Paleolimnology

Developments in Hydrobiology 15

Series editor
H. J. Dumont

Paleolimnology

Proceedings of the Third International Symposium on Paleolimnology, held at Joensuu, Finland

Edited by
J. Meriläinen, P. Huttunen and R. W. Battarbee

Reprinted from Hydrobiologia, vol. 103 (1983)

1983 **Dr W. JUNK PUBLISHERS**
a member of the KLUWER ACADEMIC PUBLISHERS GROUP
THE HAGUE / BOSTON / LANCASTER

Distributors

for the United States and Canada: Kluwer Boston, Inc., 190 Old Derby Street, Hingham, MA 02043, USA
for all other countries: Kluwer Academic Publishers Group, Distribution Center, P.O. Box 322, 3300 AH Dordrecht, The Netherlands

Library of Congress Cataloging in Publication Data

International Symposium on Paleolimnology (3rd : 1981 :
Joensuu, Finland)
Paleolimnology : proceedings of the Third International Symposium on Paleolimnology, Joensuu, Finland.

(Developments in hydrobiology ; 15)
Bibliography: p.
1. Paleolimnology--Congresses. I. Merilälnen, J.
II. Huttunen, Pertti. III. Battarbee, R. W.
IV. Title. V. Series.
QE39.5.P3I58 1981 551.48 83-13534

ISBN-13: 978-94-009-7292-6 e-ISBN-13: 978-94-009-7290-2
DOI: 10.1007/978-94-009-7290-2

Preface

The Third International Symposium on Paleolimnology was held at the beginning of September 1981 in Finland. After registration at the University of Joensuu and the opening session and welcome reception in the Town Hall, the participants moved to a hotel at Koli, a hill famed as a landmark and for its panoramic view. Here the lecture session, lasting four days, took place. Momentarily, the mosaic of the lakeland became visible between the clouds showing the hill slopes covered by mature spruce forest and scattered old birch trees, a reminder of the last slash-and-burn phase some 130 years ago.

Altogether 120 paleolimnologists participated in the meetings and most also attended one of the two post-symposium excursions to the Arctic Circle and to South Finland.

The lectures covered a wide range of paleolimnological subjects. The main topics, however, included current aspects in the dating of sediments and the history of eutrophication. The symposium was advised by the international organizing committee under the presidency of Professor Edward S. Deevey. Other members present at Koli were Dr. Maj-Britt Florin, Dr. Heikki Ignatius, Prof. Heinz Löffler, Prof. Gerbert Martinson and Dr. Magdalena Ralska-Jasiewiczowa. They also gave direction to a discussion on future plans for international activity in paleolimnology. Later in the autumn, after the end of the symposium, the participants sent special thanks to the secretaries, Mrs. Eija Meriläinen, Mrs. Tuula Nylander and Mrs. Anita Toukomaa, for solving individual problems during the symposium with smooth and rapid efficiency.

A special word of gratitude is due to the Maj and Tor Nessling Foundation and the Ministry of Education of Finland for their financial support of the symposium, as well as to others who supported the meeting and excursions in various ways. The SIL distributed information on the symposium in advance, the University of Joensuu made facilities available in Joensuu and at Koli, the Bank of Helsinki gave a flexible around-the-clock service and the town of Joensuu helped with good spirit in the welcome reception. The Geological Survey of Finland and the Maring Research Institute helped in the successful running of the excursions.

The most important people, however, were the enthusiastic participants who made "the spirit of Koli" as it was often called. In fact, that was the spirit of paleolimnology, already born in Tihany in 1967 and vigorously developed in Mikolajki in 1976. The same spirit will certainly be found in Lunz in 1986.

The Editors

View from Koli hill. The view from Koli hill is undoubtedly one of the grandest to be seen in Finland.

The imposing height of the mountain, the broad bay of Pielisjärvi, which laves its base, and the ever varying panorama which surrounds it, all combine to make a deep impression on the spectator.

In summer the landscape, as far as the eye can reach, is shrouded in a light cloud of smoke from the burning woodland, whilst in autumn, when the woods are in »the sere and yellow leaf», the entire district resembles a vast carpet, splendidly coloured in yellow and dusky green, bordered by the deep blue waters of the bay.

It is only recently that Koli hill has won the attention it deserves

Die Aussicht vom Koli-Berge. Unzweifelhaft ist die Aussicht vom Koli-Berge eine der grossartigsten in ganz Finland. Die imponirende Höhe des Berges, die weite Wasserfläche des Pielisjärvi, der seinen Fuss umspült und das abwechslungsreiche Panorama rings herum, all das übt einen mächtigen Eindruck auf den Beschauer aus.

Zur Sommerzeit ist diese Landschaft so weit das Auge reicht in einen leichten Rauchschleier, der von dem geschwendeten Boden aufsteigt, eingehüllt, und im Herbst wieder, wenn das Laub gelb wird, gleicht die Gegend einem riesigen in prachtvollen Farben von gelb und dunkelgrün spielenden Teppich, der von der dunkelblauen Wasserfläche umkränzt wird.

Erst in letzter Zeit hat man dem Koli-Berge die verdiente Aufmerksamkeit geschenkt.

Utsikt från Koli-bärget. Otvifvelaktigt är utsikten från Koli-bärget en af de mest storartade, som förekomma i Finland. Bärgets imponerande höjd, Pielisjärvis vida fjärd, som sköljer dess fot och det omväxlande panoramat rundt omkring, alt bidrar att göra ett mäktigt intryck på åskådaren.

Sommartid är landskapet, så långt ögat når, svept i en lätt slöja af svedjerök, om hösten åter, när löfven gulna, liknar näjden en jättestor matta, praktfullt färgad i gult och dunkelgrönt, kantad af den högblå fjärden.

Först på senare tider har Koli-bärget tillvunnit sig förtjänt uppmärksamhet.

Näköala Kolilta. Huoletta voipi sanoa, että Kolin näköalat ovat suurenmoisimmat, mitä Suomen luonto Kuusamoa myöden voipi tarjota. Vaaran korkeus, Pielisjärven avarat selät, jotka ulottuvat sen juurelle saakka, sekä vaihtelevat maisemat sen ympärillä antavat sille tämän ensi sijan.

Kesällä on Kolilta katsoen kaikki äärettömänä sinenä, syksyllä taas, kun kaikki nuo kasketut vaarat vesakoineen kellastuvat, leviää luonto katsojan eteen kuin suunnaton ryijy, jolla keltainen ja havuviheriä lahteilevat ja salmeilevat keskenään, selän väkevän sinen reunustamina.

Koli on vasta viime aikoina alkanut saavuttaa sitä huomiota, joka sille kohtuudella kuuluu.

Видъ съ горы Коли, несомнѣнно принадлежитъ къ числу самыхъ величественныхъ во всей Финляндіи. Высота горы, обширная водная площадь озера Пьелисъярви, омывающаго его подошву, и разнообразіе ландшафта производятъ на зрителя сильное впечатлѣніе.

Лѣтомъ, когда жгутъ лѣсъ на чищобахъ, вся мѣстность окутана легкимъ слоемъ дыма, осенью же, когда желтѣетъ листъ, она принимаетъ видъ исполинскаго ковра, сотканнаго желтымъ и темнозеленымъ цвѣтами и окаймленнаго синею полосою озера.

Туристы лишь послѣднее время стали посѣщать эту гору.

Vue du mont Koli. La vue qu'on découvre du sommet du Koli est une des plus grandioses qu'on puisse trouver en Finlande. La vaste nappe du Pielisjärvi, qui baigne le pied de l'imposante montagne, et le panorama varié qui l'entoure font une profonde impression.

En été, le paysage est voilé d'une légère fumée provenant des défrichements; mais en automne, quand le feuillage jaunit, la contrée ressemble à un vaste tapis mêlé de jaune et de vert foncé, bordé par les eaux bleues du lac.

Ce n'est que dans ces derniers temps que le mont Koli a commencé à attirer l'attention des touristes.

View from Koli hill.
Photograph and text by J. K. Inha, in: J. K. Inha, Pictorial Finland, Helsinki 1896.

Contents

PART TWO: LAKE CHANGES RELATED TO THE INFLUENCE OF MAN

PART THREE: REGIONAL PALEOLIMNOLOGY

Aspects of the developmental history and paleolimnology of lakes in Finland

Opening lecture held in Joensuu Town Hall on August 31, 1981

Pentti Alhonen

Department of Geology, Division of Geology and Palaeontology, University of Helsinki, Snellmaninkatu 5, SF-00170 Helsinki 17, Finland

The developmental history of Finnish lakes can be studied both morphologically and stratigraphically. Former water levels are indicated by raised beaches and outlet channels, and fluctuations in water level can be inferred from the lithostratigraphy of the sediments. The stratigraphic system is also the starting point for studies concerned with the changing biodynamics of lake ecosystems. Consequently, sediment stratigraphy is the fundamental basis of paleolimnology.

The geological development of all modern lakes in Finland began after the last (Weichselian) Glaciation, and therefore the main events in their history occurred during the Holocene Epoch. Of special importance in the study of the origin of the lakes is their relation to the history of the Baltic Sea. Above its highest shoreline in the supra-aquatic areas mainly of eastern and northern Finland the lake basins were formed immediately after the glaciation, but in the submerged coastal areas in southern, western, and central Finland the lakes were isolated from the various stages of the Baltic Sea as a result of isostatic uplift and tilting of the land. This crustal process is still continuing, and thus new lakes are emerging on the coast of the Gulfs of Finland and Bothnia.

Tilting caused by uneven land uplift is the main factor in the Holocene shoreline displacement of lake basins after their isolation. This is clearly seen in the geological history of our great lake complexes such as Saimaa, Päijänne, Pielinen and Näsijärvi. Their outlet channels have changed according to the transgression/regression pattern produced by a varying rate of land uplift in different parts of the lake basins (see e.g. Aario 1965; Hyvärinen 1966; and Saarnisto 1970, 1971).

Stratigraphically the isolation of a lake from some stage of the Baltic Sea is an important environmental change which is recorded by changes in the fossil diatom assemblages of the sediments. That moment is significant since it is the starting point for the typological development of the lake. This is essentially depending on the nutrient content of the geological deposits of its catchment area, although the ecosystem dynamics are also controlled by climate and by human activity in the lake surroundings.

When real paleolimnology research started in Finland 15 years ago, its first aim was to interpret the Holocene development of a eutrophic, an oligotrophic and a polyhumous (dystrophic) brown water lake (Alhonen 1967; see also Vasari 1962). The history of dystrophy is very closely related to bog development in the drainage area of a lake. High humus concentrations are typical of the majority of Finnish lakes, the average water colour value of the whole country calculated from the respective areal averages being 91 mg Pt l^{-1} (Laaksonen 1970). These brown water lakes account for more than 60% of lakes in Finland and as such they are an important subject for both limnological and paleolimnological study.

On the basis of biostratigraphical and geochemical analyses of the sediments it has become clear that both eutrophic (but non-polluted) and oligotrophic lake types have changed little with regard to their productivity. On the contrary, the main change has been environmental paludification which has led to a typological change of some lakes from mineroeutrophy to dystrophy, or more cor-

Hydrobiologia 103, 1–4 (1983).

rectly to dysoligotrophy (Alhonen 1967, 1979a, 1979b).

Analyses of a range of biological parameters have been used in paleolimnological studies of Finnish lakes. Of these the most important microfossils are diatoms and Cladocera. Attempts have been made to show trends in the limnological development of lakes and the effects of climate on their ecosystem dynamics. In comparing the planktonic/littoral ratio in the cladoceran stratigraphy of certain small lakes in south-western Finland with Holocene climatic changes it was possible to demonstrate that the water level fluctuations correlated with the known climatic history of the Holocene Epoch (see Alhonen 1970a, 1970b, 1972). Similar results are often obtained from the fossil diatom flora (e.g. Alhonen 1979a; Donner *et al.* 1978).

As it is generally known, environmental changes and the paleolimnological history of a lake can be shown from the stratigraphy of its sediments. That 'data bank' provides as much information about the lakes and their drainage basins as it does about long-term problems. On this basis cultural eutrophication and water pollution can also be traced using litho-, bio- and chemostratigraphical data (Alhonen 1979b).

We paleolimnologists know that cultural eutrophication also means distinct changes in the character of the sediments, as well as very distinct metabolic changes in the lakes. The main results of this process caused by man are increasing concentrations of nutrients and an increase in the accumulation rate of sediments. At the same time, certain indicative microfossils appear successively. I have called this the first stage (I) of the eutrophication process (Alhonen 1979b) and it is represented in the sediments by changes in fossil diatoms, cladoceran fauna and other floral or animal groups, if they are preserved.

One of the most useful ways of identifying typological changes in lakes is to trace changes in the grouping of diatoms in fossil assemblages. Nutrient enrichment is indicated by increasing alkaliphilous forms, whereas an abundance of acidophilous diatoms is clearly related to oligotrophy as can be seen from the paleolimnological results of a number of Finnish lakes.

High nutrient levels followed by increasing primary production leads, as is known, to hypolimnetic oxygen deficiency or to a permanently anaerobic hypolimnion. These second and third stages (II and III) in the eutrophication process can clearly be seen in the lithostratigraphy, first by the formation of thin sulphide layers or spots in the sediments indicating oxygen deficit in the hypolimnion during the stratification of the lake, and second, when oxygen has been completely consumed, by the formation of black sulphide gyttja. In this developmental stage of cultural eutrophication heavy pollution effects are obvious (Alhonen 1979b).

It must be kept in mind that the oxygen deficit following eutrophication depends on the volume of the hypolimnetic water. Oxygen deficiency in the hypolimnion is reached in the shallow Finnish lakes during the long winter at much lower trophic levels than elsewhere. In addition the microbial decomposition of the allochthonous humic substances increases considerably at high nitrogen and phosphorus concentrations. This is indicated, on the basis of experimental investigations and field observations, by a rapid consumption of oxygen. Data on the reaction of dystrophic waters to the introduction of industrial wastes or domestic sewage are therefore as important for the protection of polyhumous lakes in Finland as they are to an understanding of their production biology.

As mentioned above, lake sediments provide information on cultural eutrophication. Since the application of freezing techniques for sampling, observations on the relationship between sediment structure and contemporary sedimentation processes have increased our knowledge of the detailed paleolimnological history of some Finnish lakes (Saarnisto *et al.* 1977; Huttunen & Meriläinen 1978). Rhythmical laminations in Holocene lake sediments have been used for chronostratigraphy and as a methodological tool for various kinds of environmental investigation (Simola 1977, 1979, 1981). The influence of forest fires and land clearance on the developmental history of certain lakes in Finland as early as prehistoric time has been the cause of the beginning of cultural eutrophication (Tolonen 1978; see also Donner *et al.* 1978; Kukkonen 1973; Kukkonen & Tynni 1972). This can be seen from biostratigraphical evidence by the appearance of organisms indicating eutrophic conditions. However, the activities of prehistoric man, including intensive land use, slash and burn cultivation, grazing and soaking of fibre plants, especially those of hemp and flax, have also caused increasing erosion in the drainage areas of lakes. Such changes (Tolonen *et al.* 1975; Huttunen & Tolonen 1977; Huttunen 1980), recorded in the sediments by, for example, the rise of minerogenic matter in the loss-on-ignition curve and changes in other chemical components, and in the quality of seston have been commonly observed in the old-established agricultural areas of southern Finland (Tolonen *et al.* 1979).

It may be mentioned here that a reliable chronology is still a great problem in paleolimnological investigations. A detailed radiocarbon chronostratigraphy of lake sediments has proved necessary for a comparison of developmental trends in the trophic history of lakes (Tolonen 1980). A sediment chronology obtained from radioactive isotopes with a short half-life has a direct application to recent sediments where signs of pollution are apparent. This method has now been tested in Finland. Observations of annual laminations of lake sediments have also been interesting in this respect. However, their annuality should be verified microscopically by the seasonal succession of biological remains because intra-annual varved structures can be caused in many ways (see Simola & Tolonen 1981). Paleomagnetic dating of sediments complements other chronological methods.

Paleolimnology is a synthetic geoscience which builds an essential bridge between geology and limnology. It links together both terrestrial and aquatic ecological problems through time as long-term phenomena. The potentialities of lake sediments, the raw material of paleolimnology, are remarkable. Besides their use in the control of eutrophication and in plans for the restoration of polluted lakes, their investigation, including the study of chemical processes and reactions of the mud/water interface, is of great importance for environmental protection. Paleolimnology will thus contribute to the better quality of the human environment in the future.

References

Aario, R., 1965. Development of ancient Lake Päijänne and the history of the surrounding forests. Ann. Acad. Scient. Fennicae A. III. 81. 191 pp.

Alhonen, P., 1967. Palaeolimnological investigations of three inland lakes in south-western Finland. Acta Bot. Fennica 76: 1–59.

Alhonen, P., 1970. The palaeolimnology of four lakes in south-western Finland. Ann. Acad. Sci. Fennicae A. III. 105: 1–39.

Alhonen, P., 1970. On the significance of the planktonic/littoral ratio in the cladoceran stratigraphy of lake sediments. Comment. Biol. 35: 1–9.

Alhonen, P., 1972. Gallträsket: The geological development and palaeolimnology of a small polluted lake in southern Finland. Comment. Biol. 57: 1–34.

Alhonen, P., 1979a. Finnish palaeolimnology: recent advances and problems. IPPCCE Newsletter 2: 16–19.

Alhonen, P., 1979b. The sedimentary record of the cultural eutrophication and pollution of lakes in Finland. Arch. Hydrobiol. 86: 13–26.

Donner, J. J., Alhonen, P., Eronen, M., Jungner, H., & Vuorela, I., 1978. Biostratigraphy and radiocarbon dating of the Holocene lake sediments of Työtjärvi and the peats in the adjoining bog Varrassuo west of Lahti in southern Finland. Ann. Bot. Fennici 15: 258–280.

Huttunen, P., 1980. Early land use, especially the slash-and-burn cultivation in the commune of Lammi, southern Finland, interpreted mainly using pollen and charcoal analyses. Acta Bot. Fennica 113: 1–45.

Huttunen, P. & Tolonen, K., 1977. Human influence in the history of Lake Lovojärvi, S. Finland. Finskt Museum 1975: 68–105.

Huttunen, P. & Meriläinen, J., 1978. New freezing device providing large unmixed sediment samples from lakes. Acta Bot. Fennici 15: 128–130.

Hyvärinen, H., 1966. Studies on the late-Quaternary history of Pielis-Karelia, eastern Finland. Comment. Biol. 29: 1–73.

Kukkonen, E., 1973. Sedimentation and typological development in the basin of the lake Lohjanjärvi, South Finland. Geological Survey of Finland, Bulletin 261: 1–67.

Kukkonen, E. & Tynni, R., 1972. The sediment core from Lake Lovojärvi, a former meromictic lake (Lammi, South Finland). The development of Lovojärvi on the basis of its diatoms. Aqua Fennica 1972: 70–82.

Laaksonen, R., 1970. Vesistöjen veden laatu (Water quality in the water systems). Valtion painatuskeskus. Helsinki 1970. 132 pp.

Saarnisto, M., 1970. The late Weichselian and Flandrian history of the Saimaa lake complex. Comment. Phys.-Math. 37: 1–107.

Saarnisto, M., 1971. The history of Finnish lakes and Lake Ladoga. Comment. Phys.-Math. 41: 371–388.

Saarnisto, M., Huttunen, P. & Tolonen, K., 1977. Annual lamination of sediments on lake Lovojärvi, southern Finland, during the past 600 years. Ann. Bot. Fennici 14: 34–45.

Simola, H., 1977. Diatom succession in the formation of annually laminated sediment in Lovojärvi, a small eutrophicated lake. Ann. Bot. Fennici 14: 143–148.

Simola, H., 1979. Micro-stratigraphy of sediment laminations deposited in a chemically stratifying eutrophic lake during the years 1913–1976. Holarct. Ecol. 2: 160–168.

Simola, H., 1981. Sedimentation in a eutrophic stratified lake in southern Finland. Ann. Bot. Fennici 18: 23–36.

Simola, H. & Tolonen, K., 1981. Diurnal laminations in the varved sediment of Lake Lovojärvi, south Finland. Boreas 10: 19–26.

Tolonen, K., 1978. Effects of prehistoric man on Finnish lakes. Pol. Arch. Hydrobiol. 25: 419–421.

Tolonen, K., 1980. Comparison between radiocarbon and varve dating in Lake Lampellonjärvi, south Finland. Boreas 9: 11–19.

Tolonen, K., Siiriäinen, A. & Thompson, R., 1975. Prehistoric field erosion sediment in Lake Lojärvi, southern Finland and its palaeomagnetic dating. Ann. Bot. Fennici 12: 161–164.

Tolonen, K., Siiriäinen, A. & Hirviluoto, A.-L., 1979. Iron Age cultivation in SW Finland. Finskt Museum 1976: 5–66.

Vasari, Y., 1962. A study of the vegetational history of the Kuusamo district (North East Finland) during the Late-Quaternary period. Ann. Bot. Soc. 'Vanamo' 33: 443–451.

Säynäjälampi and the difficulties inherent in the dating of sediments in a hard-water lake

Ingrid U. Olsson[1], Farid El-Daoushy[1] & Yrjö Vasari[2]
[1] *Institute of Physics, University of Uppsala, Sweden*
[2] *Department of Botany, University of Helsinki, Finland*

Keywords: paleolimnology, pollen analysis, land drainage, sediment accumulation, ^{210}Pb dating, ^{14}C reservoir effect

Abstract

Previous measurements of the ^{14}C activity of lake sediments and *Potamogeton* plants from Säynäjälampi, collected in 1970, indicated a severe hard-water effect (Donner *et al.* 1971). Samples from new cores, collected in 1977, were sliced into thin sections down to 18 cm. The ^{14}C concentration in samples down to 7 cm was similar to that obtained by the Helsinki Laboratory. The corresponding apparent ^{14}C age is almost 2 000 years. Light-coloured layers, apparently due to catchment drainage, were used together with density diagrams to correlate the cores. The drainage history gives a time-scale showing that the lowest samples dated were just old enough to be pre-bomb sediments. Three samples of submerged plants were also examined, indicating some exchange with the air and an appreciable reservoir age. The activity of the plants was significantly lower in 1978 than in 1970. One sample, well below the lowest drainage stratum, indicates a reservoir age of *ca.* 1 100 years. This sample yielded an apparent age of the INS fraction which was significantly different from that of the SOL fraction. The difference was determined to *ca.* 600 years.

Total ^{210}Pb was determined via ^{210}Po using isotope dilution while ^{226}Ra was measured with the help of radon emanation. The ^{226}Ra was used to estimate the supported ^{210}Pb. The unsupported ^{210}Pb showed a higher sediment accumulation rate for the top nine cm of one core (the length investigated was 16.5 cm). The varying accumulation rate was explained by human activities, as could be seen from the heterogenous sediment composition of the cores.

Recent vegetational change is described from direct observations since 1957. Pollen analysis does not give any reliable indication of agriculture.

Introduction

Since the ^{14}C ages from deposits in northern Finland are in apparent disagreement with expected ages, samples from Säynäjälampi were studied by Donner *et al.* (1971). This study has been extended to allow more definite conclusions.

The apparent differences between the ^{14}C ages of sediment from Säynäjälampi and Maanselkä were 3 355 ± 350 years for pollen zone Hl_1b and 1 780 ± 350 for Hl_2b. The differences between Säynäjälampi and Maanselänsuo were 2 690 ± 270 years for Hl_1b and 1 550 ± 220 for Hl_4a. Further results are presented by Donner *et al.* (1971). It may be observed that the differences tend to decrease upwards in the standard profile. Since Säynäjälampi is affected by hard water a decrease of the reservoir age is natural. The data given by Donner *et al.* (1971) do not allow an accurate conclusion to be reached for the present value of the reservoir age because a) the mud does not correspond to known years and may be affected by the atomic-bomb effect, and b) the *Potamogeton* plants yield the activity of the water during the collection year (1970) and are influenced by the atomic-bomb effect which is always smoothed in water systems. Although sample 2 is

Hydrobiologia 103, 5-14 (1983).

about 3 000 years older than the terrestrial sample 1, and pollen zone Hl_1b is dated at least *ca.* 3 000 years too old, the difference between samples 1 and 2 cannot be used as a value of the reservoir age. Sample 1 consisted of 'birch, at most 5 years old'. It does not give the activity of the atmosphere during the collection year but represents an integrated value for a number of years when the excess activity was decreasing from approximately 80% (1965) to about 55% (1970). Seasonal variations were still significant in 1970. The activity of the submerged plants (sample 2) undoubtedly shows a hard-water effect but is dependent on the rate of response of the lake to the atmospheric variations of the ^{14}C content. The mud may have yielded too high an activity because of the atomic bomb effect or too low an activity if it were contaminated *in situ* by eroded material.

Lake sediments are mainly derived from autochthonous organic material and eroded material from drainage areas and watercourses with some organic remains and minor quantities from atmospheric fallout. The complex nature of the sedimentological processes that take place near the sediment-water interface and within the upper sediment layers makes dating of recent unlaminated sediments a difficult task.

Site description

Säynäjälampi (66° 10′N, 28° 58′E) is a small lake (about 1–2 m deep, 100 × 200 m in size, 263 m above sea level) in the northern part of the municipality of Kuusamo, N.E. Finland. The surrounding bedrock consists of Jatulian metabasites (Hackman & Wilkman 1925), commonly containing carbonate-quartz lenses and veins (Hackman & Wilkman 1929). On Oct. 27th, 1981 the pH of the water was 6.8.

The lake lies in a large shallow depression surrounded by till-covered hills (Fig. 1) and drains northwards through a stream which has been recently widened and deepened. It has no natural inflows, although three major ditches drain into the lake. Bedrock outcrops are frequent around the depression, and the immediate vicinity of the lake comprises land which was previously rich fen (Säynäjäsuo and Teerisuo) but has now been mostly cleared and reclaimed for agriculture. About half of

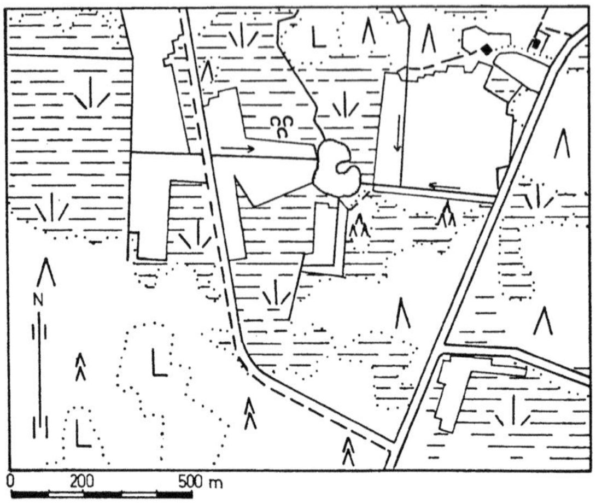

Fig. 1. Säynäjälampi and its surroundings.

the adjoining land is covered by young thickets of birch, willow and spruce.

The drainage of the fens dates from the post-war settlement period when the occupants of the nearby farms arrived. The first drainage took place in 1947–49. A second occurred in 1951–52. In the years 1964–65 nearby forests were drained to the lake.

During the first clearing phase the new fields were fertilized with potassium nitrate, containing phosphate, (about 400–600 kg ha^{-1}). In 1955–56 some parts of the surrounding fields were treated with calcium for the first time (about 3 000–4 000 kg ha^{-1}). Later, this procedure was repeated for some parts of the fields. In 1955–56 the fields were further fertilized with the so-called Y-mixture (P, K, N) (about 300–500 kg ha^{-1}). Gradually, a deficiency first of copper and somewhat later of phosphorus became apparent. Since about 1969–70 many of the western fields have not been under cultivation. Information on drainage and fertilization was obtained from a local farmer, Mr. Frans Juurtela.

Collection of material

Since 1957 Säynäjälampi has been the object of repeated Quaternary botanical studies (e.g. Vasari 1962 & 1963; Donner *et al.* 1971). From the very beginning of its history, the small lake has harboured a rich vegetation of exacting aquatic macrophytes. The vegetational development began with an alkaline phase with *inter alia Potamogeton filiformis* Pers. and *P. vaginatus* Turcz. The most luxurious phase with an abundance of great many southern species now lacking from the lake or its shores occurred in the Early Warmth Period.

The aquatic macrophytes growing in the lake today are *Nuphar lutea* (L.) Sibth. & Sm., *Hippuris vulgaris* L., *Potamogeton alpinus* Balbis, *Sparganium* sp (helophytes or plants with floating leaves) and *Utricularia intermedia* Hayne, *Potamogeton compressus* L., *P. berchtoldii* Fieber, *P. obtusifolius* Mert. & Koch, and *P. praelongus* Wulfen (submerged plants).

In October 1970, samples for the measurement of ^{14}C activity were collected from Säynäjälampi and its shores, and included submerged specimens of *Potamogeton berchtoldii, P. obtusifolius, P. alpinus* and *P. praelongus* (cf. Donner *et al.* 1971). Samples were collected for the present study in 1977 and 1978. Samples for sediment studies were collected from the ice in November 1977. The surface sediment samples were taken with a Hakala-sampler (a modification of the Kajak-sampler; Hakala 1971) 385 mm long and with an inner diameter of 44 mm. A larger sample was taken using a dry ice sampler (Saarnisto 1979). Plant material for isotope analyses was also collected in November 1977, and included twigs of grey alder (*Alnus incana* (L.) Moench) and willow (*Salix* spp.) and dry *Carex rostrata* Stokes from the shore. A sample (No. 8) consisting of submerged *Potamogeton alpinus* and *P. berchtoldii* was collected in August 1978. As this proved to be too small, other samples (9A and 9B) were collected in September the same year using a Maristo drag (Maristo 1941). Sample 9A comprised *Potamogeton compressus, P. berchtoldii, P. alpinus, Calliergon megalophyllum* Mikut., *Drepanocladus capillifolius* (Warnst.) Warnst. and *Nitella* sp. Sample 9B was composed of unidentified plant remains, probably leafless stems and roots of submerged plants. All the material in sample 9A consisted of submerged plants, without any floating leaves.

It is noteworthy that the results of collections and observations differ somewhat from year to year. Only *Potamogeton alpinus* and *P. berchtoldii* were constantly present throughout the observation years (1957, 1970 & 1978). Moreover these same species were the only pondweeds found in the uppermost gyttja samples (Vasari 1962). *Potamogeton obtusifolius* and *P. praelongus* (both lacking in the gyttja samples from the period of climatic deterioration) occurred in the first two collections, but not in 1978. *Potamogeton compressus* was found in 1957 and 1978 which would seem to indicate its continuous presence despite the lack of observations in 1970.

The finds of the two water mosses, *Drepanocladus capillifolius* and *Calliergon megalophyllum* were evidently due to the careful collection of a vast amount of material – there is little reason to suspect that they did not previously exist in the lake.

Stratigraphical observations

The larger sample collected in 1977 was used for chemical analyses and pollen counts. The chemical analyses (Ca- and Al-content of dry matter) were performed by the Kemira Oy Laboratories in Oulu in November 1978, and the pollen analyses by Mrs Raija Huttunen in early spring 1979.

The litho- and pollen stratigraphy of this frozen core (Fig. 3) together with the results of the chemical analyses are given in Fig. 2. In the lithostratigraphy attention is drawn to the darker colour of the three uppermost cm and to three whitish, somewhat irregular bands about 7 cm, 5 cm and 2 cm below the sediment surface. It is reasonable to assume that these white bands indicate the three major post-war drainage phases, 1947–49, 1951–52 and 1964–65. These figures and the approximate locations of the white bands are included in the lithostratigraphy column of Fig. 2. This assumption regarding the origin of the bands seems corroborated by the significant rise in the Al content from 3.3% at 10 cm to 5.5% at 7–6 cm. Further up, the Al content increases first to 5.9% and then to 6.1%. The purpose of the Al analyses was to estimate the amount of mineral material washed into the basin in connection with the drainage. It is interesting that despite the cultivation of the fields and some use of lime fertilizer, the Ca content of the sediment shows only a very slight increase within the uppermost 7 cm, from the precultivation value of 1.5% to 1.8–1.9% in the uppermost samples.

A calculation of the sediment accumulation rate based on these values gives the following results: 1947–51 5 mm a^{-1}, 1951–64 2.3 mm a^{-1} and 1964–77 1.5 mm a^{-1}. The first drainage seems to have had the most dramatic consequences – as is also evident from

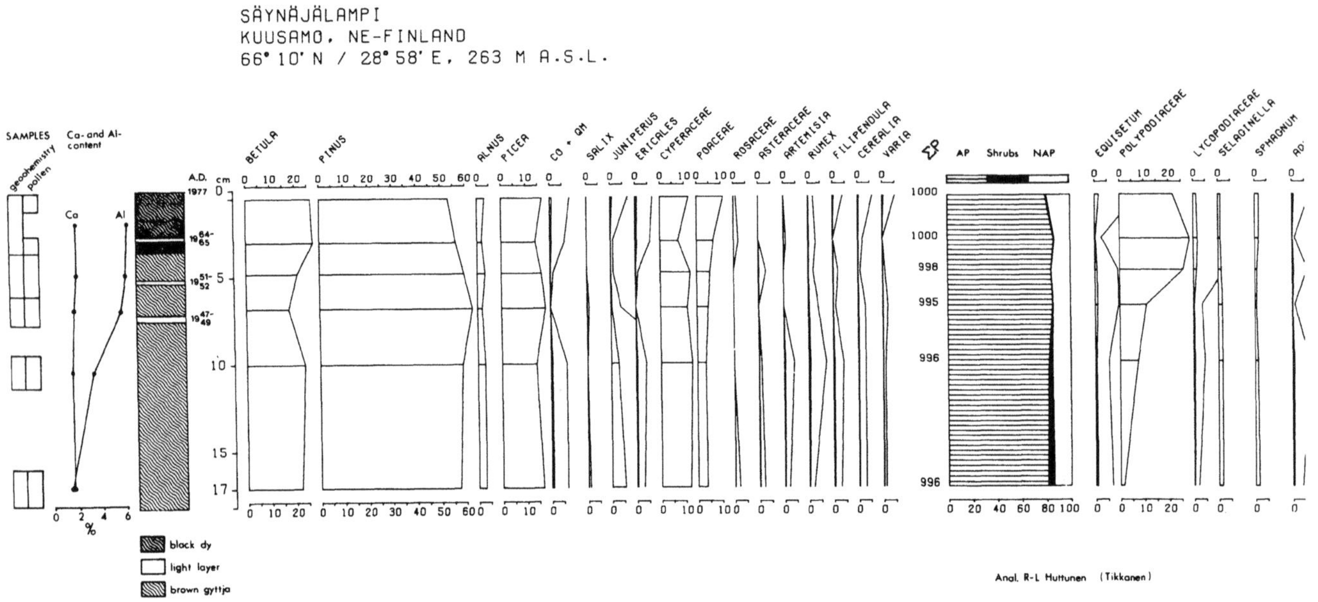

Fig. 2. Litho- and pollen stratigraphy of the frozen core from Säynäjälampi.

Fig. 3. The frozen core from Säynäjälampi.

the thickness of the white bands. The last drainage (of the forest land) had the least effect on sedimentation. The average accumulation rate during the whole 30-year period of cultivation approached 2.3 mm a^{-1}. Judging by ^{210}Pb measurements the accumulation rate was about 1 mm a^{-1} below 10 cm. This pattern closely resembles that of Hännisenlampi, eastern Finland, where Vuorinen (1978) found a change from values of 0.3–1.0 mm a^{-1} in pre-settlement sediments to 5 mm a^{-1} in the period of the most intensive land use in the 18th century. Bulk density measurements showed that below the 10 cm level the sediment in the core contains about 1.5 g cm^{-1} or 0.15 g a^{-1} dry matter whereas above this horizon the corresponding figures are about 3 g cm^{-1} or 0.7 g a^{-1}. In Lovojärvi, southern Finland, the accumulation rate during the last century was 2–3 mm a^{-1} on average, the corresponding figure for the 'pre-agricultural' years being about 0.5 mm a^{-1} (Saarnisto *et al.* 1977).

Pollen analyses

The pollen spectra (Fig. 2) are based on total sums of about 1 000 AP + NAP. It is interesting to observe that the various pollen spectra show remarkable homogeneity and that agricultural indicators play an insignificant role in this pollen diagram despite the immediate vicinity of vast fields created from land cleared after 1947.

Without a knowledge of the present vegetation it would seem hazardous to assume that there was any cultivation in the area, let alone in the immediate vicinity of the site. In principle, the diagram should contain two basically different parts, one representing the pre-cultivation period and the other representing present day cultivation. However, no such differentiation is possible.

Behre (1981) discussed the fact that pollen of cultivated species is often a very unreliable indicator of cultivation. In the Säynäjälampi diagram the only difference of any magnitude between the pollen spectra of 'pre-agricultural' and 'agricultural' periods is the increase of fern spores in the upper part of the diagram. However, this is an odd feature as ferns do not play a major role in the present vegetation. The pollen values of different trees in the surface sample correspond fairly well with the data published by Prentice (1978). A calculation based on the four uppermost pollen spectra and the average forest composition, according to the III Finnish National Forest Inventory (Ilvessalo 1960), gives almost the same R-values as Donner (1972) previously obtained for the southern Tavastia region in S. Finland. *Picea* is clearly under-represented, being the dominant tree in the higher ground nearby. *Betula* seems to be correspondingly over-represented, possibly due to the influence of young birch thickets in the vicinity of the lake, whereas the pollen values of *Pinus* well reflect the true position of pine in the forests.

Cores used for dating

The cores were 13–18 cm long. The uppermost and second white layer were easily discerned but the third layer was not observed until the seventh core was sectioned. It was also later detected in core 8 during preparation for the ^{210}Pb measurements. The cores were cut into slices soon after collection. An effort was made to correlate the cores as they were sectioned on the basis of colour changes. The slices were 0.5–1.0 cm thick. The uncertainty in each depth was estimated as 0.1–0.2 cm. The overall error in the slice volume is thus mostly around 20%. Since the light-coloured parts were presumably caused by human activity in connection with the land survey of the area after the war, the density of the sediment could be expected to help in correlating the cores. It was accordingly decided to determine the dry weight (d.w.) per unit length and also use the accumulated sediment (d.w.) as a check on whether or not levels were contemporaneous. Since normalization to dry weight is necessary for ^{210}Pb dating, core 8, selected for the ^{210}Pb measurements, was weighed first. It was seen that there were two very pronounced peaks, coincident with two of the white layers. The weighing of the other seven cores revealed that one, core 1, showed only small variations in density, expressed as dry weight per cm core, when the others exhibited peaks at least 3 times the dry weight of the sediment further down at 10–15 cm (Fig. 4). Core 1 was not used for dating since it was somewhat different from the others.

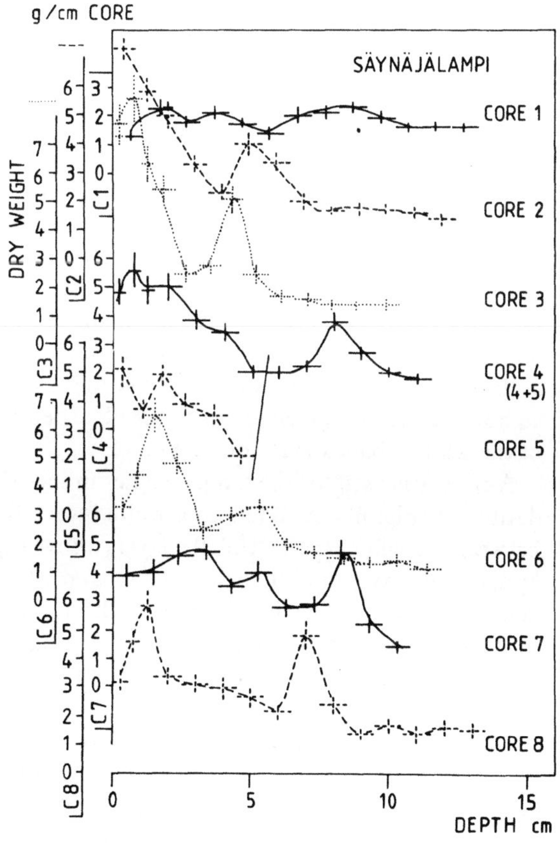

Fig. 4. Dry weight variations within the eight cores used for isotope studies.

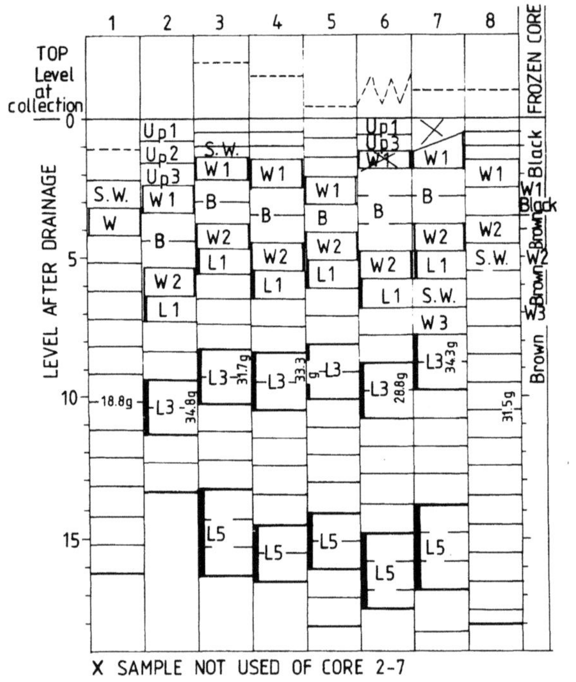

Fig. 5. Correlation between the cores based on colour and density variations of the dried sediment.

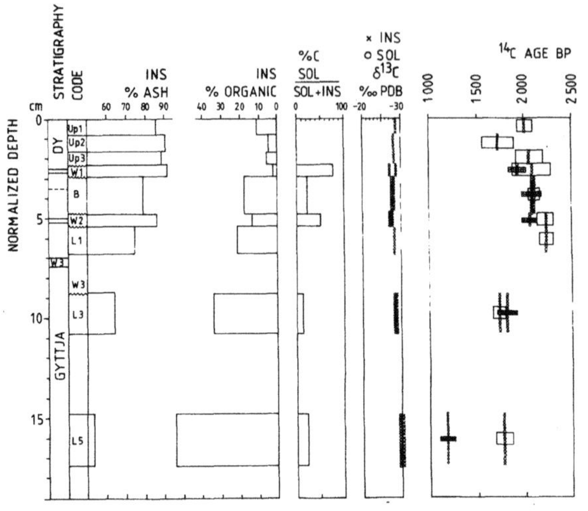

Fig. 6. Radiocarbon measurements, $\delta^{13}C$ values, organic content and relation between INS and SOL fractions.

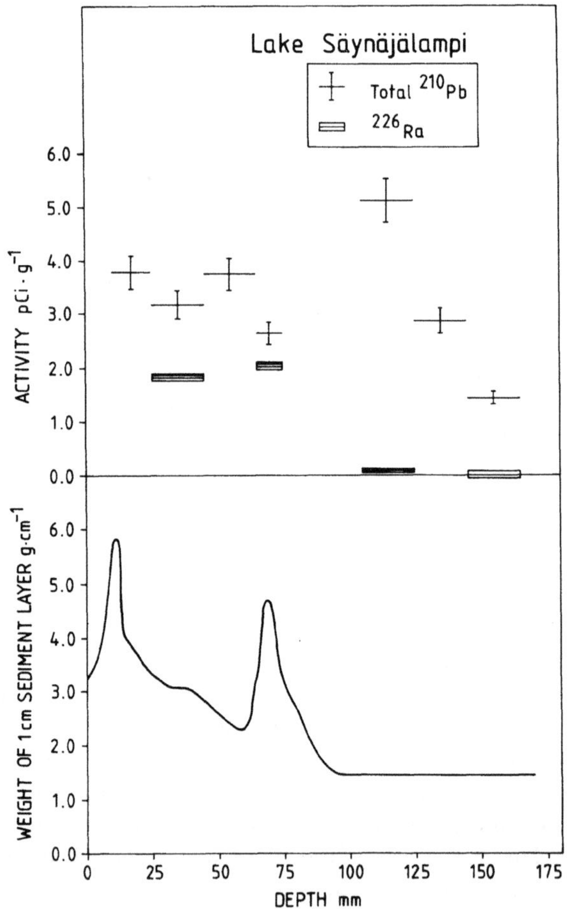

Fig. 7. Total ^{210}Pb measurements, ^{226}Ra and dry weight variations within core 8, given for the actual cross section.

The frozen core was also kept. Core 8 was selected for ^{210}Pb measurements, being one of the longest and less drained than the others. A few small animals were detected on slicing. The observations made during collection and slicing of the eight cores are indicated in Fig. 5 and are also used for correlation purposes. Cores 4 and 5 could be correlated already at this stage. The total dry weights above one selected horizon are also given in this figure. The depths of the whitish bands vary somewhat.

A code was used for the sample names in the ^{14}C dating (cf. Fig. 6). 'W' refers to a white layer, 'B', between the white layers, 'Up', the layers above the upper white layer, and 'L' below the second white layer from the top. The first figure gives the number of a specific type of layer counted from a higher level downwards (the missing numbers denote layers, which are not yet ^{14}C determined). The next figures give the cores used for the dated sample.

The ^{14}C and ^{13}C results are presented in Fig. 6 as a function of approximate depth, chosen to coincide roughly with the stratigraphy of the frozen core above 5.0 cm and with core 6 below 5.0 cm. The ^{210}Pb results are given in Figs. 7 & 8.

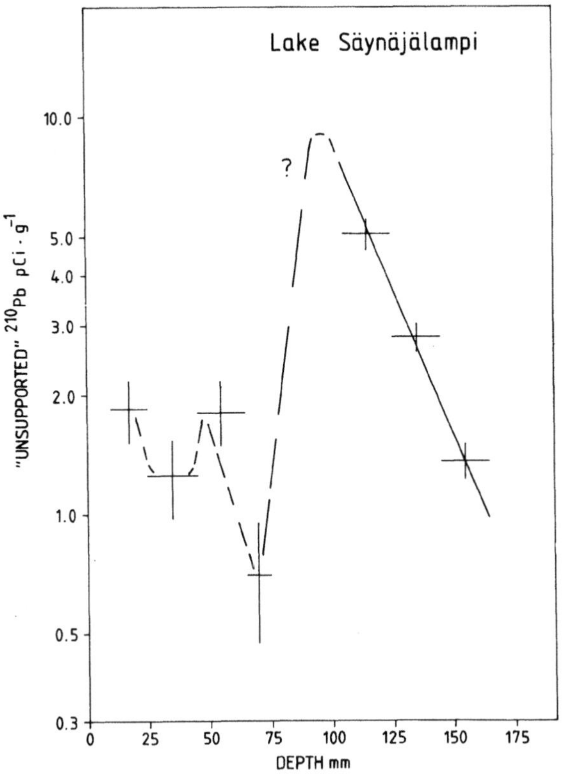

Fig. 8. Unsupported ^{210}Pb vs. depth.

^{14}C measurements

The samples were boiled with 1% HCl to remove carbonates, washed, treated overnight with 1% NaOH at 80 °C to obtain a soluble fraction (SOL), washed and again treated with HCl – this time to remove any absorbed carbon dioxide – to yield an insoluble fraction (INS).

The INS fraction of wood is more reliable than the SOL fraction containing mobile extracts, and cellulose is the most reliable fraction. The SOL fraction of lake sediments proved to be more reliable than the INS fraction especially when the organic content is low, since old organic material can be transported to the lake together with the minerogenic fraction. However, the SOL fraction is often too small to be dated with conventional equipment.

The present investigation does not reveal a significant difference between the INS and SOL fraction for the sediment (Fig. 6) except for the deepest sample, although there is an indication that the SOL fraction may be slightly younger in the case of two of the other four samples for which both this and the INS fraction were measured.

All results are normalized to $\delta^{13}C$ -25 ‰. The international standard is used and the results calculated with the half-life 5 568 years.

The apparent ^{14}C age decreases slightly with decreasing depth above the lowest white layer, and is given at most levels as over 2 000 years. It must be remembered that young samples like these, judging by both historical evidence and ^{210}Pb measurements, should manifest a deficiency in ^{14}C content because of secular variations and the industrial effect. The expected age is *ca.* 110 years (Olsson & Florin 1980). Thus, in reality, the upper part of the present sediment exhibits an apparent age of almost 2 000 years. The sediment close to the top is very little, or at least almost imperceptibly, affected by nuclear weapons.The difficulties in correlating the core slices together with the documented erosion and possible mixing or sediment disturbance (Bartlett 1951; Olsson 1972a, 1972b, 1973 & 1982) may mask the expected increase of the ^{14}C activity close to the top. The lower samples are younger than those above the lowest white layer. The SOL fraction of the lowermost sample dated yields the apparent age of about 1 180 years, and the reservoir age is therefore probably *ca.* 1 100 years, corresponding to a deficiency of about 13% of ^{14}C. Another sample, slightly higher up in the core, yields the apparent age of about 1 820 years for the SOL fraction. The INS fractions of these two samples appear to be about 1 780 and 1 740 years old, thus younger than all but one of the samples close to the top. Submerged plants from late autumn 1978 had an activity very close to the standard (Table 1) or about 30% below the contemporaneous atmospheric carbon dioxide. This is a lower activity than in 1970 (Donner *et al.* 1971) when the Δ for the sub-

Table 1. The Δ values of the submerged plants collected in 1978 from Lake Säynäjälampi.

Sample no.	Coll. month	Dating no.	$\delta^{13}C$ per mille PDB	Δ per cent (Excess over standard)
8	Aug.	U-2841	-23.1	4.2 ± 2.4
9A	Sept.	U-2842	-33.3	-1.4 ± 0.8
9B	Sept.	U-2843	-27.9	2.6 ± 0.9

merged plants was 159 ± 13 ‰ and Δ for the atmosphere approximately 5.5×10^2 ‰.

Three samples, collected in 1977, growing close to the lake were dated to yield the local ^{14}C concentration and illustrate whether there is any uptake by roots (Olsson 1979). The Δ values were 369 ± 8 (*Salix*), 367 ± 8 (*Alnus*), and 366 ± 8 ‰ (*Carex*) in agreement with those expected for the atmosphere (Nydal *et al.* 1980).

^{210}Pb measurements

The application of ^{210}Pb dating (half-life = 22.26 $\pm$ 0.22 a) particularly if combined with other suitable chronological data, could give credible sediment chronologies for time intervals which cover the last 150 a, providing that processes violating the original order of the sediments are carefully examined (Pennington *et al.* 1976; El-Daoushy 1978; Robbins 1978; Benninger *et al.* 1979; Smith & Walton 1980). ^{210}Pb is produced in the atmosphere via the decay of ^{222}Rn, an inert gas, which mainly emanates from land areas of the earth. The atmospheric or 'unsupported' ^{210}Pb is introduced to different water reservoirs, after a brief delay, through fallout processes. Besides this fraction of ^{210}Pb, which is used to construct sediment chronologies (Goldberg 1963), there is 'supported' ^{210}Pb, which appears within the sedimented material due to the *in situ* decay of ^{226}Ra, but the two may not be in secular equilibrium with each other. Bioturbation (Robbins & Edgington 1975; Fisher *et al.* 1980), physical mixing (Aller *et al.* 1980), slumping and erosion (Robbins 1978) disrupt the chronologically arranged sediment layers while chemical mobility (Koide *et al.* 1973; Eakins & Morrison 1976), grain size distribution (Megumi 1978), bulk density variations (Robbins & Edgington 1975) and varying sediment accumulation rates (Oldfield *et al.* 1978; El-Daoushy *et al.* 1982) affect the ^{210}Pb profiles so that the initial specific activity of the sediments becomes variable. A number of natural and man-made chronological tools may now be applied to provide information on specific events and supplement the ^{210}Pb results (El-Daoushy 1982a). Magnetic analyses can, for instance, be used in connection with environmental studies (Thompson *et al.* 1980).

The 'total' ^{210}Pb and the ^{226}Ra were determined using isotope dilution and radon emanation (El-Daoushy 1981a & b). Figure 7 shows the total ^{210}Pb and ^{226}Ra activities in core 8 against depth. The activity profiles demonstrate two zones of high and low radium concentration suggesting different sediment composition as described in the lower part of the same figure. The sediments were rather organic and had uniform density untill material of high inorganic content was introduced to the lake and both the density and the activity profiles assumed an irregular pattern. The 'total' ^{210}Pb was diluted and the initial concentrations reduced. Unfortunately the sediment core was not long enough to allow the application of the constant rate of supply 'c.r.s.' model (Oldfield *et al.* 1978; El-Daoushy 1982b). It is probable that part of the 'total' ^{210}Pb in the upper layers of the disturbed zone is due to the higher content of ^{226}Ra of the allochthonous but recently deposited material. On the basis of ^{226}Ra concentrations the 'unsupported' ^{210}Pb were estimated and plotted against depth on a semilogarithmic scale (Fig. 8). The sediment accumulation rate of the sediment below the disturbed zone was 1.0 ± 0.1 mm a^{-1}.

Conclusion

The deepest ^{14}C-dated sample, deposited almost 10 cm below the lowermost whitish band, yields the latest date in this series. The SOL fraction appears younger than the INS fraction. The difference is significant, 600 ± 120 ^{14}C years, for the sediment deposited one century before the first drainage phase, but not significant for the sediment deposited just before the first drainage phase and later. The general trend is that the organic content of the sediment decreases, upwards in the core, from about 50% to about 25% to 10% or even less. At the same time the Al content increases by a factor of 4 and the density by a factor of the same order although maxima are seen to be correlated to the whitish layers. The varying conditions since the drainage activities started have prevented the ^{210}Pb dating of the uppermost part, primarily because the samples examined were not deep enough to allow for the application of the 'c.r.s.' model. A sediment accumulation rate of 1.0 ± 0.1 mm a^{-1} for the pre-settlement period could be calculated in accordance with the estimates made on the basis of historical records. Pollen analysis here failed to indicate the cultivation period. The apparent ^{14}C age of the sediment is due partly to 'old' bicarbonate but, as seen from the difference between the INS and SOL fractions in sample L5, also to *in situ* contamination. When erosion occurs because of drainage old organic material is also introduced. The decrease in the ^{14}C activity from 1970–1978 for submerged plants is less than the decrease of the activity of the air but still of such magnitude that the response of the lake to the variations of the atmospheric activity is obvious. It is suggested that new submerged samples be collected within the next few years and once again some years later to study the restoration of an equilibrium between the atmosphere and the lake water. It cannot be excluded that the sample Up2, probably deposited around 1972, has a higher activity reflecting the atom-bomb peak 1963–1964 and the high level afterwards.

Acknowledgements

The authors wish to express their gratitude to Matti Saarnisto for his assistance in the collection of the cores used for this study. Sincere thanks are due to Raija Huttunen for the pollen analyses, and to Maud Söderman, Tomas Kronberg and Birgitta Hansson for assistance in the ^{14}C laboratory. We are particularly grateful to Kai Siegbahn, the Head of the Institute of Physics, Uppsala. Financial support for dating was received from the Swedish Natural Science Research Council and a grant from the Nycander Foundation. The cost of the pollen analyses was met by the Finnish National IGCP Committee.

References

Aller, R. C., Benninger, L. K. & Cochran, J. K., 1980. Tracking particle-associated processes in nearshore environments by use of $^{234}Th/^{238}U$ disequilibrium. Earth and Planet. Sci. Lett. 47: 161–175.

Bartlett, H. H., 1951. Radiocarbon datability of peat, marl, caliche, and archaeological materials. Science 114: 55–56.

Behre, K.-E., 1981. The interpretation of anthropogenic indicators in pollen diagrams. Pollen et Spores 23: 225–245.

Benninger, L. K., Aller, R. C., Cochran, J. K. & Turekian, K. K., 1979. Effects of biological sediment mixing on the ^{210}Pb chronology and trace metal distribution in a Long Island Sound sediment core. Earth and Planet. Sci. Lett. 43: 241–259.

Donner, J. J., 1972. Pollen frequencies in the Flandrian sediments of Lake Vakojärvi, south Finland. Commentationes Biol. 53: 1–19.

Donner, J. J., Jungner, H. & Vasari, Y., 1971. The hard-water effect on radiocarbon measurements of samples from Säynäjälampi, north-east Finland. Commentationes Biol. 41: 307–310.

Eakins, J. D. & Morrison, R. T., 1976. A new procedure for the determination of lead-210 in lake and marine sediments. United Kingdom Atomic Energy Authority, Harwell, AERE-R8475.

El-Daoushy, M. F. A. F., 1978. Low-level alpha and beta activity measurements and dating applications (Thesis), Uppsala University, Institute of Physics.

El-Daoushy, M. F. A. F., 1981a. Stability of two silicon detectors in low-level alpha counting. In: Methods of Low-level Counting and Spectrometry, 151–160. IAEA, Vienna.

El-Daoushy, M. F. A. F., 1981b. An ionization chamber and a Si-detector for lead-210 chronology. Nucl. Instr. Methods 188: 647–656.

El-Daoushy, F., 1982a. Dating techniques suitable for deposits younger than 500y. Eleventh Inter. Cong. on Sedimentology-symp. 35: Geochronology of Recent Deposits.

El-Daoushy, M. F. A. F., 1982b. Lead-210 and its environmental and sedimentological applications. Geol. Fören. Stockholm Förh. 104: 270–271.

El-Daoushy, F., Tolonen, K. & Rosenberg, R., 1982. The lead-210 and the moss-increment dating of two Finnish Sphagnum-hummocks. Nature 296: 429–431.

Fisher, J. B., Lick, W. J., McCall, P. L. & Robbins, J. A., 1980. Vertical mixing of lake sediments by tubificid oligochaetes. J. Geophys. Res. 85: 3997–4006.

Goldberg, E. D., 1963. Geochronology with lead-210. In: Radioactive Dating, 121–131. IAEA, Vienna.

Hackman, V. & Wilkman W. W., 1925. Suomen geologinen yleiskartta. (The general geological map of Finland.) Sheet D6, Kuolajärvi.

Hackman, V. & Wilkman, W. W., 1929. Suomen geologinen yleiskartta. Lehti D6, Kuolajärvi. Kivilajikartan selitys. Suomen geol. toimikunta – Geol. Kommissionen i Finland. Helsinki. 142 pp.

Hakala, I., 1971. A new model of the Kajak bottom sampler, and other improvements in the zoobenthos sampling technique. Ann. Zool. Fennici 8: 331–335.

Ilvessalo, Y., 1960. Suomen metsät kartakkeiden valossa. (Summary: The forests of Finland in the light of maps.) Comm. Inst. Forestalis Fenniae 52(2): 1–70.

Koide, M., Bruland, K. W. & Goldberg, E. D., 1973. Th-228/Th-232 and Pb-210 geochronologies in marine and lake sediments. Geochim. Cosmochim. Acta 37: 1171–1187.

Maristo, L., 1941. Die Seetypen Finnlands auf floristischer und vegetationsphysiognomischer Grundlage. Ann. Bot. Soc. 'Vanamo' 15(5): 1–312.

Megumi, K., 1978. A problem in ^{210}Pb geochronologies of sediments. Nature 274: 885–887.

Nydal, R., Lövseth, K. & Skogseth, F. H., 1980. Transfer of bomb ^{14}C to the ocean surface. Radiocarbon 22: 626–635.

Oldfield, F., Appleby, P. G. & Battarbee, R. W., 1978. Alternative ^{210}Pb dating: results from the New Guinea Highlands and Lough Erne. Nature 271: 339–342.

Olsson, I. U., 1972a. The pretreatment of samples and the interpretation of the results of C^{14} determinations. In: Vasari, Y., Hyvärinen, H. & Hicks, S. (Eds.), Proc. Symposium on Climatic Changes in Arctic Areas during the last 10 000 years. pp. 9–34. Acta Univ. Ouluensis. A. 3. 1972. Geol. 1.

Olsson, I. U., 1972b. The C^{14} dating of samples for botanical studies of prehistoric agriculture in northern Ångermanland. Early Norrland 1: 35–41. Stockholm.

Olsson, I. U., 1973. A critical analysis of ^{14}C datings of deposits containing little carbon. In: Proc. 8th Int. Conf. Radiocarbon Dating; Lower Hutt, Oct. 18–25, 1972. pp. 547–564. Royal Society of New Zealand, Wellington.

Olsson, I. U., 1979. The radiocarbon contents of various reservoirs. In: Berger, R. & Suess, H. E. (Eds.), Radiocarbon dating, 9th Internat. Radiocarbon Conf., Proc. pp. 613–618. Univ. Calif. Press, Berkeley Los Angeles.

Olsson, I. U., 1982. Dating non-terrestrial materials. In: Proc. 1st International Symp. on C-14 and Archaeology, in print.

Olsson, I. U. & Florin, M.-B., 1980. Radiocarbon dating of dy and peat in the Getsjö area, Kolmården, Sweden, to determine the rational limit of Picea. Boreas 9: 289–305.

Pennington, W., Cambray, R. S., Eakins, J. D. & Harkness, D. D., 1976. Radionuclide dating of the recent sediments of Blelham Tarn. Freshwater Biol. 6: 317–331.

Prentice, I. C., 1978. Modern pollen spectra from lake sediments in Finland and Finnmark, North Norway. Boreas 7: 131–153.

Robbins, J. A., 1978. Geochemical and geophysical applications of radioactive lead. In: Nriagu, J. O. (Ed.), The Biogeochemistry of Lead in the Environment. pp.285–393. Elsevier/North-Holland Biomedical Press.

Robbins, J. A. & Edgington, D. N., 1975. Determination of sedimentation rates in Lake Michigan using Pb-210 and Cs-137. Geochim. Cosmochim. Acta 39: 285–304.

Saarnisto, M., 1979. Applications of annually laminated lake sediments; a review. In: Vasari, Y., Saarnisto, M. & Seppälä, M. (Eds.), Proc. Working Session of Comm. on Holocene-INQUA. pp. 97–108. Acta Univ. Ouluensis. A. 82. 1979. Geol. 3.

Saarnisto, M., Huttunen, P. & Tolonen, K., 1977. Annual lamination in Lake Lovojärvi, southern Finland, during the past 600 years. Ann. Bot. Fennici 14: 35–45.

Smith, I. N. & Walton, A., 1980. Sediment accumulation rates and geochronologies measured in the Saguenay Fjord using the Pb-210 dating method. Geochim. Cosmochim. Acta 44: 225–240.

Thompson, R., Bloemendal, J., Dearing, J. A., Oldfield, F., Rummery, T. A., Stober, I.C. & Turner, G. M., 1980. Environmental applications of magnetic measurements. Science 207: 481–486.

Vasari, Y., 1962. A study of the vegetational history of the Kuusamo district (North East Finland) during the Late-quaternary period. Ann. Bot. Soc. 'Vanamo' 33(1): 1–140 + Appendix.

Vasari, Y., 1963. Studies on the vegetational history of the Kuusamo district (North East Finland) during the Late-quaternary period. II. Radiocarbon datings. Preliminary report. Arch. Soc. 'Vanamo' 18(2): 121–127.

Vuorinen, J., 1978. The influence of past land use on the sediments of a small lake. Pol. Arch. Hydrobiol. 25: 443–451.

The use of fallout ^{137}Cs and $^{239,240}Pu$ for dating of lake sediments

Timo Jaakkola[1], Kimmo Tolonen[2], Pertti Huttunen[3] & Soili Leskinen[1]
[1] *Department of Radiochemistry, University of Helsinki, Unioninkatu 35, SF-00170, Helsinki 17, Finland*
[2] *Department of Botany, University of Helsinki, Unioninkatu 44, SF-00170, Helsinki 17, Finland*
[3] *Department of Biology, University of Joensuu, P.O. Box 111, SF-80101, Joensuu 10, Finland*

Keywords: paleolimnology, dating, radioactive fallout, sediment, ^{137}Cs, $^{239,240}Pu$, varves

Abstract

The distribution of ^{137}Cs and $^{239,240}Pu$ in sediment core samples of the Finnish lakes Laukunlampi, Lovojärvi and Pääjärvi were determined. The sediment samples were collected using dry ice and liquid nitrogen freezing methods. The sediments of these lakes are annually laminated. A clear maximum concentration of ^{137}Cs and $^{239,240}Pu$ was found in sediment layers formed during 1962–1964, the years of maximum fallout, and the middle of the 1950's can be estimated from the ^{137}Cs and $^{239,240}Pu$ profiles. The highest concentrations, 11 500 and 820 pCi kg^{-1} dry wt for ^{137}Cs and $^{239,240}Pu$, respectively, were found in the sediment of Laukunlampi. The vertical distribution was similar for ^{137}Cs and $^{239,240}Pu$ in the lakes investigated. A slight migration of $^{239,240}Pu$ and ^{137}Cs was found and the migration of ^{137}Cs seems to be higher than that of $^{239,240}Pu$. The advantages of ^{137}Cs dating method are rapidity and simplicity. $^{239,240}Pu$ is preferable when the sample size is small. The agreement found between ^{137}Cs and $^{239,240}Pu$ dates and the annual laminae show that these fallout radio isotopes can be used for dating sediments formed during the past 25 years.

Introduction

Atmospheric nuclear weapons testing produced global radioactive fallout that has been widely studied over the past 25 years. The deposition rate of ^{137}Cs and ^{90}Sr and more recently that of $^{239,240}Pu$ has been determined regularly at several locations around the world. The annual deposition of these artificial radionuclides has been strongly dependent on the frequency of nuclear testing. The peak radionuclide fallout produced by nuclear tests occurred during 1962–1964 and a lower peak occurred in 1958–1959. Negligible amounts were deposited before 1954. In Fig. 1 the annual deposition rate of $^{239,240}Pu$ in Helsinki, Finland, is indicated. ^{137}Cs fallout follows a similar pattern.

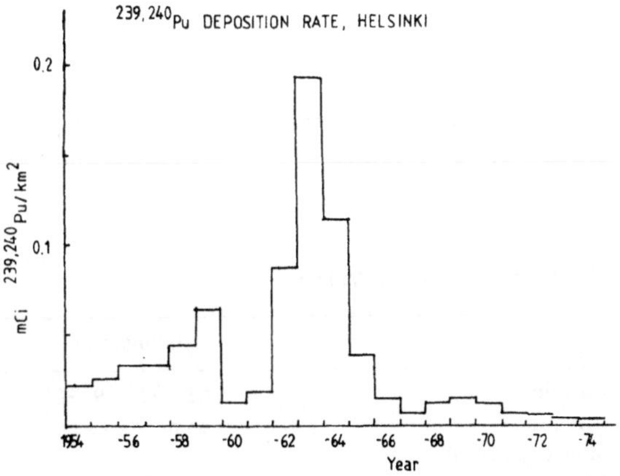

Fig. 1. The deposition rate of $^{239,240}Pu$ in Helsinki, Finland, during 1954–1974. The calculations are based on the $^{239,240}Pu$ analysis of surface air samples collected in Helsinki during 1962–1974 (Jaakkola *et al.* 1977) and on surface air concentration and deposition rate of $^{239,240}Pu$ in New York (Wrenn 1974).

The potential of ^{137}Cs in sediment core samples for dating was recognized at the beginning of the 1970's (e.g. Pennington *et al.* 1973). Since then several investigators have used ^{137}Cs and also $^{239,240}Pu$

Hydrobiologia 103, 15–19 (1983).

for the dating of recent sediments. In most of these studies the maximum ^{137}Cs and $^{239,240}Pu$ concentrations were interpreted to represent sedimentation in 1963–1964, the period of maximum fallout. In some profiles the depth at which the first considerable concentrations of ^{137}Cs and $^{239,240}Pu$ were found, was assumed to represent the year 1954.

The aim of this study has been to investigate the migration rate of ^{137}Cs and $^{239,240}Pu$ in sediment and the reliability of the dating method based on the distribution of these radio-nuclides in sediment. The investigations were carried out by analyzing annually laminated sediment core samples collected from three Finnish Lakes.

Material and methods

A description of the lakes included in this study is presented in Table 1.

The annually laminated sediment core samples were collected using the box and liquid nitrogen freezing methods developed by Huttunen & Meriläinen (1978) and Meriläinen & Huttunen (1978). The dating of the sediment was supported by both microscopic and historical analyses (Simola 1979, Battarbee 1981, Simola & Tolonen 1981, Appleby *et al.* 1979, Simola & Uimonen-Simola 1983). All the sediments studied were distinctly varved (Fig. 2).

^{137}Cs was determined directly from dried samples using NaI well crystal and a multichannel analyzer. For the analysis of plutonium samples were wet-ashed using a mixture of nitric acid and hydrochloric acid. Plutonium was separated by anion-exchange and electrodeposited on to a platinum or stainless steel disc. The ^{242}Pu isotope was used as a tracer for the determination of the chemical yield. The alpha-activity was determined using a silicon surface-barrier semiconductor detector and a multichannel pulse height analyzer. The amount of ^{239}Pu and ^{240}Pu isotopes is presented in combined form since the difference between the alpha energies of these two isotopes is too small to be separated alpha-spectrometrically.

Results

^{137}Cs and $^{239,240}Pu$ have quite a similar distribution in the sediment of Laukunlampi. Both ^{137}Cs and $^{239,240}Pu$ have a clear maximum in 1963–1965 (Fig. 3), but there seems to be some migration of both radio-nuclides downwards in the sediment. The diffusion is higher for ^{137}Cs than for $^{239,240}Pu$. The maximum $^{239,240}Pu$ concentration found during this study was 820 pCi (30.3 Bq) per kg dry weight in the sediment of Laukunlampi.

The distribution of ^{137}Cs and $^{239,240}Pu$ is also very similar in Lovojärvi (Fig. 4). The maximum content can be found in the middle of the 1960's. The sediment distribution of $^{239,240}Pu$ corresponds better to the fallout distribution than that of ^{137}Cs.

In the case of Pääjärvi the profiles of ^{137}Cs and

Table 1. Description of the lakes.

	Laukunlampi	Lovojärvi	Pääjärvi
Location	62°40′N, 29°02′E	61°05′N, 25°02′E	61°04′N, 25°05′E
Elevation (above sea level)	84 m	102 m	103 m
Maximum depth	26 m	17.5 m	87 m
Lake area (A)	0.084 km^2	0.048 km^2	13.4 km^2
Drainage area (D)	0.21 km^2	5.70 km^2	244 km^2
D/A	2.5	119	18.2
Character of the lake	kettle hole lake, stable thermal stratification during summer stagnation	eutrophic, high annual primary production	mesohumic, oligotrophic
Dominant catchment soil type	gravel and sand	loam and silt	silty moraine and clay

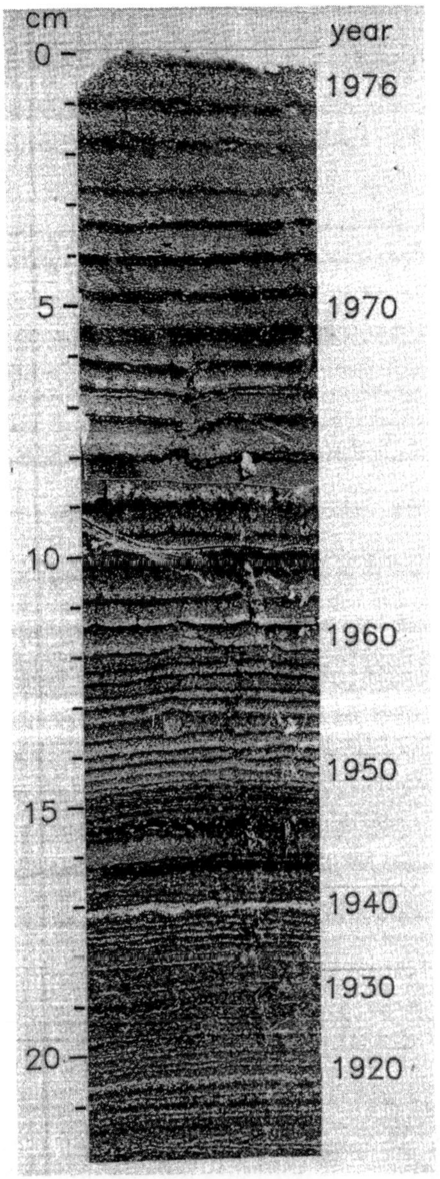

Fig. 2. A frozen core from upper laminated sediment of Laukunlampi.

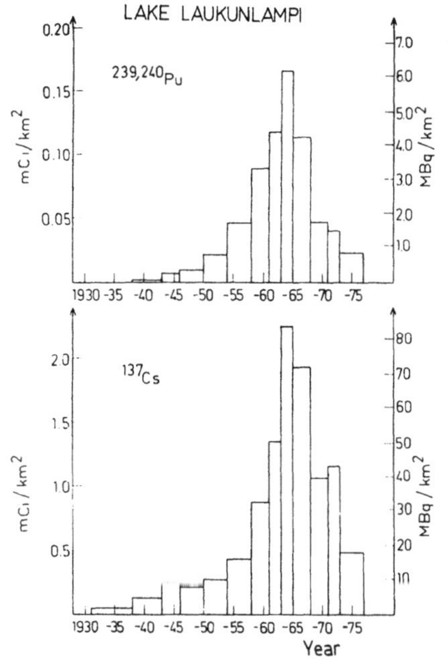

Fig. 3. Distribution of $^{239,240}Pu$ and ^{137}Cs in a sediment core from Laukunlampi (1931–1977 and 23 cm sediment).

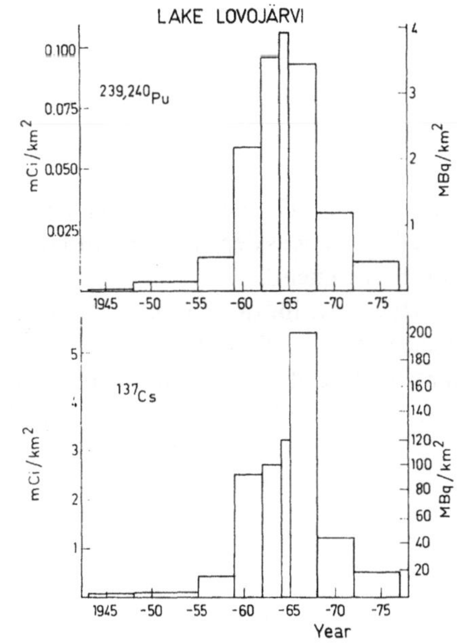

Fig. 4. Distribution of $^{239,240}Pu$ and ^{137}Cs in a sediment core from Lovojärvi (1943–1977 and 14.5 cm sediment).

$^{239,240}Pu$ are generalized because every transverse section represents several years (Fig. 5). This is due to the low sediment accumulation rate in Pääjärvi. The distribution of ^{137}Cs and $^{239,240}Pu$ is again very similar and a clear maximum can be found in the fraction representing 1960–1965. Again the diffusion rate of ^{137}Cs seems to be higher than that of $^{239,240}Pu$.

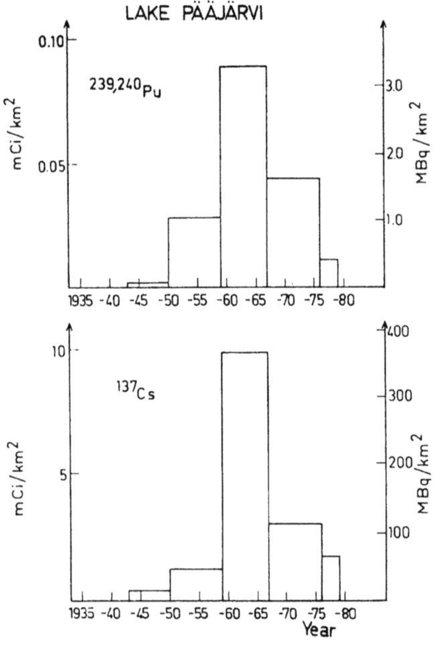

Fig. 5. Distribution of $^{239,240}Pu$ and ^{137}Cs in a sediment core from Pääjärvi (1943–1979 and 10 cm sediment).

Discussion

It is interesting to calculate the total amounts of $^{239,240}Pu$ and ^{137}Cs accumulated per unit area in the sediment cores of these three lakes and to compare them with the values of cumulative fallout. The estimated cumulative fallout $^{239,240}Pu$ was 1.6 ± 1.0 mCi km^{-2} at a latitude of 60–70 °N (Hardy *et al.* 1973). In the sediments of Laukunlampi, Lovojärvi and Pääjärvi the values are 1.9, 1.1 and 1.4 mCi km^{-2}, respectively. These values indicate that most of these radio-nuclides deposited on the lake can be found in the sediment.

The vertical distribution found in this study was almost the same for ^{137}Cs and $^{239,240}Pu$. However, the migration rate of ^{137}Cs in sediments seems to be higher than that of plutonium. The advantages of the ^{137}Cs-dating method are that it is rapid and simple. In the future the relatively short half-life of ^{137}Cs (30.1 a) will limit its use. The half-life of ^{239}Pu is 24 400 a and that of ^{240}Pu 6 540 a. The determination of $^{239,240}Pu$ is preferable when the sample size is small. The lowest measurable radioactivity of $^{239,240}Pu$ was 5 fCi = 1.85 mBq per sample.

The measurement of the ratio ^{238}Pu to $^{239,240}Pu$ gives additional information about the date of sedimentation. Before 1965 the ratio ^{238}Pu to $^{239,240}Pu$ in the atmosphere and biosphere was 0.02–0.03. After 1965 the value of this ratio increased due to the burn-up of the SNAP-9A satellite power source which contained 17 KCi of ^{238}Pu. This happened at a height of 45 km above Australia in March 1964. The ratio of ^{238}Pu to $^{239,240}Pu$ in the sediment of Laukunlampi was 0.030 ± 0.015 before 1965 and 0.059 ± 0.028 afterwards. This indicates that the plutonium found in the sediment from the years 1946–1964 really originates from the fallout deposited before the year 1964 (Table 2).

Table 2. $^{238}Pu/^{239,240}Pu$ in the sediment samples of Laukunlampi.

Year of sedimentation	$^{238}Pu/^{239,240}Pu$	
1946–1949	0.014	mean 0.030 ± 0.015*
1950–1953	0.030	
1954–1957	0.021	
1958–1961	0.026	
1961–1962	0.028	
1963–1964	0.059	
1965–1967	0.033	mean 0.059 ± 0.028*
1968–1970	0.090	
1971–1972	0.075	
1973–1976	0.037	

* Standard deviation (1 σ).

It can be concluded that it is possible to use $^{239,240}Pu$ and ^{137}Cs for reliable dating of recently formed sediment layers from lakes of different morphology hydrology. It seems to be useful also to use these radionuclides to complement the ^{210}Pb dating method. Even when the ^{210}Pb profile is disturbed due to the mixing of the surface layer (bioturbation, resuspension) the fallout nuclides, ^{137}Cs and $^{239,240}Pu$, may give useful information because about 50% of the total cumulative fallout occurred during the years 1962–1964.

Acknowledgements

We wish to thank Mr. Juha Hankkila, who carried out the radionuclide analyses of the sediment samples collected from Pääjärvi. The study was supported by The Academy of Finland, Council of Natural Sciences.

References

Appleby, P. G., Oldfield, F., Thompson, R., Huttunen, P. & Tolonen, K., 1979. ^{210}Pb dating of annually laminated lake sediments from Finland. Nature 280: 53–55.

Battarbee, R. W., 1981. Diatom and chrysophyceae microstratigraphy of the annually laminated sediments of a small meromictic lake. In: Florilegium Florinis Dedicatum (Königsson, L.-K. & Paabo, K., Ed.), Striae 14: 105–109, Uppsala.

Hardy, E. P., Krey, P. W. & Volchok, H. L., 1973. Global inventory and distribution of fallout plutonium. Nature 241: 444–445.

Huttunen, P. & Meriläinen, J., 1978b. New freezing device providing large unmixed sediments samples from lakes. Ann. Bot. Fennici 15: 128–130.

Jaakkola, T., Mussalo, H. & Tiainen, S., 1979. Plutonium in the Helsinki air during 1962–1977. In: 'Radioactive foodchains in the subratic environment', contract EY-76-C-02-3011.A003 of the US DOE, Final report, September 1979 (paper no 99).

Meriläinen, J. & Huttunen, P., 1978a. The liquid nitrogen method in sampling of the uppermost lake sediment. Pol. Arch. Hydrobiol. 25: 287–290.

Pennington, W., Cambray, R- S. & Fisher, E. M., 1973. Observations on lake sediments using fallout ^{137}Cs as a tracer. Nature 242: 324–326.

Simola, H., 1979. Micro-stratigraphy of sedimenmt laminations deposited in a chemically stratifying eutrophic lake during the years 1913–1976. Holarct. Ecol. 2: 160–168.

Simola, H. & Tolonen, K., 1981. Diurnal laminations in the varved sediment of Lake Lovojärvi, South Finland. Boreas 10: 19–26.

Simola, H. & Uimonen-Simola, P., 1983. Recent stratigraphy and accumulation of sediment in the deep, oligotrophic Lake Pääjärvi in South Finland. Dev. Hydrobiol. 00: 000–000.

Wren, M. E., 1974. In: 'Plutonium and other transuranium elements: Sources Environmental Distribution and Biomedical Effects', U.S. Atomic Energy Commission, report WASH-1359.

Caesium-137 profiles in the sediments of a partial-meromictic lake on Great Sandy Island (Fraser Island), Queensland, Australia

M. E. Longmore (McCallan)[1]*, B. M. O'Leary[2] & C. W. Rose[1]
[1] *School of Australian Environmental Studies, Griffith University, Brisbane, Australia*
[2] *Department of Physics, Queensland Institute of Technology, Brisbane, Australia*
* *Present address: Department of Biogeography and Geomorphology, Research School of Pacific Studies, Australian National University, Canberra, Australia*

Keywords: paleolimnology, caesium-137, perched lake, meromictic, diffusion, bioconcentration, sediment accumulation

Abstract

Hidden Lake is a perched, brown-water lake located in the centre of Great Sandy Island (Fraser Is.), S.E. Queensland. It is highly acid (pH 4.0), oligotrophic and is thermally and chemically stratified for most of the year. The sediments revealed a ^{137}Cs profile which departed from the temporal pattern of ^{137}Cs fallout in Brisbane and was represented by an exponential increase of ^{137}Cs towards the surface sediments from ca. 32 cm depth. The possible causes of the divergent profile are discussed, including physical and biological mixing, lag in the transport of catchment material to the sedimentary basin, diffusion, recycling and biological concentration. It is hypothesised that a combination of the last four processes, with diffusion facilitated by the highly acid conditions, are the major causes of the observed ^{137}Cs profile. Possible recycling and bioconcentration of ^{137}Cs raises questions as to the validity of this method of dating in similar environmental conditions, and as to the interpretation of other palaeochemical data. These hypotheses are to be tested against profiles obtained from ^{14}C, ^{210}Pb, $^{239/240}$Pu analyses of the sediment, and the measurement of ^{137}Cs activity in the water and biota of the lake.

Radioactive fallout in Australia

The radioisotope caesium-137 (^{137}Cs – half life 30 years) is a product of nuclear explosions since 1945 and, therefore, only a recent addition to the environment. Fallout of ^{137}Cs has occurred generally over the earth's surface since the first thermonuclear explosion in late 1952. The radioisotope is transported from the stratosphere into the troposphere in mid and polar latitudes (Davis 1961), and to the ground by rainfall. The pattern and levels of fallout are similar throughout Australia, with some variability due to latitude (Davis 1961) and meteorological conditions (Gibbs & Wilson 1965).

In Australia, nuclear fallout has been monitored since 1957 (Gibbs & Wilson 1965) and the ^{137}Cs component since 1964 (Fig. 1). The first estimated peak in ^{137}Cs fallout occurred in 1958 with the escalation of nuclear testing in the northern hemisphere, followed by a moratorium declared from late 1958 to 1961. Testing resumed and escalated in 1961/62, with a second and larger peak of ^{137}Cs fallout observed in the northern hemisphere during 1963, and in the southern hemisphere during 1964. Underground testing, following the Test Ban Treaty of 1963, led to lower levels of global fallout in the ensuing period, but the French tests in the South Pacific were principally responsible for the peak of ^{137}Cs fallout observed during 1971 in Australia. Levels have been very low, and falling, since that time. In general, levels of radioactive fallout in Australia have been smaller, by a factor of 10 or more, than in the northern hemisphere. In summary, Fig. 1 shows that the temporal variation in fallout in Australia is represented by a relatively flat curve with three small peaks of similar magnitude, followed by a period of declining fallout towards the present.

Hydrobiologia 103, 21–27 (1983).

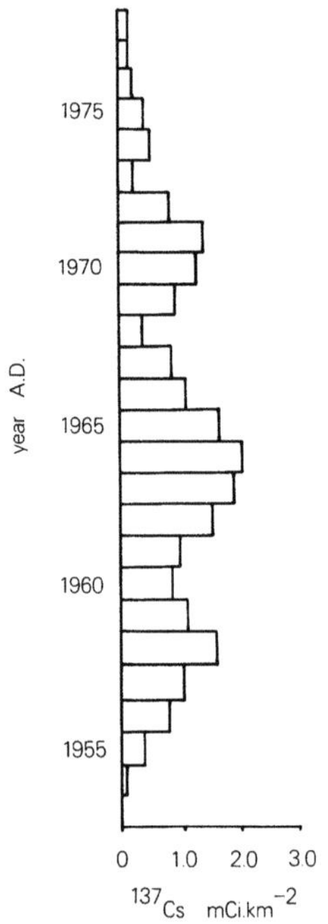

Fig. 1. Caesium-137 fallout in Brisbane 1954–1978 (corrected for radioactive decay to 1978). 1 mCi = 3.7×10^7 Bq. Caesium-137 fallout for the period 1954–1963 was estimated from extrapolated and measured ^{90}Sr data for Brisbane (Gibbs *et al.* 1965) using a ^{137}Cs/^{90}Sr ratio of 1.7 prior to 1961 and 1.5 from 1961 (Comar & Lengemann 1966). For the period 1964–1973 ^{137}Cs fallout was measured in Brisbane by the Australian Radiation Laboratory (A.R.L.) (full list of references in Bullock *et al.* 1975) and in the period 1974–1978 by A.E.R.E. (full list of references in Cambray *et al.* 1979). Data from the A.R.L. for the latter period was not available (Moroney, pers. comm.).

When ^{137}Cs reaches the ground it becomes strongly adsorbed to soil, particularly clay particles (e.g. Ritchie *et al.* 1970). As ^{137}Cs tends to move with soil particles, it has been a valuable radioactive tracer of soil movement and redistribution (e.g. Ritchie *et al.* 1974; McCallan *et al.* 1980) and has also been used as a chronological tool in lake sediment profiles (e.g. Pennington *et al.* 1973) where it mirrors the temporal variation in radioactive fallout.

While undertaking a palaeolimnological/ecological project, the use of the ^{137}Cs dating technique was attempted. Despite low levels of fallout in Australia, sufficient sedimentary material was collected, and adequately sensitive radioactive counting equipment was used, to allow the detection and measurement of ^{137}Cs in the lake sediment profile. The ^{137}Cs profile indicated that the dating technique could not be applied at this lake site. However, the measured profile raised many questions, not only as to the validity of the dating technique in certain situations, but also concerning pathways of the radioisotope in fresh water lake catchment ecosystems and the interpretation of other palaeolimnological data.

The study site

Hidden Lake (Lat. 25° 14′, Long. 153° 10′, grid ref. 9547–170088) is a perched fresh water lake ca. 100 m above m.s.l., 8–9 m deep, with a surface area of ca. 9 ha. The lake has no surface inflow or outflow. The special physical and biological characteristics of this lake and similar lakes have been described (Bayly *et al.* 1975). It is a sheltered, oligotrophic, brown water lake, high in dissolved organic acids, with a pH of ca. 4.0, a secchi disc reading of 0.75–1.5 m and low in conductivity (ca. 80 μ S). The lake is thermally stratified for most of the year, resulting in the development of anoxic conditions in the hypolimnion and the preservation of homogenous, black, highly organic (95% loss on ignition) mud to a depth of at least 3 m. The sediments are believed to be derived mainly from allochthonous sources. The catchment is steep-sided, with slopes of ca. 1 in 5, enclosed with ridges rising 20 m above the lake surface. The soils are deeply weathered podzols on siliceous sands, with a highly organic surface layer to a depth of ca. 50 cm. The vegetation consists of dry and wet sclerophyll forest (dominated by *Eucalyptus* spp.) and rainforest.

Field and laboratory methods

Four series of cores (HL–B–C–E and –F) were collected with a 1 m corer of I.D. 5.5 cm (Mackereth 1969) in 1977/78 at the deepest part of the lake (Fig. 2). The sediment/water interface appeared undisturbed. Each core series consisted of 10 to 21 cores combined at 1–2 cm intervals. An additional set of single cores was taken in 1978 at twelve random sites throughout the lake (Fig. 4), each bulked to a depth of 30 cm, dried at 105 °C, and ground through a 2 mm screen to obtain constant density and therefore constant counting geometry. The

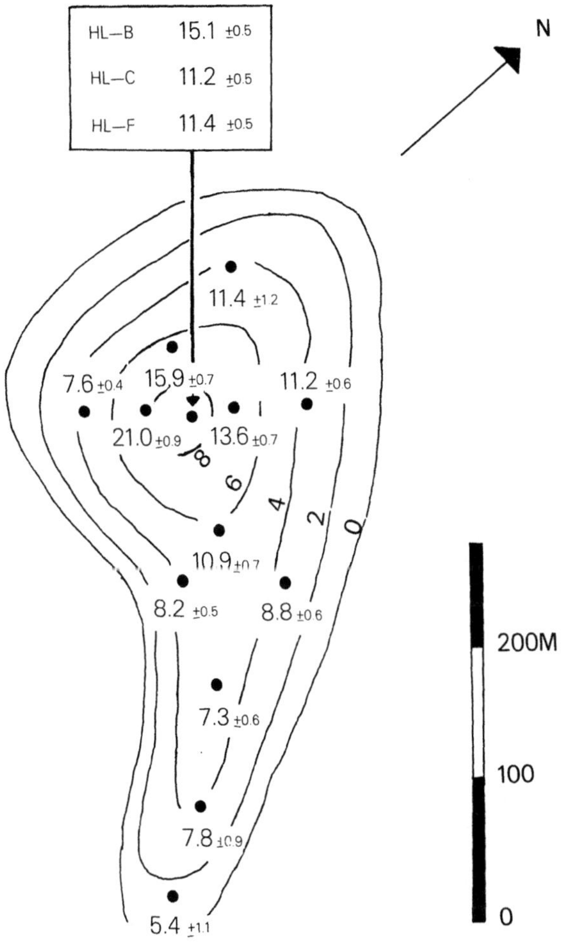

Fig. 2. Hidden Lake, Fraser Island: variation in areal concentration (mCi km^{-2}) of caesium-137 (corrected for radioactive decay to 1978). 1 mCi = 3.7×10^{7} Bq. One standard deviation is shown. Contour depths in m. HL–E sampled in same vicinity as HL–B–C and –F.

^{137}Cs activity of all sediment core samples was counted on a Ge[Li] semi-conductor detector, by measurement of the 662 keV rays emitted. The lowest specific activities which can be detected in 20 g samples by using 40 h counting times, and the 2σ limit is approximately 0.1 pCi g^{-1}. The sediment samples have a dry weight of ca. 20–30 g. The total sample is counted in a petri dish for ca. 24 h. Fuller details of the counting technique are given in McCallan *et al.* (1980).

Radiocarbon measurements were carried out on HL–F 16–18 cm and HL–F 30–32 cm and results are shown in Fig. 3.

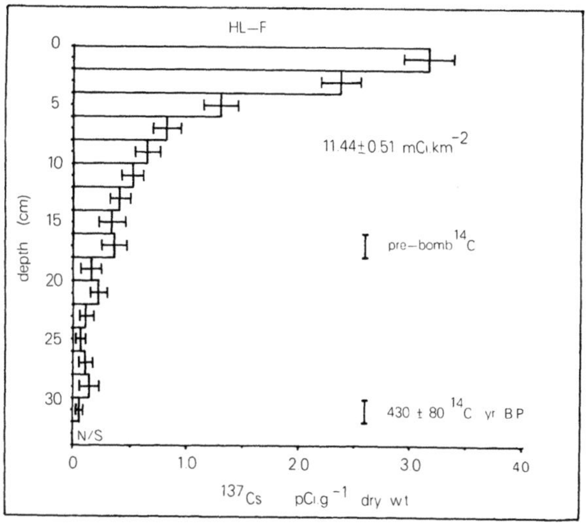

Fig. 3. Hidden Lake, Fraser Island: variation in specific activity (pCi g^{-1}) of caesium-137 with depth in lake sediment profile HL–F. 1 pCi = 3.7×10^{-2} Bq. N/S: not significant count rate. One standard deviation is shown. ^{14}C dates are also shown.

Results

The ^{137}Cs activities in all 4 core series HL–B, C, E and F (Fig. 4) show a similar distribution pattern with depth. ^{137}Cs is detectable to a depth of 30–32 cm, and increases from ca. 0.29 pCi g^{-1} (lpCi = 3.1×10^{-2} Bq (Becquerel)) at 16 cm to ca. 0.67 pCi g^{-1} at 8 cm, with an acceleration in the rate

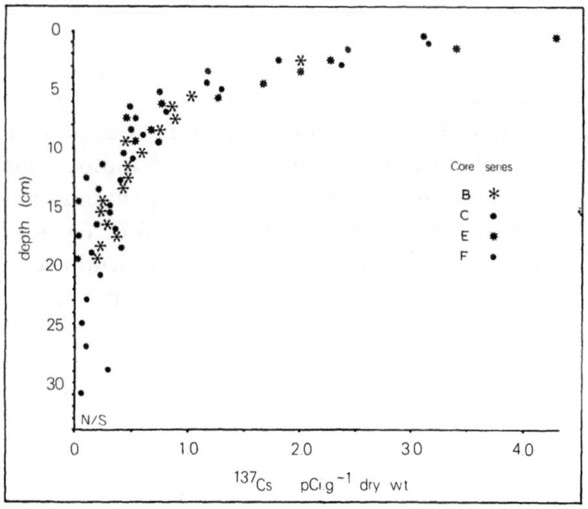

Fig. 4. Hidden Lake, Fraser Island: variation in specific activity (pCi g $^{-1}$) of caesium-137 with depth in lake sediment profiles. 1 pCi = 3.7×10^{-2} Bq. N/S: not significant count rate.

of increase to ca. 3.52 pCi g^{-1} at the surface. The general trend of the curve is an exponential increase from 32 cm to the surface.

In Fig. 3 the ^{137}Cs activities for HL–F are illustrated, along with the counting error of each 2 cm depth sample. The same basic trend in the vertical distribution of ^{137}Cs as shown by the other core series, is evident. The first detectable levels of ^{137}Cs activity occur at a depth of 30–32 cm (430 ± 80 ^{14}C a. B.P. (ANU 2678)). Towards the surface the first substantial increase in ^{137}Cs activity occurs at a depth of 16–18 cm which is ^{14}C dated as 98.3 ± 0.9% modern (ANU 2677), indicating the absence of any ^{14}C enrichment from nuclear bomb testing (Polach & Singh 1980).

The integrated ^{137}Cs activity (areal concentration) of HL–F, corrected for radioactive decay to 1978, is 11.4 ± 0.5 mCi km^{-2}. In Fig. 2 the areal concentrations (corrected to 1978) calculated for 15 cores taken at water depths varying from 1.5 to 8.5 m, ranged from 5.4 ± 1.1 to 21.0 ± 0.9 mCi km^{-2} with a mean of 11.2 ± 0.2 mCi km^{-2} and a sample standard deviation of 4.0. The areal concentrations generally increase towards the deepest part of the lake, but all values are less than the cumulative total fallout of ^{137}Cs for Brisbane (corrected to 1978) of 23.2 mCi km^{-2}.

Discussion

The vertical distribution of ^{137}Cs in the sediment cores from Hidden Lake (Figs. 3 & 4) is markedly different from expectations based on estimated and measured temporal variation of ^{137}Cs fallout in Brisbane (Fig. 1). According to sediment accumulation rates, derived from a series of 6 radiocarbon dates taken from a longer core (275 cm) dating back to 8 460 ± 100 radiocarbon years B.P. (A.N.U. 2676), the sedimentary deposits laid down in A.D. 1954 and A.D. 1964 should occur at a depth of ca. 9 and ca. 6 cm, respectively. Thus if there were no physical mixing of sediment and no diffusion, and given adequate sampling resolution, one would expect some representation of the temporal pattern of atmospheric fallout in the top ca. 10 cm of the sedimentary profile. The following discussion will be concerned with explanation of the observed ^{137}Cs profile, which diverges from these expectations.

Radionuclides are adsorbed onto, and into, organic and inorganic particles, and removed from the water by sedimentation, diffusion (Duursma & Hoede 1967; Duursma & Gross 1971; Aston & Duursma 1973; Lerman & Lietzke 1975), radioactive decay and surface outflow (Lerman & Lietzke 1975). As the lake has no surface outflow, and as the radioactive decay of ^{137}Cs is relatively slow, adsorption, sedimentation and diffusion, along with other diagenetic processes, must form the major removal mechanisms in Hidden Lake.

Environmental conditions which favour the adsorption of ^{137}Cs on to sedimentary surfaces include the presence of finer sized particles (Ritchie & McHenry 1976), the presence of clay (Ritchie *et al.* 1973), the presence of organic matter (Ritchie *et al.* 1970), a low concentration of oxygen in the water (Duursma & Gross 1971), a low water salinity (Aston & Duursma 1973) and a low concentration of hydrogen ions (Agre & Korogodin 1960; Davis 1961; Lerman & Lietzke 1975).

If adsorption is strong, there is a higher distribution coefficient between sediment and water, and the concentration gradient is therefore reduced in the interstitial water (Duursma & Gross 1971). On the other hand, increased desorption leads to the development of concentration gradients, and therefore to the diffusion of ions in the direction of lower concentrations according to Fick's Law.

Despite the absence of clay from the Hidden Lake sediment, the environmental conditions seem ideal for the adsorption of ^{137}Cs because of the fine, highly organic, colloidal nature of the sediments in persistently anaerobic waters of low salinity. It is therefore believed that the high concentration of hydrogen ions (pH ca. 4.0) may be a major controller of diffusion of ^{137}Cs in the interstitial waters. The highly acid conditions may allow the release of ^{137}Cs from the sedimentary particles, and its subsequent movement through the sediment to areas of lower concentration, deeper in the sediment and in the overlying waters. The presence of ^{137}Cs activity below 10 cm depth suggests some redistribution, possibly by diffusion, of the radioisotope within the sediments. Slow mass flow of water through the lake bottom is a possibility, and could transport ^{137}Cs downward.

In the literature, ^{137}Cs profiles have revealed the presence of the radioisotope in sediments older than A.D. 1954, and redistribution, as in Hidden

Lake, is usually represented by an elongated 'tail' into deeper sediments (Krishnaswamy *et al.* 1971; Pennington *et al.* 1973; Ritchie *et al.* 1973; Robbins & Edgington 1975). In Lake Michigan (Robbins & Edgington 1975) ^{137}Cs was repeatedly found below the ^{210}Pb-dated 1945 level in sediment cores. Movement of ^{137}Cs in the Lake Michigan cores was attributed to physical or biological mixing rather than diffusion, because of the low diffusion coefficient of ^{137}Cs (Lerman & Lietzke 1975) although this process was not completely ruled out. In sediment cores taken from Finnish lakes, ^{137}Cs activity was present in annual laminations dating back to A.D. 1931, and A.D. 1937 (Jaakola *et al.* 1983). The presence of annual laminations discounts any significant bioturbation of the sediment and suggests diffusion of ^{137}Cs.

An hypothesis of low physical or biological mixing in Hidden Lake sediments is supported by the lake being sheltered and thermally stratified for most of the year. Water currents are thus unlikely to cause sediment mixing, and one of the few organisms which can survive the anoxic conditions in the hypolimnion, *Chaoborus* spp. larvae, have not been observed in the surface sediments. In addition, the lack of ^{14}C enrichment at 16–18 cm indicates that surface sediments with ^{14}C-enriched material by bomb testing had not been mixed into these lower sediments.

The exponential increase of ^{137}Cs towards the surface sediments could be attributed to one or both of the following causes. One such explanation could be a long lag in the sedimentation (Ritchie *et al.* 1973, Stiller & Assaf 1973; Brakke, pers. comm.) where sediments from the catchment carrying the peak levels of ^{137}Cs are delayed in reaching the sedimentary basin, thus overlying and hiding the true atmospheric pattern of fallout. Due to protective tree canopies and litter layers, and the high infiltration capacity of the sandy soil minimising surface runoff, negligible erosion rates are a real possibility. This is substantiated by the slow rates of sediment accumulation on a dry wt. basis (ca. 0.03–0.04 g cm^{-2} a^{-1}) recorded in sediment traps. In the literature, slow sediment accumulation rates appear to be associated with profiles similar in shape to those of Figs. 3 & 4 (Ravera & Premazzi 1971; Ritchie *et al.* 1973).

Consideration of the integrated ^{137}Cs profiles in Hidden Lake sediments (Fig. 2) throws further light on the problem. In cores analysed in detail, at small vertical intervals, and in cores bulked down to the maximum depth of detectable ^{137}Cs, integrated totals are considerably less on average than the atmospheric fallout recorded in Brisbane. The spatial pattern of integrated totals generally follows that expected from gravity-induced flow of sedimentary material from the edges to the deepest part of the lake. In a sedimentary basin the integrated totals are expected, and are generally observed to be, greater than the atmospheric fallout, because the supplementation of direct atmospheric supply by contribution of ^{137}Cs – bearing sediments from the catchment (Krishnaswamy *et al.* 1971; McHenry *et al.* 1973; Robbins & Edgington 1975). Recorded exceptions are Lake Kinneret in Israel (Stiller & Assaf 1973) and the Lake District of England (Pennington *et al.* 1973).

The deficit of ^{137}Cs in Hidden Lake sediments suggests some net loss of ^{137}Cs deeper into the sediments or into the water. As sediment cores were sampled down to the maximum depth of detectable ^{137}Cs, the loss must be into the water column. This hypothesis is supported by the high concentration of ^{137}Cs measured in a sample of zooplankton *(Calamoecia tasmanica)* collected in 1978, which gave an activity level of 5.1 ± 0.4 pCi g^{-1}. This activity is higher than any measured in the vegetation, the soil litter layers or the surface sediments of Hidden Lake. It is probable that a portion of the ^{137}Cs in the sediment is diffusing into the water and becoming incorporated into the biotic component of the lake ecosystem. Several authors have referred to the rapid uptake of radionuclides by the biota of freshwater ecosystems, and the high concentration factors (1 000–25 000X) observed in the plankton (Il'in *et al.* 1958; Pendleton & Hanson 1958; Agre & Korogodin 1960; Eyman & Kevern 1975). Rapid removal of ^{137}Cs from the water to the sediments has also been observed by Agre & Korogodin (1960) and 97–99% of the ^{137}Cs pool has been found associated with the bottom sediments (Pendleton & Hanson 1958; Eyman & Kevern 1975).

With the characteristic rapid turnover of plankton, one would expect any ^{137}Cs incorporated into the biomass to be removed from the lake water on sedimentation, soon after the time of fallout. Nevertheless, due to the slow sedimentation rate in Hidden Lake, the surface sediment may be exposed for a sufficient time to allow decay of organic matter and release of ^{137}Cs into the water, where it is reconcentrated by the plankton. The recycling of ^{137}Cs would account for the high activities recorded in the zooplankton during a period of low radioactive fallout. Thus it is hypothesised that the remainder of the ^{137}Cs pool, not stored in the sediments, will be found in the biota of the lake. This hypothesis is consistent with the observed profiles of ^{137}Cs (Figs. 3 & 4). Diffusion of ^{137}Cs from sediment to the water body would tend to smooth out the temporal activity pattern of Fig. 1 from the sediment. Furthermore, any cycling of ^{137}Cs between the sediment and lake waters will tend to preferentially enrich the upper sediment layers which is very clearly the case in Hidden Lake (Figs. 3 & 4).

Conclusion

The ^{137}Cs profile in Hidden Lake is believed to be the result of a combination of diffusion, biological concentration from the water and at the surface sediments following sedimentation, and the time lag in sedimentation of catchment material. Diffusion and active uptake by living organisms parallels the processes postulated to explain divergent ^{137}Cs profiles in ombrotrophic peat bogs (Oldfield *et al.* 1978). Similar bog profiles, where such processes may also operate, have been described by Pakarinen & Tolonen (1977). These processes may also operate in other lakes to a greater or lesser degree depending on the nature of the sediment and the environmental conditions. It is proposed to further test the theory of ^{137}Cs diffusion in Hidden Lake sediment by comparison of the ^{137}Cs profile with that of other radioisotopes such as ^{14}C (Polach & Singh 1980) ^{210}Pb (Krishnaswamy *et al.* 1973; Robbins & Edgington 1975) and $^{239,240}Pu$ (Jaakola *et al.* 1983) which may be less mobile under these environmental conditions.

A parallel is seen with respect to mercury, and possibly other metals, where higher surface sediment concentration attributed to anthropogenic causes may be due to remobilisation, migration and enrichment of surface sediments by microbial activity in these upper layers (Sikes & Drain 1973; Elderfield & Hepworth 1975). The surface enrichment of sediments by mercury may in some cases be a natural process of diagenetic remobilisation allowing the recycling of metals (Elderfield & Hepworth 1975).

In addition, the total phosphorus profile in Hidden Lake sediment reveals a similar profile to that of ^{137}Cs. It is believed that a parallel recycling process to that postulated for ^{137}Cs, and by other authors for mercury, is occurring in the lake, with remobilisation being mainly due to the release of phosphorus from hydrous ferric oxides under the persistent anaerobic conditions (Mortimer 1971; Syers *et al.* 1973) characteristic of Hidden Lake, and biological concentration of the nutrient from the water and, after death, to the sedimentary surface.

Such postulated diagenetic processes raise questions, not only as to the validity of some radiometric dating techniques, but also as to the limitations of accurate interpretation of palaeochemical data used to reconstruct ecological processes in the recent and distant past.

Acknowledgements

Financial assistance was provided by Griffith University and A.I.N.S.E. Thanks are extended to the technicians at the two institutions, in particular R. W. Leech and J. F. Davey, for field and laboratory assistance, to the very many people with whom we discussed the work, and to C. Twang for typing the manuscript. The manuscript was critically read by J. G. Luly, R. L. Clark, G. Singh and J. Chappell whose comments were much appreciated.

References

Agre, A. L. & Korogodin V. I., 1960. The distribution of radioactive contamination in a stagnant reservoir. Med. Radiol. 5: 161–175.

Aston, S. R. & Duursma, E. K., 1973. Concentration effects on ^{137}Cs, ^{65}Zn, ^{60}Co and ^{106}Ru sorption by marine sediments, with geochemical implications. Neth. J. S. Res. 6: 225–240.

Bayly, I. A. E., Ebsworth, E. P. & Hang Fong Wan, 1975. Studies on the lakes of Fraser Island, Queensland. Aust. J. mar. Freshwat. Res. 26: 1–13.

Bullock, G. E., Duggleby, J. C., Kotler, L. H. & Wise, K. N., 1975. Strontium-90 and caesium-137 in the Australian environment during 1971, 1972 and 1973. In Fallout over Australia from Nuclear Tests. Rep. austr. Ionising Rad. Advi. Council 2. Aust. Govt. Publ. Serv.: 12–14.

Cambray, R. S., Fisher, E. M. R., Playford, K., Eakins, J. D. & Peirson, D. H., 1979. Radioactive fallout in Air and Rain. Results to the end of 1978. AERE-R9441. HMSO.

Comar, C. L. & Lengemann, F. W., 1966. Introductory paper: General principles of the distribution and movement of artificial fallout through the biosphere to man. In: Aberg, B. & Hungate, F. P. (Eds.). Radioecological Concentration Processes. Pergamon, N.Y., 1–18.

Davis, J. J., 1961. Caesium and its relationships to potassium in ecology. 1st nat. Symp. Radioecology Proc.: 539–556.

Duursma, E. K. & Gross, M. G., 1971. Marine sediments and radioactivity. In Radioactivity in the Marine Environment. Nat. Acad. Sci.: 147–160.

Duursma, E. K. & Hoede, C., 1967. Theoretical, experimental and field studies concerning molecular diffusion of radioisotopes in sediments and suspended solid particles of the sea. A: Theories and mathematical calculations. Neth. J. S. Res. 3: 423–457.

Elderfield, H. & Hepworth, A., 1975. Diagenesis, metals and pollution in estuaries. Mar. Pollut. Bull. 6: 85–87.

Eyman, L. D. & Kevern, N. R., 1975. Caesium-137 and stable caesium in a hypereutrophic lake. Health Phys. 28: 549–555.

Gibbs, W. J., Moroney, J. R., Stevens, D. J., Titterton, E. W. & Wilson, G. U., 1965. Strontium-90 in the Australian environment, 1951 to 1963. Aust. J. Sci. 28: 44–58.

Gibbs, W. J. & Wilson, G. U., 1965. Meteorological implications of measurements of strontium-90 in Australia. Aust. J. Sci. 28: 59–69.

Il'in, D. I., Moskalev, Iv. I. & Petrova, A. I., 1958. Accumulation of radioactive elements by some groups of aquatic organisms. Soviet J. Atomic Energy 3: 1013–1016.

Jaakola, T., Tolonen, K., Huttunen, P. & Leskinen, S., 1983. The use of fallout ^{137}Cs and ^{239,240}Pu for dating of lake sediments from Finland. Dev. Hydrobiol. 00: 000–000.

Krishnaswamy, D., Lal, S., Martin, J. M. & Meybeck, M., 1971. Geochronology of lake sediments. Earth planet Sci. Lett. 11. 407–414.

Lerman, A. & Lietzke, T. A., 1975. Uptake and migration of tracers in lake sediments. Limnol. Oceanogr. 20: 497–510.

Mackereth, F. J. H., 1969. A short core sampler for sub-aqueous deposits. Limnol. Oceanogr. 14: 145.

McCallan, M. E., O'Leary, B. M. & Rose, C. W., 1980. Redistribution of caesium-137 by erosion and deposition on an Australian soil. Aust. J. Soil. Res. 18: 2.

McHenry, J. R., Ritchie, J. C. & Gill, A. C., 1973. Accumulation of fallout caesium-137 in soils and sediments in selected watersheds. Wat. Resour. Res. 9: 676–86.

Mortimer, C. H., 1971. Chemical exchanges between sediments and water in the Great Lakes – speculations on probable regulatory mechanisms. Limnol. Oceanogr. 16: 387–404.

Oldfield, F., Appleby, P. G., Cambray, R. S., Eakins, J. D., Barber, K. E., Battarbee, R. W., Pearson, G. W. & Williams, J. M., 1979. ^{210}Pb, ^{137}Cs and ^{239}Pu profiles in ombrotrophic peat. Oikos 33: 40–45.

Pakarinen, P. & Tolonen, K., 1977. Distribution of lead in *Sphagnum fuscum* profiles in Finland. Oikos 28: 69–73.

Pendleton, R. C. & Hanson, W. C., 1958. Absorption of Cesium-137 by Components of an Aquatic Community. In Proc. int. Conf. Peaceful Uses Atomic Energy, 18. Geneva. United Nations, N.Y.: 419–422.

Pennington, W., Cambray, R. S. & Fisher, E. M., 1973. Observations on lake sediment using fallout Cs-137 as a tracer. Nature 242: 324–6.

Polach, H. & Singh, G., 1980. Contemporary ^{14}C levels and their significance to sedimentary history of Bega Swamp, New South Wales. Radiocarbon 22: 398–409.

Ravera, O. & Premazzi, G., 1971. A method to study the history of any persistent pollution in a lake by the concentration of ^{137}Cs from fall-out. In: Radioecology applied to the protection of man and his environment. Comm. eur. Communities int. Symp.: 1–16.

Ritchie, J. C., Clebsch, E. E. C. & Rudolph, W. K., 1970. Distribution of fallout and natural gamma radionuclides in litter, humus and surface mineral soil layers under natural vegetation in the Great Smoky Mountains, North Carolina-Tennessee. Health Phys. 18: 479–89.

Ritchie, J. C., McHenry, J. R. & Gill, A. C., 1973. Dating recent reservoir sediments. Limnol. Oceanogr. 18: 254–63.

Ritchie, J. C. & McHenry, J. R., 1976. A rapid method for determining recent deposition rates of freshwater sediments. In: Golterman (Ed.). Interactions between sediments and freshwater. Amsterdam: 203–207.

Ritchie, J. C., Sparberry, J. A. & McHenry, J. R., 1974. Estimating soil erosion from the redistribution of fallout ^{137}Cs. Soil Sci. Soc. am. Proc. 38: 137–139.

Robbins, J. A. & Edgington, D. N., 1975. Determination of recent sedimentation rates in Lake Michigan using Pb-2310 and Cs-137. Geochim. Cosmochim. Acta 39: 285–304.

Sikes, C. S. & Drain, M. P., 1973. Mercury in Sediments. Nature 244: 529.

Stiller, M. & Assaf, G., 1973. Sedimentation and transport of particles in Lake Kinneret traced by ^{137}Cs. Hydrol. Lakes Symp. 109: 397–403.

Syers, J. K., Harris, R. F. & Armstrong, D. E., 1973. Phosphate Chemistry in Lake Sediments. J. envir. Qual. 2: 1–14.

The assessment of ^{210}Pb data from sites with varying sediment accumulation rates

P. G. Appleby[1] & F. Oldfield[2]
[1] *Department of Applied Mathematics & Theoretical Physics, University of Liverpool, Liverpool L69 3BX, U.K.*
[2] *Department of Geography, University of Liverpool, Liverpool L69 3BX, U.K.*

Keywords: paleolimnology, ^{210}Pb, sediment accumulation

Abstract

The last few years have seen a dramatic growth in the use of ^{210}Pb sediment dating. Despite this, considerable doubt still surrounds the nature of the processes by which ^{210}Pb is deposited in lake sediments, and this has lead to a situation where there is a choice of dating models offering different interpretations of ^{210}Pb data. In assessing ^{210}Pb data it is therefore essential to first of all determine whether data is consistent with the assumptions of the dating model, and to then compare the ^{210}Pb chronology with independent dating evidence. We have tested ^{210}Pb data from a wide variety of sites, and our calculations indicate that the crs (constant rate of ^{210}Pb supply) model provides a reasonably accurate chronology when the total ^{210}Pb contents of cores from neighbouring locations are comparable.

Introduction

In most of the early papers on ^{210}Pb chronology (Krishnaswamy *et al.* 1971; Koide *et al.* 1973; Robbins & Edgington 1975), the methods used assumed a constant rate of sediment accumulation, and were applicable to sediment cores in which the unsupported ^{210}Pb activity declined exponentially down the core. There is, however, abundant evidence for accelerating accumulation rates in many lakes in recent times, and a need therefore to develop a reliable model for calculating ^{210}Pb dates in sites with varying rates of sediment accumulation.

Considerable doubt still surrounds the precise nature of the processes by which ^{210}Pb is deposited in lake sediments, and for this reason ^{210}Pb data from sites with varying sediment accumulation rates is interpreted by different authors in different ways. Since these processes may well vary from site to site it is unlikely that one model will be universally applicable. The purpose of this paper is to outline the principal assumptions used in ^{210}Pb chronology, and to present techniques for assessing the consistency of data with these assumptions.

^{210}Pb Chronology of lake sediments

The processes by which ^{210}Pb is delivered to catchment surfaces have been described in detail elsewhere (e.g. Krishnaswamy & Lal, 1978). The radium isotope ^{226}Ra (half-life 1622 yrs.) decays to yield the inert gas ^{222}Rn. This in turn decays (with a half-life of 3.83 days) through a series of short-lived isotopes to ^{210}Pb (half-life 22.26 yrs.). A fraction of the ^{222}Rn atoms formed by ^{226}Ra decay in soils escape from the soil particles into the interstices and diffuse through the soil into the atmosphere where they decay to ^{210}Pb. This is removed from the atmosphere by rain, snow, or dry fallout, falling either onto the land surface where it is trapped in surface soils (Benninger *et al.* 1975), or into lakes or oceans. ^{210}Pb falling into lakes is scavenged from the lake waters by sediments, and deposited on the bed of the lake. Krishnaswamy & Lal (1978) have estimated that the mean ^{210}Pb flux onto the land surface is about 0.45 pCi cm^{-2} a^{-1}. Local values of the mean ^{210}Pb flux are governed by local or regional meteorological factors, and range from 0.15 pCi cm^{-2} a^{-1} in Australia to 0.96 pCi cm^{-2} a^{-1}

Hydrobiologia 103, 29–35 (1983).

in Hokkaido, Japan.

The ^{210}Pb activity of lake sediments has two components, a supported component C_s deriving from ^{222}Rn decay within the sediment column, and an unsupported (or excess) component C deriving from the atmospheric fallout of ^{210}Pb. C_s can, for most purposes, be approximated by the ^{226}Ra concentration. In the absence of ^{210}Pb fallout, ^{210}Pb and ^{226}Ra would be in radioactive equilibrium. C is determined by subtracting C_s from the total ^{210}Pb concentration.

The unsupported ^{210}Pb concentration in each sediment layer declines with its age in accordance with the usual radioactive decay law. This law can be used to calculate the age of the sediment provided that the initial unsupported ^{210}Pb concentration when laid down on the bed of the lake can be estimated in some way.

If the erosive processes in the catchment are steady, and give rise to a constant rate of sediment accumulation, it is reasonable to suppose that every sediment layer will have the same initial unsupported ^{210}Pb concentration. In this case the unsupported ^{210}Pb concentration will decline exponentially with the cumulative dry mass of sediment. When the unsupported ^{210}Pb concentration C is plotted on a logarithmic scale, the resulting ^{210}Pb profile will be linear. The sediment accumulation rate can be determined graphically from the mean slope of the profile, or analytically by using a least squares fit procedure. In this model, sometimes referred to as the constant flux–constant sedimentation rate (cf:cs) model, the exact mechanism by which the sediment accumulated ^{210}Pb is immaterial.

^{210}Pb chronology under varying sediment accumulation rates

In many cases it is clear that rates of erosion and sedimentation have varied significantly during the past 150 years. In this event the ^{210}Pb profile may be expected to be non-linear. Many authors have observed such profiles. Figure 1 shows ^{210}Pb profiles from Lough Erne in N. Ireland which are both non-linear and non-monotonic over depths of up to 30 cm. Since changing accumulation rates may well result in variations in the initial ^{210}Pb concentra-

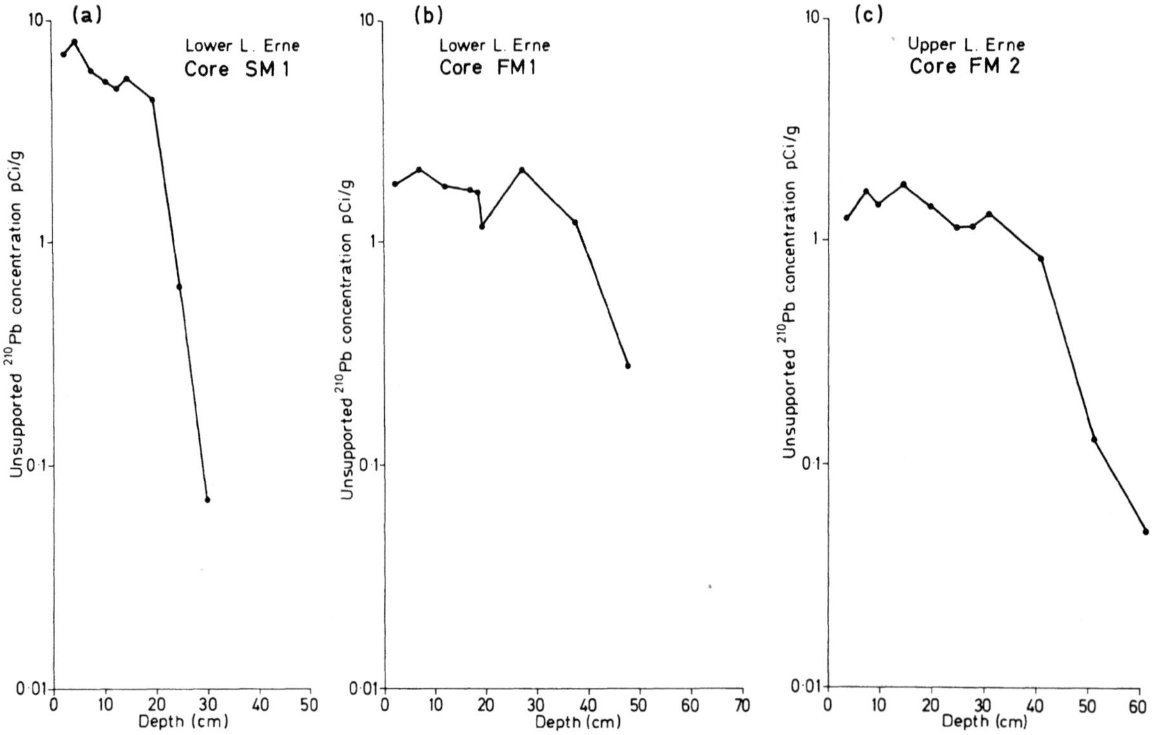

Fig. 1. Non-linear ^{210}Pb profiles from Lough Erne, N. Ireland (Oldfield *et al.* 1978).

tions of sediments, in order to construct a reliable ^{210}Pb chronology it is necessary to understand more precisely the processes by which sediment particles adsorb ^{210}Pb. The problem is further complicated by the fact that non-linear profiles may be caused by a number of other factors, including migration of ^{210}Pb through interstitial waters near the sediment water interface (Koide *et al.* 1973), mixing of near-surface sediments by physical (Petit, 1974) or biological (Robbins *et al.* 1977) processes, post-depositional redistribution of sediments either discontinuously through slumping (Edgington & Robbins 1977) or more or less continuously by sediment erosion.

There are essentially two models which are mathematically practicable for calculating ^{210}Pb dates under varying sediment accumulation rates, the constant rate of ^{210}Pb supply (c.r.s.) model and the constant initial concentration (c.i.c.) model. The physical bases for these models are discussed in more detail in Oldfield & Appleby (in press).

The constant rate of supply (or constant flux) model assumes that there is a constant fallout of ^{210}Pb from the atmosphere to the lake waters resulting in a constant rate of supply of ^{210}Pb to the sediments irrespective of any variations which may have occurred in the sediment accumulation rate. This model was proposed by Krishnaswamy *et al.* (1971). In support of this model, Benninger *et al.* (1975) have estimated that, at least in some cases, $>99\%$ of the ^{210}Pb deposited on the land surface is trapped in the soil layers. The dominant source of the ^{210}Pb in lake waters is then direct fallout onto the lake surface. Studies of the residence time of dissolved ^{210}Pb in lake waters (Schell 1977; Durham & Joshi 1980) have shown that this ^{210}Pb is rapidly transferred from the water to particulates.

If the assumptions of the c.r.s. model are satisfied, it may be shown that the cumulative residual unsupported ^{210}Pb, A, beneath sediments of age t varies according to the formula:

$$A = A(o)e^{-kt}$$

where A(o) is the total residual unsupported ^{210}Pb in the sediment column and k is the ^{210}Pb radioactive decay constant. A and A(o) are calculated by direct numerical integration of the ^{210}Pb profile. The age of sediments of depth x is then given by:

$$t = \frac{1}{k} \ln \frac{A(o)}{A}.$$

The sedimentation rate can be shown to be given directly by the formula (Appleby & Oldfield 1978):

$$r = \frac{kA}{C}$$

The ^{210}Pb supply rate is given by:

$$P = kA(o)$$

This procedure for calculating ^{210}Pb dates was first outlined by Goldberg (1963), and is set out in detail in Appleby & Oldfield (1978), and Robbins (1978).

The constant initial concentration (or constant specific activity) model assumes that an increased flux of sedimentary particles from the water column will remove proportionally increased amounts of ^{210}Pb from the water to the sediments. Under the assumptions of this model sediments will have the same initial unsupported ^{210}Pb concentration irrespective of any variations in sediment accumulation rate.

If the assumptions of the c.i.c. model are satisfied, the unsupported ^{210}Pb concentration will vary with depth in accordance with the formula:

$$C - C(o)\,e^{-kt},$$

where C(o) is the unsupported ^{210}Pb concentration of sediments at the sediment water interface. The age of a sediment layer with ^{210}Pb concentration C is therefore

$$t = \frac{1}{k} \ln \frac{C(o)}{C}$$

The calculation of ^{210}Pb dates by this procedure is illustrated in Pennington *et al.* (1976).

Assessment of ^{210}Pb data for consistency with dating models

The ^{210}Pb supply for a given core is likely to depend in a complex way on both the atmospheric input and sediment accumulation rate. The determination of a ^{210}Pb chronology will be feasible only if one or other of these factors is dominant. In considering a given data set it is therefore necessary to establish whether it conforms to either the c.r.s. model or c.i.c. model.

If the c.r.s. model is applicable, the following consequences may be noted.

1. Non-monotonic profiles may be expected in response to major changes in the accumulation rate, since faster net sediment accumulation will tend to depress initial unsupported ^{210}Pb concentrations, and vice versa.
2. Different cores from the same lake, or from the same depositional zone within a very large lake, or from different lakes within the same general area will have comparable ^{210}Pb residuals (i.e. total residual unsupported ^{210}Pb contents) despite differences in the accumulation rates.
3. The ^{210}Pb residuals of the cores should reflect the ^{210}Pb fallout from the atmosphere. Since the ^{210}Pb fallout lies in the range 0.2–0.9 pCi cm^{-2} a^{-1}, depending on the locality, the ^{210}Pb residuals should lie in the range 6–30 p Ci cm^{-2}.

All three points are well illustrated by the data from Lower Lough Erne. The profiles (Fig. 1) are non-monotonic. The ^{210}Pb residuals are virtually identical, 19.2 pCi cm^{-2} for FM1 and 20.7 pCi cm^{-2} for SM1 despite a 3-fold difference in the accumulation rates. The corresponding ^{210}Pb supply rates are 0.6 pCi cm^{-2} a^{-1} and 0.64 pCi cm^{-2} a^{-1} respectively. Table 1 summarises results from a variety of sites which satisfy the c.r.s. criteria. Figure 2 illustrates the convergence of the cumulative ^{210}Pb residuals for cores from two sites.

Table 1. ^{210}Pb parameters and sedimentation rates for cores from a variety of sites satisfying the c.r.s. criteria.

Coring site		Total residual unsupported ^{210}Pb content A(o) (pCi cm^{-2})	Unsupported ^{210}Pb conc. at surface C(O) (pCi g^{-1})	Mean sedimentation rate during (a) past 30 years r(o) (g m^{-2} a^{-1})	(b) past 100 years $\bar{r}$(g m^{-2} a^{-2})	^{210}Pb flux equivalent to ^{210}Pb residual pCi cm^{-2} a^{-1}
Ireland (Oldfield *et al.* 1978)						
Lower L. Erne	Core SM1	20.7	7.13	0.08	0.037	0.64
	Core FM1	19.2	1.82	0.31	0.12	0.60
Upper L. Erne	Core FM2	14.9	1.26	0.35	0.13	0.46
L. Augher (1977)		12.7	1.81	0.19	0.11	0.40
Wales (Elner & Wood 1980)						
Llyn Peris	Core A	38.8	1.08	1.07	0.27	1.21
	Core E	42.8	1.99	0.67	0.28	1.33
England						
Rostherne Mere	Core RMII	6.5	2.24	0.084	0.055	0.20
	Core N79	5.4	1.87	0.088	0.052	0.17
Newton Mere		5.3	4.43	0.032	0.029	0.17
Belgium (Oldfield *et al.* 1980)						
L. Mirwart	Core 1	10.5	3.29	0.092	0.14	0.33
	Core 2	10.0	2.21	0.13	0.055	0.31
	Core 3	10.7	2.34	0.14	0.12	0.33
Finland (Appleby *et al.* 1979)						
Laukunlampi		20.5	36.7	0.011	0.0072	0.64
Lovojarvi		20.1	5.55	0.078	0.059	0.63
Paajarvi		24.5	14.2	0.044	0.059	0.76
L. Michigan U.S.A. (Robbins & Edgington 1975)						
S. Margin	Core 54	3.98	10.08	0.0075	0.0069	0.12
	Core 31	5.09	6.72	0.018	0.011	0.16
S. Central	Core 11	5.37	7.10	0.020	0.019	0.17
	Core 17	6.04	11.42	0.013	0.013	0.19
N. Central	Core 105	8.14	14.5	0.014	0.010	0.25
	Core 103	8.31	15.1	0.014	0.011	0.26

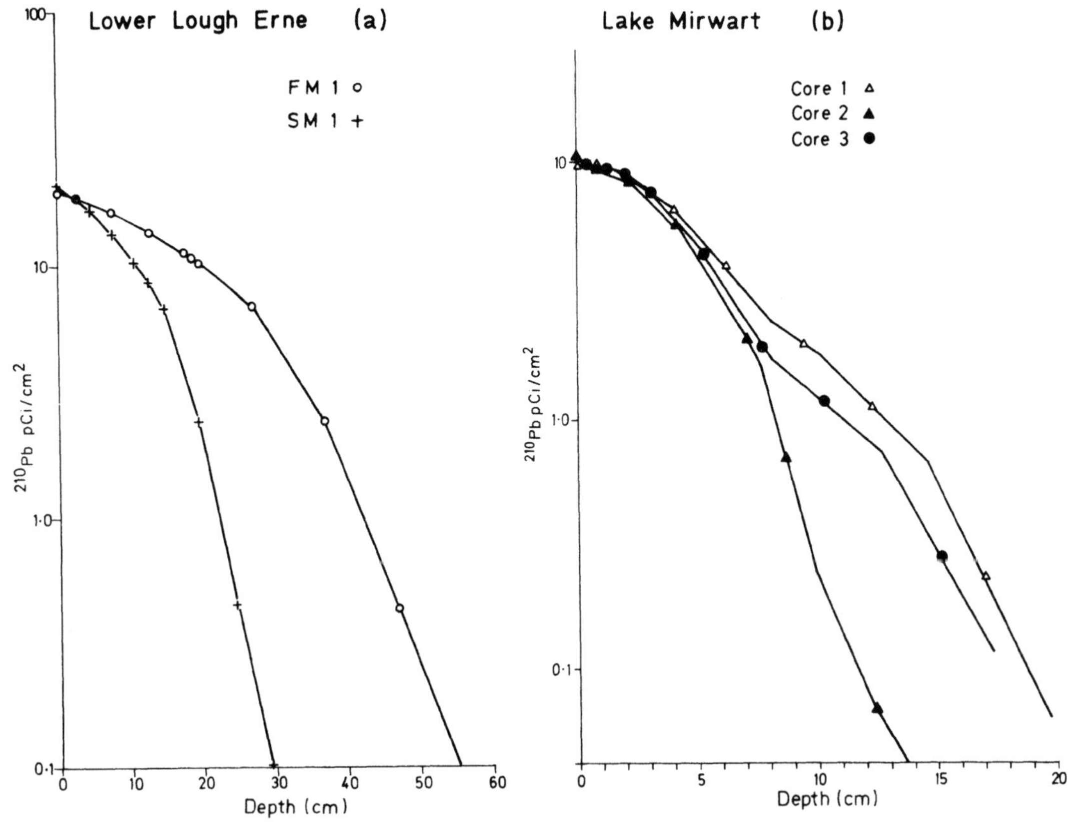

Fig. 2. Cumulative total residual unsupported ^{210}Pb contents for cores from (a) Lower Lough Erne and (b) Lake Mirwart, Belgium (Oldfield *et al.* 1980).

If the c.i.c. model is applicable, the following consequences may be noted.

1. The unsupported ^{210}Pb concentration must show a monotonic decline with depth.
2. The total cumulative residual unsupported ^{210}Pb in sediment cores from the same lake should vary roughly in proportion to the mean sediment accumulation rate.

In view of the efficiency at which ^{210}Pb is scavenged from lake waters by particulates, it is unlikely that this model will be widely applicable except possibly at sites where sediment focusing is a major factor.

In order to assess whether one or other of the models is generally valid, we have plotted the ^{210}Pb residuals A(o) and surface ^{210}Pb activities C(o) for about 50 cores against the corresponding mean sedimentation rates $\bar{r}$ (Fig. 3). If the c.r.s. model were generally valid, there should be no significant relation between A(0) and $\bar{r}$. On the other hand, since the ^{210}Pb activity is inversely proportional to the sediment accumulation rate the graph of C(o) against $\bar{r}$, plotted on log-log paper, should approximately follow a line making an angle of 45° with each axis. If the c.i.c. model were generally valid, A(o) should be proportional to $\bar{r}$, and there should be no significant relation between C(o) and $\bar{r}$. The graphs clearly support the c.r.s. model. The great majority of the cores have a ^{210}Pb residual in the range 6–30 pCi cm^{-2}. The average value of 17.7 pCi cm^{-2} corresponds to a mean ^{210}Pb supply rate of 0.55 pCi cm^{-2} a^{-1}. This compares well with estimates of the mean ^{210}Pb fallout.

Figure 3(a) includes data from Lake Michigan in the U.S.A. and Lough Neagh in N. Ireland which appears to be consistent with the c.i.c. model. In the case of the Lake Michigan data (from Robbins & Edgington 1975), six of the cores (see Table 1) have

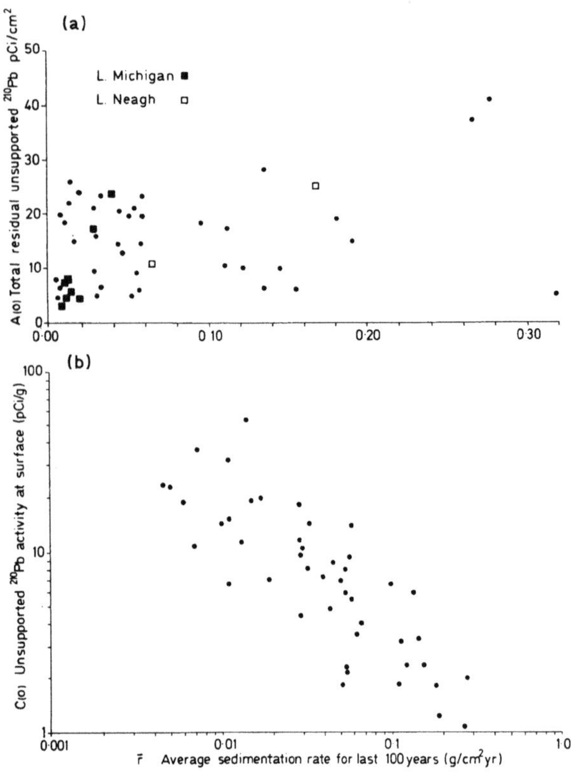

Fig. 3. ^{210}Pb parameters vs mean sediment accumulation rate $\bar{r}$ during the past 100 years for about 50 cores from various localities. Fig. 3(a) plots the total residual unsupported ^{210}Pb content A(o) vs $\bar{r}$. Fig. 3(b) plots the unsupported ^{210}Pb concentration at the surface C(o) vs $\bar{r}$.

^{210}Pb residuals which correspond more or less to the measured atmospheric ^{210}Pb fallout of ~ 0.2 pCi cm^{-2} a^{-1}. The two remaining cores, however, 29 and 100A, have ^{210}Pb residuals well in excess of these values. Both sets of results can be interpreted in terms of post-depositional sediment focusing (Oldfield & Appleby, in press).

When the ^{210}Pb profiles satisfy the assumptions of either model, errors in the chronology may still arise as a result of vertical mixing processes in the near surface sediments. Robbins *et al.* (1977) have modified the constant flux-constant sedimentation rate model to allow for such processes. A modified c.r.s. model which takes account of sediment mixing is set out in Oldfield & Appleby (in press).

Summary

Perhaps the most important conclusion that we would reach is that there is no single model that will give a reliable ^{210}Pb chronology in all cases, and that each data set must be evaluated independently for consistency with one or other of the dating models. A tentative procedure for evaluating data is as follows:

I Linear Profiles
All models give the same chronology

II Non-linear Profiles

(a) If the ^{210}Pb residuals are comparable with the known atmospheric flux, or with the ^{210}Pb residuals or nearby cores, the c.r.s. model would appear to be applicable. In our experience, ^{210}Pb dates calculated in this way have generally been consistent with independent dating evidence.

(b) If the ^{210}Pb residuals do not satisfy the requirements of the c.r.s. model, but there is independent evidence that the primary sediment accumulation rate has been constant, the c.i.c. model will be applicable. This case may occur in situations where sediment focusing takes place.

(c) If the ^{210}Pb residuals do not satisfy the requirements of either model, a ^{210}Pb chronology cannot be reliably established. We have found that ^{210}Pb dates calculated in these circumstances have in most cases been in conflict with independent dating evidence.

Acknowledgements

We wish to thank Dr. R. W. Battarbee for permission to use his data from Lough Erne and Loch Augher, Mr. R. Nelms for permission to use his data from Rostherne mere, and Mr. J. P. Smith for permission to use his data from Newtonmere.

References

Appleby, P. G. & Oldfield, F., 1978. The calculation of lead-210 dates assuming a constant rate of supply of unsupported ^{210}Pb to the sediment. Catena 5: 1–8.

Appleby, P. G., Oldfield, F., Thompson, R., Huttunen, P. & Tolonen, K., 1979. ^{210}Pb dating of annually laminated lake sediments from Finland. Nature 280: 53–55.

Benninger, L. K., Lewis, D. M. & Turekian, K. K., 1975. The use of natural ^{210}Pb as a heavy metal tracer in the river estuarine system. In: T. M. Church (ed.). Marine Chemistry in the Coastal Environment. Am. Chem. Soc. Symp. Ser. 18: 202–210.

Durham, R. W. & Joshi, S. R., 1980. Recent sedimentation rates, ^{210}Pb fluxes, and particle settling velocities in Lake Huron, Laurentian Great Lakes. Chem. Geol. 31: 53–66.

Edgington, D. N. & Robbins, J. N., 1976. Pattern of deposition of natural and fall-out radionuclides in the sediments of Lake Michigan and their relation to Limnological processes. In: Nriagu, J. O. (Ed.). Environmental Biogeochemistry, 2. Ann. Arbor Science, M.I.: 705–729.

Elner, J. & Wood, C., 1980. The history of two linked but contrasting lakes in N. Wales from a study of pollen, diatoms and chemistry in sediment cores. J. Ecol. 68: 95–121.

Goldberg, E. D., 1963. Geochronology with ^{210}Pb. In: Radioactive Dating. Int. Atom. Energy Ag. Vienna: 121–131.

Koide, M., Bruland, K. W. & Goldberg, E. D., 1973. Th-228/Th-232 and Pb-210 geochronologies in marine and lake sediments. Geochim. Cosmochim. Acta 37: 1171–1187.

Krishnaswamy, S., Lal, D., Martin, J. M. & Meybeck, M., 1971. Geochronology of lake sediments. Earth planet. Sci. Lett. 11: 407–414.

Krishnaswamy, S. & Lal, D., 1978. Radionuclide Limnochronology. In: Lerman, A. (Ed.). Lakes, Chemistry, Geology & Physics. Springer Verlag, N.Y.: 153–177.

Olfield, F. & Appleby, P. G., in press. Empirical testing of ^{210}Pb dating models for lake sediments. In: Haworth, E. Y. & Lund, J.W. G. (Eds.). Lake Sediments and Environmental History. Leicester Univ. Press.

Oldfield, F., Appleby, P. G. & Petit, D., 1980. A re-evaluation of lead-210 chronology and the history of total lead influx in a small South Belgian Pond. Ambio 9: 97–99.

Oldfield, F., Appleby, P. G. & Battarbee, R. W., 1978. Alternative ^{210}Pb dating: results from the New Guinea Highlands and Lough Erne. Nature 271: 339–442.

Pennington, W., Cambray, R. S., Eakins, J. D. & Harkness, D. D., 1975. Radionuclide dating of the recent sediments of Blelham Tarn. Freshwat. Biol. 6: 317–331.

Petit, D., 1974. Pb-210 et isotopes stables du plomp dans des sediments lacustres. Earth planet. Sci. Lett. 23: 199–205.

Robbins, J. A., 1978. Geochemical and geophysical applications of radioactive lead. In: Nriagu, J. O. (Ed.). Biogeochemistry of Lead in the Environment. Elsevier Scientific, Amsterdam: 285–393.

Robbins, J. A. & Edgington, D. N., 1975. Determination of recent sedimentation rates in Lake Michigan using ^{210}Pb and Cs-137. Geochim. Cosmochim. Acta. 39: 285–304.

Robbins, J. A., Krezoski, J. R., Mozley, S. C., 1977. Radioactivity in sediments of the Great Lakes: Post-depositional redistribution by deposit-feeding organisms. Earth planet. Sci. Lett. 36: 325–333.

Schell, W. R., 1977. Concentrations, physico-chemical states and mean residence times of ^{210}Pb and ^{210}Po in marine and estuarine waters. Geochem. Cosmochim. Acta. 41: 1019–1031.

Mineral magnetic studies of lake sediments

A brief review

F. Oldfield[1], C. Barnosky[2], E. B. Leopold[2] & J. P. Smith[3]
[1] *Department of Geography, University of Liverpool, Liverpool, U.K.*
[2] *Quaternary Research Centre, University of Washington, Seattle, Wash., U.S.A.*
[3] *Department of Geography, The Polytechnic, Wolverhampton, U.K.*

Keywords: paleolimnology, magnetic minerals, sediment sources

Abstract

This review outlines the origin and environmental significance of magnetic minerals in lake sediments. Attention is drawn to situations where the patterns of mineral magnetic variation is a reflection of processes other than changing erosion rates and fire incidence. The use of mineral magnetic techniques in sediment source tracing, palaeoclimatic studies and the reconstruction of particulate pollution history is illustrated by means of case studies from Britain and N. America. The value of magnetic susceptibility as an on-site core logging technique is shown by reference to data from Lake Washington.

Introduction

In a paper presented to the second Symposium on Paleolimnology, Oldfield *et al.* (1978a) considered the 'mineral' magnetic properties of lake sediments and illustrated their use in core correlation and in the study of erosion rates and processes. Several other papers have appeared since 1978 exploring further the use of magnetic parameters in core correlation and in the study of sedimentation on a whole lake/whole catchment basis (e.g. Bloemendal *et al.* 1979; Dearing *et al.* in press; Dearing 1982, 1983; Oldfield *et al.* 1980 & in press). The basis for the links with erosion rates and processes has also been further investigated by Thompson & Morton (1979). Realization of the significance of fire in the production of magnetic minerals persistent both in the soil (Oldfield *et al.* 1978), and in sediments (Rummery *et al.* 1979) has led to the use and evaluation of magnetic measurements in sedimentary studies of fire history (Rummery 1983). The present account includes a brief summary of magnetic minerals in lakes and goes on to consider applications of sediment magnetism complementary to those illustrated in the papers of Dearing (1983), Rummery (1983) and Thompson (1983).

Hydrobiologia 103, 37–44 (1983).

Magnetic minerals in lake sediments

The present summary is based on measurements from some six hundred sediment cores, from over sixty lakes. Most data sets are from Holocene sediments formed under cool temperate conditions, though late-glacial and full-glacial samples have been analysed from at least ten of the sites and earlier Pleistocene material from two.

Moreover, the sites include records from perhumid tropical and semi-arid sub-tropical environments. Several generalizations are provisionally indicated:

1. Catchment lithology, more specifically primary magnetic mineral content, exercises a major control on the overall type and concentration of magnetic minerals in lake sediments. Catchments rich in basic igneous rocks, for example basalts, are likely to give rise to high magnetic concentrations in the sediment record, provided there is a significant allocthonous minerogenic particulate input. This is as true of lakes in semi-arid areas (e.g., Lake Lenore, Washington, U.S.A.), as of cool temperate 'oceanic' lakes (e.g., Lough Neagh, Northern Ireland). Only under conditions of minimal catchment size, rapid autocthonous organic sedimentation and/or reduced erosional loss will such

lithologies fail to yield high concentrations of magnetic minerals in the sediment.

2. In areas of bedrock lacking primary magnetic minerals (e.g., the limestones of the Jura region, Eastern France), the magnetic mineralogy of the sediments is dominated by the erosive input of secondary magnetic minerals formed in the soil by the process summarized in Mullins (1977) and Longworth *et al.* (1979). The case of the Lac d'Annecy (Dearing 1979) illustrates this well. Magnetically intermediate lithologies will tend to yield magnetic minerals from both the substrate and the soil in varying proportions, according to rock type and denudational regime (see below).

3. In no case among the ones studied so far do we have conclusive evidence for the authigenic formation of persistent magnetic minerals in the sedimentary record; on the contrary, strong circumstantial evidence for detrital origin is present in most cases. Authigenic magnetic mineral formation is, however, documented by other authors (Jones & Bowser 1978) and may be significant in some extreme, highly polluted contexts.

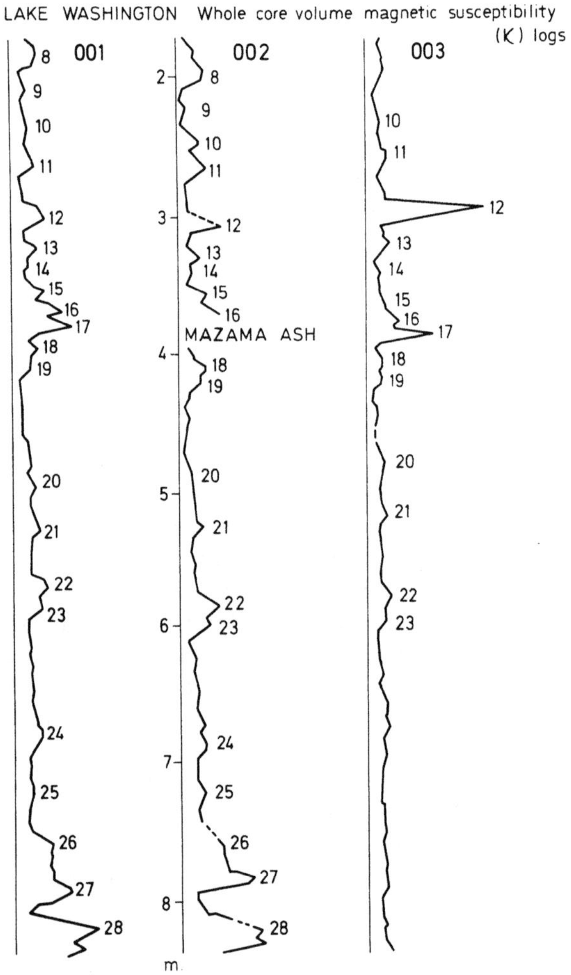

Fig. 1. Whole core volume susceptibility (K) measurements on parts of three unextruded gravity cores from Lake Washington. A sequence of correlated peaks is shown.

4. Rarely, the main contribution to the initial susceptibility in sediments (χ) (cf. Oldfield *et al.* 1978a) arises not from the ferrimagnetic or anti-ferromagnetic content, but from high concentrations of paramagnetic iron deposited by precipitation from iron-rich ground water (cf. Bloemendal, 1982). In all iron-rich sediments, some paramagnetic contribution is to be expected and this becomes relatively more significant as ferrimagnetic concentrations decline. Since paramagnetic minerals cannot carry a magnetic remanence, samples from such material are marked by a very low SIRM/χ ratio. We have encountered early Flandrian sediments of this type at several sites.

5. Atmospheric magnetic mineral inputs are likely to be significant only in areas affected by volcanic ash falls or particulate pollution from fossil fuel combustion. The former has been documented in Oldfield *et al.* (1978a, 1980) and is illustrated in Figures 1 and 4. The latter may be inferred from recent studies of ombrotrophic peat (Oldfield *et al.* 1978b, 1981) and is explored more fully below.

6. Climate influences magnetic mineralogy in lake sediments through its effect on weathering and hydrological regimes. Since magnetic mineralogy can exert a strong influence on the intensity of natural remanence in sediments, this provides a *geochemical* explanation for a relationship between climatic change and natural remanent magnetization quite different from the *geophysical* linkages sometimes inferred. The theme of magnetic mineralogy and climatic change is considered briefly in section 5 below.

The role of susceptibility logging in core selection

Implicit in the establishment of detailed susceptibility-based core correlation schemes is the potential value of this approach in the preliminary appraisal of a set of cores from a given lake. Figure 1 shows part of the susceptibility logs from three 8–11 m long gravity cores of Postglacial age from the Madison Park 'Trench' area of Lake Washington. Each log was constructed by passing unextruded core sections through a 7 cm diameter coil sensor coupled to a portable battery driven digital susceptibility meter supplied by Bartington Instruments. All the logging could be completed within the space of a normal working day. Twenty-one correlatable 'peak' features can be identified over the 8.5 m of log here portrayed. Visual markers in the form of the Mount Mazama volcanic ash (feature 17) and a striking laminated band (the trough between 27 and 28) reinforce the proposed correlation, though most of the features find no apparent expression in the visual stratigraphy. Whereas cores 001 and 002 correlate extremely well over the whole span shown here; core 003 is anomalous from feature 23 downwards, either as a consequence of unanticipated variations in depositional environment or a failure to retrieve a valid stratigraphic sequence during the coring operation. Clearly, rapid scans of this type, which neither delay nor preempt any further type of study, may be of widespread use now that relatively cheap, portable equipment is readily available. In the present case, as in the earlier studies of Lough Neagh (Thompson *et al.* 1975) the detailed nature of the correlations shown is an expression of the frequency with which the depositional system has been perturbed (here presumably through natural, rather than human agencies) by events in the large and magnetically rich catchment. The availability of purpose made, portable, sensors makes this type of logging feasible as a precursor to further study in many paleolimnology projects.

Magnetic minerals and terrestrial sediment sources

Elucidating the dominant sources of sediments and identifying shifts in sediment source are important concerns from several points of view. Mineral magnetic studies are particularly promising in this connection, since the mechanisms of magnetic 'enhancement' in soils, first outlined systematically by Le Borgne (1955) and subsequently by Mullins (1977), give rise to variations which are often diagnostic of source type. They will, for example, often provide a basis for differentiating between surface soil derived materials and those derived from deeper in the substrate by such mechanisms as shoreline erosion and channel scour. Oldfield *et al.* (1979) and Walling *et al.* (1979) were able to identify the main source of suspended stream sediments in the instrumented Permian sandstone catchment of the Jackmoor Brook in South-west England, and Dearing (1979) was able to establish detailed source-sediment linkages in Lake Frisa (a basalt based lake on Mulll, North-west Scotland), using a similar approach. The example used here is drawn from an estuarine situation, the Rhode River, an arm of Chesapeake Bay. Its intimate association with closely monitored catchments yielding sediment to the head of the River makes this an ideal context within which to study source-sediment linkages in all phases of the system, as well as their change through time.

It is the relationship between magnetic parameters, particle size and the primary and secondary (soil derived) magnetic components which provides the basis for distinguishing top soil from substrate derived materials.

Table 1 is based on an extensive magnetic survey of the potential sediment sources in the catchment. Differentiation between these on the basis of a range of magnetic parameters is clearly shown. Figure 2 plots these same parameters down core for one of some 30 analysed profiles from the Estuarine sediment. Between 120 cm and 90 cm, the assemblage of characteristics changes from low χ (4 to 6), very low SIRM (<100), low SIRM/χ (<20) and high 'S' (<-0.4), to high χ (8 to 20), high SIRM (400–1100), high SIRM/χ (~50) and low 'S' (-0.7 to -0.9). In the lower part, the parameters are comparable to those for substrate samples, whilst in the upper part of the core, there is perfect correspondence with surface derived silts and clays. ^{210}Pb analyses (J. Donaghue, pers. commun.) confirm that this shift in sediment type took place within the last 200 years. Further studies are in progress, designed to document in greater detail the present processes and patterns of material flux in the Rhode River ecosystem and to establish the spatial and temporal variations in sediment types and sources

Table 1. Magnetic characteristics of potential sediment sources in the Rhode River catchment.

Substrates and beach sands	Coarse surface material	Fine surface material
Low χ (<8)	χ (4–8)	χ (6–20)
Low SIRM (<200)	SIRM (200–600)	SIRM (500–1 500)
Low SIRM/χ (<30/~10)	SIRM/χ (~40)	SIRM/χ (~50–60)
'S' (-0.1 to -0.5)	'S' (-0.6 to -1.0)	'S' (-0.6 to -1.0)

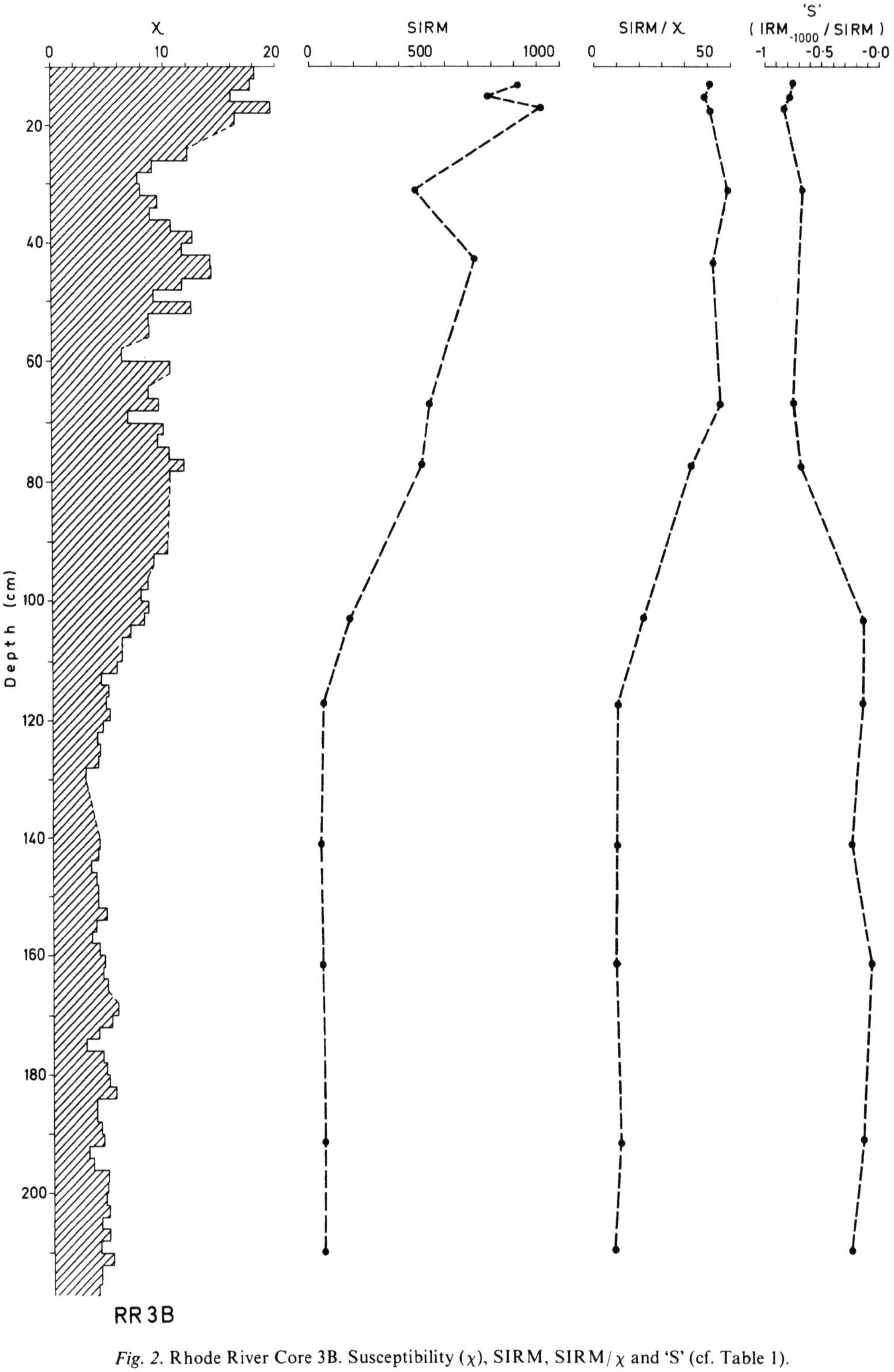

Fig. 2. Rhode River Core 3B. Susceptibility (χ), SIRM, SIRM/χ and 'S' (cf. Table 1).

revealed by variations in magnetic mineralogy of the type shown in Figure 2. By virtue of the widespread formation of distinctive secondary soil minerals the present type of study is practical in a wide range of environmental contexts.

The mineral magnetic expression of climate change

Figure 3 plots volume susceptibility for a series of stratigraphically consecutive piston cores from Battleground Lake in southern Washington State. Each extruded core section was laid in a shallow glass-fibre trough and passed through the magnetic susceptibility coil with readings being recorded every 2 cm. The 12 m of sediment span Late-Wisconsin and Postglacial time, with a basal ^{14}C date of 14 840 ± 220 (QL-1539). The lake is in a closed crater basin with a small low-rimmed catchment in basalt. Below 7 m the high susceptibility values reflect a substantial detrital input of ferrimagnetic basalt from the catchment. Above 7 m the values are consistently low save where volcanic ash bands are recorded. The major change from ca. 7.5 to 7 m corresponds with the lithostratigraphic boundary, the sudden increase in loss-on-ignition values and the change in pollen assemblage from *Artemisia-Picea-Pinus* to *Pseudotsuga-Alnus rubra-Abies* dominance. The change in magnetic mineral concentrations clearly identifies the Late-Wisconsin/Postglacial boundary and reflects a diminution in allocthonous detrital input from the catchment with the development of more stable soils and complete vegetation cover. Across the transition, additional magnetic parameters indicate a growing proportion of haematite or alternatively stable single domain magnetite in the inwashed material, as a result of weathering. A shift to proportionally more secondary soil-derived, as against primary substrate-derived minerals is suggested by an increase in the 'viscous' magnetic component (cf. Mullins 1977 and below). Thus the magnetic mineralogy preserves a record of the climatic amelioration and of the associated change in sedimentation, weathering and pedogenic regimes.

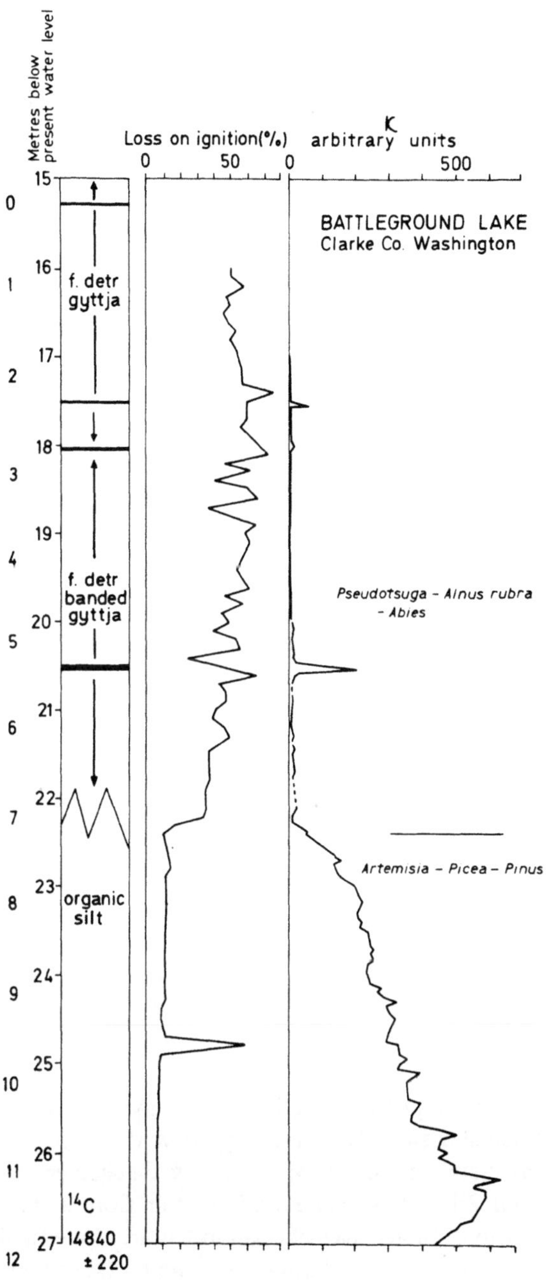

Fig. 3. Battleground Lake: stratigraphy, loss-on-ignition and whole core volume susceptibility.

Magnetic minerals and particulate pollution

Oldfield *et al.* (1978b, 1981) have shown that, as a result of fossil fuel combustion since the mid-19th century, ombrotrophic peat bogs have trapped increasing concentrations of atmospherically derived magnetic minerals in the form of spherules present for example in fly-ash. These are often directly associated with heavy metals (cf. Revitt *et al.* 1981). Figure 4 shows the location of 4 Mackereth minicores from

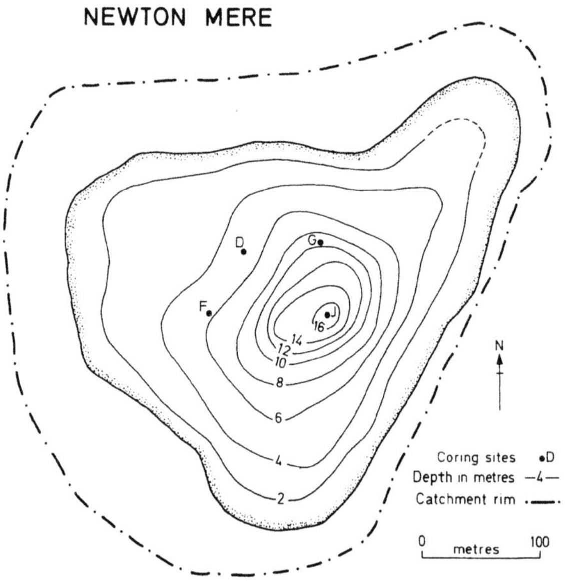

Fig. 4. Newton Mere, Cheshire, U.K. showing bathymetry, catchment limits and coring locations.

Newton Mere, a closed glacial kettle hole lake in rural south Cheshire, England, between the industrial zones of South Lancashire, Deeside, and the West Midlands. The surface area of the catchment is comparable to that of the lake, there are no inflowing streams and the recent sediments are organic gyttjas. For Core G, the magnetic parameters χ, SIRM, SIRM/χ and 'S' (Fig. 5) are plotted alongside Pb and Cu concentrations, and the dry mass sedimentation rate derived from ^{210}Pb determinations, which, using the c.r.s. model (Appleby & Oldfield 1983), give the ages as listed alongside. The same magnetic and chemical parameters are plotted for cores D & F together with total Zn and Ni concentrations. For Core J the 4 previously noted magnetic parameters are plotted and, in addition, the 'quadrature' susceptibility (see below). In all cores, χ and SIRM increase in the top 18–30 cm. The ^{210}Pb chronology dates this to the last ~70 years, the timespan over which the main increase in magnetic spherule flux to dates peat surfaces in North Europe takes place (Oldfield *et al.* 1981). A closely parallel increase in copper concentrations is recorded along with less clearly marked increases in lead (especially cores D & G) and Zinc. The SIRM/χ ratios in the upper parts of each core are comparable to those in measured fall-out samples despite the differences in ratio between cores in the earlier sediments. The 'S' ratios trend into the range characteristic of contemporary atmospheric input from earlier values much too low and indicative of a high fine grained haematite component derived from the Triass rich drifts of the catchment. The quadrature component is a function of the presence in the sediments of very small magnetic crystals at the stable single domain, superparamagnetic size transition. The percentage plotted here has been calculated by measuring the magnetic susceptibility using a low frequency (χ_{LF}) and a high frequency (χ_{HF}) Bartington Instrument sensor. Whereas the former measures the total susceptibility, the latter measures only the in-phase component. The 'viscous', out of phase, 'quadrature' component can be calculated as a percentage using the formula:

$$\frac{\chi_{LF} - \chi_{HF}}{\chi_{LF}} \times 100.$$

Viscous grains are common in soils (Mullins 1977) and in soil derived sediments. We have never found a large viscous component in any of the industrial/domestic fossil fuel combustion derived particulates. The trend in the top 18 cm of Core G can thus be confidently interpreted as an increased contribution from fossil derived atmospheric particulate input. In the most recent sediments in all four cores the magnetic mineral input is dominated by this contribution.

Conclusion

By concentrating on aspects of magnetic mineral flux to lake sediments not fully considered in the previous symposium paper (Oldfield *et al.* 1978a) we have shown that, in many cases, simple variations in the concentration of primary magnetic minerals in allocthonous catchment-derived materials do not provide an adequate model of mineral magnetic variations in lake sediments. Depending on the environmental context, mineral magnetic variations can be used to aid the elucidation of a diversity of problems.

Acknowledgements

The work numerised in this paper has been funded by NSF, NERC, the Royal Society, the Leverhulme

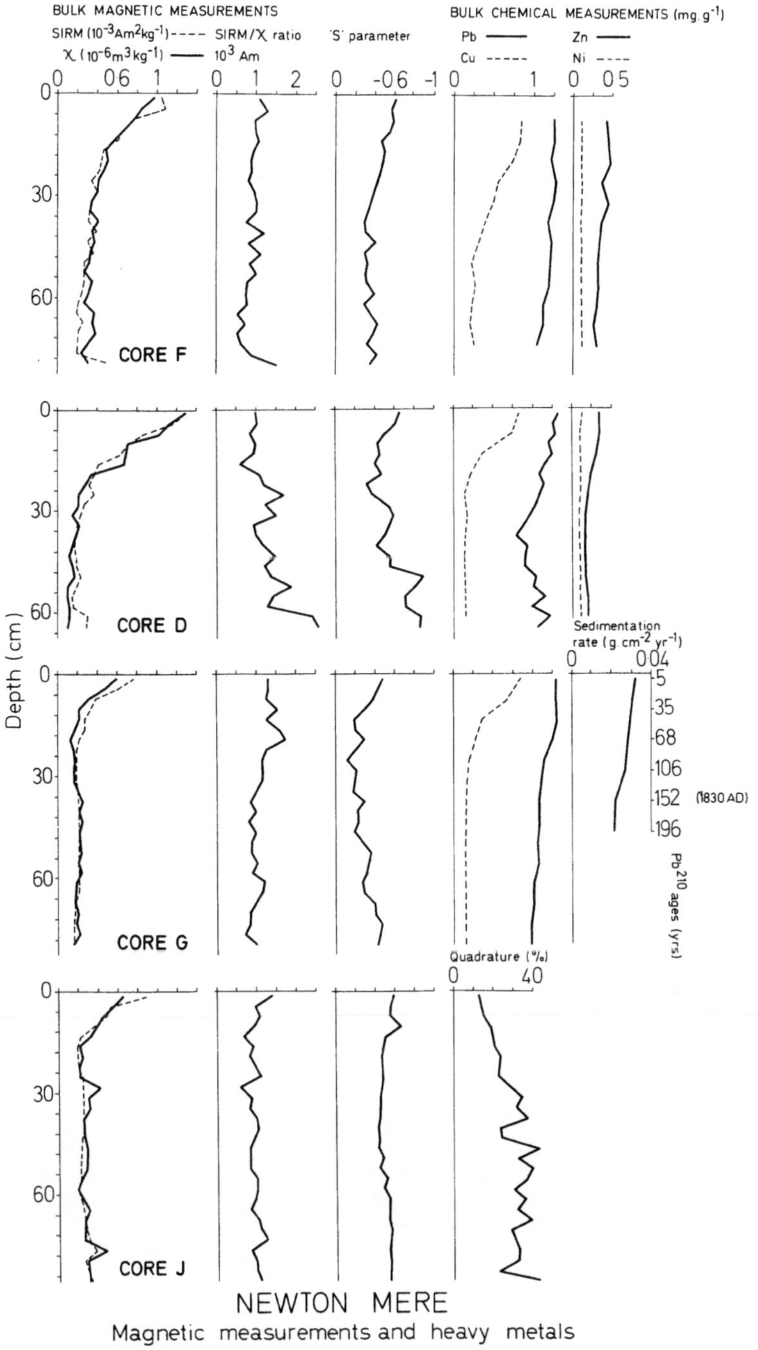

Fig. 5. Magnetic measurements and heavy metal analyses for Mackereth minicores from Newton Mere.

and Nuffield Foundations, the Institute of Hydrology, the Universities of Liverpool and Washington (Seattle) and Wolverhampton Polytechnic. Dr. J. Donaghue provided samples from Rhode River Core 3B and Dr. R. Thompson has been of crucial importance to all stages of the work. G. W. Bartington of Bartington Instruments has provided instrumentation used in several of the case studies outlined.

References

Appleby, P. G. & Oldfield, F., 1983. The assessment of ^{210}Pb Data from sites with varying sediment accumulation rates. Devel. Hydrobiol. 00: 000–000.

Bloemendal, J., Oldfield, F. & Thompson, R., 1979. Magnetic measurements used to assess sediment influx at Llyn Goddiondnon. Nature 280: 50–51.

Bloemendal, J., 1982. Unpublished Ph.D. Thesis, University of Liverpool 000 pp.

Dearing, J. A., 1979. The application of magnetic measurements to studies of particulate flux in lake-watershed ecosystems. Unpublished Ph.D. Thesis, University of Liverpool 000 pp.

Dearing, J. A., 1982. Core correlation and total sediment influx. In Berglund, B. (ed.). Paleohydrological changes in the temporate zone in the last 15 000 years. IGCP 158B. Lake and mire environments. Project guide Vol. 3. Lund., 1–23.

Dearing, J. A., 1983. Changing patterns of sediment accumulation in a small lake in Scania, S. Sweden. Devel. Hydrobiol. 00: 000–000.

Dearing, J. A., Elner, J. K. & Happey-Wood, C. M. (Mrs. D. G. E. Wood), 1981. Recent sediment flux and erosional processes in a Welsh upland lake catchment based on magnetic susceptibility measurements. Quat. Res. 16: 356–372.

Jones, B. F. & Bowser, C. J., 1978. The mineralogy and related chemistry of lake sediments. In: Lerman, A. (ed.). Lakes, Chemistry, Geology, Physics. Springer-Verlag, New York.: 179–236.

Le Borgne, E., 1955. Susceptibilité Magnetiqué anormale du sol superficiel. Ann. Geophys. 11: 399–419.

Longworth, G., Becker, L. W., Thompson, R., Oldfield, R., Dearing, J.A. & Rummery, T. A., 1979. Mössbauer effect and magnetic studies of secondary iron oxides in soils. J. Soil Sci. 30: 93–110.

Mullins, C. E., 1977. Magnetic susceptibility of the soil and its significance in Soil Science: A Review. J. Soil Sci. 28: 223–246.

Oldfield, F., Dearing, J. A., Thompson, R. & Garrett-Jones, S. E., 1978a. Some magnetic properties of lake sediments and their possible links with erosion rates. Pol. Arch. Hydrobiol. 25: 321–331.

Oldfield, F., Thompson, R. & Barber, K. E., 1978b. Changing atmospheric fallout of magnetic particles recorded in recent ombrotrophic peat sections. Science 199: 679–680.

Oldfield, F., Rummery, T. A., Thompson, R. & Walling, D. E., 1979. Identifications of suspended sediment sources by means of magnetic measurements: some preliminary results. Wat. Resour. Res. 15: 211–218.

Oldfield, F., Appleby, P. G. & Thompson, R., 1980. Paleoecological studies of lakes in the Highlands of Papua New Guinea. The chronology of sedimentation. J. Ecol. 68: 457–478.

Oldfield, F., Tolonen, K. & Thompson, R., 1981. Magnetic measurements of particulate atmospheric pollution history in dated Finnish peat profiles. Ambio 10: 185–188.

Oldfield, F., Appleby, P. G. & O'Garra, A. T., in press. Evidence from lake sediments for recent erosion rates in the Highlands of Papua New Guinea.

Revitt, M., Bryan Ellis, J. & Oldfield, F., in press. Variations in heavy metals of stormwater suspended solids in a separate sewer system. In: Yen, B. C. (ed.). Urban Storm Drainage. Proc. 2nd int. Conf. Urban Storm Drainage.

Rummery, T. A., Bloemendal, J., Dearing, J., Oldfield, F. & Thompson, R., 1979. The persistence of fire-induced magnetic oxides in soils and lake sediments. Ann. Geophy. 35: 103–107.

Rummery, T. A., 1983. The use of magnetic measurements in interpreting the fire histories of lake drainage basins. Devel. Hydrobiol. 00: 000–000.

Thompson, T. A., 1983. Global Holocene magnetostratigraphy. Devel. Hydrobiol. 000: 000–000.

Thompson, R., Battarbee, R. W., O'Sullivan, P. E., Oldfield, F., 1975. Magnetic susceptibility of lake sediments. Limnol. Oceanogr. 20: 687–698.

Thompson, R. & Morton, D. J., 1979. Magnetic susceptibility and particle-size distribution in recent sediments of the Loch Lomond Drainage basin, Scotland. J. Sed. Pet. 49: 801–812.

Global holocene magnetostratigraphy

R. Thompson
Department of Geophysics, University of Edinburgh, Edinburgh, Scotland

Keywords: paleolimnology, magnetism, dating, Australia, North America, Japan, Europe, Near East

Abstract

Paleolimnomagnetic records from five regions of the world have been combined with historical magnetic field observations in order to produce regional geomagnetic master curves.

Introduction

During the past 10 000 years the earth's magnetic field has been slowly changing. These secular changes in intensity and shape of the field have been produced by the ponderous yet turbulent fluid circulation of the liquid part of the earth's core. The variations in direction of the earth's magnetic field have been found to be recorded in the magnetization of some lake sediments. By measuring the remanence of these often weakly magnetized sediments using sensitive magnetometers, and by radiocarbon dating the organic fraction of the sediments, the past secular variations of the geomagnetic field have been determined at several localities around the world. Paleomagnetic records are now available from lakes in Australia, Japan, North America, Europe and the Near East. Work is also in progress on lake sediment cores from East Africa and South America and a detailed 2 000 year magnetic record is available from Japanese and Chinese archaeomagnetic studies.

In this paper the available well-dated paleolimnomagnetic records are gathered together, processed mathematically and converted to a common time scale to form regional master curves. (Figs. 1-4). The chronologies of the master curves are all based on the radiocarbon dating method combined with a tree ring calibration procedure (Clark 1975). Apart from historical measurements of the geomagnetic field, which have been used to construct the most recent part of the master curves, paleomagnetic measurements from lake sediments have been used exclusively to reconstruct the past geomagnetic changes.

Regional master curves

Regional geomagnetic master curves are presented for five different regions of the world. They have been derived from type lake sediment cores chosen, for each region, on the quality of their paleomagnetic records and their ^{14}C chronologies. The paleomagnetic data from these type cores have been processed using the mathematical methods described below. The final geomagnetic master curves resulting from these analyses are presented in Figs. 3 & 4. Whenever a single lake core did not cover as complete a time span as possible within a region, a second core was selected from a nearby lake to extend the type core record. The lakes and cores selected for the five regions are as follows: South Australia, Bullenmerri core BC extended by Keilambete core KF (Barton & McElhinny 1981); North America, St. Croix core 75 (Banerjee *et al.* 1979); Western Europe, Lomond core LLRP1 extended by Windermere core W3 (Thompson &

Hydrobiologia 103, 45-51 (1983).

Turner 1979); Eastern Europe, Pääjärvi core P4 (Huttunen & Stober 1980) extended by Lovojärvi core D (Tolonen *et al.* 1975); Near East, Kinneret core K8 (Thompson & Stiller, unpublished). The original direction vs depth data for these cores are plotted in Figs. 1 & 2. Smooth curves have been fitted to these data sets and converted to the tree ring calibrated time scale in order to form the re-

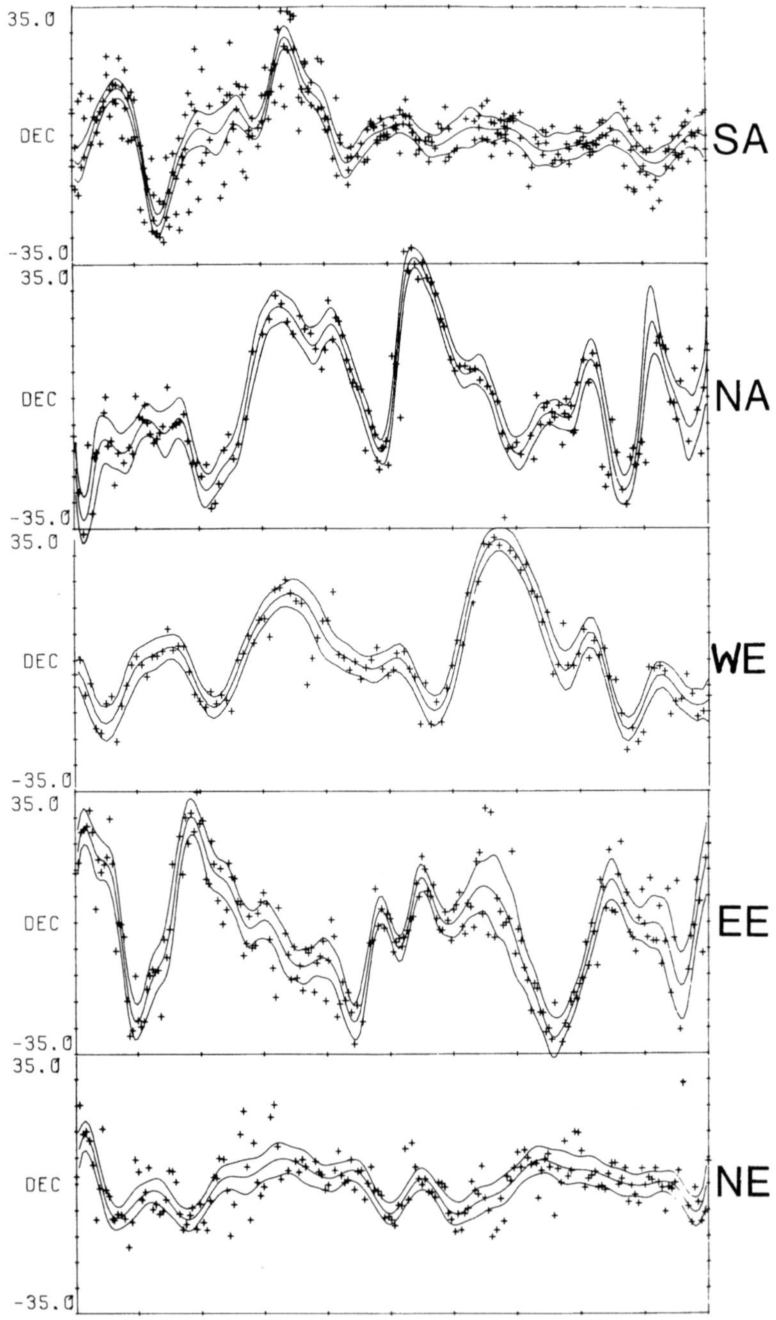

Fig. 1. Spline fit and 95% confidence bands for declination vs. depth data of type core records. Top of cores at left of diagram. Number of data points, number of spline pieces and depth range in metres as follows: SA: South Australia, 319, 30, 4.94 m; NA: North America, 156, 34, 18.40 m; WE: Western Europe, 120, 23, 2.80 m; EE: Eastern Europe, 211, 33, 4.96 m; NE: Near East, 180, 22, 4.45 m.

gional master curves presented in Figs. 3 & 4. Inspection of the curves reveals that magnetic direction changes have been quite different from one region to another. Although there are a few similarities in form of the various records of Figs. 3 & 4, which may suggest common geomagnetic origins, the overall picture is one of a complex Holocene field dominated by local dynamic processes.

The part of each regional master curve corresponding to the last 350 years is based on a spherical

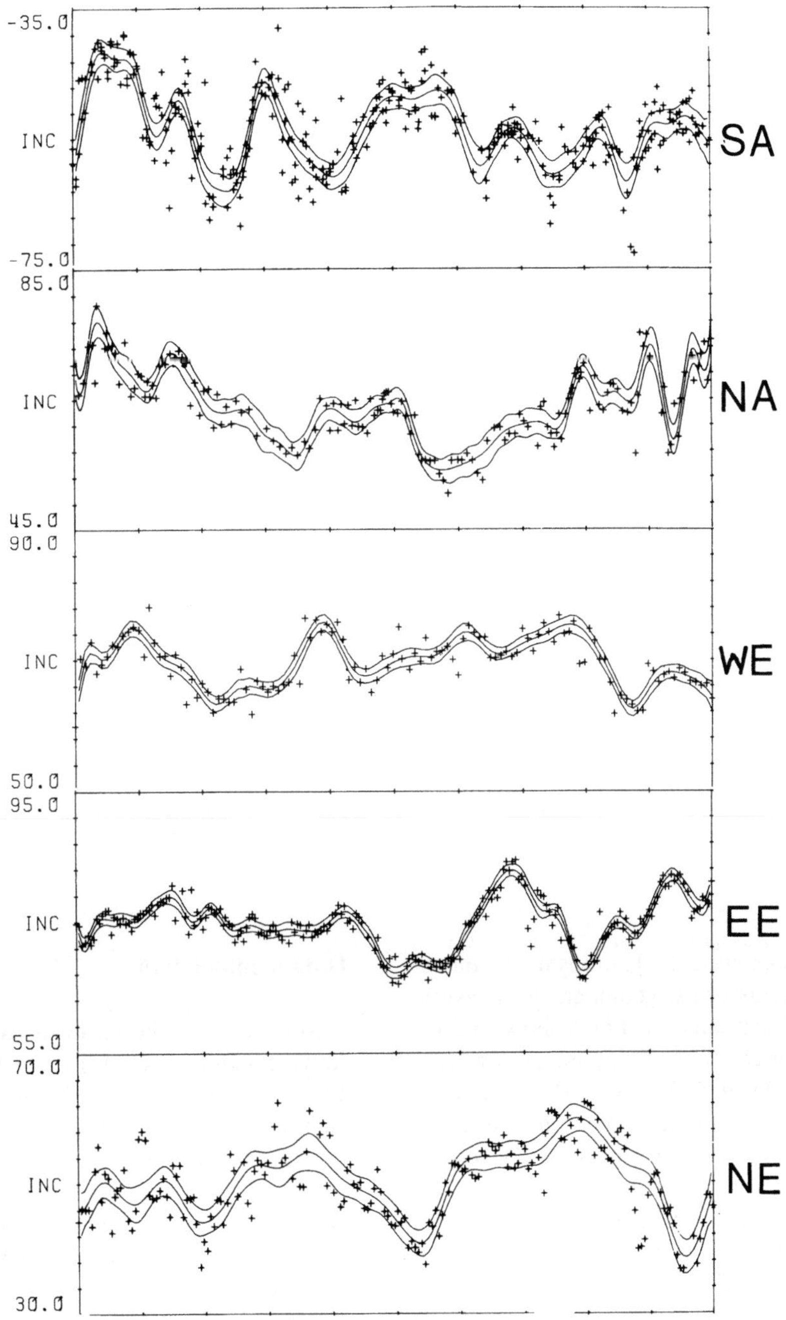

Fig. 2. Spline fit and 95% confidence bands for inclination vs. depth data of type core records. Core, data points, spline pieces and depth range as in Fig. 1.

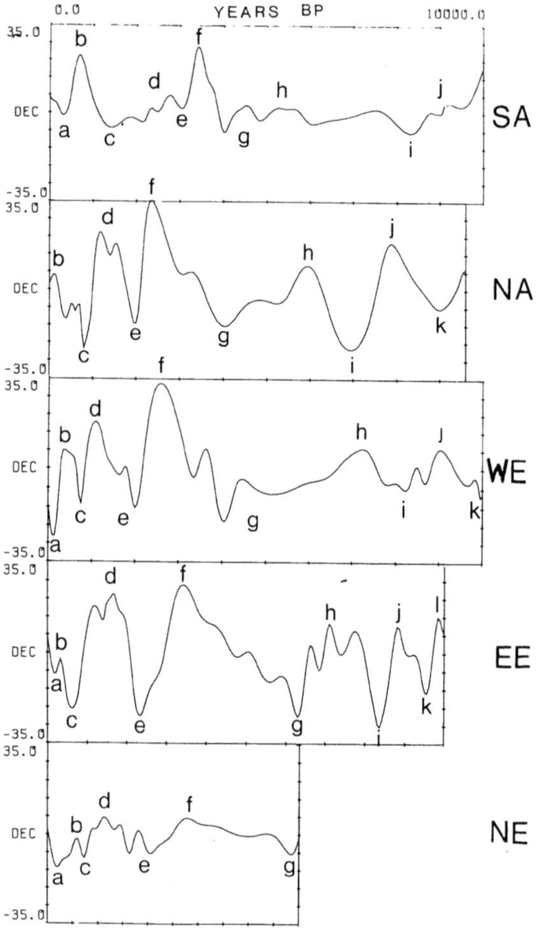

Fig. 3. Regional declination master curves. Tree ring calibrated time scale in years BP. (O BP = 1950 AD). Letters identify the turning points listed in Table 1.

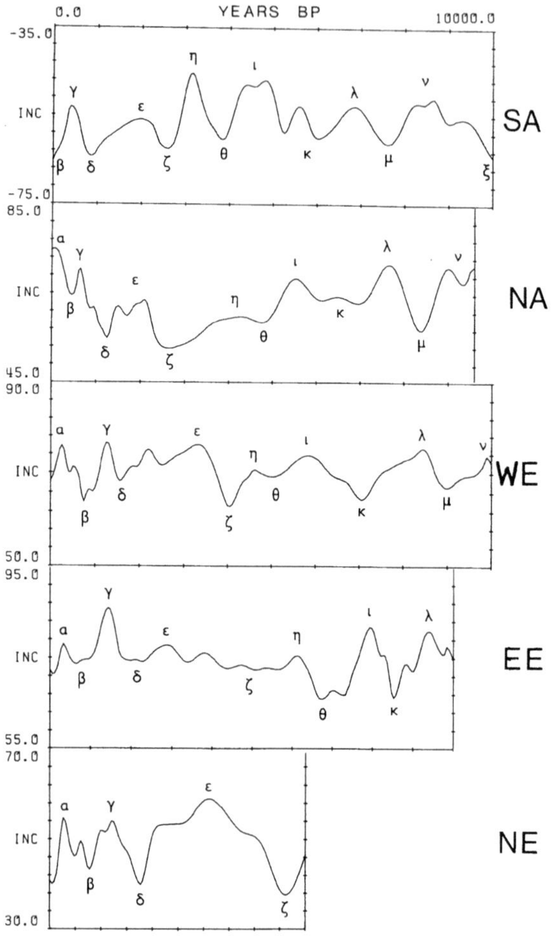

Fig. 4. Regional inclination master curves. Time scale as in Fig. 3. Letters identify the turning points listed in Table 1.

harmonic analysis model (Thompson & Barraclough 1981) of historical geomagnetic measurements (Fig. 5). Each curve in Fig. 5 shows a local change in virtual geomagnetic pole position from AD 1600 until AD 1975. The westerly declination maximum and the inclination maximum recorded in Europe around 1810 AD and 1710 AD, respectively, can be seen to have occurred in a region which extended as far south as Cape Town but not as far east as the Caspian Sea. Clockwise motion of the virtual geomagnetic pole is seen over the whole world except for an area centred on the Southern Indian ocean, where anticlockwise looping took place.

Data manipulation

Starting with the series of paleomagnetic direction measurements of Figs. 1 & 2 and their associated ^{14}C age/depth pair series, clearly defined mathematically reproducible procedures were followed in order to produce the regional master curves of Figs. 3 & 4. These mathematical procedures involved detrending the paleomagnetic data sets, curve fitting and then finally converting the depth scales to a common time-scale.

The specific operations performed on the paleomagnetic data to calculate the smooth curves from the data points of Figs. 1 & 2 were: (1) rotate data

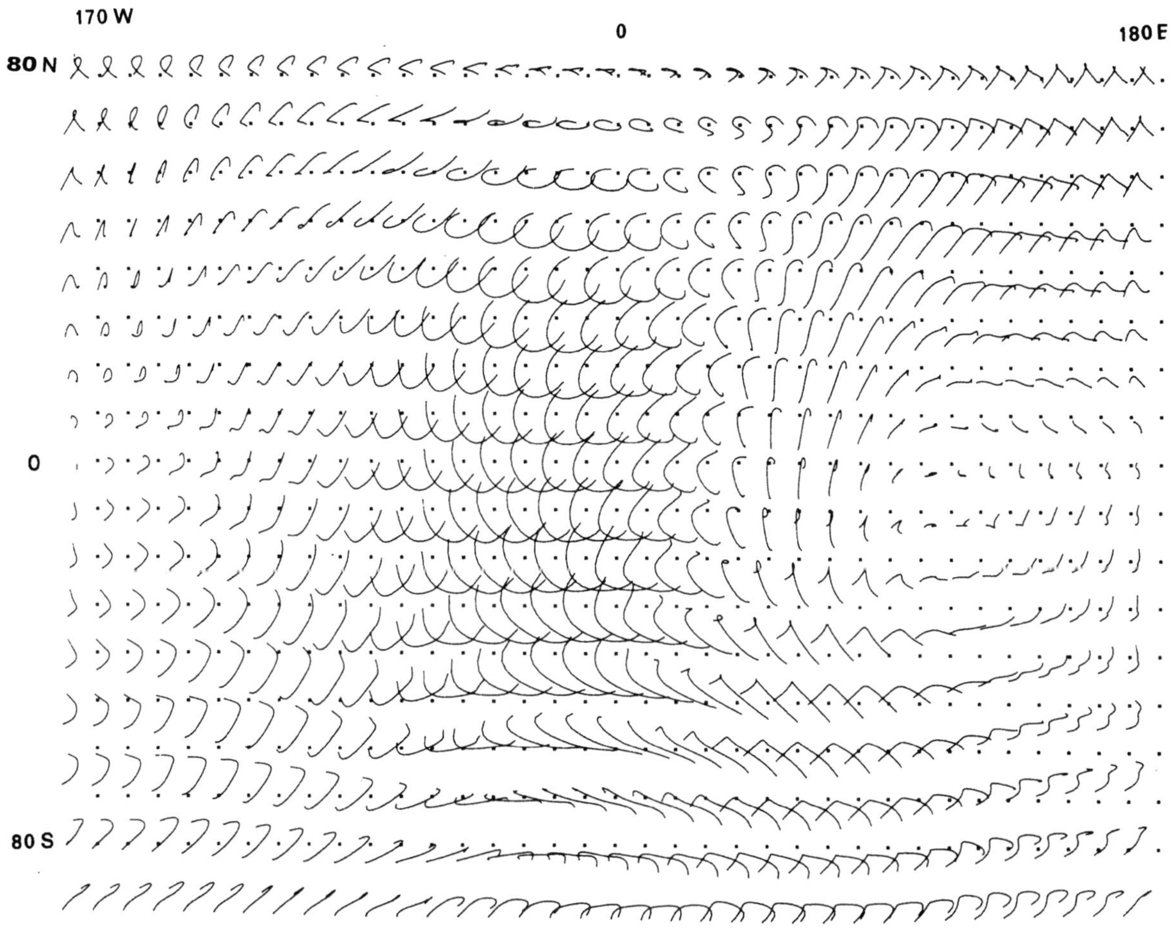

Fig. 5. Historical local virtual geomagnetic pole paths from AD 1600 to AD 1975 on a 10° latitude and longitude grid. The well known clockwise looping observed at London (Bauer 1895) can be seen in the left central part of the diagram. The region of low secular variation in the Pacific is seen at the centre of the spiral in the upper part of the diagram.

by firstly setting the mean declination to zero and secondly setting the mean inclination to zero; (2) linearly detrend the rotated declinations and inclinations using the robust weighted least squares curve fitting method described by Thompson & Clark (1981); (3) fit least squares cubic splines to the detrended data using the above curve fitting method, basing the degree of smoothing on cross validation (Thompson & Clark 1981).

Magnetostratigraphic features

Estimated ages for the main secular variation features of the regional master curves of Figs. 3 & 4 are tabulated in Table 1. Only the major magnetic features which are likely to be of regional significance are included in the tabulation. The alphabetic system of labelling the broad features or turning points from the top of the core downwards follows that originally proposed by Thompson & Turner (1979). The features are labelled in order to aid their recognition and in order to simplify discussions. The labelling is not meant to imply that any of the features have been produced by the same underlying magnetic source. Some features are poorly developed and particularly difficult to label. For example, the South Australian declination turning points SAc to SAe and SAg to SAi are of very low amplitude. Inclination will probably be of more

Table 1. Magnetostratigraphic ages.

	SA	NA	WE	EE	NE	EA
Declination						
a	300	–	140	160	220	0
b	680	100	450	300	700	700
c	1300	750	600	600	850	1200
d	2000	1200	1000	1400	1300	1650
e	2800	2000	2000	2200	2300	2200
f	3500	2400	2600	3100	3200	3100
g	4500	4000	4900	5700	5600	4400
h	5500	5900	7100	6500	–	5100
i	8300	7000	8300	7600	–	7300
j	9000	7900	9100	8000	–	–
k	–	9000	10000	8700	–	–
l	–	–	–	9000	–	–
Inclination						
α	–	50	240	300	300	–
β	–	420	650	600	800	400
γ	400	750	1150	1300	1400	760
δ	900	1200	1650	1900	2000	1000
ϵ	1900	2300	3100	2600	3600	1300
ζ	2600	2900	3800	4600	5300	1550
η	3200	3700	4300	5500	–	1750
θ	3600	4400	5000	6400	–	2800
ι	4600	5300	6000	7200	–	4100
κ	6000	6600	7100	7800	–	4600
λ	6800	7700	8300	8600	–	5100
μ	7900	8400	8800	–	–	5600
ν	8600	9600	9700	–	–	6600
ξ	10000	–	–	–	–	–

SA	South Australia	(35°S 140°E)
NA	North America	(45°N 90°W)
WE	Western Europe	(55°N 05°W)
EE	Eastern Europe	(60°N 30°E)
NE	Near East	(30°N 35°E)
EA	Eastern Asia	(35°N 140°E)

a to l: declination turning points; α to ξ: inclination turning points. Ages tabulated ^{14}C years BP. The pre 2 000 BP EA magnetostratigraphic features are taken from Horie *et al.* (1981). The EA ages are rather poorly known, based here on a linear interpolation between the basal tephra layer and the archaeomagnetic features recorded in the upper sediments.

value in Australian Holocene magnetostratigraphic work. Some features e.g. WEe and WEg have double peaks which again can be difficult to label. For this type of double turning point the overall feature has been identified, rather than the more extreme peak, as broad overall features can generally be recognised more easily. The ages in Table 1 have been based on all available dating information. It is therefore recommended that the ages in Table 1, rather than the ages of the turning points, as indicated by the time scales of Figs. 3 & 4, are used in magnetostratigraphic dating studies. The curves of Fig. 5 form a global model which can be used for dating sediments deposited since AD 1600. The global model is summarized by a series of spherical harmonic coefficients which allow declination and inclination values to be calculated at 50 year intervals for any location in the world.

The procedure of labelling and dating the magnetic features is subjective so gauging the true accuracy of the ages in Table 1 is very difficult. However, bearing in mind the problems of labelling broad features and the uncertainties in ^{14}C-dating of lake sediments, the typical error in the estimated ages of the secular variation features of the regional master curve is probably around 5%.

By matching paleomagnetic features found in new sedimentary sequences with those of the regional master curves (Figs. 3–5) paleomagnetic measurements can be used as a chronological tool. The geographic areas over which the regional master curves of Figs. 3 & 4 will extend are yet to be determined. However, a number of observations suggest that the regions will be at least 1 000 km across. These observations include the reproducibility of paleolimnomagnetic records across Western Europe, the similarity of archaeomagnetic records between China and Japan (Wei *et al.* 1981) and the character of the present magnetic field. Furthermore, inspection of Fig. 5 indicates that the diameter of regions experiencing similar secular changes is again in the order of a few thousand kilometres.

References

Banerjee, S. K., Lund, S. P. & Levi, S., 1979. Geomagnetic record in Minnesota lake sediments – Absence of the Gothenburg and Erieau excursions. Geology 7: 588–591.

Barton, C. E. & McElhinny, M. W., 1981. A 10 000 year geomagnetic secular variation record from three Australian Maars. Geophys. J. R. Astr. Soc. 67: 257–278.

Clark, R. M., 1975. A calibration curve for radiocarbon dates. Antiquity 49: 251–266.

Horie, S., Yaskawa, K., Yamamoto, A., Yokoyama, T. and Hyodo, M., 1981. Paleolimnology of Lake Zigaki, Arch. for Hydrobiol.

Huttunen, P. & Stober, J., 1980. Dating of palaeomagnetic records from Finnish lake sediment cores using pollen analysis. Boreas 9: 193–202.

Thompson, R. & Barraclough, D. R., 1981. Cross validation, cubic splines and historical secular variation. Abs. in IAGA Bulletin 45: 120.

Thompson, R. & Clark, R. M., 1981. Fitting polar wander paths. Phys. Earth Planet Interiors 27: 1–7.

Thompson, R. & Turner, G. M., 1979. British geomagnetic master curve 10 000–0 yr. B.P. for dating European sediments. Geophys. Res. Lett. 6: 249–252.

Tolonen, K., Siiriäinen, A. & Thompson, R., 1975. Prehistoric field erosion sediment in Lake Lovojärvi, S. Finland and its paleomagnetic dating. Ann. Bot. Fenn. 12: 161–164.

Wei, Q. Y., Li, D. J., Cao, G. Y., Zhang, W. X. & Wang, S. P., 1981. The polar wandering path for the last 6 000 years. Abs. in International Association of Geomagnetism and Aeronomy (IAGA) Bulletin 45: 259.

The use of magnetic measurements in interpreting the fire histories of lake drainage basins

T. A. Rummery
Department of Geography, University of Liverpool, P.O. Box 147, Liverpool, L69 3BX, U.K.

Keywords: paleolimnology, secondary ferrimagnetic oxides, S.I.R.M., magnetic susceptibility, laminated sediments

Abstract

The magnetic parameters S.I.R.M. and magnetic susceptibility have been used to try and establish the fire histories of lake drainage basins. The technique is demonstrated using sediments from three lakes: Llyn Bychan (N. Wales), a lake with a recently burnt catchment, Lake Biscarrosse (S.W. France), a lake with a well documented fire history, and Lake Laukunlampi (E. Finland), a lake with laminated sediments and a long, but unknown fire history.

Introduction

The magnetic enhancement of soils by burning was first demonstrated by Le Borgne (1955). He surmized that under the action of reducing gasses weakly magnetic forms of iron reduce to strongly ferrimagnetic magnetite on cooling. Longworth *et al.* (1979), however, have shown that the end product of the burning process is often non-stoichiometric magnetite. Rummery *et al.* (1979) demonstrated that forest fires produce magnetically enhanced secondary ferrimagnetic oxides in the soil with magnetic susceptibility (χ) and saturated isothermal remanent magnetisation (S.I.R.M.) values up to two to three orders of magnitude greater than unburnt soils.

This paper attempts to demonstrate the formation of secondary ferrimagnetic oxides in burnt soils and their persistence over time in the soil and lake sediment environment, were their presence can be used as a means of identifying past forest fires. The technique may therefore be used as a tool in examining and recreating past fire episodes and histories in lake catchments.

Methods

The sediments from three lakes were studied. Those from Llyn Bychan, Wales and Lake Biscarrosse, France were sampled using a 1 m Mackereth corer. After measurements of continuous whole-core magnetic susceptibility, the sediment was extruded in 1 or 2 cm slices. The sediments from Laukunlampi were sampled using a freezing technique (Huttunen & Merìlainen, 1978), and subsampled every 0.25 cm. Replicate samples were subsampled at 0.5 cm intervals.

All samples were prepared for single sample magnetic measurements of magnetic susceptibility (χ) and saturated isothermal remanant magnetisation (S.I.R.M.) (cf. Oldfield *et al.* 1978). Sediments from lakes Biscarrosse and Laukunlampi were analyzed for pollen and charcoal (Faegri & Iversen 1964; Swain 1973). The lake Biscarrosse core BP6 was dated by the ^{137}Cs method.

Results

Llyn Bychan is a small lake in an area of magnetically favourable lithology of slates and siltstones

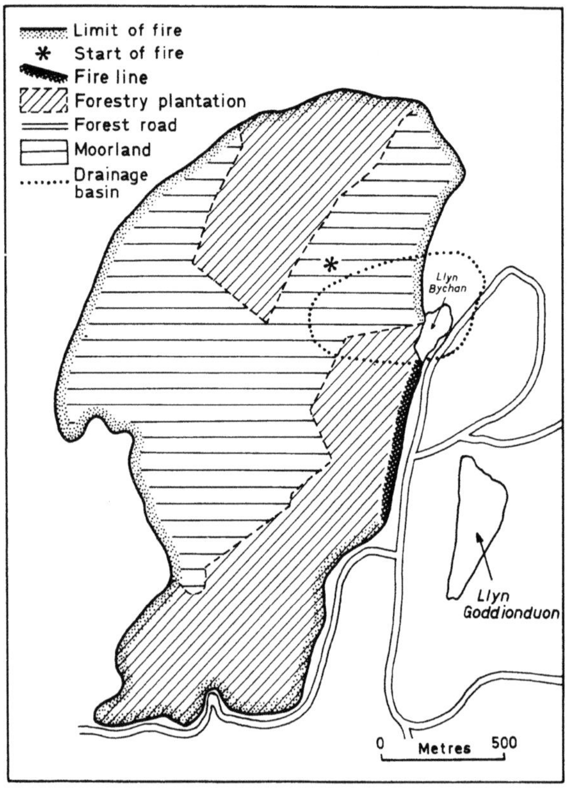

Fig. 1. The area burnt in the 1976 forest fire, N. Wales.

10 km north-west of Betws-y-Coed, Wales. In August 1976, half of the lake catchment was burnt in a forest fire (Fig. 1).

The effects on the magnetic properties of χ and S.I.R.M. in the soil by the heating in the fire is demonstrated in Table 1. Both the χ and S.I.R.M. values of the burnt soils are typically 2–3 orders of magnitude greater than the values for the unburnt soils, indicating the formation in the burnt soils of magnetically enhanced secondary ferrimagnetic oxides. Rapid erosioin of the soils followed the fire removing magnetically enhanced oxides and transporting them into the lake systems. The lake sediments were sampled 3 months after the fire and the continuous core χ traces for 5 cores are shown in Fig. 2. Cor LB1 was taken from the deepest part of the lake; the other cores are more marginal. The high values at the top of LB1 and the low values at the top of the marginal cores suggests that the magnetically enhanced material was initially concentrated in the deeper parts of the lake. If single samples are measured defined units can be used and Fig. 3 plots single sample S.I.R.M. against depth for core LB6 a paired core to LB1. LB6 has S.I.R.M. values of around 200×10^{-6} G.Oe.cm^3 g^{-1} up to 4 cm and S.I.R.M. values in the topmost sediments reaching $4\,270 \times 10^{-6}$ G.Oe.cm^3 g^{-1}. Three years after the fire event it was found that the magnetically enhanced layer was present over the whole surface of the sediment.

Table 1. Magnetic parameters for burnt and unburnt soils.

	S.I.R.M. -10^{-6} G.Oe.cm^3 g^{-1}	$\chi \times 10^{-6}$ G.Oe.cm^3 g^{-1}	S.I.R.M./χ
a. Burnt			
BY1	23051	399	57
BY5/1	16733	1168	14
BY5/2	108390	3264	33
BY9	11528	100	115
BY10/1	40024	713	56
BY10/2	175736	4516	39
TB3	7320	378	19
b. Unburnt			
BY20/2	211	2.0	106
BY20/3	156	2.0	78
BY19/2	200	3.0	67
BY19/3	47	1.0	47
BY19/4	49	1.0	49
BY17/2	100	0.7	142
BY17/3	207	0.3	69

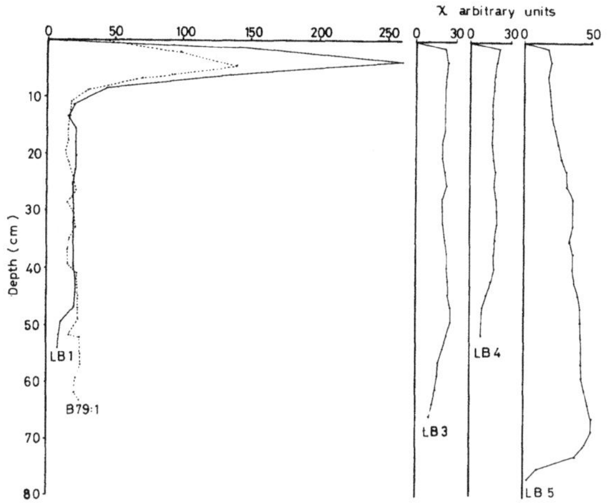

Fig. 2. Continuous core χ results for 5 cores from Llyn Bychan.

In order to evaluate the persistence of fire-induced marker layers in lake sediments, Lakes Biscarrosse and Sanguinet, in the sandy Landes region of south-west France were cored. Forests in the catchment areas of these lakes were severely damaged by extensive fires in the 1940s and especially in 1949. Details of fires and of the magnetic characteristics of the sediments are reported in Rummery (1981). Here some of the data for L. Biscarrosse are summarised (cf. Fig. 4).

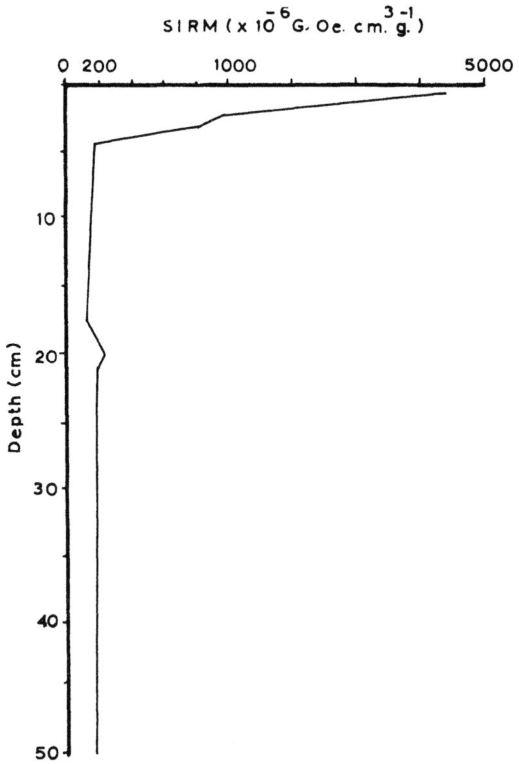

Fig. 3. Single sample S.I.R.M./depth plot for core LB6 from Llyn Bychan.

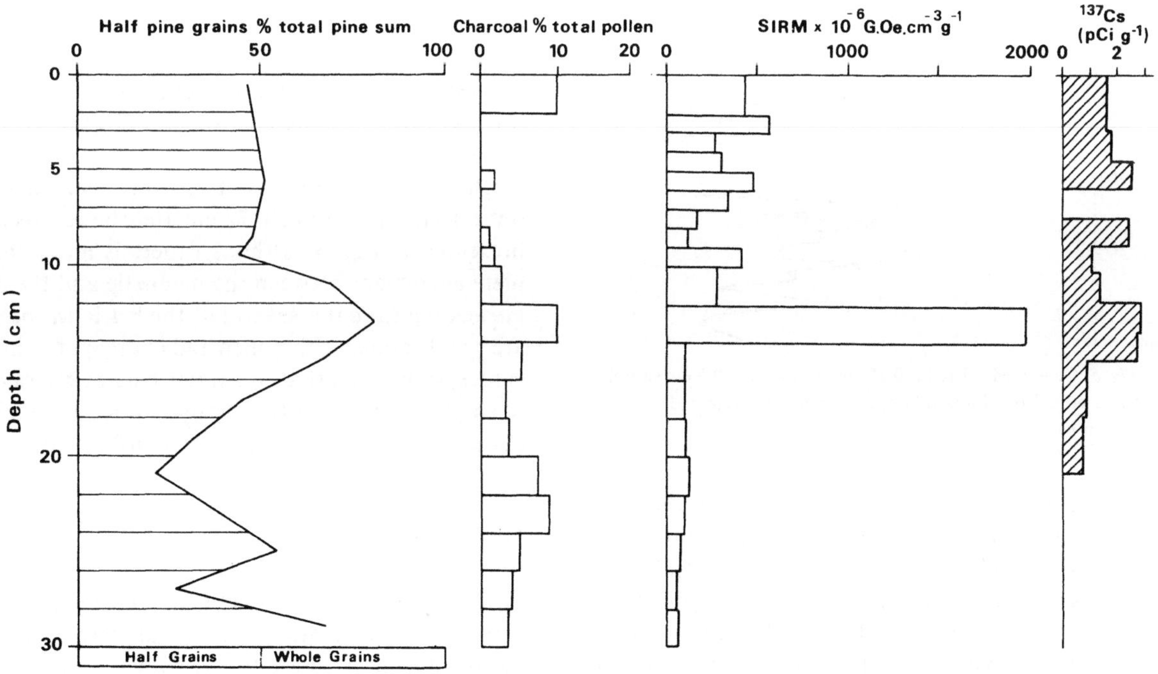

Fig. 4. Broken pine pollen ratio, charcoal, S.I.R.M., and ^{137}Cs for core PB6, Lake Biscarrosse, France.

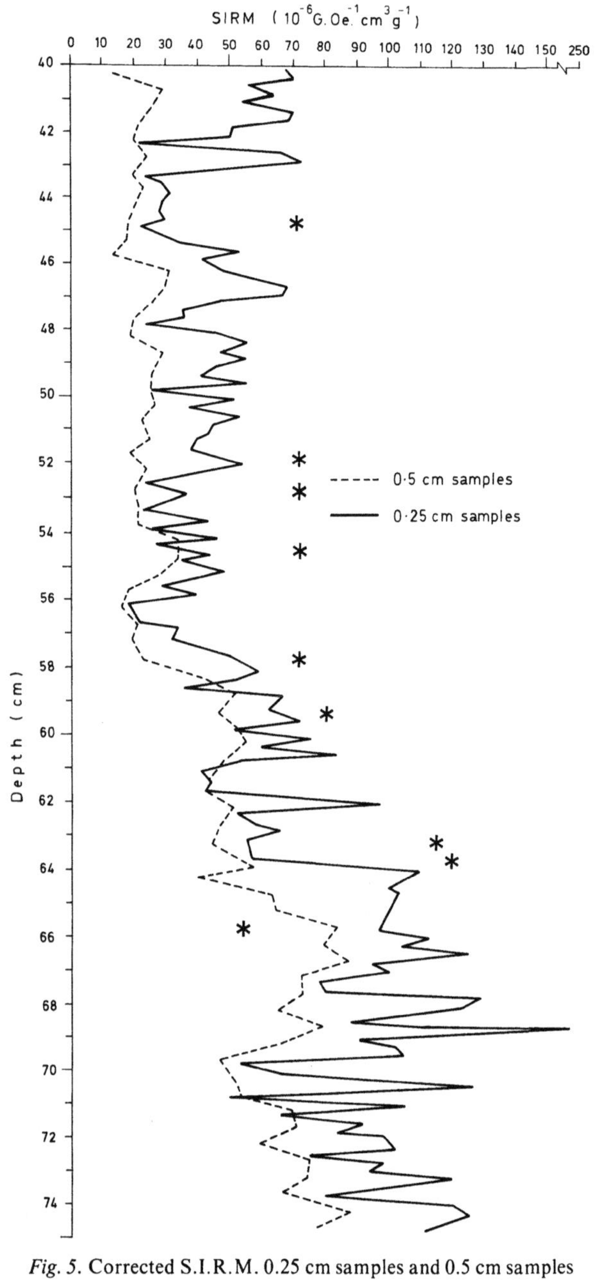

Fig. 5. Corrected S.I.R.M. 0.25 cm samples and 0.5 cm samples for a core from Laukunlampi, E. Finland. Rejected samples shown by a star.

A strong peak in S.I.R.M. was found in core BP6 (Fig. 4). In order to assess its relationship to the 1949 fire, pollen and charcoal analysis and ^{137}Cs dating was carried out. Charcoal may be considered to be the most direct evidence of fire and Oldfield (1978) demonstrated that a large broken to whole pine pollen ratio is also indicative of an ecological disturbance such as a forest fire. It can be seen (Fig. 4) that the charcoal maximum at 12–14 cm, the maximum pine pollen breakage ratio of 80% and the S.I.R.M. peak are all synchronous. In addition, the ^{137}Cs profile indicates that the peak falls slightly before the 1954 onset of ^{137}Cs fallout (Digerfeldt *et al.* 1975) and compatible therefore with a date during the previous decade of forest fires or with the extremely severe 1949 fire.

In an attempt to test the applicability of the technique on a longer time-scale the annually laminated sediments of Laukunlampi, a small kettlehole lake in North Karelia, E. Finland, were examined (Battarbee 1982). Initial measurements of S.I.R.M. showed a number of sharp peaks in the upper 1 m of sediment. Replicate measurements of 40–75 cm at 0.5 cm intervals (Fig. 5) failed to reproduce some of the peaks and these values have consequently been discarded. The S.I.R.M. data for both 0.25 and 0.5 cm samples are shown in Fig. 5

The S.I.R.M. profile can be divided into two parts, the lower part from 56–75 cm with an average time interval between S.I.R.M. peaks of 20–21 years (using varve counts), and the upper part from 40–56 cm with an average time interval between peaks of 10 years. The time interval between S.I.R.M. peaks in the lower part may be consistent with a rotational slash and burn system of agriculture and the reduced time interval between the peaks in the upper part, if caused by fire, may suggest the more intensive exploitation of the forests during the nineteenth century by this system of agriculture.

Charcoal and pollen analysis has been carried out on samples from 59–72 cm and the results are presented in Fig. 6. Although there is not a complete agreement between the magnetic and the palaeoecological data, several of the S.I.R.M. peaks are synchronous with pollen and charcoal fire indicators, suggesting that fire layers may be restricted to single samples. Further analysis is required for a definite history of this site to be constructed.

Discussion

The high level of magnetic enhancement in the soils from the formation of secondary ferrimagnetic oxides during a forest fire is shown in the Llyn Bychan catchment. Erosion of the oxides from the

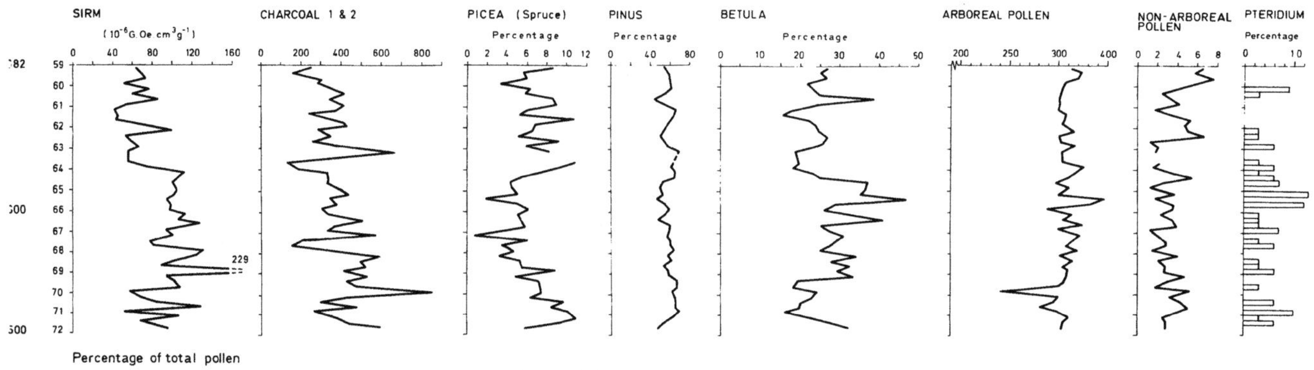

Fig. 6. S.I.R.M., charcoal, and pollen fire-indicators for Laukunlampi sediment 59–72 cm.

soil into the lake sediments leads to the formation of a magnetically distinct layer in the sediment which provides an easily identifiable fire level, a fixed datum for future studies and a feature upon which a core correlation scheme can be based. The Landes study demonstrates the persistence over time of the magnetically enhanced oxides in the lake environment and the use of the S.I.R.M. peak as a fire indicator by virtue of its synchroniety with known pollen and charcoal fire indicators. Construction of a multiple core correlation scheme using magnetic features and the rapid identification of a core suitable for the more time-consuming pollen and charcoal analysis is also demonstrated. In the case of Laukunlampi, the relationship between fires and S.I.R.M. peaks is ambiguous. This may be because the catchment area of the lake has an unfafourable magnetic lithology for this kind of study. The Laukunlampi soils were poor in convertible forms of iron but the substrate was rich in primary magnetic minerals. Correspondingly, in the lake sediments there is difficulty in differentiating between fire formed S.I.R.M. peaks and those formed by large inputs of primary magnetic minerals from the substrate.

In general the advantages of using magnetic measurements to identify past forest fires are numerous. The technique is rapid, undestructive and can differentiate between fires burning within the catchment from those outside. Movement of the oxides is primarily within the drainage basin and it is therefore likely that local fires will be preferentially recorded. The technique can also differentiate between areas in the catchment where the temperatures of the fire were high enough to form magnetically enhanced secondary ferrimagnetic oxides. A soil temperature of at least 400 °C, a rapid heating and cooling of the soil and a reducing soil atmosphere is required for the maximum formation of these oxides. A differentiation may therefore be made between canopy fires, cool ground fires (often controlled burns) and hot ground fires resulting from accidents and natural forest fires. A thorough review of the conditions needed to induce maximum magnetic enhancement in bedrock is given in Oldfield *et al.* (1981). It is suggested that the technique should be used to compliment rather than replace other more time-consuming techniques for identifying fire records.

Acknowledgements

I would like to thank P. Huttunen for the pollen and charcoal work carried out on the Laukunlampi sediments.

References

Battarbee, R. W., in press. Diatom and chrysophyceae microstratigraphy of the annually laminated sediments of a small meromictic lake in N. Karelia, Finland. Striae.

Digerfeldt, G., Battarbee, R. W. & Bengtsson, L., 1975. Report on annually laminated sediment in Lake Jälarsjön, Nacka, Stockholm. Geol. För. Stockh. Förh. 97: 29–40.

Faegri, K. & Iversen, J., 1964. Textbook of Pollen Analysis. Blackwell scientific Publications, Oxford.

Huttunen, P. & Meriläinen, J., 1978. New freezing device providing large unmixed sediment samples from lakes. Ann. bot. fenn. 15: 128–130.

Le Borgne, E., 1955. Susceptibilité magnetique anormale du sol superficial. Ann. Geophys. 11: 399–419.

Longworth, G., Becker, L. W., Thompson, R., Oldfield, F., Dearing, J. A. & Rummery, T. A. 1979. Mössbauer effect and magnetic studies of secondary iron oxides in soils. J. Soil Sci. 30: 93–110.

Oldfield, F., 1978. Pollen analysis of recent sediments from two lakes near Halifax Nova Scotia, Canada. Pollen Spores 20: 167–175.

Oldfield, F., Dearing, J., Thompson, R. & Garret-Jones, S., 1978. Some magnetic properties of lake sediments and their possible links with erosion rates. Pol. Arch. Hydrobiol. 25: 321–331.

Oldfield, F., Thompson, R. & Dickson, D. P. E., 1981. Artificial magnetic enhancement of stream bedload: a hydrological application of supermagnetism. Phys. Earth planet. Interiors 26: 107–124.

Rummery, T. A., Bloemendal, J., Dearing, J., Oldfield, F. & Thompson, R. 1979. The persistence of fire-induced magnetic oxides in soils and lake sediments. Ann. Geophys. 35: 103–107.

Rummery, T. A., 1981. The Effects of Fire on Soil and Sediment Magnetism. Unpubl. Ph.D. Thesis. Univ. Liverpool.

Swain, A. M., 1973. A History of fire and vegetation in north-eastern Minnesota as recorded in lake sediments. Quat. Res. 3: 383–396.

Changing patterns of sediment accumulation in a small lake in Scania, southern Sweden

J. A. Dearing
Department of Geography, Coventry (Lanchester) Polytechnic, Coventry, England

Keywords: paleolimnology, sedimentation pattern, magnetic susceptibility, Scanian lakes, palaeomagnetism

Abstract

Magnetic susceptibility measurements have been used to correlate synchronous depths in ca. 50 1 m. cores of recent sediment from a small lake (area ca. 55 ha) in Scania, southern Sweden. The three-dimensional picture of sediment accumulation which emerges provides a basis for studying sediment deposition patterns through time. Using a palaeomagnetic chronology, the results show that the pattern and rate of accumulation have dramatically altered during the past 350 years, thus making semiquantitative studies of downcore sediment properties in a single core problematical. Possible reasons for these phenomena are briefly discussed.

Introduction

Palaeoecological studies often use time-consuming or costly methods of analysis, and consequently those based on lake sediments are frequently restricted to single or duplicate cores. Conventional sampling strategies in small lakes choose cores from the deepest or central zone of the lake bed, a choice not usually based on prior knowledge of accumulation patterns, but on generally accepted notions of constant and conformable accumulation in relatively deep water away from shallow margins. Recent studies of spatial variations in sediment accumulation rates (Bloemendal *et al.* 1979; Davis 1976; Dearing *et al.* 1981) do not support the universal application of the concept 'sediment focusing' (cf. Lehman 1975) and illustrate the complex nature of sedimentation in many small lakes. The present paper describes sediment accumulation patterns for two periods in the recent history of the small eutrophic kettle-hole lake Havgardssjö in Scania, S. Sweden (Fig. 1, inset) and discusses possible reasons for the observed phenomena. The study stems from a larger multi-disciplinary project aimed at reconstructing palaeoenvironments in the lake catchment, and utilises the data obtained from a network of cores correlated by magnetic susceptibility (Oldfield *et al.* 1978; Thompson *et al.* 1975; Thompson *et al.* 1980) which form the basis of ongoing total sediment influx calculations.

The lake basin is of simple form with an island, which since a water level lowering of 1.5–2.0 m during the nineteenth century has been connected to the shore. Statistics for the present lake are listed in Table 1, and it is thought that the water body does not undergo thermal stratification but remains fairly well mixed throughout the year. The catchment supports an intensive agricultural land use based on cereals and sugar beet, with some pasture and deciduous woodland. Soils are derived from thick deposits of till and glacial clay and are artificially drained over much of the catchment. Consequently, there is little surface water in the catchment and no well-defined inflow streams.

Hydrobiologia 103, 59–64 (1983).

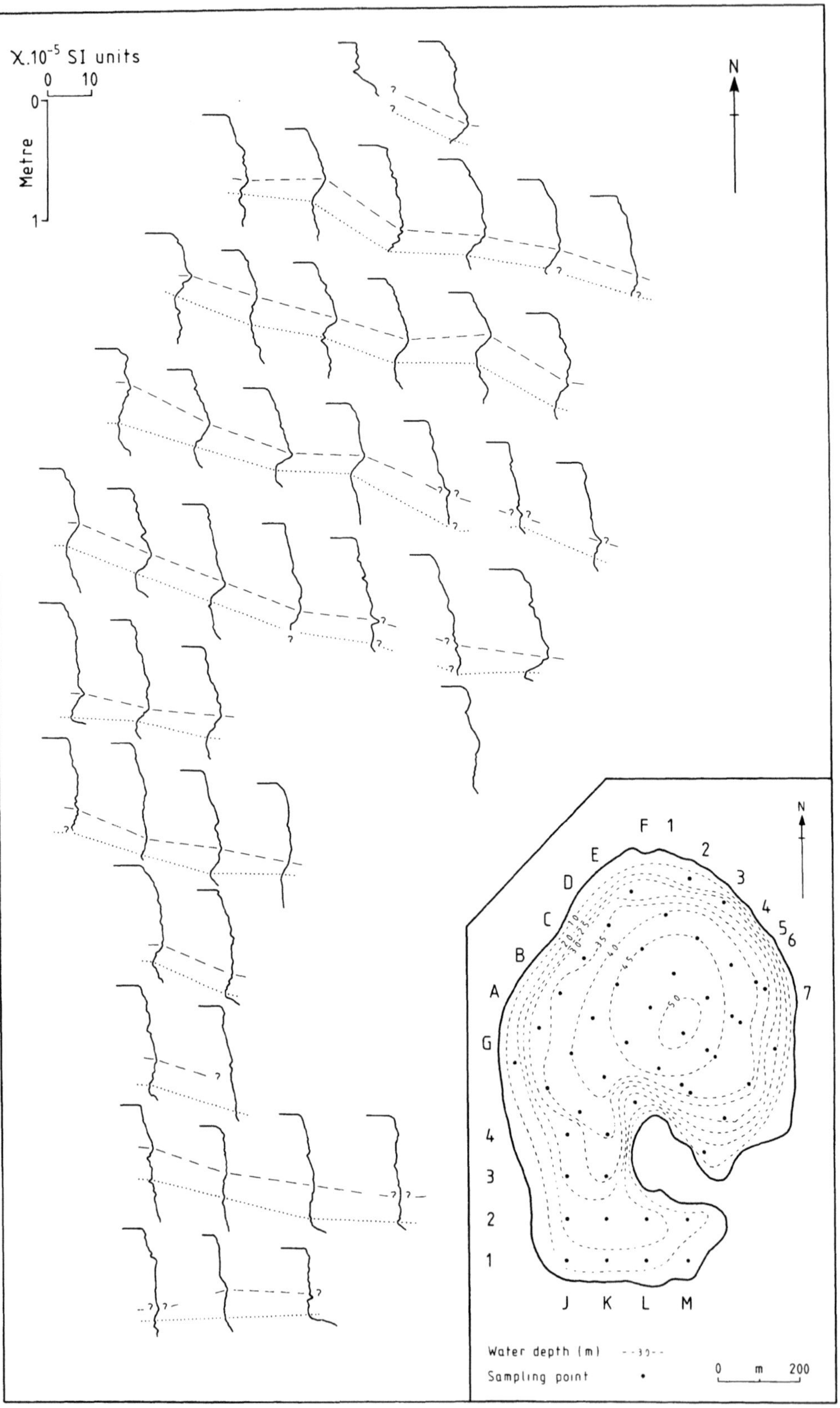

Fig. 1. Magnetic susceptibility profiles for 47 1 m sediment cores, showing inter-core correlations between peak values (dashed lines) and between trough values (dotted lines). Dotted lines also mark approximate position of stratigraphic boundary. Inset: underwater contours and sampling grid.

Methods

Sediment cores were systematically sampled on a ca. 100 m × 100 m grid basis using a simple piston corer operated with rods from an anchored boat and using 1.3 m plastic core tubes. The grid was formed by stretching floating rope between the shore and the peninsular, Turestorpsö, or between the shore and temporary rafts made of polystyrene floats anchored to the lake bed, and marking 100 m intervals on each transect with small buoys or 6 m wooden poles driven into the sediment (Fig. 1, inset). Fifty per cent of the cores were sampled in clear plexiglass tubes which showed that mud-water interfaces remained intact. All cores were carefully stoppered with notched or perforated bungs at the mud-water interface to avoid mixing during transit and magnetic measurement.

Magnetic susceptibility was measured on a Digico long-core susceptibility bridge (Molyneux & Thompson 1973), at 2 cm intervals. Relative declination was measured both continuously (Molyneux & Thompson 1973) and on 5.4 cm³ orientated samples extruded from a 4 m sequence of similarly orientated 1 m. Russian cores using a Digico low-speed balanced flux gate spinner (modified from Molyneux 1971). All measurements were undertaken at the Palaeomagnetic Laboratory, Department of Geology, Lund University. Calcareous bedrock excluded the possibility of using radiocarbon dating (cf Deevey *et al.* 1954).

Results

The susceptibility profiles are of similar form (Fig. 1) with a major peak and trough in values discernible in all cores except for some near the lake margin and in the extreme shallower southern part of the lake. The synchroneity of the susceptibility peak and trough features seems certain in view of a visible stratigraphic change from dark brown gyttja to light brown clayey gyttja which coincides with the susceptibility trough in all cores which have so far been extruded and examined internally. The single sample relative declination data shows characteristic fluctuations (Fig. 2) which can be assigned ages by comparison with historical records of secular variation (Aitken 1974) and declination 'master curves' based on radiocarbon dated sedi-

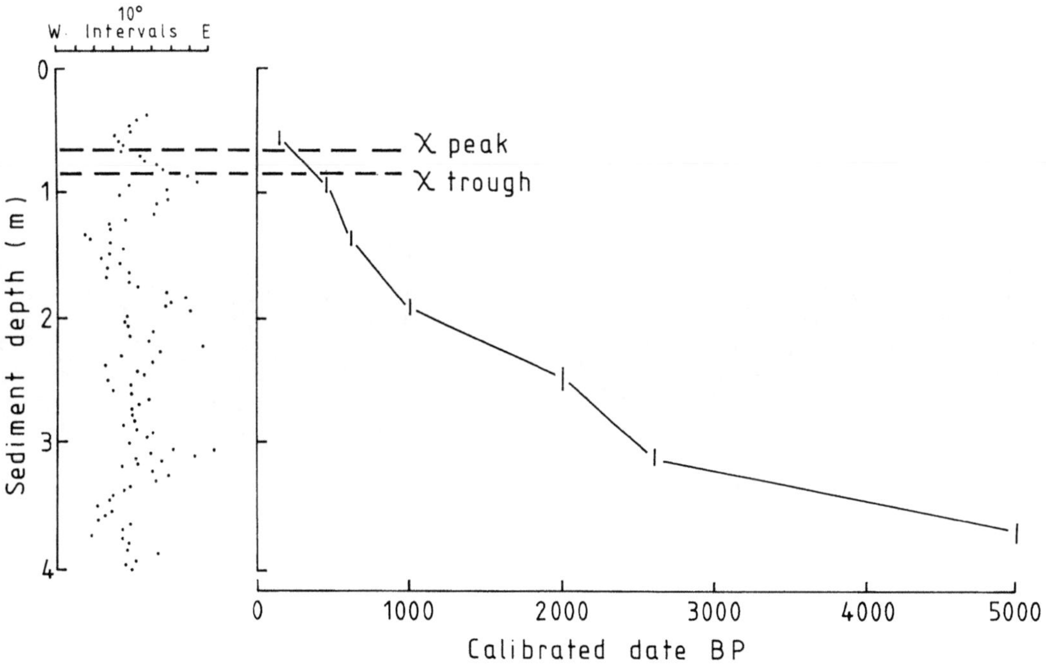

Fig. 2. Relative declination data for a central core, with derived depth-age curve 150–4900 BP and position of susceptibility (χ) peak and trough (cf. Fig. 2).

ment cores (Thompson & Turner 1979). The designated age of the lowest declination swing at 4 900 BP is supported by pollen analyses of sediments at 4 m which describe a spectrum characteristic of pre-*Ulmus* decline (ca. 5 000 BP) deciduous forest, with herb pollen frequencies registering ca. 20% of the total pollen sum, and *Ulmus* pollen frequencies accounting for 6–7% of total arborealpollen. From the derived depth–age curve (Fig. 2) the susceptibility peak and trough features can be dated to ca. 1712 AD and ca. 1550 AD, respectively. Using the mud-water interface as the third synchronous level, mean sedimentation rates can be calculated for the periods ca. 1550–ca. 1712 AD and ca. 1712–1979 AD at those sites which are represented by cores exhibiting common susceptibility features. These accumulation rates are shown in Fig. 3 by means of isolines. The earlier period (Fig. 3a) shows a rather uniform pattern of accumulation with maximum values occupying small areas in semi-central regions of the lake in water depths generally not exceeding 4.5 m. A finger of relatively low accumulation rates extends north-west from the tip of the peninsular. Measured accumulation rates range from ca. 0.5 mm a^{-1} to ca. 2.0 mm a^{-1} with a mean value of 1.23 mm a^{-1}. In the later period (Fig. 3b) the accumulation pattern is more complex with low isolines extending out from the margins into the central zone of the lake. Isolines of higher accumulation exist in the north-east corner and southern half of the lake, with highest isolines represented by sedimentation in single cores from water depths of 3–4.5 m. Measured accumulation rates range from ca. 1.2 mm a^{-1} to ca. 3.6 mm a^{-1} with a mean value of 2.25 mm a^{-1}.

Discussion

The generally higher levels of accumulation rate in the later period could possibly reflect the low wet density values (g DW ml^{-1}) of the uppermost levels

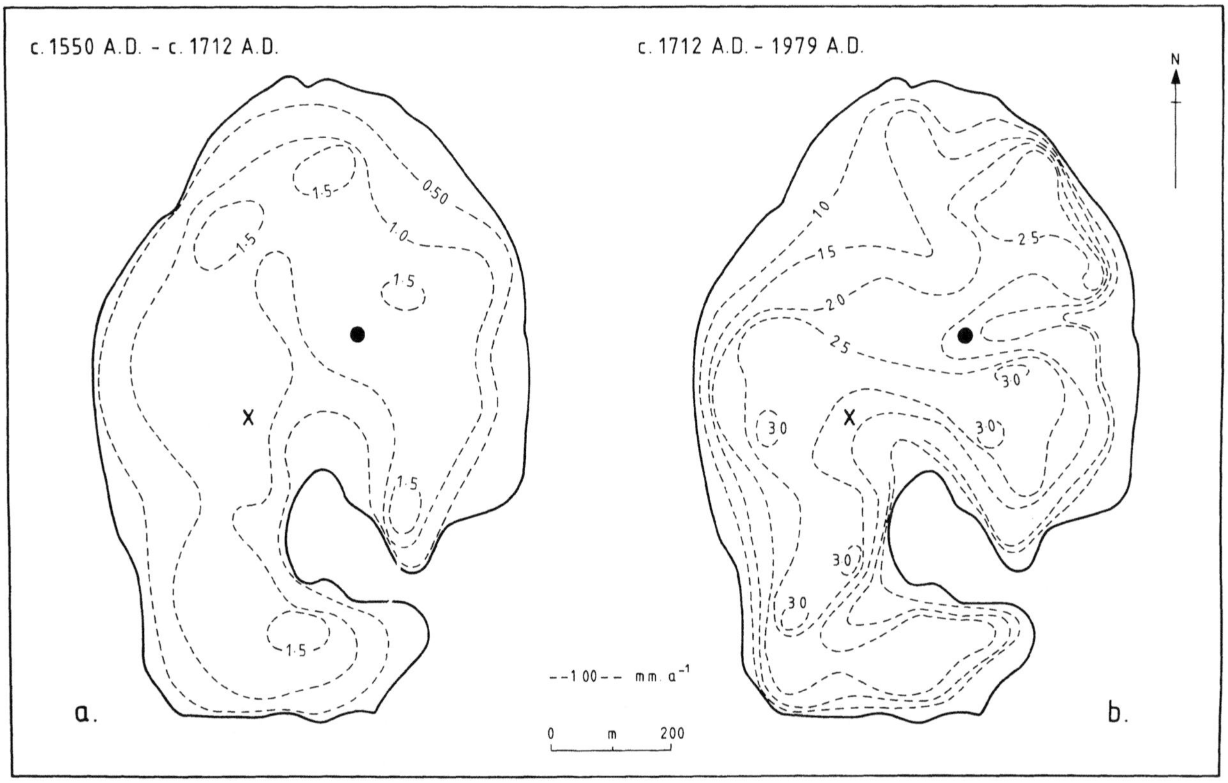

Fig. 3. Sediment accumulation patterns for two periods a) ca. 1550 AD–ca. 1712 AD and b) ca. 1712 AD–1979 AD. ● marks deepest zone, X marks core with mean sedimentation rate during both periods.

of sediment. There the values are 50–80% lower than the general range of values for levels in the uppermost metre of sediment. However the mean thickness of the section of sediment above the susceptibility peak is 60 cm and adjustment for the different wet density values would effectively reduce this value by only ca. 5–10 cm, a drop in the mean accumulation rate to 1.87 mm a^{-1} (cf. 1.23 mm a^{-1} in the lower zone). The higher accumulation rates cannot be explained by increased water content alone and seem likely to reflect increased sediment flux to the lake bed. Disucssion of the possible causes for this phenomenon lies outside the scope of the present paper, but preliminary studies of total sediment influx and the history of farming in the catchment indicate progressive soil erosion from 1600 AD to the present time (Oldfield *et al.*, in press).

Different patterns of sediment accumulation are caused by changes to the internal dynamics of the limnic system. The observed patterns could be artefacts of complex slumping-redeposition cycles where different zones of the lake are periodically subject to gain or loss of sediment as thresholds for sediment slope stability are transgressed. But the close correlation between widely spaced cores suggests that sedimentation at the majority of sampling sites is regular, and that sediment columns are not solely the product of localised mass sediment movement and deposition. The more recent pattern could be due to the varying thickness of the uppermost labile sediment caused by short term and local turbulence and currents. But, as described above, no core has more than 10 cm of labile sediment and observed differences in this thickness are not large enough to explain the range of sedimentation rates over the lake bed. A more plausible explanation lies with the changes in water movements caused by environmental factors, such as wind exposure. Progressive deforestation and hedge removal in the catchment over the last few centuries (unpublished documentary evidence) has had the effect of increasing the exposure of the lake to wind. Havgårdssjön is bounded to the south, east and north-east by wooded hills rising to ca. 30 m above the lake level, and the lake is most exposed to the west, especially the north-west. A breakdown of wind direction statistics for west Scania in the period 1931–1960 AD (Taesler 1972) for winds originating from west points and east points of the compass gives figures of 48% and 33%, respectively. Observations by Tyge Brahe who kept records for the isle of Ven (lying between Scania and Denmark) during the period 1582–1597 AD showed figures of 40% and 37%, respectively (La Cour 1876), a reduced dominance by westerlies as a result of the greater incidence of anticyclones during the 'Little Ice Age' (Lamb 1977). Conditions were cooler then and it can be assumed that there was a longer annual period of ice cover on Scanian lakes. A combination of more westerlies and greater exposure to them in more recent times is a possible explanation for the observed changes in sedimentation pattern. A wind generated current along the north-west/south-east axis of the central part of the lake may result in the periodic occurrence of a body of relatively turbulent water extending from shore to shore, and at times reaching the deepest lake bed, in which the deposition of fine material is restricted. This material remains in suspension until it drifts laterally into less turbulent areas in the sheltered southern half of the lake, and to a lesser extent the north-east corner, where it can settle out of suspension. A more sheltered and deeper (perhaps by 2.0 m) lake in the earlier period which had annual ice cover might have seen longer periods of relatively calm water in which sediments could deposit conformably over much of the lake bed. The increased depth and shelter of the lake then may have led to the development of a thermocline, with sediment moving from the margins to the centre after overturn (cf. Davis 1968, 1973).

While the results support the idea that sediment generally moves from shallower to deeper water (Davis 1968) as demonstrated by highest sediment accumulation rates away from the margins, they show that focusing of sediment to the deepest or most central area does not stricity occur. In addition, accumulation rates at sites over most of the lake bed are not constant through time, or even of constant proportions of the contemporaneous mean accumulation rate over the whole lake bed. These last observations have implications for studies aimed at interpreting downcore fluctuations in relative or absolute data. Semi-quantitative comparisons of such fluctuations could only be made with any confidence in cores which possessed mean (or some constant proportion of the mean) accumulation rates throughout the whole period since 1550 AD. Although this study is limited to the examination of just two periods it reveals that none of the cores has the mean accumulation rates for both periods. The core which comes closest to having both mean rates is A3 which was sampled 230 m away from the deepest zone of the lake (Fig. 4) in a water depth of 4.1 m. The sedimentation rate for C4, the core from the deepest zone, is very similar to the mean rate in the early period (cf. 1.1 mm a^{-1} and 1.23 mm a^{-1}), but is only ca. 70% of the mean rate in the later period (cf. 1.6 mm a^{-1} and 2.25 mm a^{-1}). If pollen grains sedimented in a similar manner to the rates and patterns described here, an absolute count in C4 would underestimate mean values within the later period by ca. 30%, thus giving an erroneous and misleading picture of pollen deposition over the two periods.

On the present evidence sedimentation patterns may have altered in many small to medium sized lakes during the last few centuries as a result of deforestation, accelerated erosion and infilling, climatic change or water level lowering. It seems unwise for palaeoecological studies based on recent sediments from such lakes to proceed without information about the spatiality of sedimentation as can be obtained through the correlation of widely sampled cores (cf. Dearing 1982).

Acknowledgements

This work was undertaken while the author was the holder of a Royal Society Fellowship at the Department of Quaternary Geology, Lund University, Sweden. Sincere thanks go to Prof. Björn Berglund, Dr. Gunnar Digerfeldt, Dr. Göran Bylund, and Magnus Thelaus of that institution for their support and guidance, and to Dr. Bob Jones of the Department of Geography, Coventry Polytechnic for pollen analyses.

Table 1. Lake and catchment statistics.

Site	55°29′N 13°22′E
Altitude	50.8 m a.s.l.
Lake surface area	54.7 ha
Catchment area	141 ha
Max. water depth	c 5.5 m
Mean water depth	c 2.5 m
Annual water depth fluctuation	±0.5 m
Water residence time	4.6 y
Water volume	$1.4 \cdot 10^6 m^3$
pH	8.20–8.75
Mean rainfall	594 mm a^{-1}
Mean annual temp. °C	8 °C

References

Aitken, M. J., 1974. Physics and Archeology. Clarendon Press, Oxford, pp. 291.

Bloemendal, J., Oldfield, F. & Thompson, R., 1979. Magnetic measurements used to assess sediment influx at Llyn Goddiondon. Nature 280: 50–53.

Davis, M. B., 1968. Pollen grains in lake sediments: Redeposition caused by seasonal water circulation. Science 162: 796–799.

Davis, M. B., 1973. Redeposition of pollen grains in lake sediment. Limnol. Oceanogr. 18: 44–52.

Davis, M. B., 1976. Erosion rates and land use history in southern Michigan. Envir. Conserv. 3: 139–148.

Dearing, J. A., 1982. Core correlation and total sediment influx. In: Berglund, B. Palaeohydrological changes in the temperate zone in the last 15 000 years. IGCP 158B. Lake and mire environments. Project guide Vol 3. Lund.: 1–23.

Dearing, J. A., Elner, J. K. & Happey-Wood, C. M., (Mrs D. G. E. Wood), 1981. Recent sediment flux and erosional processes in a Welsh upland lake-catchment based on magnetic susceptibility measurements. Quat. Res. 16: 356–372.

Deevey, E. S., Gross, M. S., Hutchinson, G. E. & Kraybill, H. L., 1954. The natural C-14 Content of materials from hard-water lakes. Proc. nat. Acad. Sci., Washington, 40: 285–288.

La Cour, 1876. Tyge Brahes meteorologiske dagbog, holdt paa uraniborg for aarene 1582–1597. K. Danske Vidensk. Selsk.: Collectanea Danica, Copenhagen.

Lamb, H. H., 1977. Climate: Present, Past and Future, 2. Climate History and the Future. Methuen, London.

Lehman, J. T., 1975. Reconstructing the rate of accumulation of lake sediment: the effect of sediment focusing. Quat. Res. 5: 541–550.

Molyneux, L., 1971. A complete result magnetometer for measuring the remanent magnetisation of rocks. Geophys. J.T. astr. Soc. 32: 429–434.

Molyneux, L. & Thompson, R., 1973. Rapid measurement of the magnetic susceptibility of long cores of sediment. Geophys. J.R. astr. Soc. 32: 479–481.

Oldfield, F., Dearing, J. A., Thompson, R. & Garret-Jones, S., 1978. Some magnetic properties of lake sediments and their possible links with erosion rates. Pol. Arch. Hydrobiol. 25: 321–331.

Oldfield, F., Battarbee, R. W. & Dearing, J. A., in press. Sedimentary evidence for environmental change. Geograph. J.

Taesler, R., 1972. Klimatdata för Sverige, SMHI/SIB. Publ. Stockholm.

Thompson, R., Battarbee, R. W., O'Sullivan, P. E. & Oldfield, F., 1975. Magnetic Susceptibility of lake sediments. Limnol. Oceanogr. 20: 687–698.

Thompson, R., Bloemendal, J., Dearing, J. A., Oldfield, F., Rummery, T. A., Stober, J. C. & Turner, G. M., 1980. Environmental applications of magnetic measurements. Science 207: 481–486.

The use of X-ray radiographic methods in studying sedimentary properties and rates of sediment accumulation

Valter Axelsson
Department of Physical Geography, University of Uppsala, Box 554, S-751 22 Uppsala, Sweden

Keywords: paleolimnology, X-ray radiography of sediment cores, sedimentary properties, sediment accumulation

Abstract

X-ray radiographs of unextruded sediment cores are used for the documentation and the interpretation of primary and secondary sedimentary structures, and for the correlation of synchronous laminations between different cores. The vertical variation in bulk density, water content and void ratio, and the increase of solids with increasing sediment depth are calculated from the recorded film density along the radiographs. The X-ray radiographic technique for the study of sedimentary properties is fast and non-destructive. The technique is especially valuable when studying the uppermost part of the sediment cover and the processes of sediment redistribution and sediment accumulation. Studies are being undertaken in several Swedish lakes and coastal bays, where mono- and stereoradiographs are being used to classify bottom and sediment types, to interpret changes in the depositional environment, and to calculate past and present rates of sediment accumulation.

Introduction

The development of lake and sea basins is reflected in the composition and sequence of the sediments which are deposited on their floors. The adverse environmental effects of man's activities (e.g. soil erosion and water pollution) are also recorded in the sediment deposits. However, the environmental changes which are recorded within the depositional sequence are very difficult to interpret. This is partly due to the lack of knowledge about the processes and rates of sediment redistribution and accumulation. There is, therefore, a great need for improved methods of analysis of the uppermost portion of the sedimentary sequence in particular, and for the calculation of past and present rates of sedimentation.

X-ray radiography of sediment cores is a fast, non-destructive scanning and recording technique which may reveal structures that are not revealed by other methods of analysis. The information about primary and secondary sedimentary structures that is provided by X-ray radiographs simplifies the determination of sedimentary properties, and facilitates the interpretation of changes in the depositional environment. This information is obtained before the sediment core is distorted through extrusion from the coring tube, and the technique is, therefore, especially valuable when studying the uppermost, often very soft part of a deposit (Axelsson & Händel 1972).

Absorption of X-rays is dependent upon the wavelenght of the radiation, the density and thickness of the sample, and the atomic number of the absorbing material. This means that inhomogeneities in the sediments will cause variable darkening of the X-ray film. The qualitative information given by the radiographic image may be quantified by densitometric evaluation.

The first application of radiography in the study of sediments was documented by Hamblin (1962). Valuable reviews on radiographic techniques and

Hydrobiologia 103, 65–69 (1983).

their applications in sedimentology are given by Bouma (1969) and by Krinitzsky (1970). The aim of the present paper is to give a brief description of a new, radiographically based method of sediment analysis and to examplify the use of radiographic methods in studying sedimentary properties and rates of sedimentation.

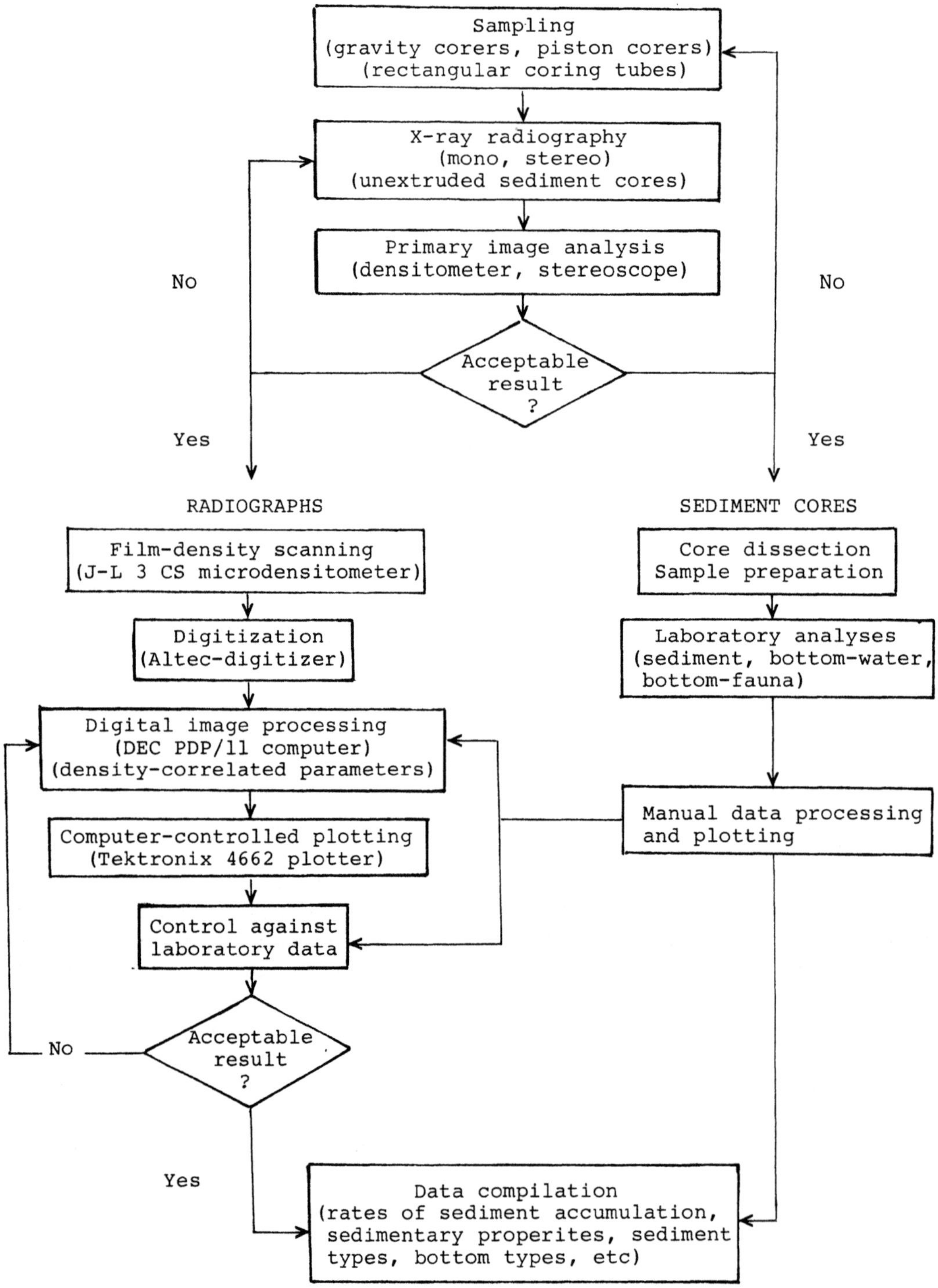

Fig. 1. Flow diagram of radiographically based sediment analysis.

A radiographic technique for the study of sedimentary properties

The technique includes the development of methods for sediment sampling, as well as for X-ray radiography of sediment cores, image analysis of radiographs, and the radiographic determination of density-correlated sedimentological parameters. The principal components of the radiographic technique are illustrated in the flow diagram, Fig. 1.

Samples for radiography should, if possible, be of uniform thickness and of rectangular or square

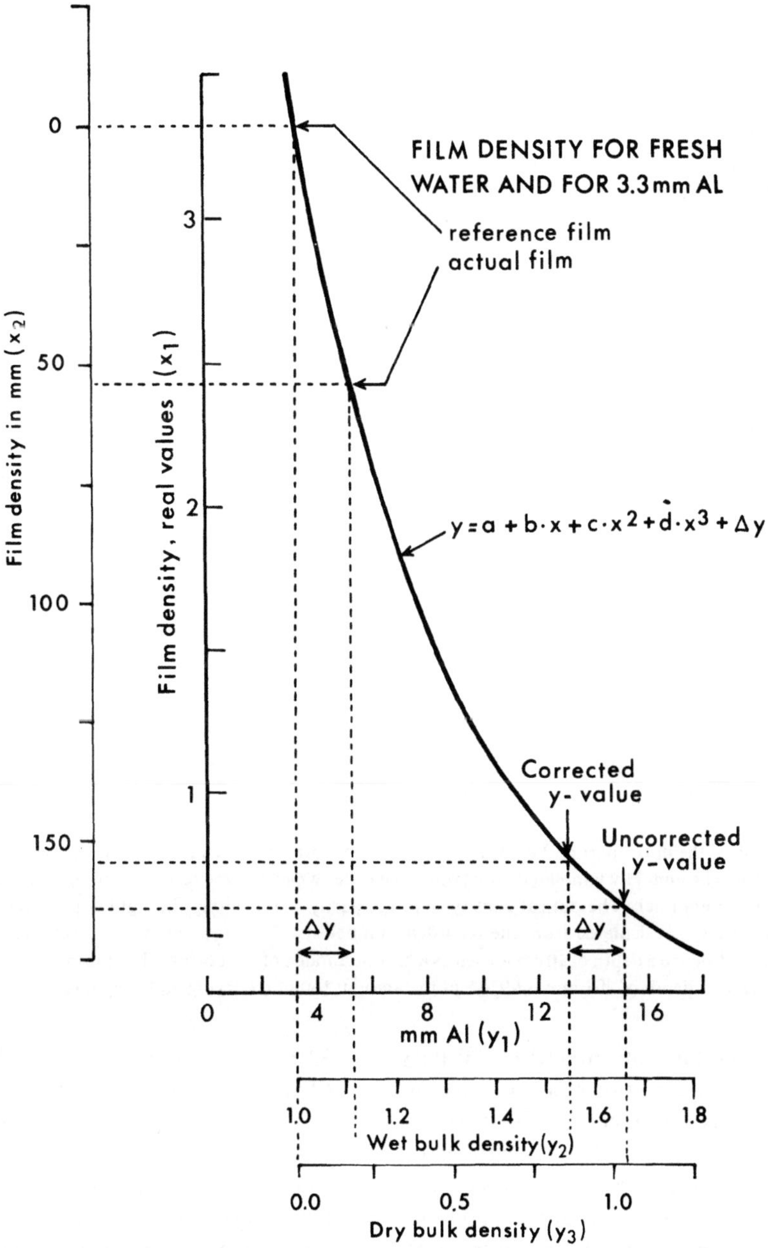

Fig. 2. The film density curve for the aluminium wedge and for corresponding bulk densities of the sediment layers on the reference film ($\Delta y = 0$). After calculating the Δy value (which varies with the exposure) the given relationship may be used to calculate the probable bulk densities (y) from recorded film densities (x). The constants a, b, c, and d are dependent on the characteristic curve of the X-ray film and on the selected density-correlated parameter.

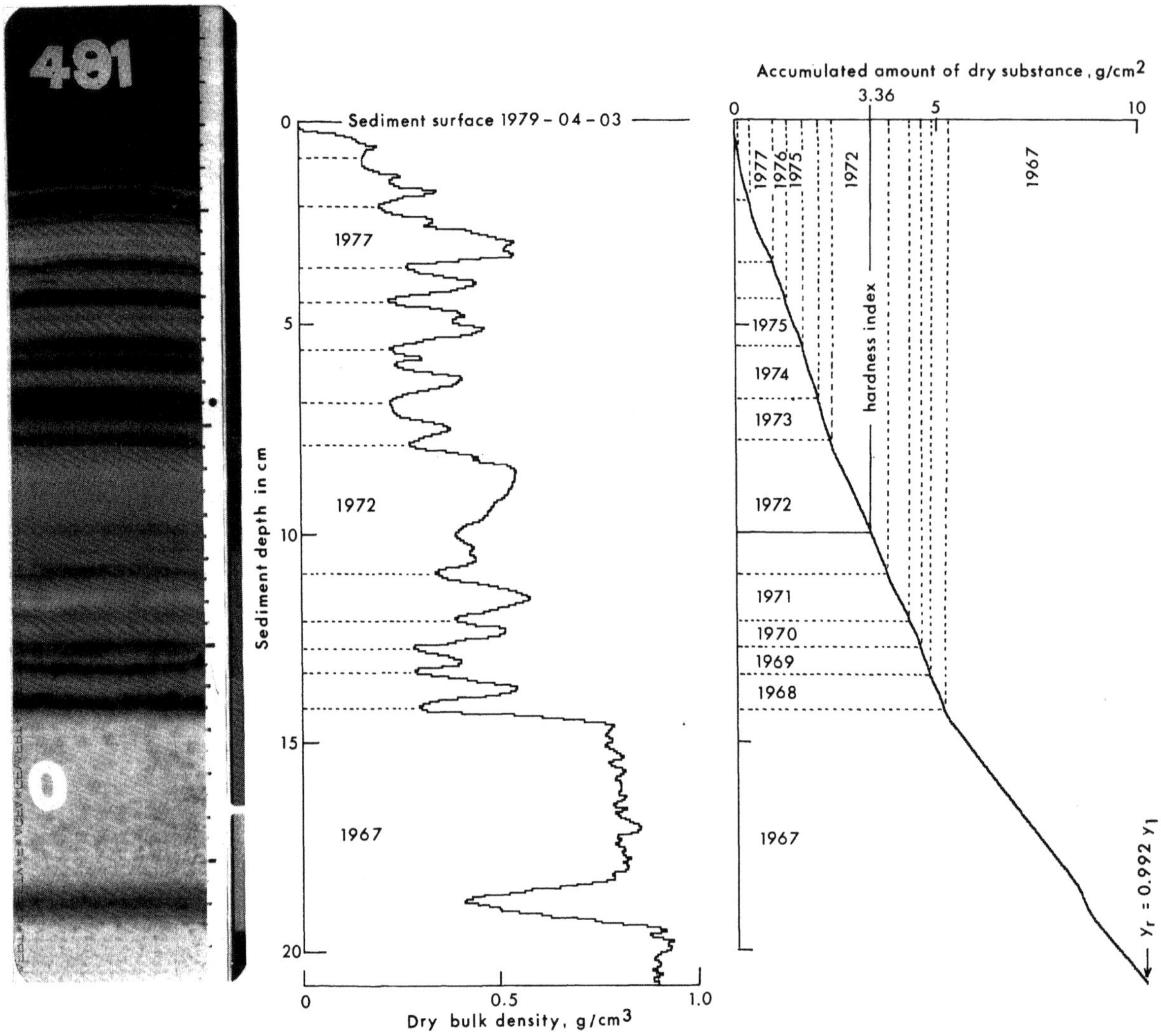

Fig. 3. X-ray radiograph of the upper part of the unextruded sediment core 491 with adjoining bottom water from the Nylandsfjärden Bay in northern Sweden. The annually laminated core (with turbidities which are the result of the submarine slides which occurred in 1967) is rectangular and 3 cm thick. The optical density steps are partly visible along the right edge of the radiograph. The vertical variation in dry bulk density is calculated from the recorded film density along the centerline of the radiograph. The accumulated amount of dry substance down to a depth of 10 cm is used as a hardness index of the bottom. The radiographically calculated amount of dry substance (y_r) down to a depth of 208 mm is 99.2% of the amount found by drying and weighing the deposits (y_l).

cross-section, to facilitate interpretation. Gravity and piston corers with exchangeable, rectangular coring tubes, specially designed for X-ray radiography have been constructed. The coring tubes are made of transparent acrylic glass which has a small coefficient of absorption of X-rays at low and moderate levels of radiation energy. This permits observation and radiography of the cores to be undertaken before the contents are extruded. The inside section of the tubes which are normally used is 30 mm × 60 mm and the wall thickness is 4 mm (Axelsson & Håkanson 1978).

The unextruded surface-sediment cores with adjoining bottom water are radiographed in an upright position, together with a 7-stepped, aluminium wedge. The thinnest (on the radiographs darkest) step of the wedge exhibits the same film density on the radiographs as fresh water, and the succeeding steps as sediments with the wet bulk densities 1.05, 1.1, 1.2, 1.3, 1.4, and 1.6 respectively.

The wet bulk density (y) at level z is given by:

$$y_z = a + \frac{x - b}{c}$$

where a is the density of the pore water, x the thickness, in mm, of the aluminium wedge which gives the same film density as the sediment core at level z, b the thickness, in mm, of the aluminium wedge which gives the same film density as the bottom water in the coring tubes, and c a factor which varies with the composition of the dry substance and with the energy of radiation (Axelsson 1979). For 30 mm thick, silty, sediment cores with fresh bottom water, X-rayed through a Beryllium window at 50 kV, the value of b is 3.3 and the value of c approximately 18.

The vertical variation in density-correlated sedimentological parameters is calculated from the recorded variations in the film density along the radiation images of the sediment cores and the aluminium wedge. The film-density curves are digitized and used (together with the relationship, illustrated in Fig. 2) to calculate the vertical variations in wet bulk density, dry bulk density (Fig. 3), water content, porosity, void ratio, effective overburden pressure, the increase of solids with increasing sediment depth (Fig. 3), and in the plotting of sedimentation-compression curves.

The examples illustrated in Figs. 2 and 3 are based on the following data: Philips K 140 Be X-ray apparatus set at 50 kV and 2 mA, focal spot size 0.6 × 0.6 mm, focal distance 100 cm, Agfa-Gevaert X-ray film Structurix D7 Pb, developing time 5 minutes at 20 °C in G 230 developer.

Results and applications

Stereoradiographs of unextruded sediment cores facilitate the interpretation of changes which occur in the depositional environment. They may also be used to study the position of larger, embedded particles, the three-dimensional shape of sedimentary structures, and the mixing activity of bottom fauna.

The X-ray radiographic technique for the calculation and recording of density-correlated sedimentary properties is fast and accurate, and is particularly applicable to sediments with a low bulk density. The accuracy of the calculations increases with decreasing density of the deposits. For bulk density values lower than 1.20 the standard deviation from the true values is generally less than 1%. The technique is, therefore, especially valuable when studying the uppermost, often very loose, part of a sediment deposit, and the processes of sediment redistribution and sediment accumulation. It may also be used for the correlation of synchronous laminations between different cores.

The radiographically calculated curves, that show the vertical variation in dry bulk density (Fig. 3), reflect important sedimentary properties. They are therefore particularly useful for sediment classification. The accumulated amount of solids, down to a sediment depth of 10 cm, is used as a hardness index (Fig. 3) for the classification of lake- and sea-bottoms. The sedimentation-compression curves, that relate the void ratio to the effective overburden pressure, are very useful in studying the processes of resuspension and pore-water transport, as well as in analysing changes in the depositional environment.

Studies are being undertaken in several Swedish lakes and coastal bays, where both mono- and stereoradiographs are being used to document primary and secondary sedimentary structures. The rate of sediment accumulation is determined from measurements of the X-ray radiographs of sediment cores and by the dating of dried subsamples by radioactive methods (i.e. ^{137}Cs, ^{210}Pb). Film density curves are used to quantify the amount of dry substance between datable sediment layers (Fig. 3). The radiographs form a valuable base for the calculation and prediction of sedimentation rates by comparing radiographs of cores from the same basin which were obtained at different times. In some of the basins, which are being examined the contemporary rate of sedimentation has been calculated by comparing radiographs of cores taken at yearly intervals.

References

Axelsson, V., 1979. Sjöars sediment studeras med röntgenteknik. Ymer 99: 133–144.

Axelsson, V. & Håkanson, L., 1978. A gravity corer with a simple valve system. J. Sed. Petrol. 48: 630–633.

Axelsson, V. & Händel, S. K., 1972. X-radiography of unextruded sediment cores. Geogr. Ann. 54A: 34–37.

Bouma, A. H., 1969. Methods for the study of sedimentary structures. John Wiley & Sons, Inc., New York, 458 pp.

Hamblin, W. K., 1962. X-ray radiography in the study of structures in homogeneous sediments. J. Sed. Petrol. 32: 201–210.

Krinitzsky, E. L., 1970. Radiography in the earth sciences and soil mechanics. Plenum Press, New York, 163 pp.

An automated technique for the accurate positioning of sediment core sites and the bathymetric mapping of lake basins

Richard W. Battarbee[1], Colin Titcombe[1], Karl Donnelly[2] & John Anderson[1]
[1] *University College London, 26 Bedford Way, London, WCIH OAP England*
[2] *Donnelly, Connolly & Poots, Engineers and Surveyors, Iveagh Court, Harcourt Road, Dublin 2, Ireland*

Keywords: paleolimnology, sediment core positioning, hydrographic survey, computer contouring, morphometric calculation

Abstract

Automated techniques for sediment core site positioning, hydrographic survey, and contour mapping are described. The computer contouring system can be extended to calculate values for morphometric variables such as maximum length (l_{max}), maximum width (w_{max}), lake perimeter (l_0), lake area (A) and lake volume (V). The techniques are exemplified using results from Augher Lough, Northern Ireland.

Introduction

Bathymetric maps of lake basins are essential for the calculation of many morphometric variables used in limnology (cf. Wetzel 1975; Håkanson 1981) and they are routinely used for navigation and as a guide to the location of water sampling and sediment coring sites. This paper describes how sample site positions, hydrographic survey, contour map production and calculation of morphometric data can be made quickly and accurately using automated and computer-based techniques. The methods are exemplified by reference to a research project on Augher Lough, a small eutrophic lake in County Tyrone, Northern Ireland. The project aims to assess reasons for spatial and temporal variations in sediment accumulation and sediment diatom concentration, and involves data from about 30 accurately located and stratigraphically correlated 1 m sediment cores.

Site positioning

The survey procedure takes advantage of the wide range of theodolite distancers now available for survey work (Short 1981). In our study we used a shore-based Geodimeter 710, an instrument capable of the dynamic fixation, by polar co-ordinates, of retro-reflecting prisms mounted or hand-held in a small boat. The average range of the instrument is

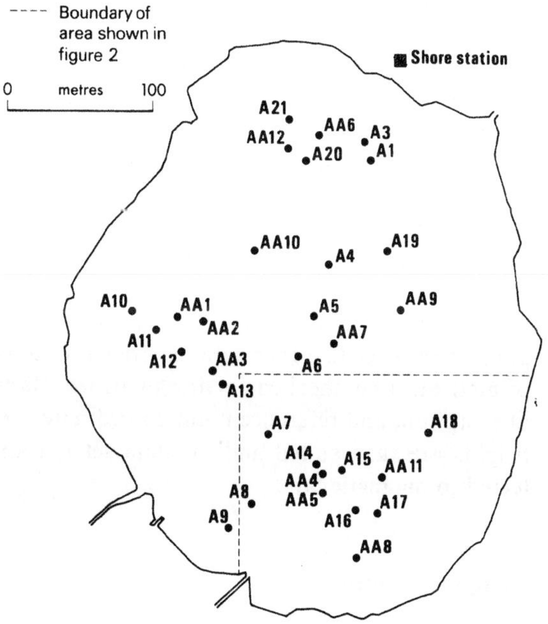

Fig. 1. 32 core sites for Augher Lough located using a shore-based electronic distancing theodolite.

Hydrobiologia 103, 71–74 (1983).

3 km although ranges up to 5 km are possible if the number of prisms is increased. Measurements are presented in digital form and values for horizontal and vertical angle, slope and horizontal distance, and height difference can be displayed as necessary. Corrections for scale factor can be keyed in. Corrections for curvature and refraction are built into a later model. The instrument can be connected to an automatic data logger which records angles and slope distance and will accept additional station information. With solid-state recorders the data can be transferred directly to magnetic tape. Static targets such as buoy locations, water sampling sites, and core sites can be fixed with extreme accuracy (±5 mm). The 32 core sites fixed in this way for Augher are shown in Fig. 1.

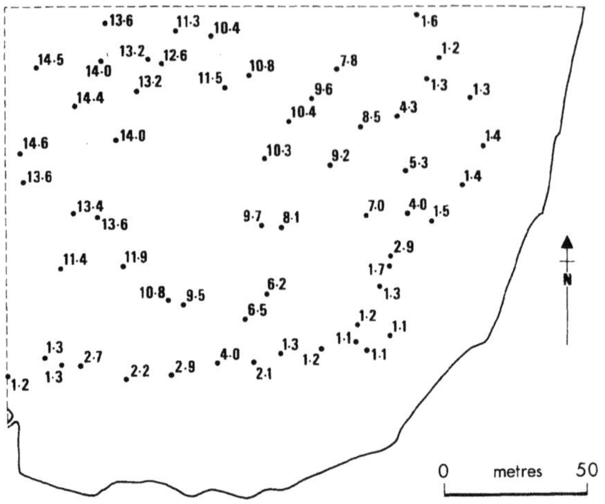

Fig. 2. SE corner of Augher Lough (see Fig. 1) showing computer plotted spot-depths recorded on site using echo sounding and an electronic distancing theodolite.

Hydrographic survey

Using this system hydrographic survey work involves recording water depth values for a series of accurately located sites and is most conveniently carried out using continuous tracking and an echo sounder with an analogue trace. Errors in continous tracking are greater than for a static target; with a boat speed of two knots down-the-line distancing errors would be +2–3 m if closing and -2–3 m if retreating. Tracking across the line of sight provides better distance accuracy. Depth readings from the echo sounder are transmitted to the shore station by radio telephone either after each fix is made or at timed intervals (e.g. every 5 seconds). After an adequate number of depth points have been fixed the data recorded on site are printed out and checked for irregularities. Shore line, station, and reference point co-ordinates and heights are then added and the data set is transferred to magnetic tape.

Computer contouring

The position and value of each surveyed point can be plotted directly using the magnetic tape data and a rectangular co-ordinatograph linked to a desk-top computer (cf. Fig. 2). Contours can be fitted manually, or more conveniently by computer. The contouring algorithm involves three stages. First, gradients in x and y positive directions are calculated at the original data points. This is achieved by a natural neighbour interpolant technique (Sibson 1982). A tile or Theissen polygon is generated for each of the original data points (Fig. 3), and the tile for each point is formed by the perpendicular bisectors between the point and its neighbouring points. The areas of these tiles provide one of the inputs to the gradient calculation. Another important input is the subtile area of each tile and these are calculated by simulating its re-

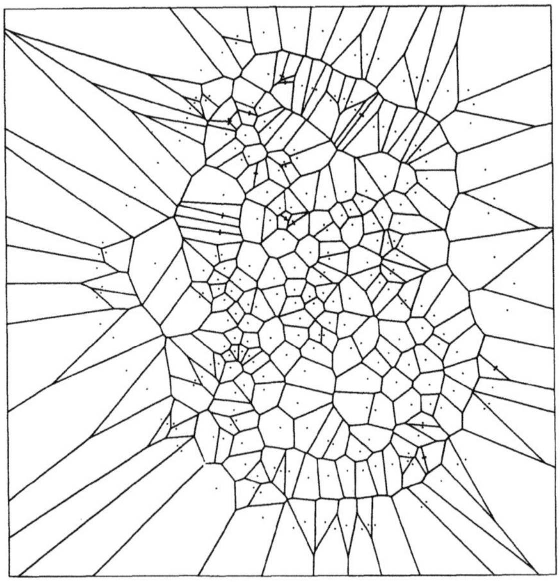

Fig. 3. Theissen polygons and the location of depth data points used in the tessellation.

moval from the tessellation. When the point is removed its neighbours form subtiles by capturing portions of its areas. The subtile area of each neighbour is expressed as a ratio of the deleted tile's area thus providing a formal description of the degree of association between the deleted point and its neighbours.

The second stage of the process involves overlaying a regular grid on to the tessellation. Each point on the grid is introduced into the tessellation in turn. Subtile areas and inter-point distance data are used to calculate depth and x, y gradients for each grid point. On completion of these calculations height and gradient data are available for the whole grid. Thus, for a 6 × 6 grid, one hundred and eight data values are calculated.

The regular grid data is processed in stage three. Cells on the grid are used in turn to generate contours and contouring takes place within triangular units. Each cell in the grid is subdivided into sixteen triangles. Height data at the vertices and mid-points of each triangle side are derived using hollow array operators (Sibson 1981). These arrays express heights of the triangle corners and edges as linear combinations of the height and gradients on the corners of the cell units. The mathematical theory on which the hollow array operators are based is founded on a piecewise quadratic approach.

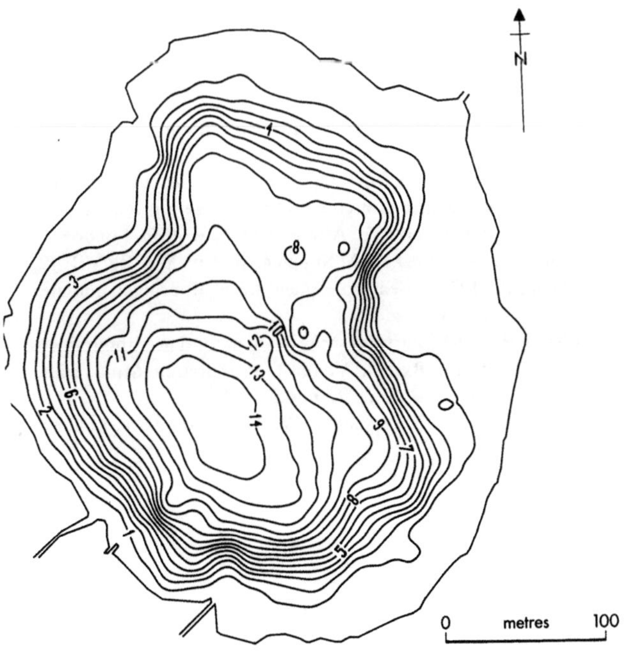

Fig. 4. Computer contoured bathymetric map of Augher Lough.

Contouring within triangle units is performed by a Fortran subroutine extracted from a Harwell report (Marlow & Powell 1976). The subroutine uses triangle mid-point and vertex heights to define a conic. Requested contours are then generated for the part of the conic which lies within the triangle. Continuity and smoothness of contours at triangle and grid cell boundaries is ensured by the common usage of the quadratic function, the function being continuously differentiable across triangle boundaries. Contours defined in this way at 1 m intervals are shown in Fig. 4.

Morphometric calculations

Maximum length (l_{max}), maximum width (w_{max}), lake perimeter (l_0) and lake area (A) are easily computed once an accurate lake outline has been digitized, and lake volume (V) can be computed in different ways.

Maximum length (l_{max}) across a lake may be obtained by calculating the distance between each point on the shore and all other points, the longest distance being retained. Within a computer program the maximum length is initialized to zero and updated every time a greater distance is encountered. The accuracy of this strategy increases as more points are used to define the lake shoreline.

The maximum width (w_{max}) of the lake is the greatest distance between the shores perpendicular to the line of maximum length. This is obtained by generating a series of perpendicular bisectors at increments along the line of maximum length. The bisector with the longest portion of its length inside the lake gives the maximum width. Increased accuracy is achieved by decreasing the increment between the perpendicular bisectors.

Lake perimeter (l_0) may be calculated by summing distances between successive points around the shoreline. The calculation becomes more accurate as more points are used to define the lake outline.

Lake area (A) may be computed by triangulation. The lake surface is divided into triangular units and their areas summed. Triangles are formed between a base point and adjacent pairs of shoreline points. The technique is widely used and works for very irregular shapes. If the outline turns back on itself triangles have negative areas. When the trian-

Table 1. Morphometric data for Augher Lough calculated using computer-based techniques.

Area (A)	92751.33 m^2
Perimeter (l_0)	1213.16 m
Maximum length (l_{max})	389.59 m
Maximum width (w_{max})	329.72 m
Volume (V)	0.486×10^6 m^3

gle areas are summed the signs cancel out to give the correct area measurement. Islands, and lakes within islands, may be included in the outline definition. Land areas within the lake are therefore not included in the final area measurement. Greater accuracy is again obtained by increasing the number of points used to define the outline. By transforming digitized shorelines between coordinate systems (i.e. mm to yards, m, km or miles) area calculations can be made in any units desired.

For single basin lakes (such as Augher) calculation of lake volume (V) is easy and is based upon the measurement of areas between closely spaced contours. Volume is calculated by either plotting depth against contour area and integrating the area under the curve or by using contours to define a series of truncated irregular cones. Volumes between successive portions of irregular cones are summed to give total volume. In more complex basins with depressions and ridges on the lake floor the solution is less simple. However, since the contouring strategy defined above uses small triangles as the basic unit for contouring and since the contouring problem is treated in piecewise quadratic form, contour string configurations are limited in their complexity. Area calculations can therefore be made within individual triangles and a running count maintained for areas between contour levels. This facility is a planned enhancement for the contouring program, but at present, in these complex cases, volume is obtained rather inefficiently by operating on a large file containing all the completed contour strings.

Table 1 lists values for morphometric parameters for Augher from computer-based measurements.

Discussion

Accuracy in the positioning of sediment core sites is essential especially when grid coring (e.g. Bloemendal *et al.* 1979), when attempting to relate the position of a core to cores taken previously from a particular site, and when it is necessary to avoid known areas of sediment disturbance. For lakes without winter ice-cover, electronic distancing allows sites to be rapidly and precisely fixed.

This site fixing method is most useful when used for hydrographic survey. It eliminates problems of maintaining constant boat velocity and direction when using the transect survey technique (cf. Håkanson 1981) and allows the data to be automatically transferred to magnetic tape for computer plotting. If the spot-depth data so produced is contoured by computer, a bathymetric map can be objectively and rapidly constructed and the program can be extended to compute values for morphometric variables. In addition, if sufficient stratigraphic data are available palaeo-bathymetric maps can be produced in the same way and values for whole-basin sediment volumes can be computed for successive time periods.

References

Bloemendal, J., Oldfield, F. & Thompson, R., 1979. Magnetic measurements used to assess sediment influx at Llyn Goddionduon. Nature 280: 50–51.

Håkanson, L., 1981. A manual of lake morphometry. Springer-Verlag, New York, 78 pp.

Marlow, S. & Powell, M. J. D., 1976. A Fortran subroutine for plotting the part of a conic that is inside a given triangle. UKAEA Harwell Rep. AERE-R 8336, HMSO, London.

Short, M., 1981. 1981 EDM Supplement. Civ. Eng. Surv.:1–12.

Sibson, R., 1981. A seamed element for contouring. Computer J. 24: 378–382.

Sibson, R., 1982. A brief description of natural neighbour interpolation. In: Barnet, V. (ed.). Interpreting Multivariate Data. J. Wiley & Sons, London: 21–36.

Wetzel, R. G., 1975. Limnology. W. B. Saunders Co., London.

Recent climatic changes recorded by the salinity of pore waters in the Dead Sea sediments

M. Stiller, P. Kaushansky & I. Carmi
Isotope Department, The Weizmann Institute of Science, 76 100 Rehovot, Israel

Keywords: paleolimnology, Dead Sea climate, pore-waters, diffusion, sediments, salinity, porosity

Abstract

The salinity of lakes is subject to variations imposed by climatic changes. These variations are recorded in the salinity profile of pore waters. Meromictic lakes, such as the Dead Sea, are a special case where waters which underlie the mixolimnion reflect salinity variations. In a sediment core from Dead Sea shallow waters, the salinity profile exhibited a minimum at about 1.9 m depth. It is shown by a diffusion model that this minimum can be attributed to lower salinities which prevailed at the sediment water interface for several decades around the turn of this century. No such minimum was observed in a sediment core from the deepest part of the lake where, during the last two centuries, the overlying brines had a constant salinity.

Introduction

In closed lakes with large watersheds, the lake level is a sensitive indicator of fluctuations in its water balance caused by climatic and man-made changes.

The levels of the Dead Sea, a terminal, chloride, hypersaline lake, have been reconstructed back to the early 19th century (Klein 1965: Fig. 1), and found to be fairly well correlated with rainfall intensity in Jerusalem (Klein 1981). A historical record of surface densities of the Dead Sea since the early 19th century (Fig. 1) shows that low values characterize higher lake levels and vice versa (Nissenbaum 1970; Assaf & Nissenbaum 1977). Also, the meromictic structure of the Dead Sea is seen from density profiles measured in 1864, 1919, and the 1930's and from a detailed study in 1959–1960 (Neev & Emery 1967). In this paper, we show that salinity fluctuations in the upper water mass imposed by climate variations over the watershed, are recorded in the pore waters of underlying sediments.

Methods

Two sediment cores were taken from the Dead Sea (Fig. 1), shortly after the destruction in February 1979 of its meromictic structure (Steinhorn *et al.* 1979). Pore waters were separated from 5 cm sediment slices by centrifugation at 15 000 rpm and 23 °C. The density ρ (g cm^{-3}) of the pore waters was measured by the method of Kaushansky and Starobinetz (1982), and is reported at 25 °C. The salinities of the pore waters, s (g kg^{-1}), were calculated from the relation between the density and salinity $\beta = (1/\rho)\partial\rho/\partial s$, with $\beta = 7.4.10^4$ g kg^{-1} at 25 °C. (Steinhorn 1980).

The sediments were vacuum dried at 40 °C. The weight of pore water was obtained by adding the weight of salt in the dried sediment to the weight lost by vacuum drying, and the porosity calculated from $p = (W_{pw}\rho_{pw}^{-1})(W_{pw}\rho_{pw}^{-1} + W_s\rho_s^{-1})^{-1}$, where W_{pw}, W_s, ρ_{pw} and ρ_s are the weight and densities of, respectively, pore water (pw) and sediments (s). ρ_s was taken to be 2.65 gcm^{-3}.

Hydrobiologia 103, 75–79 (1983).

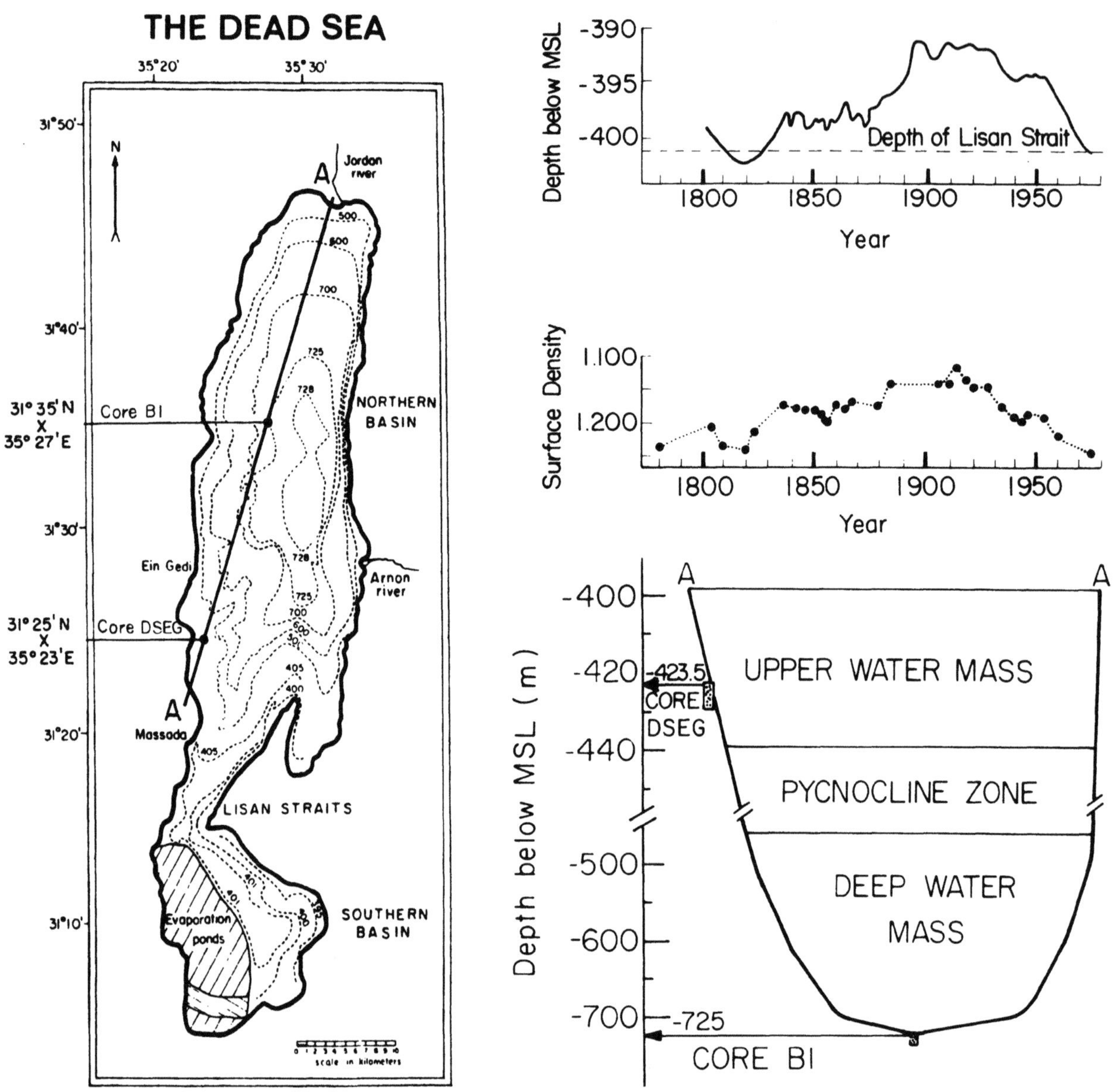

Fig. 1. Left: Bathymetric map of the Dead Sea with locations of core DSEG and core B1. Right: top - fluctuations of the level of the Dead Sea (Klein 1965); middle - fluctuations of its surface density (Nissenbaum and Assaf 1977); bottom - transect across line A-A showing the meromictic structure in 1959/60 and the depth of sediment cores.

Results

Core DSEG was taken from a shallow part of the Dead Sea (Fig. 1). The densities of the entire profile (Fig. 2) are lower than that of the brine at the sediment-water interface (SWI) which, at that time, was 1.233 gcm^{-3}. The minimum probably reflects the downward diffusion of overlying waters of considerably lower densities than nowadays (Fig. 1). In core B1, taken from the deepest part of the lake (Fig. 1), large halite crystals were found at 80 cm depth (Fig. 2); its pore waters have a rather constant density of 1.23203 $\pm$ 0.00036 g cm^{-3} (Fig. 2), which is within the range of the monimolimnion densities between 1975 and 1979: 1.23179 $\pm$ 0.00012 (Steinhorn 1981).

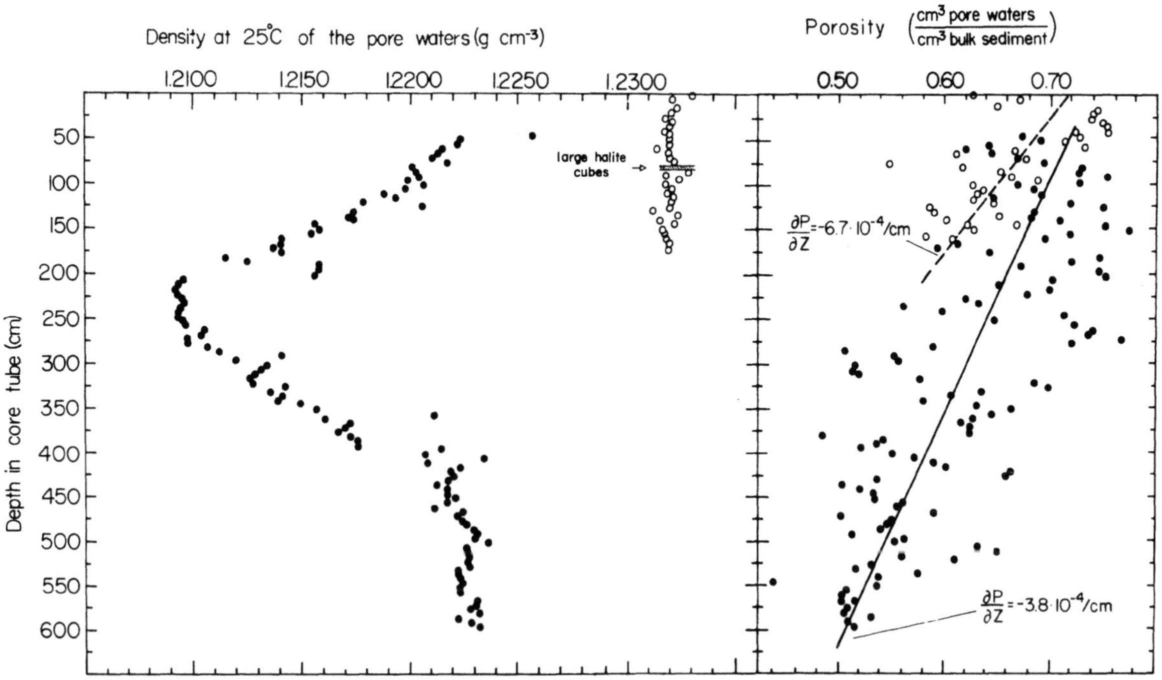

Fig. 2. Densities of pore waters at 25 °C and porosity of the sediments. Core DSEG (●), 550 cm, taken with a Mackereth (1958) corer, 19 September, 1979. Core B1 (○), 176 cm, taken with a Benthos 2171 corer, 19 December, 1980.

Time dependent changes in the concentration of dissolved constituents in pore waters of sediments are described by (Stiller *et al.* 1975):

$$\partial(pC)/\partial t = \partial/\partial z(pD_b\partial C/\partial z) - \partial(UpC)/\partial z \qquad (1)$$

where p is porosity of the sediment (cm^3 pore water cm^{-3} bulk sediment), C is concentration (gcm^{-3}), D_b is coefficient of diffusion in the bulk sediment (cm^2 bulk sediment s^{-1}), U is advective velocity (cms^{-1}) and z is depth in sediment (cm). In solving eq. (1), a coordinate system fixed in time and space was used and compaction currents and upwelling were neglected. The present Dead Sea and its pore waters are a mixture of chlorides of approximately 6M (Mg^{2+} 44, Na^+ 40, Ca^{2+} 17 and K^+ 7.8 gl^{-1} in March 1977; Steinhorn 1979). The model assumes the pore water salinity, S (gl^{-1}), to be a 6M NaCl brine with $D = 1.6\ 10^{-5}\ cm^2s^{-1}$ at 25 °C (Harned and Owen 1960). Eq. (1) thus simplifies to:

$$p\partial S/\partial t = \partial/\partial z(pD_b\partial S/\partial z) \qquad (2)$$

Molecular diffusion is hindered by the tortuosity of the sediments, and porosity is considered to be a measure of this property. We used an empirical relation suggested by Lerman (1978) for the molecular diffusion coefficient in the bulk sediment: $D_b = p^nD$. When the porosity is low, e.g. less than 0.75, as is the case with Dead Sea sediments (Fig. 2), the chosen value of n becomes significant.

We solved eq. 2 numerically to a depth of 10.5 m, from 1805 to the present, with upper boundary values constant for periods of one year. The time and depth increments were, respectively, 5 days and 5 cm so that the modulus was far less than 0.5, hence satisfying the stability requirement (Carslaw and Jaeger 1959).

During the last two centuries, the depth of the sediment-water interface (SWI) for core DSEG varied between 20 and 31 m relative to the lake surface. The salinities at SWI were deduced (Fig. 3) as follows. a) 1805–1821: they were assumed to be identical to that of surface waters; b) 1821–1935 (rising lake levels): a linear gradient existed between the surface and a depth of about 60 m, as seen in historical profiles (Assaf & Nissenbaum 1977). The salinity at the SWI was calculated by interpolation to this depth; c) 1935–1959 (declining lake levels): we allowed the salinity of the SWI to gradually reach that of the surface waters. d) 1959–1979 (declining lake levels): the upper layer of uniform salinity was 40 m in 1959 (Neev & Emery 1967), and gradually deepened to 175 m in 1978 (Steinhorn

1982). Therefore, for this period we used measured surface salinities, or salinities at SWI whenever available.

In 1805, the meromictic structure had already existed for about 100 years (Stiller & Chung 1981), but lake levels were low (-399 m to -401.05 m; Klein 1981) and the salinity of the mixolimnion was probably rather high. Therefore, for core DSEG the chosen initial salinity in 1805 was 330 gl^{-1}, i.e. close to the 340 gl^{-1} of the monimolimnion.

The calculated and measured salinity profiles of core DSEG are compared in Fig. 3. During sampling, part of the top of the core was lost and the profiles fit the data best by assuming this loss to be 25 cm.

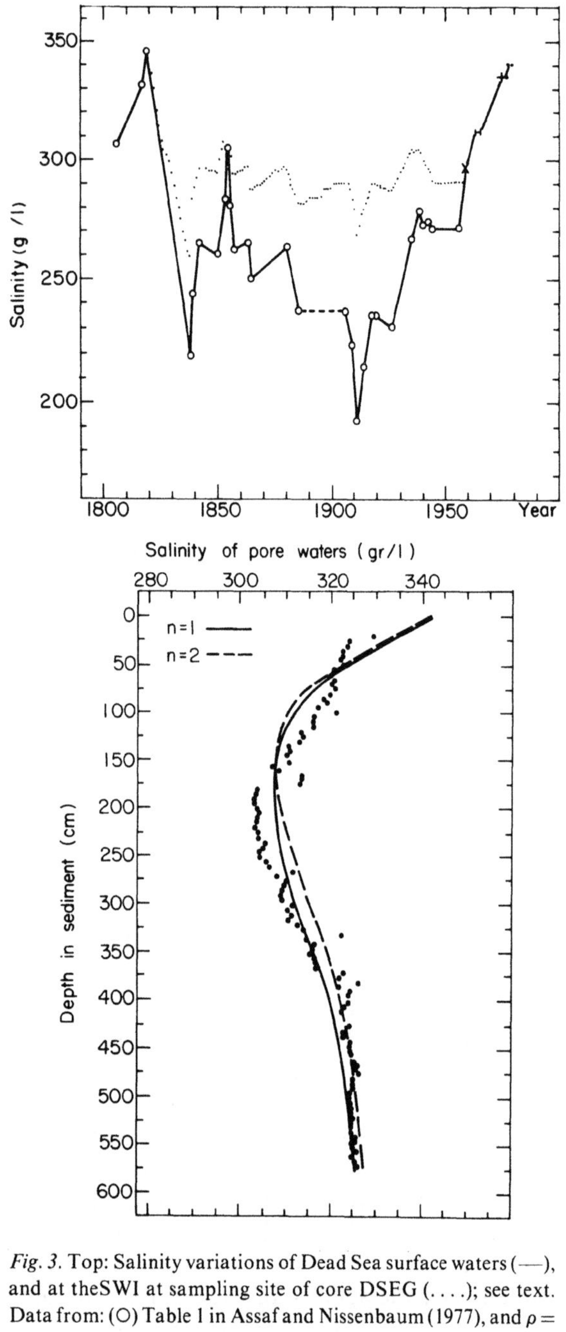

Fig. 3. Top: Salinity variations of Dead Sea surface waters (—), and at theSWI at sampling site of core DSEG (. . . .); see text. Data from: (O) Table 1 in Assaf and Nissenbaum (1977), and $\rho =$ 1.227 17 °C in 1817 as measured by Gay Lussac; (X) Neev and Emery (1967); (H) Nissenbaum (1969), (+) Beyth (1977), (●) Steinhorn (1981), Bottom: Measured pore water salinities (●●●●) in core DSEG and calculated salinity profiles.

Discussion

The model presented above fits the measured salinity profile in core DSEG rather well. Of the two diffusion profiles presented, the one with $n = 1$ (i.e. $D_b = pD$), fits the data better (Fig. 3). This may be due to the rather coarse texture of the sediments (50% to 60% are 2μm–63μm grains, Garber 1980), so that tortuosity is relatively small and, consequently, the diffusion less hindered.

The uniformity of the profile in core B1 suggests that the deep waters have maintained a constant brine concentration for quite a long period of time. However, the presence of halite crystals does not exclude a process of halite dissolution within the pore waters which could wipe out any memory of lower salinities. Halite crystals were not found in core DSEG, possibly because during its deposition transitions from saturation to undersaturation in the mixolimnion were fast enough so that any precipitated halite redissolved rapidly before being buried in the sediment.

From ^{14}C ages in core DSEG: 2130 and 7710 BP at depths of 105 and 545 cm (Kaufman, pers. comm.), and the porosity gradient (Fig. 2), we estimate that the effect of compaction on the salinity distribution in the sediment is one order of magnitude smaller than the effect of diffusion. The ^{14}C ages also indicate that between 1805 and 1979 the SWI could have moved upwards by 10 cm at most. Upwelling, if any, must be small, because its incorporation into eq. 2 would generate a profile with the minimum shifted upwards, impairing the fit with the measured salinity. For an exact model, profiles of the different chloride components of the Dead Sea brines have to be measured and compared to diffusion profiles calculated for each with the respective cross coupling terms (Lasaga 1979). Rigorously, diffusion calculations cannot be performed for salinity as an entity. Our approximation which regards salinity as a pure 6M NaCl solution and also scarcity of data in the historical density record, render refinements such as incorporation of compaction and upwelling, unjustified.

References

Assaf, G. & Nissenbaum, A., 1977. The evolution of the Upper Water Mass of the Dead Sea 1819–1976. In: Green, D. C. (Ed.). Desertic Terminal Lakes. Proc. Symp., Utah State Univ. pp. 61–72.

Beyth, M., 1977. Recent evolution of the Dead Sea brines. Geol. Survey Israel, Internal Rep. MG/3/77, 11 p.

Carslaw, M. S. & Jaeger, J. C., 1959. Conduction of heat in solids. Oxford Univ. Press, London, 510 p.

Garber, R., 1980. The sedimentology of the Dead Sea. Ph.D. Thesis, Rensselaer Polytechnic Institute, 157 pp.

Harned, H. S. & Owen, B. B., 1960. The physical chemistry of electrolytic solutions. Reinhold Publ. Co., New York, 607 p.

Klein, Z., 1965. On the fluctuations of the level of the Dead Sea since the beginning of the 19th century. Hydrol. 7, revised ed. Hydrol. Serv., Jerusalem, 83 pp.

Klein, Z., 1981. Variations of the Dead Sea level during the last nine hundred years. Teva Vehaaretz 23: 212–218 (in Hebrew).

Kaushansky, P. & Starobinetz, S., 1982. A temperature gradient column for the rapid measurement of the density of brines. Mar. Chem. 11: 289–292.

Lasaga, A. C., 1979. The treatment of multi-component diffusion and ion pairs in diagenetic fluxes. Am. J. Sci. 279: 324–346.

Lerman, A., 1978. Chemical exchange across sediment-water interface. Ann. Rev. Earth Planet. Sci. 6: 281–303.

Mackereth, F. J. H., 1958. A portable core sampler for lake deposits. Limnol. Oceanorgr. 3: 181–191.

Neev, D. & Emery, K. O., 1967. The Dead Sea, depositional processes and environments of evaporites. Geol. Surv. Israel Bull. 41, 147 pp.

Nissenbaum, A., 1969. The geochemistry of the Jordan River -Dead Sea system. Ph.D. Thesis. UCLA, 288 pp.

Nissenbaum, A., 1970. Chemical analyses of the Dead Sea and Jordan River water 1778–1830. Israel J. Chem. 8: 281–287.

Steinhorn, I., 1980. The density of Dead Sea water as a function of temperature and salt concentration. Israel J. Earth Sci. 29: 191–196.

Steinhorn, I, 1982. A hydrographical and physical study of the Dead Sea during the destruction of its long-term meromictic stratification. Ph.D. Thesis, Weizmann Inst. Sci., Israel, 323 pp.

Steinhorn, I., Assaf, G., Gat, J. R., Nishry, A., Nissenbaum, A., Stiller, M., Beyth, M., Neev, D., Garber, R., Friedman, G. M. & Weiss, W., 1979. The Dead Sea: Deepening of the mixolimnion signifies the overturn of the water column. Science 206: 55–57.

Stiller, M., Carmi, I. & Münnich, K. O., 1975. Water transport through Lake Kinneret sediments traced by tritium. Earth Planet. Sci. Lett. 25: 297–304.

Stiller, M. & Chung, Y. C., 1983. Radium 226 in the Dead Sea: a possible tracer for the duration of meromixis. Limnol. Oceanogr. (in press).

Trace element associations as indicators of sediment accumulation in lakes

Lidiya V. Sergeeva
Institute of Lake Research, Academy of Sciences of the U.S.S.R., Petrovskaya nab. 2, Leningrad 197046, U.S.S.R.

Keywords: paleolimnology, geochemistry, trace elements

Abstract

A geochemical study of lacustrine sediments was undertaken as part of a major paleolimnological project concerned with the impact of man on lakes. Factor analysis was applied to the geochemical data obtained from a core from one lake. Three principal factors which explained most of the variance of the initial data were identified. Factor I is related to the organic:mineral component ratio of the sediment, and can be considered as an indirect index of change in lake trophic status through time. Factor II is related to the granulometric composition of the sediment and matches evidence for changes in the hydrological regime of the lake. Factor III reflects changes in redox potential and is more closely related to processes occurring within the lake than other factors. Specific geochemical associations of elements are connected with each factor and are the factor indicators. The correlation of these associations are analyzed as geochemical indices of the variability of sediment accumulation conditions in time.

Introduction

Geochemistry is of particular importance in paleolimnological investigations. The statistical treatment of geochemical data not only assists in the analysis of problems, but also helps to identify new geochemical indices (Sergeeva 1981; Tarnovsky 1981).

Materials and methods

In the present report the results of factor analysis are applied to geochemical core data (Lawley & Maxwell 1967). These data are part of a larger body of data collected by a group of different specialists during the investigation of sediment cores from six lakes in the Latgalian elevation, Pribaltica (Jakovleva *et al.* 1981).

The lakes have all been formed in postglacial time and their watersheds are composed of unconsolidated glacial and fluvioglacial material. Lake areas are small, ranging from 0.087 to 0.57 km^2. Although the lakes are situated in similar natural conditions, they differ with respect to trophic level and the impact of man on the catchments. This is manifested by the varying ratio of ploughed areas to natural forest within the watersheds. The sediment cores have similar stratigraphies: sandy and sandy-clayey terrigenous carbonate and noncarbonate material from the base, above which a thick layer of organogenic silts with a varying content of organic matter (20–80% dry matter) occurs. This indicates that sediment accumulation has been mainly determined by variations in biogenic factors through time, and this should be reflected by the distribution of chemical elements in the cores.

The concentrations of trace elements (Ti, Zn, V, Ga, Pb, Mn, Cu, Ni, Co), and the content of organic matter (OM) and mineral matter (MM) were determined. Trace elements were analyzed by emission spectral analysis, and organic matter was

Hydrobiologia 103, 81–84 (1983).

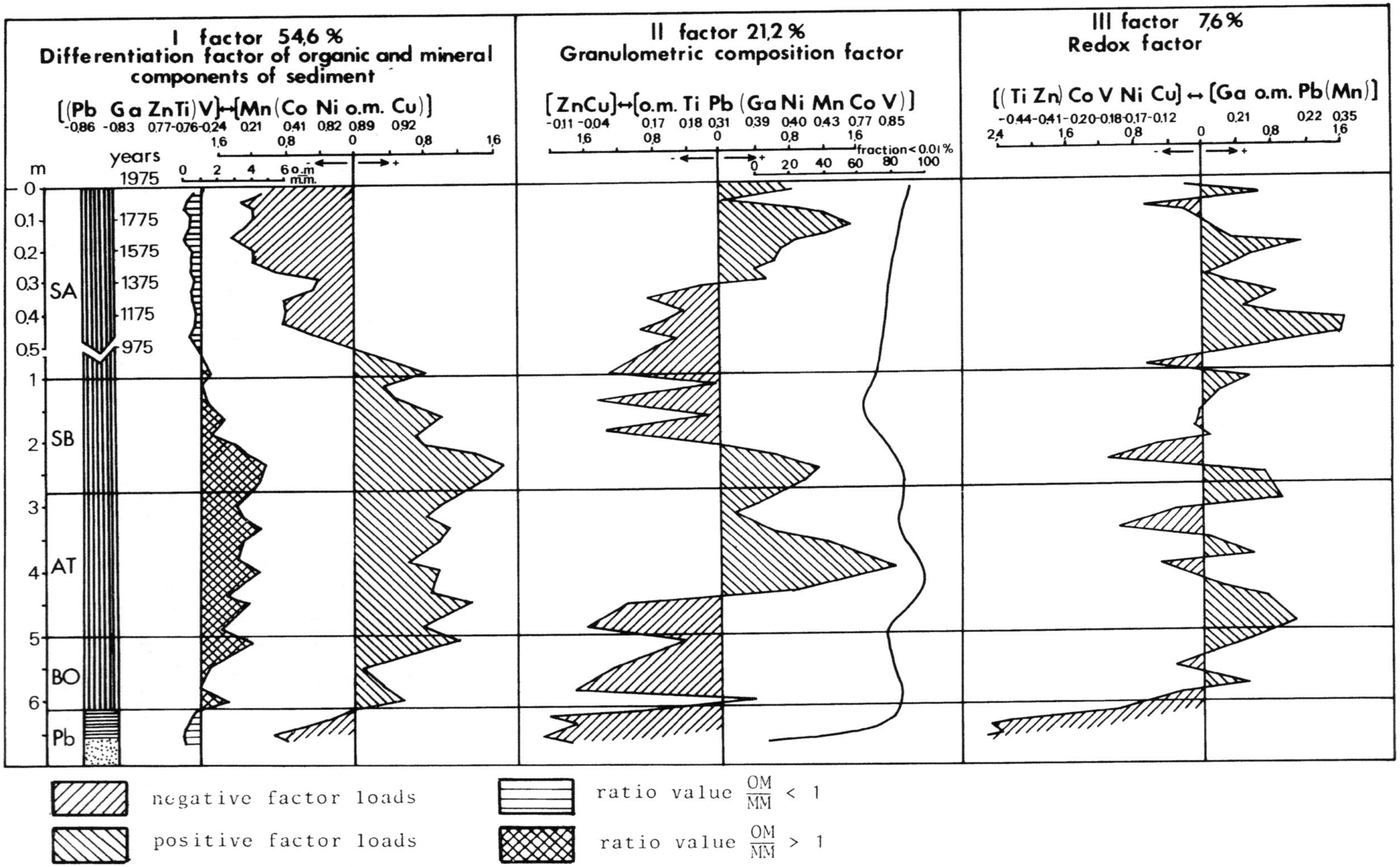

Fig. 1. Graph of the results of factor analysis of trace element concentration in a sediment profile from lake Vorkalu. [. . .] associations of elements, (. . .) elements characterized by a factor load significant at the 5% level.

found by loss on ignition at 500° C. The ash weight indicated the mineral matter (100% - OM).

Results and discussion

Factor analysis of geochemical data enables a number of causative factors which account for the variability of the whole data complex in sediment cores to be found. The results of this analysis for the postglacial sediments of eutrophic lake Vorkalu are shown in Fig. 1. The quantitative evaluation of the importance of each factor is represented by the amount of the total variance explained by each factor expressed in percent. The correlation of the factor with each chemical element is also evaluated quantitatively in terms of the value of factor load. The factor loads are regarded as correlation coefficients between the factor and the content of the elements. They can have either a negative or positive sign, and these allow the objective identification of the specific geochemical associations related to each factor selected.

The graphs (Fig. 1) show variations in factor loads through time. The first factor (I) is responsible for 54.6% of the total variation. The elements connected with it which have significant load values are Co, Ni, Cu and OM (positive) and Pb, Ga, Zn, Ti (negative). The correlation matrix indicates that the positive geochemical associations are primarily related to organic matter ($r = +0.41$–0.80). Factor I may then be assumed to reflect the differentiation of organic and mineral components of the sediment. Changes in the intensity of this factor throughout the profile can be explained by changes in the ratio of the volume of organic matter capable of being preserved under certain conditions of sediment accumulation to that of mineral substances deposited simultaneously (OM/MM). As is seen in Fig. 1, the changes of the OM/MM index in the profile conforms well with the variability of factor I.

The second factor (II) explains 21.2% of the total variation and is related to the granulometric composition of the sediments. Elements of the iron family (Mn, Ni, Co, V) and gallium (Ga) are most closely associated with it. Figure 1 shows that the variability of this association correlates well in time with that of the pelite fraction content in the sediment, which has maximum values generally coinciding with the pelagial. Significant factor loads on these elements are explained by considerable variations of their content in sediments. It may be that high factor values could indicate periods of high water during the history of the lake. The changes revealed with the help of factor II completely coincide with the reconstruction of lake level regime based on diatom analysis.

The third factor (III) accounts for only a small amount of the variation (7.6%) and can be explained by changes in redox potential in the sediments. This is indicated by the opposite signs of factor loads on manganese which, according to its redox potential is the most sensitive element to changes in oxidation-reduction. It is to be expected that positive loadings coincide with sediments rich in silts and organic matter (sediments BO-SA). The boundaries and characteristics of the climatic periods of the Holocene by Blytt-Sernander have been defined by V. I. Khomutova according to the results of palynological analysis (Jakovleva et al. 1981). Maximum factor loads for this association occur in sediments of the early Subatlantic period where vivianite crystals have been discovered. These data allow the development of the lake to be traced. Trophic level and water level fluctuations are related to alterations of climatic conditions. During the Preboreal period mineral sediments accumulated in shallow water since the humid climate and the large quantity of glacial material, not stabilised by vegetation, allowed erosive processes in the watershed to dominate. From the beginning of the Boreal period and until the onset of the Subatlantic period the changes in ratio of mineral to organic matter in the sediment are closely connected with water level fluctuations. Sediments with the highest content of organic matter occur during the Atlantic and during the beginning of the Boreal period when climatic conditions were most warm and humid and water levels were highest. The granulometric composition of the sediments became coarser and their organic content decreased due to a reduction in water level resulting from climatic changes during the Subboreal and the beginning of the Subatlantic period. The accumulation of fine silts begins from the middle of the Subatlantic period. However, the changes in sediment composition of this period do not conform with the relationships found earlier since the increase in the terrigenous component of the sediments coincides with a rather high water level. The beginning of this process, assuming a constant sediment accumulation rate is believed to date from about 800–900 AD. The increase in the terrigeous component at this time is probably due to the beginning of ploughing in the drainage basin and a consequent intensification of erosion. More humid climatic conditions contributed to the maximum erosive development between the 14th and 18th centuries. Factor II indicates that high lake levels were characteristic of this time. Analogous processes do not occur in lakes with watersheds not exposed to human impact. These data suggest that human impact in the watershed disturbs natural geochemical processes.

References

Lawley, D. N. & Maxwell, A. E., 1967. Factor analysis as statistical method (in Russian). Moscow, Izdatelstvo 'Mir', pp. 144.

Sergeeva, L. V., 1981. Trace element associations as indices of sediment accumulation in Latvian lakes. In: Paleolimnological approach to study of human impact on lakes (in Russian). pp. 52–61. Leningrad, Izdatelstvo 'Nauka'.

Jakovleva, L. V., Khomutova, V. I., Drabkova, V. G. & Sergeeva, L. V., 1981. Postglacial history of lacustrine sediment accumulation in Latvian water bodies and its change under influence of human factor. In: Paleolimnological approach to study of human impact on lakes, pp. 36–51. Leningrad, Izdatelstvo 'Nauka'. (in Russian).

Tarnovsky, A. A., 1981. Geochemistry of bottom sediments of present lakes (in lakes of the Karelian Isthmus), Leningrad, Izdatelstvo of Leningradian State University. (in Russian).

Transfer functions for estimating paleoecological conditions (pH) from East African diatoms

F. Gasse[1] & F. Tekaia[2]
[1] *Ecole Normale Supérieure, 92260, Fontenay-aux-Roses, France*
[2] *Laboratoire de Statistique, 4 Place Jussieu, 75230, Paris Cedex 5, France*

Keywords: paleolimnology, diatoms, East Africa, transfer functions, pH-indicators

Abstract

Our purpose is to establish the quantitative relationship between recent diatom floras and ecological parameters, in order to extrapolate the results to the past. The parameter pH is here considered as an example.

This work is based on the study of about 160 diatom samples from East Africa and of their corresponding biotopes. We propose some statistical methods to interpret the data.

Correspondence analysis allows us to define the pH-indicator species. The regression calculations allow pH values to be calculated using the percentage of the diatom species in a sample.

Introduction

Our goal is to use fossil diatoms to quantify paleoecological conditions in East African lakes. This objective necessitates the following steps:

a) studying the relationships between recent diatom communities and their corresponding biotopes of a statistically representative number of samples taken from East Africa;

b) establishing an equation (transfer function) which would allow the value of an ecological parameter to be calculated, from the relative percentages of diatom species in a sample;

c) extending the correlation to fossil samples once transfer functions are established for recent diatoms.

This paper is intended to demonstrate the methods of approach used rather than to present final results. The parameter pH is exemplified since so far it appears to be the environmental factor indicated best by diatom floras.

Material studied and data collection

East African lakes are numerous and diversified. During the Quaternary they underwent wide fluctuations in water-level and water chemistry which have been recorded by changes in fossil diatom assemblages (Richardson & Richardson 1972; Holdship 1976; Harvey 1976; Richardson *et al.* 1978; Gasse 1975, 1977, 1980).

The modern samples were collected from 98 stations (lakes, swamps, peat-pogs, rivers and thermal springs) situated between 12 °N and 12 °S latitude, 26 °E and 43 °E longitude, and ranging from 155 m below sea-level to 4 000 m.a.s.l. Detailed description of the localities will appear elsewhere (Gasse 1983). In most cases, pH, conductivity and temperature were measured in the field when the diatoms were collected. The pH ranges from about 5 to 10.9, and the conductivity from about 10 to 50 000 μS cm^{-1}. Chemical analyses of the corresponding waters were published by Talling & Talling (1965), Talling (1976), Kilham (1971), and Gasse (1975). The water temperature varies from a few degrees above zero in the mountains to 35 °C in the deserts.

Hydrobiologia 103, 85–90 (1983).

It reaches 55 °C in thermal springs.

Two principal chemical types of water can be distinguished: the sodium-bicarbonate-carbonate type, corresponding to the majority of the samples, were the factors pH, alkalinity and conductivity are strongly and positively correlated between themselves, and the sodium-chloride type which has a pH near 7.

The following analyses are based on 156 contemporary diatom samples including phytoplankton, periphyton and mud. For each sample, a systematic inventory was established and the percentage of each taxon was evaluated by counting 300 to 1 000 valves distributed on four slides. 579 species and varieties were identified. The check-list and autecological data will appear elsewhere (Gasse 1983).

Statistical methods used

As considerable progress has been made in the establishment of transfer functions between micro-organism assemblages and ecological parameters (Kipp 1976; Roux 1979; Bryson & Kutzbach 1974), we attempt here to apply such methods to diatoms.

Correspondence analysis (CA)

The CA method (Benzecri 1973; Lebart *et al.* 1977; Benzecri 1980) aims at synthethizing the information contained in a data matrix and visualizing the relationships between the elements of two sets i (e.g. taxa percentages) and j (e.g. samples). The main properties and examples for defining diatom assemblages are given by Gasse & Tekaia (1979, 1982).

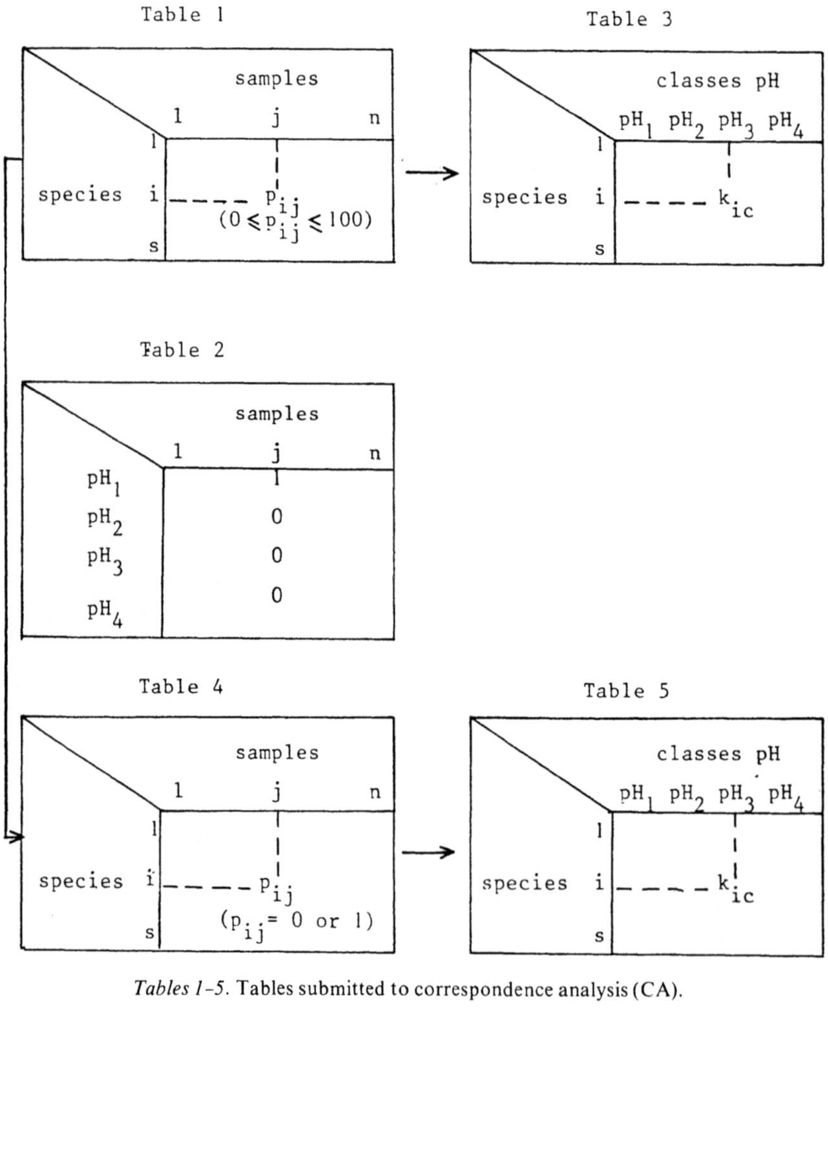

Tables 1–5. Tables submitted to correspondence analysis (CA).

We present here the types of tables which have undergone factor analysis and indicate for each table whether it is considered as a principal table (defining a factorial axis), or as a supplementary element of another table. The supplementary element does not participate in the definition of the axes, but is placed with regard to the axes defined by the principal table.

In Tables 1–5, n and s refer to the number of samples and species, respectively. Table 1 is the initial table where the intersection of the row i and the column j is the percentage of species i in the sample j. In Table 2, sample j is also defined by its pH. One can consider breaking down this parameter into classes as follows: pH_1, 5–6.9 (37 samples); pH_2, 7–7.9 (36 samples); pH_3, 8–8.6 (38 samples); pH_4, 8.7–10.9 (37 samples).

If the pH of sample j is equal to 5.5 (class pH_1), the value is coded as 1 000, meaning that this parameter is 1 in this class, where it is found, but 0 in all others. Table 2 will be placed as supplementary to Table 1.

Table 3 is constructed from Table 1. If k_{ic} is designated as the value found at the intersection of the species i and the pH class c, k_{ic} is the sum of the percentages p_{ij} of species i in the samples j situated in the class c of pH. Table 3 will be analyzed as a principal table. Table 1 is added to it as the supplementary table.

Table 4 is the table of presence absence; it is obtained from Table 1 by replacing the values that are not 0 by 1. If the species i is present in the sample j, $p_{ij} = 1$, if not $p_{ij} = 0$.

Table 5 is constructed from Table 4 in the same manner as Table 3 from Table 1. In this case, k_{ic} is the number of times that the species i is present in the samples found in the class c of pH. This table will be analyzed as a principal table and Table 4 will be added to it as a supplementary element.

Regression analysis

The goal of regression analysis (Cazes 1978; Benzecri 1978) is to establish a transfer function that uses the distribution of the species in a sample to determine the approximate value of an ecological parameter. The species play the role of independent variables, and the ecological parameter which is to be estimated is the dependent variable.

In our case, the number of species is high (579). Therefore, we cannot carry out regressions from Table 1 because it would give illusory results. It is known that the quality of a regression is based on the coefficient of multiple correlation R; the more this coefficient approaches 1, the better the regression. In cases where the independent variables are numerous, a high coefficient is obtained even if some are independent of the variable to be explained. It is therefore necessary to reduce the number of independent variables so that a significant result can be obtained. To do this a CA of the initial table (Table 1) is carried out (Cazes 1978; Roux 1979) and the 579 species are reduced to a number of factors. Each sample is defined by its co-ordinates on the factors which are now considered as the independent variables. An advantage of proceeding in this manner is that the new variables are not correlated between themselves, and the problem of an unstable coefficient of multiple correlation is avoided. On the other hand the inconvenience of this procedure is that it involves a table of 7 dimensions (if 7 factors are kept), and only a part of the initial formation is retained. It is necessary to ensure that the most important part of the information has been kept on the factorial space obtained. It is for this reason that we have analyzed Table 3 with Table 1 as a supplementary table. CA of Table 3 defines only 3 factors, and allows all the information to be maintained.

Establishing a transfer function

Starting with the results obtained from CA and the regression analysis, it is possible to establish a transfer function which is represented by the following formula:

$$y_{(j)} = \sum_{i}^{s} a_i p_{ij} + \bar{y} + e_j$$

where: s is the number of species

$y_{(j)}$ is the estimated value of the parameter y in sample j

p_{ij} is the percentage of the species i in the sample j

$\bar{y}$ is the mean of the parameter y for all the samples

e_j is the difference between the estimated value and the measured value of the parameter y in sample j

a_i is the coefficient of species i

$$a_i = \frac{1}{100}\left(\frac{b_1}{\sqrt{\lambda_1}} G_1(i) + \frac{b_2}{\sqrt{\lambda_2}} G_2(i) + \frac{b_3}{\sqrt{\lambda_3}} G_3(i)\right)$$

where λ_1, λ_2, λ_3 are the relative inertias, respectively, to the 1st, 2nd and 3rd factorial axis constructed by CA of Table 3.

$G_1(i)$, $G_2(i)$ and $G_3(i)$ are the coordinates of the species i on the 1st, 2nd and 3rd factorial axis from Table 3, respectively.

b_1, b_2 and b_3 are the regression coefficients obtained from Table 3 with Table 1 as the supplementary table.

Results

pH-indicator species

The selection of indicator species is carried out by treating the data with CA. We considered the species percentages and analysed Table 2 followed by Table 1 as a supplementary table (CA-percentage). We eliminated those species which were present in less than 3 samples and those with a percentage of less than 5% in any sample. The CA was then conducted with 245 taxa and 156 samples. Figure 1 shows the projection of the species points and pH-classes on the factorial plane 1–2. The species characterizing each class of pH are situated close to the corresponding pH point. In Table 6, we have selected some of the species which have a strong influence on the definition of the factorial axes.

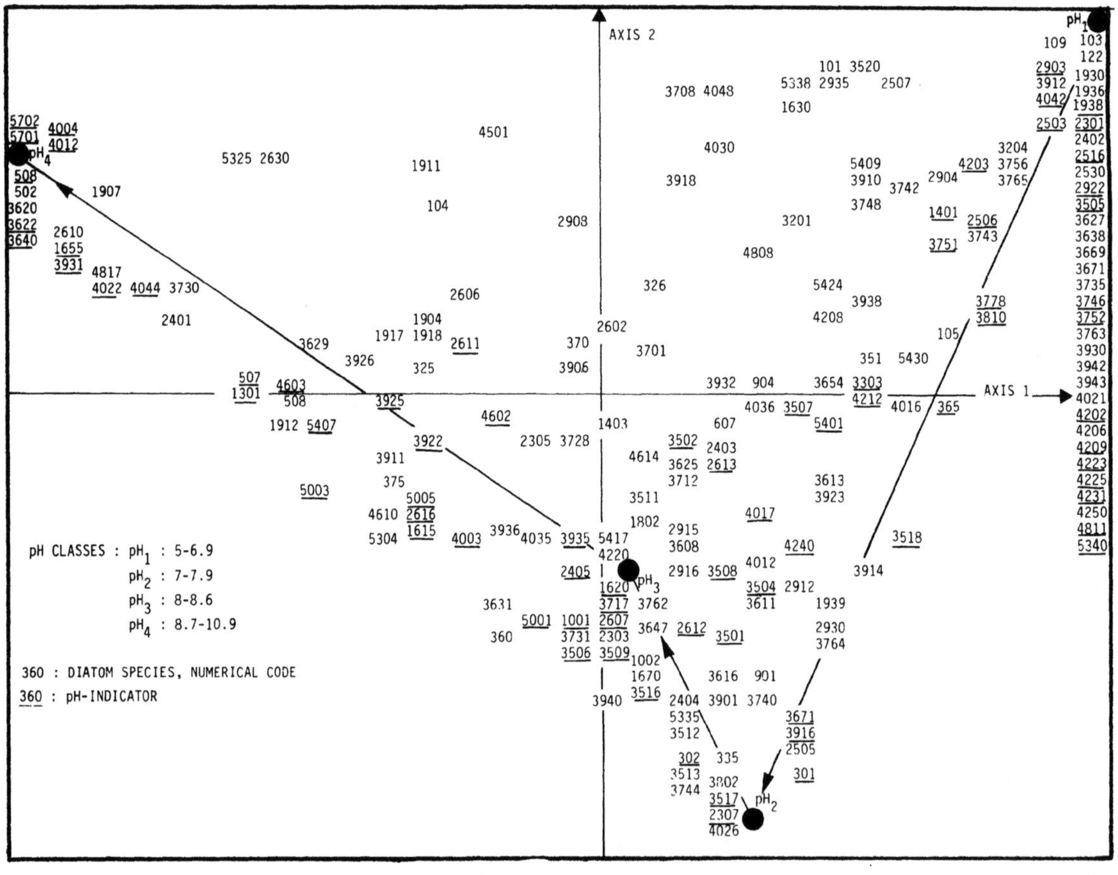

Fig. 1. Correspondence analysis (CA) based on Table 1 (initial table, 245 taxa, 156 samples), and Table 2 (supplementary table, 156 samples, 4 pH classes). Projection of the species points and pH classes on factorial plane 1–2. The underlined species are those having an important contribution in the definition of the factorial axes. Their taxonomy is given in Table 6. On the right, the points corresponding to the species indicated in the vertical column were clustered round point pH_1.

Table 6. pH-indicator diatoms in East Africa.

Class	pH	Taxa
CLASS pH_1	pH : 5–6.9	*Pinnularia graciloides*
		Pinnularia appendiculata
		Pinnularia obscura
		Pinnularia microstauron
		Pinnularia cardinalis
		Eunotia praerupta
		Eunotia af. pseudoveneris
		Melosira distans v. africana
		Cymbella fonticola
		Navicula bryophila
		Navicula subtilissima
		Navicula tantula
		Stauroneis nana
		Surirella ovata
		Gomphonema olivaceum
		Diploneis elliptica
	pH near 7	*Pinnularia tropica*
		Pinnularia acrosphaeria
		Pinnularia borealis
		Eunotia lunaris
		Eunotia flexuosa
		Cocconeis thumensis
		Stauroneis anceps
		Cocconeis diminuta
		Nitzschia linearis
		Nitzschia umbonata
		Navicula perpusilla
		Achnanthes minutissima
		Navicula symmetrica
		Amphora tenerrima
		Mastogloia elliptica
		Pinnularia interrupta
		Nitzschia kutzingiana
CLASS pH_2	pH : 7–7.9	*Synedra acus + v.*
		Cyclotella stelligera
		Gomphonema parvulum
		Navicula salinicola
		Navicula iranensis
		Navicula cryptocephala
		Nitzschia elegantula
		Nitzschia palea
		Melosira distans
		Melosira varians
		Amphora coffaeaformis
		Melosira agassizii
		Fragilaria pinnata
CLASS pH_3	pH : 8–8.6	*Melosira italica + v.*
		Melosira granulata
		Melosira nyassensis
		Fragilaria brevistriata
		Fragilaria leptostauron
		Amphora ovalis
		Melosira granulata v. angustissima
		Epithemia zebra
		Cocconeis placentula
		Nitzschia amphibia
		Campylodiscus clypeus
	pH near 8.6	*Stephanodiscus astraea*
		Rhopalodia gibba
		Synedra rumpens v. neogena
		Nitzschia lancettula
		Nitzschia fonticola
		Fragilaria lapponica
		Fragilaria construens
		Fragilaria pinnata v. trigona
		Stephanodiscus damasii
		Stephanodiscus hantzschii
		Synedra berolinensis
		Nitzschia frustulum
		Cyclotella meneghiniana
		Chaetoceros muelleri
CLASS pH_4	pH : 8.6–10.9	*Anomoeoneis sphaerophora + v.*
		Rhopalodia gibberula
		Cyclotella ocellata
		Epithemia argus
		Nitzschia subrostrata
		Nitzschia vitraea
		Nitzschia sigma
		Nitzschia estohensis
		Nitzschia latens
		Nitzschia pusilla
		Anomoeoneis costata
		Navicula damasii
		Navicula irregularis
		Navicula elkab
		Thalassiosira faurii
		Thalassiosira rudolfi
		Cyclotella iris

These species seem to display a clear preference for a given pH or pH class. If they are abundant, and if several taxa characteristic of the same pH class are associated in a sample, they can be considered as good pH-indicators for the investigated area.

A CA taking into account the presence-absence of taxa was also carried out. Most of the species situated around the pH_1 point in Fig. 1 also characterize this class by their presence. But the effects of pH above 7 do not show up clearly. Many species are able to tolerate a wide range of pH, but the species preference, deduced from the CA percentage, is a more sensitive indicator than the species tolerance.

Transfer function

We have carried out regressions using the methods described above, and successively considered presence-absence (CA with principal Table 5, supplementary Table 4) and species percentage (CA with principal Table 3, supplementary Table 1) situations. The coefficients of multiple correlation, calculated by the method of least squares, are $R = 0.745$, and $R' = 0.857$, respectively. The best result is obtained by taking into account species percentage.

We applied the transfer function established above to the 156 recent samples. The difference between the estimated values and the measured values is $<\pm0.3$ in 43% of the cases, $<\pm0.5$ in 65% of the cases, $<\pm1.0$ in 90% of the cases.

Large differences (<1.2) are observed in only two cases which correspond to the highest measured values (10.3, 10.9) that were underestimated by our calculation.

Before using the transfer function on fossil samples the stability of the formula must be tested by applying it to new modern samples. In addition other improvements might be added; we are especially hoping to reduce the number of taxa necessary in its application.

Acknowledgements

We are grateful to Dr. J. Talling, Professor R. B. Wood, Professor J. Kalff, Dr. P. Kilham, and Dr. F. A. Perrott for having collected and sent numerous recent diatom samples. We would also like to thank Professor J. P. Benzecri for his criticism and comments. This work was supported by the Centre National de la Recherche Scientifique and the Ecole Normale Supérieure de Fontenay-aux-Roses.

References

Benzecri, J. P., 1973. La taxinomie. L'analyse des données 1. Dunod, Paris. 615 pp.

Benzecri, J. P., 1978. Problèmes statistiques et méthodes géométriques. Cah. Anal. Données 3: 131–146.

Benzecri, J. P., 1980. Pratique de l'analyse des données. Analyse des correspondances. Dunod, Paris. 315 pp.

Bryson, R. A. & Kutzbach, J. E., 1974. On the analysis of Pollen-Climate Canonical Transfer Functions, Quater. Res. 4: 162–174.

Cazes, P., 1978. Méthodes de régression III. L'analyse des données. Cah. Anal. Données 3: 385–391.

Gasse, F., 1975. L'évolution des lacs de l'Afar Central (Ethiopie et T.F.A.I.) du Plio-Pléistocène à l'Actuel. Thesis, University of Paris 1: 390 pp., 3: 59 pl.

Gasse, F., 1977. Evolution of Lake Abhé (Ethiopia and T.F.A.I.) from 70 000 b.p. Nature 265: 42–45.

Gasse, F., 1980. Late Quaternary changes in lake levels and diatom assemblages on the Southeastern margin of the Sahara. Paleoecology of Africa 12: 333–350.

Gasse, F., 1983. Diatoms from East Africa. Bibliotheca Phycologica, in press.

Gasse, F. & Tekaia, F., 1979. La paléoécologie des diatomées. Cah. Anal. Données 4: 81–94.

Gasse, F. & Tekaia, F., 1982. Tentative definition, comparison and interpretation of fossil diatom assemblages from Eastern Africa. Proceedings of VIth Symp. on Recent and Fossil diatoms, Budapest, 1980. Acta Geol. Ac. Scient. Hung, 25: 135–140.

Harvey, T. K., 1976. The paleolimnology of Lake Mobutu Sese Seko, Uganda-Zaire: the last 28 000 years. Ph. D. Thesis, Duke University, U.S.A. 103 pp.

Holdship, S. A., 1976. The paleolimnology of Lake Manyara, Tanzania; a diatom analysis of a 56 meter sediment core. Ph. D. Thesis, Duke University, U.S.A. 112 pp.

Kilham, P., 1971. Biogeochemistry of African lakes and rivers. Ph. D. Thesis, Duke University, U.S.A. 199 pp.

Kipp, N. G., 1976. New transfer function for estimating past sea-surface conditions from sea-bed distribution of planktonic foraminiferal assemblages in the North Atlantic, Geol. Soc. Am. Mem. 145: 3–41.

Lebart, L., Morineau, A. & Tabard, N., 1977. Techniques de la description statistique, Dunod, Paris.

Richardson, J. L., Harvey, T. J., Holdship, S. A., 1978. Diatoms in the history of shallow East African lakes. Pol. Arch. Hydrobiol. 25: 341–343.

Richardson, J. L. & Richardson, A. E., 1972. History of an African Rift lake and its climatic implications. Ecol. Monog. 42: 499–534.

Roux, M., 1979. Estimation des paléoclimats d'après l'écologie des Foraminifères. Cah. Anal. Données 4: 61–79.

Talling, J. F., 1976. Phytoplankton: composition, development and productivity. In: J. Rzoska (ed.). 'The Nile, Biology of an ancient River', Junk, The Hague. pp. 385–402.

Talling, J. F. & Talling, I. B., 1965. The chemical composition of African lake waters. Int. Rev. ges. Hydrobiol. 50: 421–463.

Interpretation of lake quality from contemporary diatom assemblages

Pertti Huttunen & Jouko Meriläinen
University of Joensuu, P.O. Box 111, SF-80101, Joensuu 10, Finland

Keywords: paleolimnology, diatoms, assemblage indicators, Finland

Abstract

Diatom assemblages from the topmost sediment of 49 lakes were ordinated by means of a program for detrended correspondence analysis and reciprocal averaging (Decorana). Five lake groups were separated, each having more or less characteristic diatom assemblages and water quality. Shifts from one group to another caused by factors such as acidification, artificial drainage, diffuse nutrient loading and sewage are discussed.

Introduction

An important early work on diatom assemblages from Finland and Russian Karelia was published by P. T. Cleve (1891). This and many other studies revealed his character as an astute diatomist: instead of only reporting diatom floras he also grouped and presented the diatoms of analyzed samples in relation to water quality. Forty years later R. W. Kolbe (1932) presented a paper outlining the principles of diatom ecology.

pH has been one of the most discussed characteristics of lake quality and a number of diatom workers have been paid much attention to the dependence of diatoms on the hydrogen ion concentration of the water. Hustedt's classical work on the diatoms of Java, Bali and Sumatra, published in the years 1937–39, emphasizes the classification of diatoms on the basis of the kind of water in which their occurrence is optimal. Cholnoky introduced fresh ideas to the ecology of diatoms, the main concepts being combined comprehensively into an ecological monograph (Cholnoky 1968). In the meantime many other authors have been attracted to studies on the ecology of diatoms and they have been primarily concerned with the preferences and tolerances of diatoms with respect to a range of water characteristics (e.g. Mölder 1946, 1964; Jørgensen 1948; Järnefelt 1956; Hustedt 1957; Round 1957; Foged 1958). Nygaard (1956) applied pH for diatoms to the interpretation of early changes in the ecology of a Danish forest lake. He introduced a quantitative treatment of Hustedt's pH system and embodied his ideas in the indices α, ω and ϵ. Because Nygaard's concept has also inspired our investigation the indices are repeated here:

$$\text{index } \alpha = \frac{\text{acid units}}{\text{alkaline units}}$$

$$\text{index } \omega = \frac{\text{acid units}}{\text{number of acid taxa}}$$

$$\text{index } \epsilon = \frac{\text{alkaline units}}{\text{number of alkaline taxa}}$$

The acid and alkaline units are calculated by adding the relative frequency of acidobiontic/alkalibiontic taxa multiplied by five to the relative frequency of acidophilous/alkaliphilous taxa. The ecological significance of the acidobiontics is weighted by a arbitrary factor of 5.

Nygaard compared the present diatom assem-

Hydrobiologia 103, 91–97 (1983).

blages and pH ranges of some Danish brown-water lakes. In spite of limited material and using pH data collected by different investigators and originating from varying times of year the regression between the pH of the water and diatom assemblages was rather good. The method was used to reconstruct the trophic history of lake Gribsø which was shown to have experienced a gradual increase of dystrophication (Nygaard 1956).

One of the present authors (JM) tested Nygaard's indices, based on the contemporary diatom assemblages from 14 Finnish lakes, against the lake pH data (Meriläinen 1967). The article was supplemented by another work on three meromictic lakes, one in Finland and two in Norway (Meriläinen 1971). The regression calculated from the $\log_{10}$ of index α predicted pH of the water quite well and the accuracy was suggested to be around one pH unit. The risks of comparing the indices ω and ϵ were pointed out if they were calculated by different authors. In these studies the importance of measuring pH during a comparable successional phase of the water body, most preferably during the autumnal full circulation was emphasized (cf. Charles 1982; Norton *et al.* 1981).

The present research programme builds on these earlier experiences but in addition to pH the present study embraces almost fifty different lake characteristics.

Material and methods

The study is based on diatom assemblages originating from the topmost sediment of 49 lakes situated in eastern Finland. The sediment samples are estimated to represent material accumulated during the last 5 to 10 years before sampling. All sites belong to a larger program that began in 1977 and that has been introduced in more detail elsewhere (Meriläinen *et al.* 1981; Meriläinen *et al.* 1982). Most lakes are located on the poor granite-gneiss complex and some on the more fertile Karelidic schists. The size of the lakes varies from five to several hundred hectares. A great number of lake properties, such as productivity, land use and percentage of peatland in the catchment vary considerably. Lakes with a short residence time and shallow unstratified lakes have been avoided. Water samples were taken during the period of autumnal full circulation and supplementary samples during distinct thermal summer stratification. Waters were analyzed using the standard methods of the Water Research Institute of Finland. Sediments samples were collected in late winter through the ice using a Kajak sediment sampler. Five subsamples were obtained from the deepest part of each lake and the three topmost centimetres were homogenized. Most sediment analyses were carried out by the Geological Survey of Finland. Diatom analyses have been made according to Battarbee (1979). Altogether 302 taxa from 49 lakes were included in the analysis. The Cornell Ecological Program Nr 40, Decorana, a program for detrended correspondence analyses and reciprocal averaging was used for data ordination (Hill 1979). The first axis is calculated by reciprocal averaging and the second and higher axes are calculated avoiding the systematic relation of any kind between these and the first axis. So the arch effect (horseshoe effect) that commonly occurs with reciprocal averaging or principal components analysis is avoided (Hill & Gauch 1980).

Results

Detrended correspondence analyses and reciprocal averaging of diatom assemblages ordinate the lakes to at least five rather distinct groups (see Fig. 1). The characteristic taxa of each group are also presented in Fig. 1. Those diatoms, such as *Anomoeoneis serians, Cymbella hebridica, Eunotia faba, E. tenella, Frustulia rhomboides* + v. *saxonica, Melosira lirata* f. *seriata, Navicula mediocris* and *N. subtilissima,* that are usually found in most unproductive environments characterize Group 2. Species that in Finnish waters are often correlated with human impact, for instance *Asterionella formosa, Cyclotella pseudostelligera* and *C. stelligera* v. *tenuis* are found more or less frequently in Group 4.

On the basis of the physico-chemical properties of both catchments and lake-water the five groups are distinctly different. Many of the properties, such as conductivity, calcium content, field percentage and the pH of the water support the separation of the groups on algal criteria (cf. Table 1).

Some of the relationships between the compo-

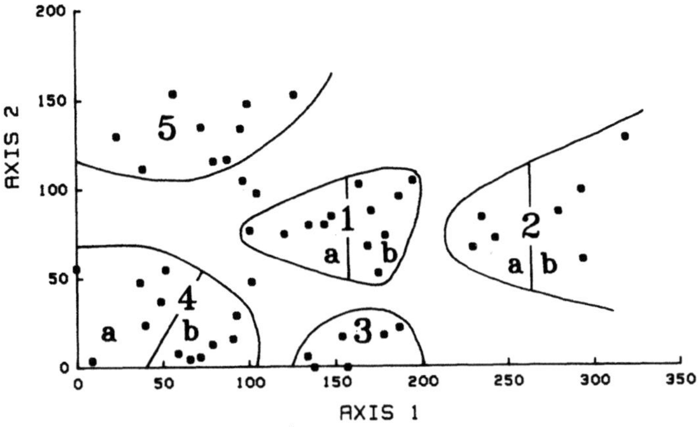

5

Achnanthes minutissima v. cryptocephala
Anomoeoneis exilis
Asterionella formosa
Cyclotella iris
C. comta
C. kuetzingiana v. radiosa
C. pseudostelligera
C. stelligera v. tenuis
Fragilaria construens
F. pinnata

4

Cyclotella stelligera
Fragilaria brevistriata
F. construens
F. pinnata
Melosira ambigua
Navicula pupula

a

Achnanthes lanceolata
Amphora ornata
Cyclotella comta
Fragilaria lapponica
Navicula dicephala
N. lanceolata

b

Melosira distans v. perglabra
M. distans v. tenella
M. italica ssp. subarctica
Neidium affine

1

Cyclotella kuetzingiana v. radiosa
Eunotia lunaris
Frustulia rhomboides v. saxonica
Melosira lirata

a

Achnanthes minutissima v. cryptocephala
Cyclotella comta
C. stelligera
Fragilaria pinnata
F. virescens
Melosira distans v. tenella

b

Anomoeoneis serians
A. serians v. brachysira
Cymbella gracilis
Eunotia pectinalis
E. veneris
Frustulia rhomboides
Melosira distans v. perglabra
Navicula mediocris
Pinnularia interrupta

3

Cyclotella stelligera
Eunotia pectinalis
Fragilaria construens
Frustulia rhomboides
F. rhomboides v. saxonica
Melosira ambigua
M. distans v. perglabra
M. distans v. tenella
M. italica
M. italica v. valida

2

Anomoeoneis serians
Cymbella gaeumanni
C. hebridica
C. ventricosa
Eunotia faba
E. rhomboides
E. tenella
E. veneris
Frustulia rhomboides
F. rhomboides v. saxonica
Melosira lirata f. seriata
Navicula mediocris
N. subtilissima
Pinnularia gibba
P. interrupta
P. viridis

a

Fragilaria constricta
Melosira distans v. alpigena
Nitzschia palea
Pinnularia hemiptera

b

Eunotia bactriana
Tabellaria binalis

Fig. 1. The ordination of diatom assemblages of the 49 lakes by Decorana analysis and the characteristic diatoms for each Group.

Table 1. The characterization of lake groups.

	Group 1					Group 2				
	n	$\bar{x}$	SD	max	min	n	$\bar{x}$	SD	max	min
Catchment (ha)	12	2120.63	3743.59	11142.0	33.2	7	83.86	63.88	210.8	20.0
Field (%)	12	8.81	18.32	65.06	.00	7	.93	1.86	5.12	.10
Peatland (%)	12	19.42	15.55	45.4	2.5	7	22.82	20.98	54.8	.5
Area (ha)	12	122.63	222.34	643.0	16.4	7	35.63	14.81	61.2	16.8
Secchi (m)	12	3.52	1.93	8	1.2	7	6.54	3.08	10	1.5
Colour (Pt mgl^{-1})	12	53.3	32.36	120	5	7	17.1	27.82	80	5
Conductivity (mSm^{-1})	12	3.08	1.01	4.60	1.53	7	1.35	.39	2.10	.89
pH	12	6.51	.32	7.09	5.95	7	5.25	.57	6.10	4.61
Alkalinity ($mvall^{-1}$)	12	.10	.07	.21	.01	7	.01	.01	.03	.00
N-total (μgl^{-1})	12	312.5	81.59	450	150	7	176.1	94.73	360	95
P-total (μgl^{-1})	12	9.6	6.54	29	5	7	6.0	3.65	13	2
Na (mgl^{-1})	10	1.32	.48	2.3	.6	5	.62	.41	1.3	.2
K (mgl^{-1})	10	.65	.29	1.2	.3	5	.40	.12	.6	.3
Ca (mgl^{-1})	12	2.18	.94	3.7	.6	7	.66	.38	1.1	.2
Mg (mgl^{-1})	12	.69	.28	1.1	.1	7	.29	.13	.5	.1
Cl (mgl^{-1})	12	2.17	2.52	9.9	.6	7	.80	.37	1.3	.4
SO_4 (mgl^{-1})	12	6.30	5.78	24.4	3.3	7	3.81	.83	5.3	2.8

nent score axes and water chemistry give useful information that can be applied to paleolimnological work. In general, axis 1 corresponds relatively well with the nutrient status of the lake. There is also a good correlation between axis 1 and the pH of water in the autumn (Fig. 2a) and axis 1 and conductivity (Fig. 2b). Axis 2 seems to reflect the depth of the productive water layer, measured by transparency and water colour. Otherwise there is no clear dependence between the water parameters and higher axes.

Discussion

A reliable comparison of the lakes with each other is possible only if the measurements of lake properties take place under comparable circumstances. Meriläinen (1967) preferred to sample the lakes for water analyses during the autumnal full circulation. The same procedure was followed in the present study although other periods have been used in related investigations (e.g. Norton *et al.* 1981). Figure 3 illustrates the pH of the 151 lakes

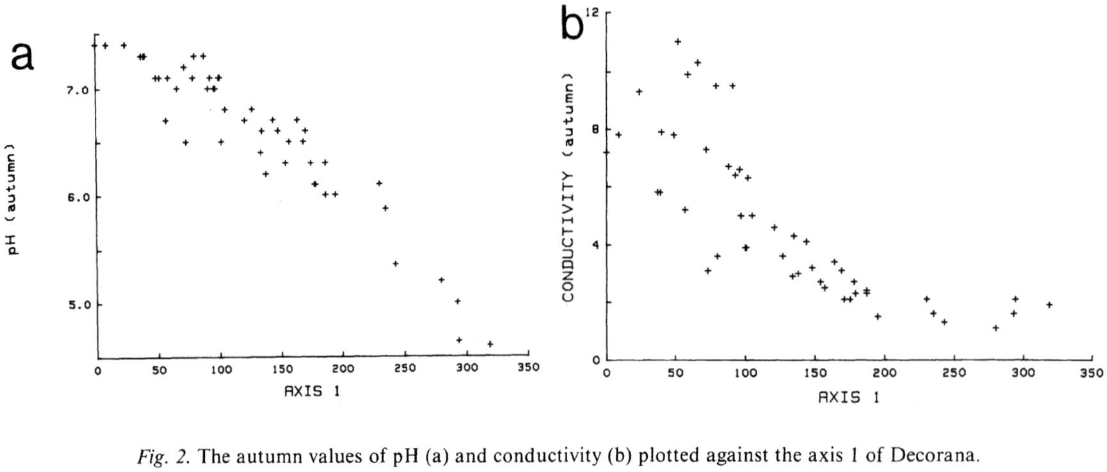

Fig. 2. The autumn values of pH (a) and conductivity (b) plotted against the axis 1 of Decorana.

roup 3				Group 4					Group 5				
x̄	SD	max	min	n	x̄	SD	max	min	n	x̄	SD	max	min
7158.97	7044.39	16036.0	1003.2	12	534.56	538.84	2080.0	78.4	9	442.60	584.24	1924.8	21.0
1.56	1.75	3.53	.00	12	20.42	10.70	46.40	6.68	9	6.65	5.88	18.57	.05
43.60	22.19	80.0	15.0	12	11.90	8.88	31.4	2.0	9	3.73	3.37	10.0	.0
306.23	285.19	825.0	86.0	12	137.10	342.38	1220.0	3.6	9	72.31	61.90	195.2	6.8
1.47	.33	2	1.0	11	2.63	1.24	4	1.0	9	4.42	1.34	7	2.8
148.3	31.89	200	110	11	52.2	41.97	140	15	9	18.3	8.29	30	10
2.70	.23	3.00	2.40	12	8.37	1.64	11.00	5.80	9	5.69	1.86	9.30	3.60
6.25	.19	6.50	6.00	12	7.15	.13	7.39	6.98	9	7.14	.26	7.46	6.70
.05	.02	.07	.02	12	.32	.10	.57	.19	9	.30	.14	.60	.15
465.0	51.28	530	420	12	440.8	163.90	870	260	9	310.0	123.79	590	190
32.8	11.09	50	18	12	21.1	10.02	42	12	9	11.4	11.35	41	4
1.44	.25	1.8	1.1	8	2.81	.34	3.4	2.5	4	1.85	.90	2.9	1.0
.66	.18	.9	.5	8	1.84	.34	2.3	1.3	4	1.35	.42	1.8	.9
2.08	.42	2.6	1.6	12	7.26	2.04	10.6	4.9	9	3.92	1.24	6.2	2.5
.78	.16	1.0	.6	12	1.86	.70	2.9	.1	9	1.84	2.23	7.6	.6
1.52	.37	2.2	1.2	12	3.18	1.14	5.5	1.7	9	2.09	.87	3.5	1.1
4.58	2.66	9.7	2.1	12	12.95	3.75	18.3	7.1	9	7.07	2.69	11.2	3.2

during the autumnal full circulation and deviation from this during the summer stratification. The summer values are considerably higher in a number of lakes and lower in some of them. In general, the former represents lakes that have small catchments with a relatively greater field area and less peatland; their water is also more transparent and the total iron content of the water is lower. The highest summer values (measured in daytime) correspond with higher productivity and the lowest summer

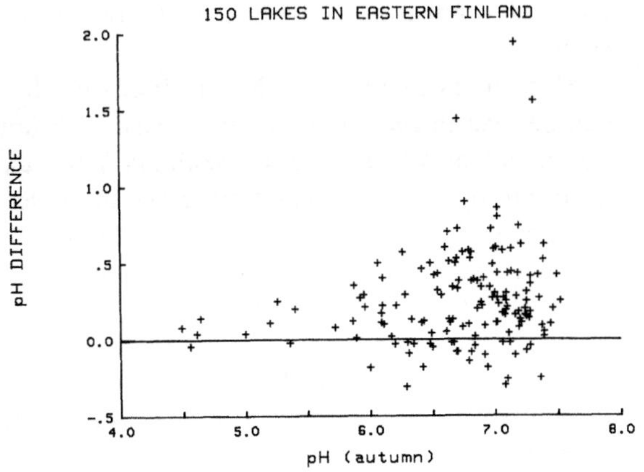

Fig. 3. The autumn pH against pH difference (summer pH–autumn pH) in the 150 lakes.

values with the inflow of allochthonous material (humic water with high concentration of iron).

The structural characteristics of the diatom assemblages ordinate the 49 lakes into five groups that differ also from each other with respect to their physico-chemistry (Fig. 4). Furthermore, the data indicate that lakes can shift from one group to another, especially from Group 1 lakes, that represent typical undisturbed lakes in E. Finland, to the others. Without doubt some of the lakes in other groups have remained in the same category for a long time, perhaps since their isolation. But at present and during recent times different types of human activity have caused an increase in the movement of lakes away from Group 1 to other groups.

It is suggested that acidification causes a shift towars Group 2, the more or less acid clear-water lakes. Furthermore, it is expected that acid precipitation will accelerate this process and that the methods presented here can be used to study the changes of the lakes due to acidification, especially in the case of those lakes that are situated in extreme acid environments where conventional indices (e.g. Nygaard 1956) are not easily applied.

The shift from mesohumic oligotrophic lakes (Group 1), towards Group 3 is assumed to correlate with the acceleration of artificial drainage of forest and peatland that increase the humic load to lakes.

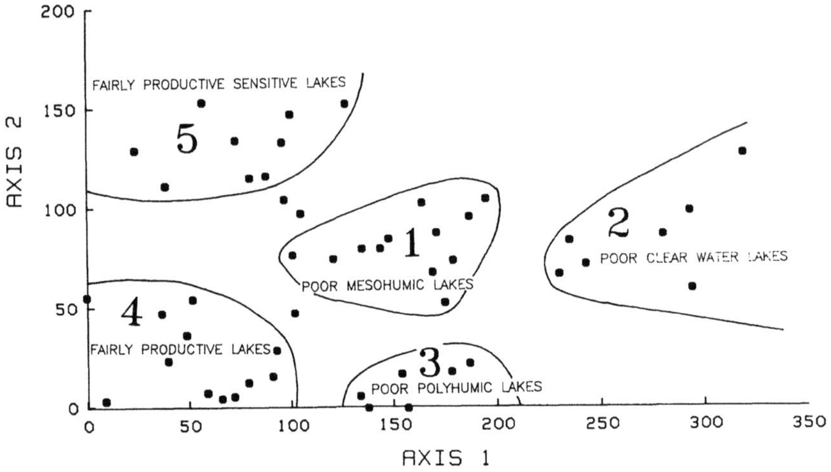

Fig. 4. The grouping of lakes according to Decorana analysis.

As a result the lakes become very coloured reducing the thickness of the euphotic zone. During the past 20 years most (70%) of the bogs in southern Finland have been drained to improve forest production which, if our assumption is correct, has greatly increased the number of lakes in Group 3.

The lakes of Group 4 are also coloured but they are slightly more productive than the previous ones. Fields are more abundant around these in comparison to the others. It is likely that soil erosion and leaching have shifted lakes from Groups 1 and 3 towards Group 4. Therefore these lakes are here called fairly productive lakes or lakes of harmonic eutrophication.

Finally, there is a group of lakes (Group 5) characterized by the fact that they have been disturbed or are sensitive to disturbance. The origin of this state is heterogenic: earlier they were used as a soaking and retting place for hemp and flax or, more recently, as a receiver of sewage waters. In addition, some have extreme morphometric properties and a good protection against the wind, the characteristics of meromictic lakes or basins inclined to meromixis. A sudden increase of nutrients can shift a lake, if deep enough, from any group to Group 5. In coloured lakes precipitation of humic matter with nutrients and sedimentation is a common phenomenon that is expected during the change.

It has been shown (Alhonen 1983) that some Finnish oligotrophic and eutrophic lakes have changed little during their whole history. No doubt, all groups are represented in Finland through the whole post-glacial history of the landscape. But during the earliest stages of development Group 4, lakes with harmonic eutrophication, was much larger because of the closer contact of the water with minerogenic material and richer, non-leached lake surroundings. Climatic changes connected with the migration of spruce caused the deterioration of the soil, and an increase of humic matter in lakes. In our model it is represented by the shift towards the poor polyhumic lakes probably via poor clear waters. This dystrophication was finally stopped by clearance of forests and the creation of agricultural land that greatly dictated the development of Finnish lakes from then to the present stage.

Many of the main trends that are discussed above can be seen in the history of each basin. Without doubt not only the diatom assemblages but also other biological assemblages preserved in the sediment can be treated by these methods and used to interpret lake changes and this is one of the later aims of the study. Each sediment zone represents an independent lake at a point in time. The fossil assemblage when correlated with modern reference data reveals a clear insight into the ecology of the paleo-lake.

The different groups identified here are separated by differences in diatom assemblage structure. The characterization of the groups was based on modern species composition and differences in physicochemistry. The next step of the study is to test the hypotheses presented here by analyzing core assemblages from selected lakes.

Acknowledgements

Assistance with lake selection was carried out by Ms. Kaija Kiiskinen (née Pirttiala) who, together with Mr. Teuvo Lappalainen, collected the samples and carried out most of the diatom counts. Their work was financed by the Maj and Tor Nessling Foundation and by the Academy of Finland.

References

Alhonen, P., 1983. Aspects of the developmental history and paleolimnology of lakes in Finland. Devel. Hydrobiol., this volume.

Battarbee, R. W., 1979. Diatoms in lake sediments. In: Berglund, B. E. (Ed.). Paleohydrological changes in the temperate zone in the last 15 000 years. Subproject B, Vol. II, pp. 177-225.

Charles, D. F., 1982. Studies of Adirondack Mountain (N.Y.) lakes: Limnological characteristics and sediment diatom-water chemistry relationships. Doctoral Dissertation, Indiana Univ. Bloomington.

Cholnoky, B. J., 1968. Die Ökologie der Diatomeen in Binnengewässern. Weinheim. 669 pp.

Cleve, P. T., 1891. The Diatoms of Finland. Acta Soc. Fauna Flora Fenn. 8(2): 1-68 + 3 pl.

Foged, N., 1958. The diatom in the basalt area and adjoining areas of archean rock in West Greenland. Meddel. Grønland 156(4): 1-146.

Hill, M. O., 1979. Decorana. A Fortran program for detrended corespondence analysis and reciprocal averaging. Ecology and Systematics. Cornell Univ. Ithaca.

Hill, M. O. & Gauch, M. G., 1980. Detrended correspondence analysis: an improved ordination technique. Vegetatio 42: 47-58.

Hustedt, F., 1937-39. Systematische und ökologische Untersuchungen über die Diatomeen-Flora von Java, Bali und Sumatra nach dem Material der Deutschen limnologischen Sunda-Expedition. I. Arch. Hydrobiol. Suppl. 15: 131-177, 187-295, 393-506, 638-790 + 28 pl.; Suppl. 16: 1-155, 274-394.

Hustedt, F., 1957. Die Diatomeenflora des Flusssystems der Weser im Gebiet der Hansestadt Bremen. Denkschrift über den heutigen ökologischen Zustand der Wasserstrassen im bremischen Staatsgebiet auf Grund der Kieselalgen Flora mit Beziehung auf die Abwasserverhältnisse. Abh. Naturwiss. Ver. Bremen 34: 181-440.

Järnefelt, H., 1956. Zur Limnologie einiger Gewässer Finnlands. XVI. Mit besonderer Berücksichtigung des Planktons. Ann. Zool. soc. 'Vanamo' 17: 1-201.

Jörgensen, E. G., 1948. Diatom communities in some Danish lakes and ponds. Kong. Dansk. Videnska. selsk. biol. skr. 5(2): 1-140.

Kolbe, R. W., 1932. Grundlinien einer allgemeinen Ökologie der Diatomeen. Ergebn. d. Biol. 8: 221-348.

Meriläinen, J., 1967. The diatom flora and the hydrogen ion concentration of the water. Ann. bot. fenn. 4: 51-58.

Meriläinen, J., 1971. The recent sedimentation of diatom frustules in four meromictic lakes. Ann. bot. fenn. 8: 160-176.

Meriläinen, J., Huttunen, P. & Pirttiala, K., 1981. Research strategy for the ecology of diatoms. The diatom study of the 151 lakes, in the eastern Finland. In: Ross, R. (Ed.). Proceedings of the sixth Symposium on Recent and Fossil diatoms. pp. 413-423. Otto Koetltz. Koenigstein.

Meriläinen, J., Huttunen, P. & Pirttiala, K., 1982. The effect of land-use on the diatom communities in lakes. Hydrobiol. 86: 99-103.

Mölder, K., 1946. Die geographische Verbreitung der Algen in Estland nebst einem Verzeichnis der Konjugaten. Ann. bot. soc. 'Vanamo' 21(6): 1-34.

Mölder, K., 1964. Abhängigkeit der Diatomeen von Felsgrund in Mittelfinnland. Ann. bot. fenn. 1: 104-109.

Norton, S. A., Davis, R. B. & Brakke, D. F., 1981. Responses of northern New England lakes to atmospheric inputs of acid and heavy metals. Completion report Project A-048-ME. Land and Water Resources Center. Univ. of Maine, Orono. 90 pp.

Nygaard, G., 1956. Ancient and recent flora of diatoms and Chrysophyceae in lake Gribsø. Folia Limn. Scand. 8: 32-94, 253-262 + 12 pls.

Round, F. E., 1957. The distribution of Bacillariophyceae on some littoral sediments on the English Lake district. Oikos 8: 16-37.

Bacterial decomposition of organic matter in lacustrine sediments

V. G. Drabkova
Institute of Lake Research, Academy of Sciences of the U.S.S.R., Petrovskaya nab. 2, Leningrad 197046, U.S.S.R.

Keywords: paleolimnology, organic matter, bacteria, aerobic decomposition, oxygen consumption

Abstract

The investigations were carried out in lakes situated in the forest zone (Karelian Isthmus, Baltic zone, South Ural) and in the forest-steppe (South Ural) of the USSR. The lakes differ in their bioproductivity and in the intensity of their human influence. The amount of organic matter accumulating in the sediments is closely related to production and decomposition processes in the trophogenic layer. Processes of organic matter transformation were found to be most active in the uppermost sediment. The quantity of bacteria shows no correlation with the organic matter content of the sediment. Increase in organic matter, up to 70–80% of dry weight, is often accompanied by a decrease in bottom bacteria. The intensity of aerobic decomposition of the labile organic matter can be judged from the oxygen demand of the sediments. However, it is important to differentiate between chemical and biological oxidation processes. The quantity of bottom bacteria correlates closely with the value of oxygen consumption only in cases of high sediment redox potentials.

Introduction

Production and decomposition processes in the trophogenic zone are closely related to the amount of organic matter entering the sediments (Ohle 1956). The amount of organic carbon entering the sediments is inversely proportional to the lake depth and directly proportional to the concentration of organic carbon in the upper water layers. According to Hargrave (1975) organic carbon sedimentation can be expressed as a linear function of the ratio of carbon supply (autochthonous and allochthonous organic matter) to the mixed-layer depth. Hargrave showed a correlation between the surface sediment organic content and annual organic carbon sedimentation. Values were similar in those lakes where terrestrial debris accounts for a large fraction of the organic input and in deep lakes where the sinking material has already been extensively transformed. The greatest changes in organic content after sedimentation occur in shallow lakes, suggesting that decomposition processes in the sediments of these lakes are intensified. The surface sediments contain a large quantity of bacteria whose biomass may be 2.5–7.8% of the total amount of organic matter in the mud, and the active mud layer with high numbers of microorganisms may be 4–12 cm deep.

Methods

The amount of unstable organic matter in sediments can be estimated from measurements of oxygen consumption. The oxygen consumption of mud can be determined by using bell-jars to isolate sections of the mud-water interface or by using sediment cores (Hayes & MacAulay 1959), Gambarjan (1962), Rybak (1966), Romanenko & Romanenko (1969). Due to its simplicity the latter method has

Table 1. Organic matter, redox potential, bacteria number (N), oxygen consumption by surface sediments, sediment (P_1) and aquatic (P_2) bacteria production (gCm^{-2} day^{-1}) in lakes of different geographical zones of the USSR (June–August).

Zone	Lake	Organic matter content (% dw)	Redox potential (mV)	N (cells × 10^6 g^{-1})	Oxygen consumption (gO_2 m^{-2} day^{-1})	P_1		P_2
						by respiration	by P/B coefficient	
Tundra	Seidozero	19–23	+200–+310	80–85	0.21–0.31	0.025	0.013	0.030
North taiga	Vjalozero	38	-	165	0.95	0.085	0.065	0.120
	V. Senozero	20	-	165	0.50	-	-	-
Middle taiga	Krasnoye	26	-200–+300	125–200	0.50–0.79	0.065	0.045	0.175
South taiga	L. Snidzinjas	30–60	-180–+200	120–580	0.52–1.22	0.093	0.055	0.162
(Pribaltica)	Snidzinjas	30–54	-300–+140	180–620	0.52–1.47	-	-	-
South taiga	B. Kisegach	51	+250–+465	25–150	0.41–0.90	-	-	-
(South Ural)	B. Terenkul	56	+250–+400	40–120	0.32–1.27	0.026	0.022	0.206
	Argajash	71	+250–+420	60–155	0.33–0.80	0.072	0.062	0.234
	B. Ishkul	51	+390–+420	60–205	0.39–1.10	0.032	0.022	0.132
Forest-steppe	Atkul	23	-155–+405	760–780	0.86–0.92	-	0.077	0.335
	Selezjan	37	-130–+280	400–1300	0.71–2.36	-	0.148	0.422
	Butash	23	-180–+360	3100–11600	1.1 -1.52	-	0.360	0.255
	Kunkul	22	-200–+365	2400–8000	0.78–1.29	-	0.211	0.324
	B. Shantropay	22	-185–+375	4360–8400	0.75–1.08	-	0.267	0.333
	Gorkoye	32	-150–+250	1100–3500	0.48–0.82	-	0.256	0.500

found widest application. The oxygen consumption values obtained by these methods allows the intensity of the aerobic decomposition of organic matter to be estimated. However, many authors have shown that oxygen consumption values are also considerably influenced by chemical oxidation and zoo- and phytobenthos respiration.

Results and discussion

Since the oxygen consumption of sediments is not only related to bacterial respiration, the number of bacteria in sediments does not always correlate with the rate of oxygen consumption. This interrelationship exists only where decomposition rates are low, bacterial numbers are low, and the redox potential is rather high (Hayes & MacAulay 1959). Edwards and Rolley (1965) studied the oxygen consumption of sediments in heavily polluted rivers in England and found no correlation between it and the quantity of bacteria.

As in the water column, the main mass of microorganisms in the sediments is represented by heterotrophic bacteria, therefore a change in organic matter concentration should result in a change in bacterial abundance. However, bacteria abundance does not normally correlate with the total amount of organic matter in sediments. Furthermore, according to the data of Kuznetzov (1950), the increase of organic content up to 70–80% of the dry weight hinders bacteria development. A comparison of the organic matter content and the total number of bacteria in sediments of lakes of different geographical zones has also shown that there is no direct correlation between these values. The greatest content of organic matter is typical for sediments of the southern taiga lakes of the South Ural. Here, independently of the lake bioproductivity, the organic content (measured as loss on ignition) amounts on average to 51–71% of the dry sediment. The values for lake muds in the northern and middle taiga region and in the forest-steppe zone were similar. In the muds of the northern taiga they were 20–44%, in the middle taiga, 22–44%, and in the forest-steppe, 22–39%. The least number of bacteria was found in sediments of the lakes of the northern taiga, tundra and southern taiga (South Ural) where values from $70\text{–}80 \times 10^6$ cells g^{-1} fresh mud are common. In the muds of the middle taiga zone the number of bacteria is more than double. The greatest number of bacteria (up to 11.6×10^9 cells g^{-1} fresh mud) was recorded in the forest-steppe lakes (Table 1). An inverse tendency is observed in the correlation between the organic matter content in sediments and the quantity of bacteria.

In the same way, there is little direct correlation between oxygen consumption and the organic matter content of muds. In southern taiga lakes of the South Ural the oxygen consumption of the mud during the growing season was on average $0.4\text{–}0.8\ gO_2\ m^{-2}\ day^{-1}$, while in lakes of the north and middle taiga values were $0.5\text{–}1.0\ gO_2\ m^{-2}\ day^{-1}$. These values were somewhat higher in the lakes of the forest-steppe, viz. $0.9\text{–}1.45\ gO_2\ m^{-2}\ day^{-1}$. Such discrepancy between the organic matter content and the intensity of oxygen consumption of the muds is easily explained since the latter value mainly characterizes microbiological processes, whose intensity, as shown above, does not correlate with the concentration of organic matter. In all the lakes investigated, excluding those of the forest-steppe, a direct dependence between number of bacteria and the intensity of oxygen consumption of the muds has been noted, the most distinct dependence being observed in northern lakes. For the southern taiga lakes of the South Ural and the northern lakes where the redox potential of the sediments was very high (Table 1) the oxygen consumption of the muds is commonly connected with the respiration of bottom microflora. The only exception is the deep-water lake B. Kisegach (depth 30 m) where chemical oxidation increases in importance in the central regions. In lakes of the middle and south taiga (Pribaltica), the proportion of chemical oxidation in oxygen consumption of muds amounts to 30–50%.

In the forest-steppe lakes, where the intensity of oxidation processes increases and thus the redox potential diminishes (to −150 mV), the importance of microflora in sediment oxygen consumption decreases noticeably, and there is no connection between bacteria number and oxygen consumption values in the sediments of these lakes. In this case the intensity of oxygen consumption by muds cannot serve as an index of heterotrophic microflora activity. For the sediments of the lakes where the microflora activity is primarily responsible for the oxygen consumption, it is possible to calculate bacterial production on the basis of this value as well as the coefficient of energy exchange K_2 for near-bottom bacterioplankton. The utilization of coefficient K_2 of near-bottom bacterioplankton for bottom bacteria is fully justified, because the intensity of bacteria respiration, calculated per cell was found to be similar in water and in the sediments of the lakes. A second method of calculation was also used. It was based on the data of bottom bacteria biomass and the P/B coefficient of near-bottom bacterioplankton. Both calculation methods gave analogous results, and almost identical results were observed for the lakes with a high sediment redox potential (Table 1). For the forest-steppe lakes only the second method of bacterial production calculation could be used. The lowest production of bacteria in muds was recorded in the northern-taiga lakes whilst the highest was found to occur in those of the forest-steppe. The highest ratio of sediment bacteria production to aquatic bacteria production was recorded for lakes of the tundra and northern taiga and in lakes of the forest-steppe (Table 1). In the lakes of the tundra and northern taiga this is probably due to low water temperatures, restricting significant decomposition in the water column. In the lakes of the forest-steppe the increase in the importance of bottom bacteria in the general recycling of organic matter was found to be correlated with the mass development of macrophyte vegetation and periphyton. In the lakes of the middle and south taiga the bottom microflora production in the summer period does not normally exceed 30–45% of bacterioplankton production, whilst during the ice-cover period the production of bottom bacteria appeared to be almost equal to that of bacterioplankton.

An analogous correlation has been found between the intensity of aquatic and sediment bacteria respiration. In lakes of the middle and south taiga the intensity of oxygen consumption by bottom microflora accounts for 22–45% of that of the bacterioplankton. In the forest-steppe lakes this value also fluctuates within similar ranges, except for lake Butash where this value amounts to 80% or higher.

References

Edwards, R. W. & Rolley, H. L., 1965. Oxygen consumption of river muds. J. Ecol. 53: 1–19.

Gambarjan, M. E., 1962. Method for analyzing the intensity of the destruction of organic matter in sediments of deep lakes (in Russian). Microbiology 31: 895–898.

Hargrave, B. T., 1975. The importance of total and mixed-layer depth in the supply of organic material to bottom communities. Symp. Biol. Hung. 15: 157–165.

Hayes, F. R. & Anthony, E. H., 1959. Lake water and sediment. VI. The standing crop of bacteria in lake sediments and its place in the classification of lakes. Limnol. Oceanogr. 4: 299–315.

Kuznetzov, S. I., 1950. Microbiological characteristics of the destruction of organic matter in mud sediments (in Russian). Transaction of the laboratory of sapropel sediments 4: 15–28.

Ohle, W., 1956. Bioactivity production and energy utilization of lakes. Limnol. Oceanogr. 1: 139–149.

Romanenko, V. I. & Romanenko, V. A., 1969. Destruction of organic matter in sediments of the Rybinsk reservoir. In: Physiology of water organisms and their role in the organic matter circulation (in Russian): 24–31.

Rýbak, J. I., 1966. Method for analyzing the microstratification in the near-bottom water layers. Bulletin de l'Academie Polonaise des sciences 14: 321–325.

Relationship between sedimentary pigments and primary production: evidence from core analyses of twelve Italian lakes

Piero Guilizzoni, Giuliano Bonomi, Gaetano Galanti & Delio Ruggiu
C.N.R. Istituto Italiano di Idrobiologia, Largo V. Tonolli, 50/52, 28048 Pallanza (NO), Italia

Keywords: paleolimnology, sedimentary pigments, primary production, core analyses

Abstract

A statistically significant correlation of the type $y = ax^b$ between sedimentary plant pigments and contemporary algal primary production has been found in a study of the recent trophic evolution of twelve Italian lakes. The equation is used to assess baseline production levels in periods (50–70 years ago) when human influence was low.

Introduction

The record of pigment degradation products laid down in lacustrine sediments allows historical changes in lake primary production to be determined, adding greatly to our understanding of the natural development of lakes and the process of eutrophication. Such an understanding is necessary for enlightened lake management.

Wetzel (1970) has documented the usefulness of sedimentary pigment analysis in the back-calculation of primary productivity through the entire post-glacial period.

In a previous study (Adams *et al.* 1978) data on sedimentary pigments and recent primary productivity from several northern Italian lakes were gathered. A good correlation was found between pigment concentration per unit organic matter in the 0–10 cm sediment depth and contemporarily derived data on primary productivity from six lakes. In this regard a more detailed study was made on Lake Mergozzo, situated near Lake Maggiore (Guilizzoni *et al.* 1981).

When surficial sedimentary phaeophytin from recent sediments of six cores from Lake Mead, Nevada, were regressed against annual primary production data, the correlation was highly significant (.001 level), indicative of the potential application of this method (Adams, pers. comm.).

In this paper data are presented from additional twelve northern Italian lakes with very different trophic and morphometric characteristics (Guilizzoni *et al.*, in press). Analyses of sediments were made to evaluate as accurately as possible the relationships between plant pigment derivatives in the surface sediment layers and corresponding primary production levels. The results shown are a part of those gathered during a more comprehensive study designed to identify trophic changes from both core data and limnological data (Guilizzoni *et al.*, in press).

Material and methods

From summer 1978 to 1980 a crust freezer (Adams *et al.* 1978) was used for collecting sediment cores of about 1 m from the deepest zone of the lakes. In Lake Como an additional core was taken at a depth of 40 m in a zone close to the town of Como. Sections of the crust were removed every 2 cm in the first 10 cm of the core, every 5 cm down to 20 cm, and every 10 cm in the remaining part. Extraction of the chlorophyll derivatives and carot-

Hydrobiologia 103, 103–106 (1983).

enoids was performed following Sanger & Gorham (1972). Sedimentary pigment degradation units ($SPDU_{665}$ and $SPDU_{445}$) relative to the content of organic matter in the sample were defined as arbitrary units, according to Wetzel (1970). Reproducibility of the results (C.V. % on 10 replicates) was 4.8 and 13.5 for $SPDU_{665}$ and $SPDU_{445}$, respectively.

Results and discussion

Depth distributions of chlorophyll and carotenoid degradation products for the lakes are shown in Fig. 1. The trend of pigment concentration within the sediments of the various lakes is similar. In general, from the oldest layers to a level of about 20–30 cm, all the lakes, with the exception of Lake Mezzola which has a short residence time, show a gradual increase in pigment concentration, followed by a sharper rise in the most recent sediments. As this rise is mainly due to recent eutrophication, the pigment concentrations in the lower part of the curve may be considered as the expression of a baseline trophic status for the lakes considered. Using this criterion a ranking of the lakes from the most to the least productive would be: Alserio, Varese, Como (40 m depth), Segrino, Pusiano, Endine, East and West Annone, and the deep lakes Como (400 m depth), Maggiore, Garda, and Iseo.

There are large differences in the pigment concentration between the lakes, the mean value of $SPDU_{665}$ in the surface sediment varies from ca. 1 300 for Alserio to ca. 60 for lakes Garda and Pozzo di Riva. Although the deep Lake Garda is the least productive among the investigated lakes, its core analysis displays a clear recent increase of algal pigments and it consequently suggests an increase in lake productivity.

The primary production data for the single lakes, as shown in Fig. 1, were obtained in different years within the last ten years but for Lake Maggiore older data are also included. Because of bioturbation it is difficult to establish a year to year correspondence between deposited pigments and primary production. We have therefore averaged pigment values from two or three different levels around the one theoretically corresponding (on the basis of the calculated sediment accumulation rates) to the production data. Data from Lake Mergozzo (Guilizzoni *et al.* 1981) are also included in the regression analysis.

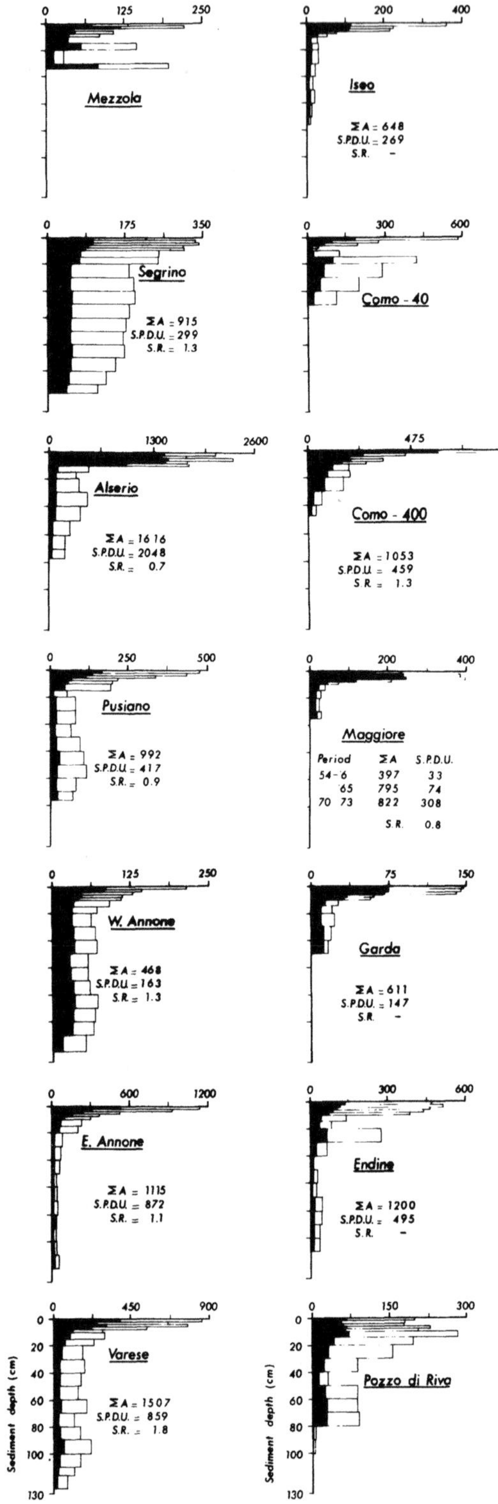

Fig. 1. Depth distribution of chlorophyll (black) and carotenoid (white) derivatives in sediment cores. Average daily primary production (Σ A, in mg C . m^{-2} . d^{-1}), SPDU (chlorophyll plus carotenoid derivatives/g organic weight) and S.R. (sediment accumulation rate, in cm . a^{-1}) are also shown.

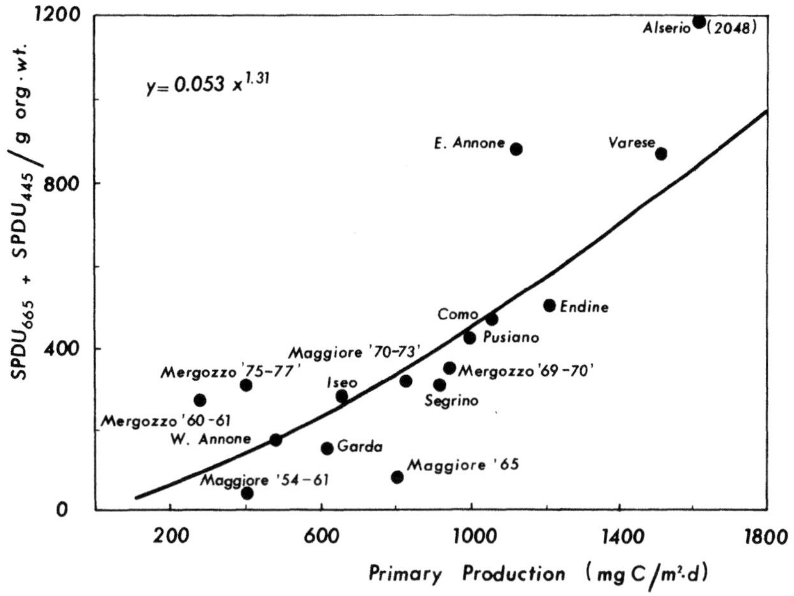

Fig. 2. Relationship between phytoplankton primary production and chlorophyll plus carotenoid derivatives in lake sediments.

Significant regressions ($P \leqslant 0.01$) of the type $y = a x^b$ were found between sedimentary pigments and primary production. The equation shown in Fig. 2, takes into account the sum of the two categories of pigments and was used to calculate approximate present primary production values for lakes Pozzo di Riva and Mezzola, lakes for which algal production data were not available. The values obtained were 521 mg C . m^{-2} . d^{-1} and 490 mg C . m^{-2} . d^{-1}, respectively. Although tentative, these values are in accord with the meso-eutrophic condition of the lakes as indicated by chemical and biological studies (Baudo *et al.* 1979; de Bernardi *et al.*, in press).

On the basis of accumulation rates calculated using the ^{137}Cs method (Premazzi 1978 and pers. comm.) the equation of Fig. 2 can be used to back-calculate production levels for the period prior to 50–70 years ago, when human impact on the lakes was much less severe than at present (Table 1). These values should correspond to baseline productivity levels, associated with general morphometric and climatic characteristics.

From Table 1 it can be seen that the past annual production of the large, deep lakes Garda, Iseo, Como and Maggiore was in the range of 31–58 g C . m^{-2}, corresponding to a daily average production of 85–160 mg C . m^{-2}. These values are in good agreement with the expected values for oligotrophic lakes. The table also shows that, of the shallow lakes, Endine and East and West Annone were essentially oligotrophic, whereas the rest were mesotrophic or eutrophic.

Despite differences in factors that may be expected to affect the preservation of pigments in sediments and to cause variations in the amount of allochthonous pigments, the correlation seems to be of general applicability to Italian subalpine lakes. From a practical point of view, the difference

Table 1. Recent measured annual (g C . m^{-2} . a^{-1}) algal production (a) and calculated past production (b) prior to 50–70 years ago. Per cent difference is also indicated.

Lake	(a)	(b)	$\frac{(a)-(b)}{(a)} \cdot 100$
Varese	550 (1979)	159	71
E. Annone	407 (1972–73)	72	82
W. Annone	171 (1972–73)	83	51
Pusiano	362 (1972–73)	108	70
Alserio	590 (1972–73)	295	50
Segrino	340 (1972–73)	169	50
Endine	438 (1973)	57	87
Garda	223 (1970–72)	31	86
Como -400 m	384 (1970–71)	58	85
Iseo	237 (1971–72)	31	87
Maggiore	300 (1970–73)	44	85

between present and past production rates indicates the changing degree of human impact on the lakes in the last few decades; on the other hand, the calculated values for the past indicate what could be achieved, in terms of productivity reduction, following nutrient abatement with the use of suitable treatment plants.

Acknowledgements

We thank Prof. M. S. Adams (Dept. of Botany, University of Wisconsin-Madison, U.S.A.), who provided detailed information about pigment sediment analysis and reviewed the manuscript. Thanks are also due to Prof. S. R. Brown (Queens University, Canada) for his kindness in critical reading of the text. The courtesy of Dr. G. Premazzi (Biology Division, C.C.R. Euratom, Ispra, Italy) for determining accumulation rates, is greatly appreciated.

This study was made within the Finalized Project: 'Environmental Quality', Subproject: 'Water', Theme: 'Eutrophication', of the Italian National Research Council.

References

Adams, M. S., Guilizzoni, P. & Adams, S., 1978. Sedimentary pigments and recent primary productivity in northern Italian lakes. Mem. Ist. Ital. Idrobiol. 36: 267–285.

Baudo, R., Galanti, G., Guilizzoni, P. & Varini, P. G., 1979. Relationships between heavy metals and aquatic organisms in Lake Mezzola hydrographic system (Northern Italy). 1. Hydrochemistry. Mem. Ist. Ital. Idrobiol. 37: 163–185.

de Bernardi, R., Giussani, G., Guilizzoni, P. & Mosello, R., (in press). Indagini conoscitive per una caratterizzazione limnologica dei 'Piccoli laghi' lombardi. Documenti dell'Ecologia – Regione Lombardia, Assessorato per l'Ecologia e i Beni Ambientali.

Guilizzoni, P., Bonomi, G., Galanti, G., Ruggiu, D. & Saraceni, C., 1981. Relazione tra l'evoluzione trofica del Lago di Mergozzo ed il contenuto in pigmenti vegetali, sostanza organica, carbonio e azoto dei suoi sedimenti. Mem. Ist. Ital. Idrobiol. 39: 119–145.

Guilizzoni, P., Bonomi, G., Galanti, G. & Ruggiu, D., (in press). Basic trophic status and recent development of some Italian lakes as revealed by plant pigments and other chemical components in sediment cores. Mem. Ist. Ital. Idrobiol. 40.

Premazzi, G., 1978. Metodo di valutazione della rata di sedimentazione. pp. 181–195. Atti del Convegno sull'Eutrofizzazione in Italia, Roma 3–4 Ottobre 1978.

Sanger, J. E. & Gorham, E., 1972. Stratigraphy of fossil pigments as a guide to the postglacial history of Kirchner marsh, Minnesota. Limnol. Oceanogr. 17: 840–854.

Wetzel, R. G., 1970. Recent and postglacial production rates of a marl lake. Limnol. Oceanogr. 15: 491–503.

The mythical concept of eutrophication

M. C. Whiteside
Department of Zoology & Graduate Program in Ecology, University of Tennessee, Knoxville, TN 37916, U.S.A.

Keywords: paleolimnology, lake ontogeny, lake succession, eutrophication

Abstract

Paleolimnologists frequently invoke the term 'eutrophication' for describing apparent enrichment phases in the history of a lake. I argue that this term is often used incorrectly and that alternative explanations can serve as more accurate descriptions. Increased organic content in the sedimentary record may result from increased nutrient availability (eutrophication), but it can also reflect decreased residence time of water, or changes in biotic interactions, or changes in lake morphometry.

Additionally, I argue that 'eutrophication' is an inappropriate term for describing the aging process of lakes. Lake ontogeny is the preferred term, as it does not imply directional changes in nutrients, nor in community structure.

Introduction

Development of the concept of eutrophication has been adequately reviewed by several authors (Hutchinson 1969 & 1973; Rodhe 1969). It is agreed that the earliest definitions referred to nutrient status – oligotrophy meaning nutrient-poor, and eutrophy meaning nutrient-rich. During eutrophication the supply of nutrients available to primary producers increases. This can be accomplished by increased allochthonous inputs, or by internal mechanisms that allow more nutrient to become available (Likens 1972). Implicit in this concept is the biological consequence of increased nutrient concentrations, i.e. increased primary production. An 'increase' in trophic status usually involves secondary changes in the physio-chemical nature of the lake, and changes in the flora and fauna.

Many of the biological responses to changes in nutrient input or availability become preserved in the sedimentary record (Frey 1969). It is this sedimentary record that paleolimnologists use to describe historical changes in lakes. Among the clues that paleolimnologists use is organic carbon content. Increased carbon content is frequently regarded as evidence for increased trophic status. Type and numbers of plant and animal remains are also used as evidence of lake conditions.

Organic carbon content per unit sediment (usually expressed as % loss on ignition) is generally considered the most convenient and simplest measure for estimating historical changes in productivity. A typical pattern of carbon change for north-temperate lakes is shown in Fig. 1. This curve shows the initial increase in organic carbon content following glacial retreat (Fig. 1, Zone A), a period of constancy that has been termed 'trophic equilibrium' by Hutchinson (1973) (Fig. 1, Zone B), and a period reflecting agricultural activity (Fig. 1, Zone C). Zone C may have decreased organic carbon, or increased content depending on allochthonous inputs. The same rationale for interpreting these patterns has been applied to lakes of other climatic regions (Cowgill & Hutchinson 1966; Tsukada 1967). Perhaps in most cases this interpretation is correct, but in other instances the convenience of this argument may cause us to overlook alternative explanations.

Hydrobiologia 103, 107–111 (1983).

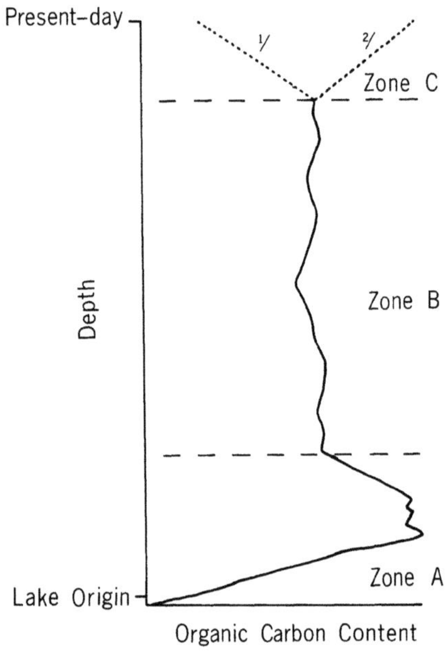

Fig. 1. A typical organic carbon profile found in north-temperate lake sediments. Zone A represents the initial stages following glacial retreat. Zone B is a period of stable carbon content - a stage termed by Hutchinson (1969) as 'trophic equilibrium'. Zone C reflects the activities of modern man; curve 1 shows increased soil erosion, while curve 2 reflects an increase in allochthonous carbon within the lake basin.

Alternative explanations

Organic carbon

Interpretation of carbon profiles is complicated because of changes in sediment accumulation rates during lake development, and uncertainty as to the source of carbon (allochthonous vs. autochthonous carbon). Accumulation rates can be established through radiocarbon dating, but the origin of sedimentary carbon poses greater difficulties. Mackereth (1966) argued convincingly that the typical carbon pattern (Fig. 1) was the result of allochthonous carbon inputs, presumably refractory dissolved organic carbon (DOC). Ideally, the key to trophic determinations using sedimentary carbon would be through particulate organic carbon analyses (POC) - not DOC. DOC:POC is approximately 10:1 for unpolluted lakes and enrichment increases POC to reduce the ratio (Wetzel 1975).

Another difficulty with regard to sedimentary carbon is that paleolimnologists usually assume the quantity of organic carbon to be a function of a nutrient input, but changes in autochthonous carbon accumulation can occur independently of changes in nutrient availability. This seemingly improbable state may be more common in the sedimentary record than currently recognized. At least two sets of conditions could make this possible:

(1) Schindler (1978) has shown that phosphorus turnover (nutrient inputs) and lake-water residence time (flow rates) are good predictors of chlorophyll biomass. The rate of nutrient inputs is very important in determining algal biomass; however, the kinetics of nutrient assimilation is mediated by residence time. This relationship is not unlike the response of algae in chemostats. So, although many of the assumptions are violated, lakes behave somewhat as chemostats, where nutrient levels and dilution rates determine algal biomass. A decrease in dilution rate can increase algal biomass without increasing the rate of productivity. The lesson for paleolimnologists is that increased biomass, i.e., autochthonous sedimentary carbon, is not necessarily linked to nutrient enrichment. It is possible to have constant nutrient inputs accompanied by increased residence time (e.g. climatic dry periods) when primary productivity would remain constant, while biomass would increase. In the sedimentary record this could be perceived as a period of 'eutrophication'. Conversely, a decrease in residence time with no accompanying change in nutrient status would be reflected by lesser amounts of autochthonous carbon in the sediments. This could be interpreted as a period of 'oligotrophication' (Hutchinson 1973).

(2) The second set of conditions involves ecological interactions among secondary producers. Shapiro and his co-workers (Shapiro *et al.* 1975) have shown experimentally that low algal biomass can be maintained in the presence of high nutrients, providing large, herbivorous zooplankters were present. In the sedimentary record this would be reflected by low carbon, 'oligotrophic' conditions accompanied by high numbers of zooplanktonic remains.

There are other conditions that are yet insufficiently understood that could also affect the nature of organic carbon deposition. Additionally, the role of littoral productivity in lake ontogeny is poorly understood. Wetzel (1979) argues that sedimentary records of most lakes are greatly influenced by littoral zone productivity. A greater understanding of the relationship between littoral and limnetic productivity may help to resolve several perplexing questions arising from sediment analyses.

Biotic remains

Remains of plants and animals that become incorporated into the sediments provide us with direct evidence of past conditions within a lake. In dealing with these remains it is important to realize that the sedimentary matrix incorporates biota over time and space. The interpretations that we apply to these data cannot invoke biotic interactions that require knowledge of time-space relationships. It makes no sense to argue nutrient/food limitations, or competitive interactions to account for changes in numbers of remains found at a particular point in time. These arguments are difficult to establish for modern-day biota. Thus, sedimentary information regarding plants and animals is most accurately interpreted at the ecosystem or community level of ecology; attempts to establish arguments at lower levels of integration is necessarily unfounded speculation.

Another difficulty with biotic remains from sediments is to resolve whether the changes have resulted from variations in nutrient status, or from biotic interactions. We cannot assume, *a priori*, that shifts in populations or communities are totally dependent on nutrient regimes; they could be more closely linked to biotic interactions (Hrbáček 1961; Brooks 1969; Zaret 1981). Kerfoot (1981) has argued effectively that predation could account for changes in *Bosmina* morphologies during the postglacial history of Frains Lake, Michigan. It is possible that observations such as the disappearance of *Daphnia*, or changes in abundance of smaller zooplanktonic forms (*Bosmina* or *Chydorus sphaericus*) are more closely related to predatory interactions than to nutrient enrichment.

In this section I have argued that alternative explanations should be considered carefully by paleolimnologists, rather than simply invoking nutrient enrichment, or 'eutrophication'. Indeed, alternative explanations could answer some troublesome observations. For example, one perplexing problem has been the recovery of diverse assemblages of Chydoridae (Cladocera) from sediment levels that have high organic carbon content. These levels have been attributed to periods of increased productivity, and, in this manner, higher trophic status has been linked to increased diversity. This interpretation is directly counter to empirical observations, i.e., diversity of present day lakes is inversely correlated to trophic status (Margalef 1964). Specifically, Whiteside and Harmsworth (1967) have shown a strong negative correlation between chydorid diversity and lake trophy. Why then, do we consistently find increased faunal diversity associated with increased carbon content in the sedimentary record? It is quite possible that one of the alternatives would apply to this situation. In fact, the most credible explanation for the chydorid/sedimentary carbon correlation can be argued from the work of Devol & Wessmar (1978). They show that productivity of aquatic macrophytes is more closely associated with lake morphometry than to nutrient status. This, coupled with the fact that a substantial portion of macrophyte carbon becomes incorporated into the sediments (Wetzel 1979), would result in an increase in chydorid habitat and an increase in sedimentary carbon without any change in the nutrient status of a lake.

Concepts of Lake history

The terms 'eutrophication,' 'succession,' and 'ontogeny' are used frequently when describing the historical development of lakes, and it is not un common for these concepts to be synonymized. I believe that these terms are not synonymous.

Paleolimnologists use the term eutrophication for describing the process of alleged enrichment. Succession is defined as the universal process of directional change in a community with time, and it can be recognized by the progressive changes in the species composition of the community (Krebs 1978). During succession communities are generally thought to progress from less mature states (Pioneer) to more mature states (Climax). The extremes of these states have characteristic structural and functional biological properties (Smith 1980). Ontogeny was originally defined as the biological development of the individual. The ontogeny of a community may emphasize biological interactions, or the 'biocoenosis' (Goulden 1969), but generally this concept includes all lake-basin developments (Wetzel 1975).

There would be no basis for confusion if the three concepts were not related. The most obvious relationship is their historical dimension. If a scientist were to analyze lake ontogeny within a deterministic framework, then lake ontogeny would become synonymous with lake succession. Thus, the change of community properties with time could be interpreted as either ontogeny or succession, depending upon one's philosophy of community organization.

The aging of lakes has been called 'natural' eutrophication, and, if used in this sense, is synonymous with succession and ontogeny. The oft redrawn figure showing eutrophication proceeding with community succession (Lindeman 1942) and the early work of Deevey (1955) helped solidify this conceptual synonymy.

Many limnologists now accept the mistaken notion that all lakes naturally proceed from a pioneer state of oligotrophy to a more advanced state of eutrophy. Even if all lakes followed this pattern – which they do not (Alhonen 1967) – it would be incorrect to draw the analogy between terrestrial and aquatic succession. Margalef (1964) pointed out the fallacy of this argument. The traditional characteristics of pioneer and mature communities (Smith 1980) do not correspond to oligotrophic and eutrophic lakes. In fact the most diagnostic criteria, i.e., species diversity, food-web complexity, and the production to biomass ratio suggest that oligotrophy is the more mature biotic community.

Conclusions

These observations lead to some general conclusions:

(1) Paleolimnologists should be very cautious in applying the term eutrophication when interpeting data from lake sediments. There are alternative explanations for data generally perceived as being evidence for nutrient enrichment. We should not be too hasty in applying readily acceptable dogma to historical events.

(2) Ontogeny is a more desirable term for describing lake history because it does not imply traditional successional concepts, nor progressive lake enrichment.

(3) Eutrophication should be used only to describe nutrient enrichment, or the consequences of nutrient enrichment – this may occur during several phases in a lake's history, or it may be rare during ontogeny.

References

Alhonen, P., 1967. Paleolimnological investigations of three inland lakes in south-western Finland. Acta bot. fenn. 76: 1–59.

Brooks, J. L., 1969. Eutrophication and changes in the composition of the Zooplankton. In: Eutrophication, Causes, Consequences, Correctives. Washington, D.C.: National Academy of Sciences, 1970, pp. 236–255.

Cowgill, U. M. & Hutchinson, G. E., 1966. The history of the Petenxil basin. In: The History of Laguna de Petenxil. A Small Lake in northern Guatemala. Mem. Connecticut Acad. Arts Sci.: 121–126.

Deevey, E. S. Jr., 1955. The obliteration of the hypolimnion. Mem. Ist. ital. Idrobiol. Suppl. 8: 9–38.

Devol, A. H. and Wissmar, R. C., 1978. Analysis of five North America lake ecosystems. V. Primary production and community structure. Verh. int. Ver. Limnol. 20: 581–586.

Frey, D. G., 1969. Evidence for eutrophication from remains of organisms in sediments. In: Eutrophication, Causes, Consequences, Correctives. Washington, D.C.: National Academy of Sciences, 1970, pp. 594–613.

Goulden, C. E., 1969. Developmental phases of biocoenosis. Proc. National Acad. of Science 62: 1066–1073.

Hrbáček, J., 1962. Species composition and the amount of the zooplankton in relation to the fish stock. Rozpravy Čsav, Řada Mat. a přir. Ved. 72: 1–116.

Hutchinson, G. E., 1969. Eutrophication, Past and Present. In: Eutrophication, Causes, Consequences, Correctives. Washington, D.C.: National Academy of Sciences, 1970, pp. 17–26.

Hutchinson, G. E., 1973. Eutrophication. Am. Sci. 61: 269–279.

Kerfoot, W. C., 1981. Long-term replacement cycles in Cladocera communities: A history of predation. Ecol. 62: 216–233.

Krebs, C. J., 1978. Ecology: The Experimental Analysis of Distribution and Abundance. 2nd Ed. Harper & Row, New York, 678 pp.

Likens, G. E., 1972. Eutrophication and aquatic ecosystems. In: Nutrients and Eutrophication: The Limiting-Nutrient Controversy. Limnol. Oceanogr. Special Symposium. 1: 3–13.

Lindemann, R. L., 1942. The trophic-dynamic aspect of ecology. Ecology 23: 399–418.

Mackareth, F. J. H., 1966. Some chemical observations on post-glacial lake sediments. Phil. Trans. Soc. Lond. B., Biol. Sci. 250: 165–213.

Margelef, R., 1964. Correspondence between the classic types of lakes and their structural and dynamic properties of their populations. Verh. int. Ver. Limnol. 15: 169–175.

Rodhe, W., 1969. Crystallization of eutrophication concepts in northern europe. In: Eutrophication, Causes, Consequences, Correctives. Washington, D.C.: National Academy of Sciences, 1970, pp. 50–64.

Schindler, D. W., 1978. Factors regulating phytoplankton production and standing crop in the world's freshwaters. Limnol. Oceanogr. 23: 478–486.

Shapiro, J., Lamarra, V. & Lynch, M., 1975. Biomanipulation: An ecosystem approach to lake restoration. In: Brezonik, P. L. (Ed.) Sympo. Water Qual. Management through Biological Control (Proceedings). Univ. of Florida, Gainesville. 164 pp.

Smith, R. L., 1980. Ecology and Field Biology (3rd Ed.). Harper & Row, N.Y. 835 pp.
Tsukada, M., 1967. Successions of Cladocera and benthic animals in Lake Nojiri. Japan. J. Limnol. 28: 107–123.
Wetzel, R. G., 1975. Limnology. Saunders, W. B., Philadelphia. 743 pp.
Wetzel, R. G., 1979. The role of the littoral zone and detritus in lake metabolism. Arch. Hydrobiol. Beih. 13: 145–161.
Whiteside, M. C. & Harmsworth, R. V., 1967. Species diversity in chydorid (Cladocera) communities. Ecology 48: 644–667.
Zaret, T. M., 1981. Predation and freshwater communities. Yale University Press. New Haven, 187 pp.

Paleolimnological reconstruction of the effects of atmospheric deposition of acids and heavy metals on the chemistry and biology of lakes in New England and Norway

Ronald B Davis,[1] Stephen A. Norton,[2] C. Thomas Hess[3] & David F. Brakke[4]
[1] *Department of Botany & Plant Pathology, and Institute for Quaternary Studies, University of Maine at Orono, Orono, ME 04469, U.S.A.*
[2] *Department of Geological Sciences, University of Maine at Orono, Orono, ME 04469 U.S.A.*
[3] *Department of Physics, University of Maine at Orono, Orono, ME 04469, U.S.A.*
[4] *Huxley College, Western Washington University, Bellingham, WA 98225, U.S.A.*

Keywords: paleolimnology, acidic precipitation, lake acidification, heavy metal pollution, diatoms, cladocerans

Abstract

Sediment cores from nine lakes in southern Norway (N) and six in northern New England (NE) were dated by ^{137}Cs, ^{210}Pb and in NE also by pollen, and were analyzed geochemically and for diatoms. Cores from two N and three NE lakes were analyzed for cladocerans. ^{137}Cs dating is unreliable in these lakes, probably due to mobility of Cs in the sediment. In Holmvatn sediment, an up-core increase in Fe, starting ca. 1900, correlates with geochemical indications of decreasing mechanical erosion of soils. Diatoms indicate a lake acidification starting in the 1920's. We propose that soil Fe was mobilized and runoff acidified by acidic precipitation and/or by soil acidification resulting from vegetational succession following reduced grazing. Even minor land use changes or disturbances in lake watersheds introduce ambiguity to the sedimentary evidence relating to atmospheric influences. Diatom counts from surface sediments in 36 N and 31 NE lakes were regressed against contemporary water pH to obtain coefficients for computing past pH from subsurface counts. Computed decreases of 0.3–0.8 pH units start between 1890 and 1930 in N lakes already acidic (pH 5.0–5.5) before the decrease. These and lesser decreases in other lakes start decades to over a century after the first sedimentary indications of atmospheric heavy metal pollution. It is proposed that the acidification of precipitation accompanied the metal pollution. The delays in lake acidification may be due to buffering by the lakes and watersheds. The magnitude of acidification and heavy metal loading of the lakes parallels air pollution gradients. Shift in cladoceran remains are contemporary with acidification, preceding elimination of fishes.

Introduction

Southern Norway and northern New England (Vermont, New Hampshire, and Maine) contain numerous lakes which, because of the low buffering capacity of their watershed soils and waters are sensitive to acidic precipitation (Overrein *et al.* 1980; Hendrey *et al.* 1980). The degree and chronology of acidification and associated pollution of these lakes by heavy metals from the atmosphere is poorly known, due to the scarcity of long-term series of direct measurements of chemical and biological parameters. Consequently, we are examining the 'records' of these parameters in lake sediments spanning the past few hundred years.

Lakes sampled and methods

The results come from 36 SNSF-project lakes (described by Wright *et al.* 1977) in Norway, and 31 lakes in northern New England (described by Norton *et al.* 1981). In brief, the Norwegian lakes (New England in parentheses) range in pH from 4.4–7.1 (4.4–7.0), alkalinity 0–6 (0–10) ppm as $CaCO_3$, conductance 7–59 (15–47) $\mu S\ cm^{-1}$ at 20° (25°) C,

Hydrobiologia 103, 113–123 (1983).

elevation 150–1 246 (78–1 123) m, A_0 0.2–6.0 (0.01–7.7) km^2, and depth at coring site 8.5–61.0 (2.4–22.1) m. Cores, 0.4–0.8 m long, were taken from each lake using a 6.25 cm stationary-piston corer (Davis & Doyle 1969). Cores were sectioned in 0.5 cm increments by upward extrusion. At all the lakes, the top ('surface') increment was analyzed for diatoms. For cladocerans, the surface increment from 26 Norwegian and 18 New England lakes was analyzed. Entire cores from nine Norwegian and six New England lakes were studied for chronology, geochemistry, and diatoms; and two Norwegian and three New England cores for cladocerans. For the study of cores, we chose lakes whose watersheds were relatively free of disturbance for at least 150 years, so we could distinguish the effects of changing atmospheric inputs.

Each core section was thoroughly mixed, and split for the different analyses. Geochemical analyses were carried out on volumetric samples which were dried at 110 °C to obtain percent water, ignited at 550 °C to obtain percent 'organic matter', dissolved as in Buckley and Cranston (1971), and analyzed by atomic absorption flame and flameless spectrophotometry to obtain percent inorganics. Chronological analyses of three types were done. (1) ^{137}Cs concentration was measured by gamma-ray spectral analysis using a calibrated Ge(Li) detector with low background lead shield. Dry sediment was counted for 10^4 or 2×10^4 seconds, and the spectrum analyzed by Compton continuum subtraction (Covell 1956). (2) ^{210}Pb concentration was determined by alpha-ray spectral analysis of ^{210}Po. The efficiency of chemical separation onto Ag discs was determined by ^{208}Po tracer. The nuclides were extracted by the method of Eakins and Morrison (1978). (3) Pollen analyses were made on the New England cores using standard methods. Historical records provided dates of initial settlement, lumbering, and agriculture in the surrounding lowlands. Changes in pollen stratigraphy reflecting these activities were used for dating the sediment. For diatom analyses, 0.5 cm^3 subsamples of sediment were digested in chromic acid, and the washed residue settled onto cover slips by the method of Battarbee (1973). Six-hundred or more valves were counted per subsample. For cladoceran analyses, 2.0 cm^3 subsamples of sediment were digested in hot KOH, washed and mounted on slides (Brakke 1980). Each animal part was tabulated separately until 200 individuals had been counted derived from the most abundant part for each species (Brakke 1980). A total of 500–600 separate parts usually represented 200 individuals.

Results and discussion

Holmvatn

Holmvatn has a surface area of 1.8 km^2, and is located in Nissedal, Telemark, Norway at 671 m. It has a granitic watershed with rocky barrens, juniper-heath, bog, and scattered stands of conifers or birch. Haying and sheep grazing were practiced until 1946 (Johannessen & Wright 1980). There is no record of farms, lumbering, or heather burning in the watershed which has only one cabin. In the mid-1970s: pH $\cong$ 4.7, conductance $\cong$ 22 $\mu S\ cm^{-1}$ at 20 °C, and TOC $\cong$ 2.2 $mg\ l^{-1}$ (estimated equivalent color 22 Pt units) (Wright *et al.* 1977). In March, 1978 a sediment core was recovered in 26.5 m of water.

Sediment chronology. The ^{210}Pb concentrations are given in Fig. 1. The CRS model (Krishnaswami *et al.* 1971) has been used to determine the date-depth curve (Fig. 2) down to 10.25 cm (ca. 1894). Below this, the CRS curve is increasingly sensitive to estimated levels of supported Pb-210 and undetected unsupported Pb-210 resulting in erroneously old dates. We use the CIC model which employs a linear–log regression of unsupported Pb-210 to determine the portion of the date–depth curve from 10.25 to 17.25 cm (ca. 1840), and beyond 17.25 cm we extrapolate 50 years (dashed in Fig. 2). Errors given for the CRS portion of the curve are based on counting errors and for the CIC portion, on error estimates for the regression.

The ^{137}Cs concentration profile (Fig. 1) resembles the fallout curve from 1954 to 1966, but lacks the recent period of low concentration since the sharp decrease of nuclear weapons testing in the atmosphere (Toonkel 1980). Using the initiation of significant fallout (1954) as a stratigraphic marker would place 1954 at 11 cm or deeper where sediments are ca. 1900 and older according to ^{210}Pb. The peak Cs-137 concentrations occur at 2.5–4.0 cm which date to 1955–1968 by Pb-210. This brackets the 1963–64 ^{137}Cs fallout peak, but the

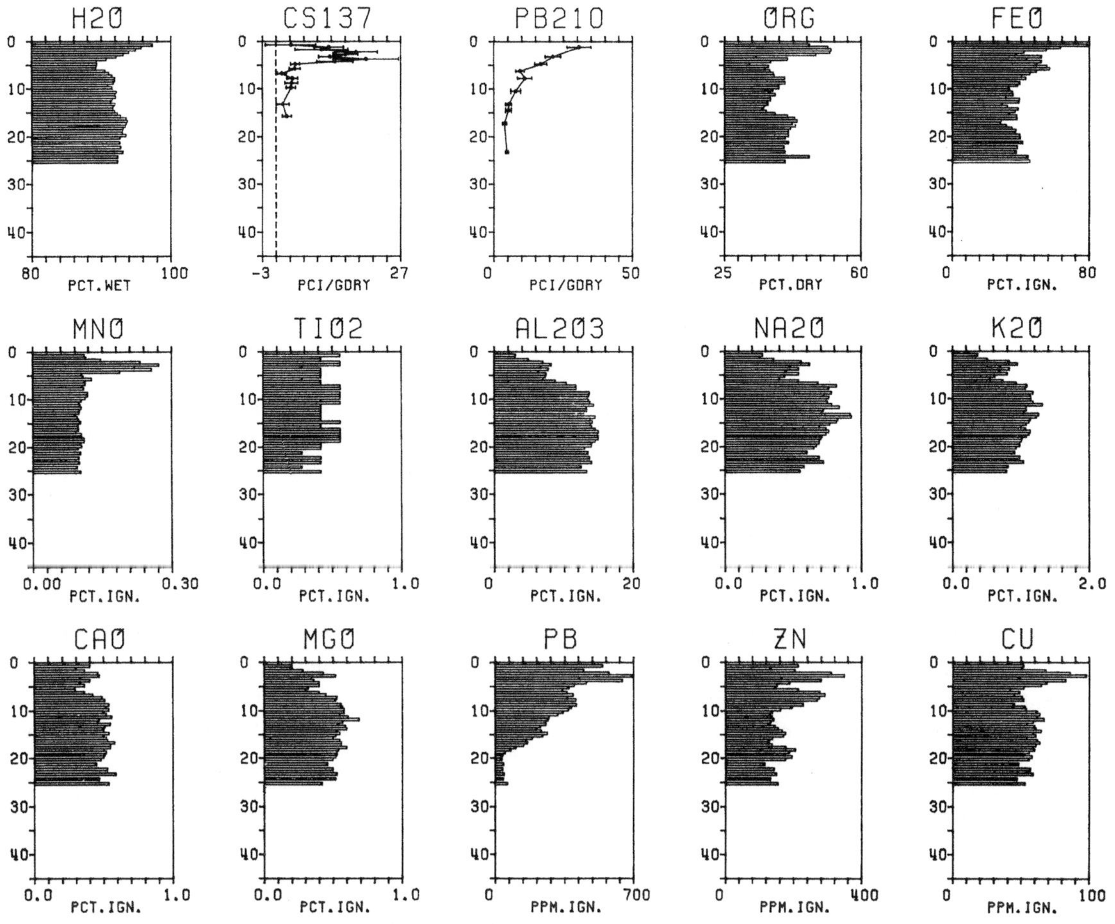

Fig. 1. Chemical composition of the core from Holmvatn, Norway. Depth in cm.

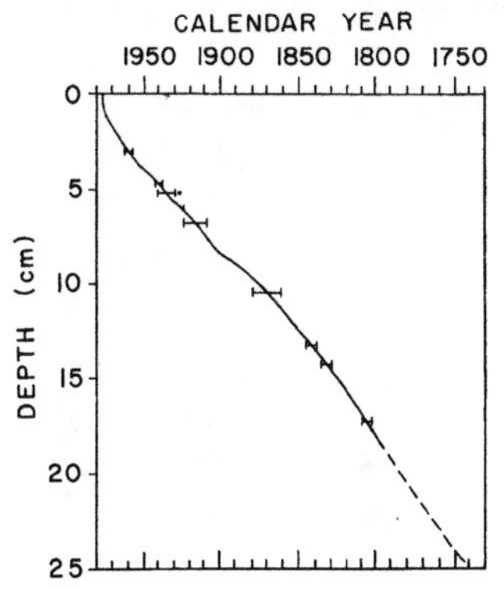

Fig. 2. Date–depth curve for the core from Holmvatn, based on ^{210}Pb analysis. Bars are one standard deviation.

peak concentrations in the sediment may be an artifact of dilution by increasing Fe concentrations near the surface. Other chemically similar lakes dated by us by both methods show a similar lack of a period of low ^{137}Cs concentrations in the uppermost sediments (and may have a pronounced concentration of ^{137}Cs at the surface), and also have ^{137}Cs in sediment several decades too old according to ^{210}Pb. If the ^{210}Pb dating is correct, then this suggests that ^{137}Cs dating may not be useful or reliable in these types of lakes.

Geochemistry. Holmvatn sediment is rich in organic matter, resulting in a high water content. This, and the approximate positive relationship between water and organics in the profile (Fig. 1) is due to the sponginess of organic detritus (Norton *et al.* 1981). Iron (FeO) increases upward irregularly from 28–39% below 9.5 cm to 49–57% (3–6 cm) to 80% in the top cm. The great enrichment at the top is probably due to diagenetic precipitation there and recent sedimentation of FeO(OH) and/or $Fe(OH)_3$ floc, some of which dissolves upon burial. This enrichment reduces the percentages of all the other solids.

The up-core percentage increase of FeO starting at 9.5 cm ca. 1903) is correlated with a percentage decrease in SiO_2, shortly (8.5 cm) joined by decreasing Al_2O_3, and, finally starting in the 1920s (7–8 cm) decreasing Na_2O, K_2O, CaO, and MgO. Deposition rates of these constituents were calculated using the CIC and CRS ^{210}Pb dating models. The resultant deposition rate profiles roughly parallel the percentage profiles. This suggests that the following occurred: (1) increasing leaching of Fe from the watershed with subsequent in-stream or in-lake precipitation, focusing and sedimentation, and (2) decreasing mechanical erosion and sedimentation of inorganic detritus.

Decreasing mechanical erosion and increasing Fe mobility could result from changing land use. Increasing Fe mobility could also result from acidic precipitation. Changing land use in the watershed during 1900–1930 is not indicated (or denied) by Johannessen and Wright (1980), but grazing and haymaking ceased in 1946, culminating a period of decrease in such practices which started at some unknown earlier time. According to Drabløs *et al.* (1980), sheep grazing of the subalpine outfields in the Holmvatn region decreased greatly around the turn of the century. Grazing of such outfields increased in the region since the 1920's and is greater now than in the 19th century, but this renewal probably did not occur at Holmvatn. The expected results of reduced grazing are increases in coniferous forest (now 30% of Holmvatn watershed), heath, and raw humus topsoil. This can mobilize Fe by increasing organic ligands (Beck *et al.* 1974) and reducing soil redox potential, and by acidifying the soil (Norton 1973). The process can also acidify the runoff (Rosenqvist *et al.* 1980). In addition or alternatively, in very poorly buffered terrain as around Holmvatn, acidified precipitation can increase the leaching of Fe and acidify the runoff (Norton 1981). At Holmvatn, we may be seeing the combined effects of changing land use and acidic precipitation.

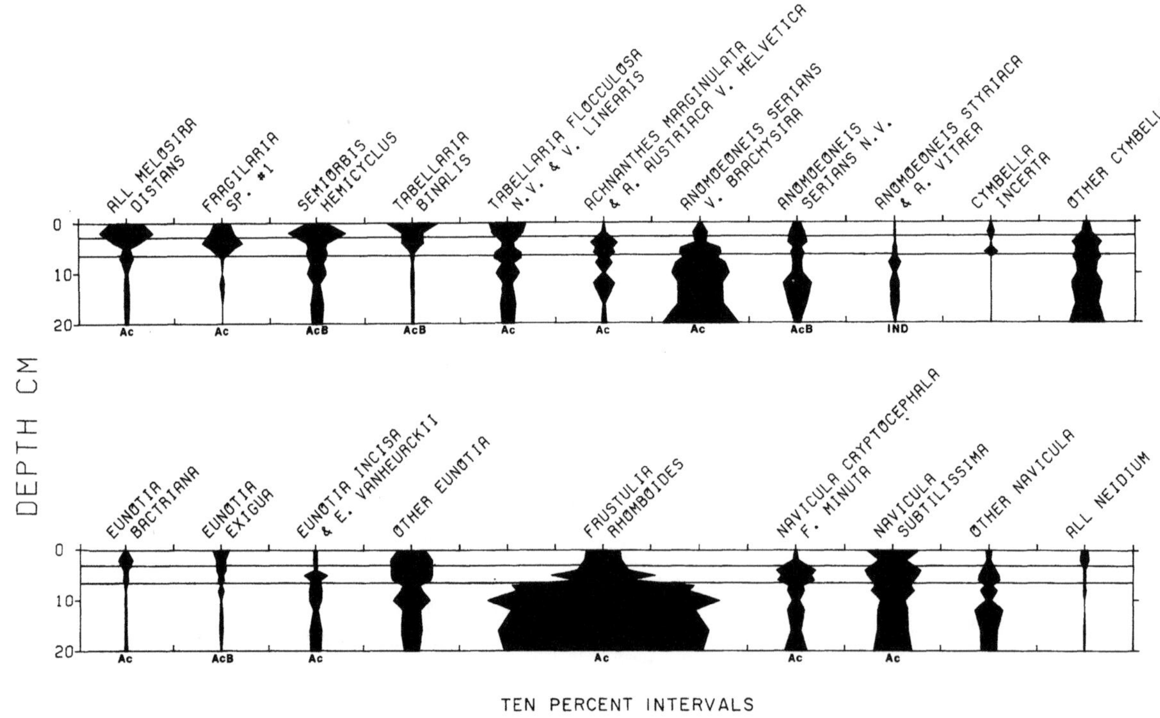

Fig. 3. Diatom stratigraphy in the core from Holmvatn. Only the most abundant taxa are given. AcB = acidobiontic; Ac = acidophilic; IND = indifferent. Zonation lines at 6.75 and 3.25 cm are based on a cluster analysis. Analyst: F. Berge (lab standardized taxonomy).

Drabløs *et al.* (1980) have argued that on a regional basis (there may be individual exceptions), the recent acidification of lakes in southern Norway is due mainly to acidic precipitation.

Lead concentration increases up-core from 18.5 cm (ca. 1835) to values 14X background. The top 2 cm are strongly modified by the FeO peak. We interpret the long term increase as caused by atmospheric deposition of Pb. Zinc starts to increase at 9 cm (ca. 1910). The short-lived decrease at 4–6 cm and the decrease at 0–2 cm are largely artifacts of the FeO peaks there. A Cu increase starts at 5 cm (ca. 1945), a bit earlier if the effect of FeO abundance is considered. The increases in Zn and Cu relative to Pb are less than is commonly observed (Norton, unpublished) and may indicate that Holmvatn has always been fairly acidic; this condition would keep Zn and Cu more mobile than Pb.

The coincident MnO, Pb, Zn, and Cu peaks are not causally related. Concentrations in the sediment relate primarily to atmospheric loading, sediment focusing, and surface water pH. If the enrichment of the surface by FeO were absent, Pb, Zn, and Cu surface concentrations would be about triple those shown (Fig. 1).

Diatom remains. A cluster analysis of the diatom counts indicates that the major changes occurred or began at 6.75 cm (ca. 1927) and 3.25 cm (ca. 1963). The first change correlates (with a slight delay) with the shift in geochemistry. The diatom stratigraphy (Fig. 3) includes up-core decreases in the acidophilic *Frustulia rhomboides* and *Anomoeoneis serians brachysira,* and the indifferent *A. styriaca* and *A. vitrea.* Increases occur in the acidobiontic *Semiorbis hemicyclus, Tabellaria binalis,* and *Eunotia exigua,* and in several acidophilic *Eunotia* species, *Melosira distans,* an unnamed species of *Fragilaria* similar to *F. bicapitata,* and in *Navicula cryptocephala* var. *minuta* which then decreases above 3.25 cm. These results indicate that Holmvatn was quite acidic prior to ca. 1927 when it began to become more acidic, and that acidification intensified at ca. 1963. No systematic change in the deposition rate of diatom frustules accompanied the acidification.

pH calibration of sedimentary diatom assemblages

Regressions relating the diatom assemblages in surface sediments to the pH's of the upper waters were computed for the 36 Norwegian and the 31 New England lakes. The regression coefficients were then applied down-core to infer water pH in past times.

Surface-sediment diatom data. The relative abundances of diatom taxa in the Norwegian lakes are clearly related to pH (Fig. 4). At the higher pH's, *Cyclotella* species, most *Fragilaria* species, certain *Achnanthes* species, *Anomoeoneis vitrea,* most *Navicula* species, and all the *Nitzschia* species are most abundant. At the lower pH's, there are greater abundances of most varieties of *Melosira distans, Semiorbis hemicyclus, Tabellaria binalis, Anomoeoneis serians* var. *serians,* and the *Eunotia* species. *Frustulia rhomboides* has a less steeply sloping relationship with pH than most *Eunotia* species. The acidophilic *Anomoeoneis serians* var. *brachysira* changes little in relative abundance below pH 6.4. *Navicula cryptocephala* var. *minuta* and *Navicula subtillissima* are relatively most abundant at low pH. These relationships strongly support our interpretation of the diatom stratigraphy at Holmvatn. In New England, the relationships between the abundances of the taxa and pH, given for 25 of the lakes by Norton *et al.* (1981) are not changed greatly by the additional six lakes.

Regressions of the surface-sediment data. The diatom data were regressed against pH in four different ways: (1) multiple linear regression of Hustedt's (1939) pH-preference groups, i.e. acidobiontic, indifferent, and alkaliphilic; to allow for some 'independent' variation, the acidophilics were not regressed; there were no alkalibiontics; (2) simple linear regression of the log of Nygaard's (1956) alpha index; (3) multiple linear regression of 16 common taxa; and (4) multiple linear regression of the taxa in the first principal component (PCA). Cluster analyses of the two data sets indicated distinct breaks between the diatom assemblages from lakes pH >6.2 versus those pH ≤6.2. Accordingly, each of the four types of regressions was run in two ways: (1) for lakes of all pH's, and (2) for lakes pH ≤6.2 (limited range). The F-ratios of the regressions were highly significant ($p < 0.001$). The regression coefficients were applied down the cores from the nine Norwegian and six New England lakes. If, at any lake the pH in 1978 was >5.7 or the reconstruction of pH based on the full or limited range regres-

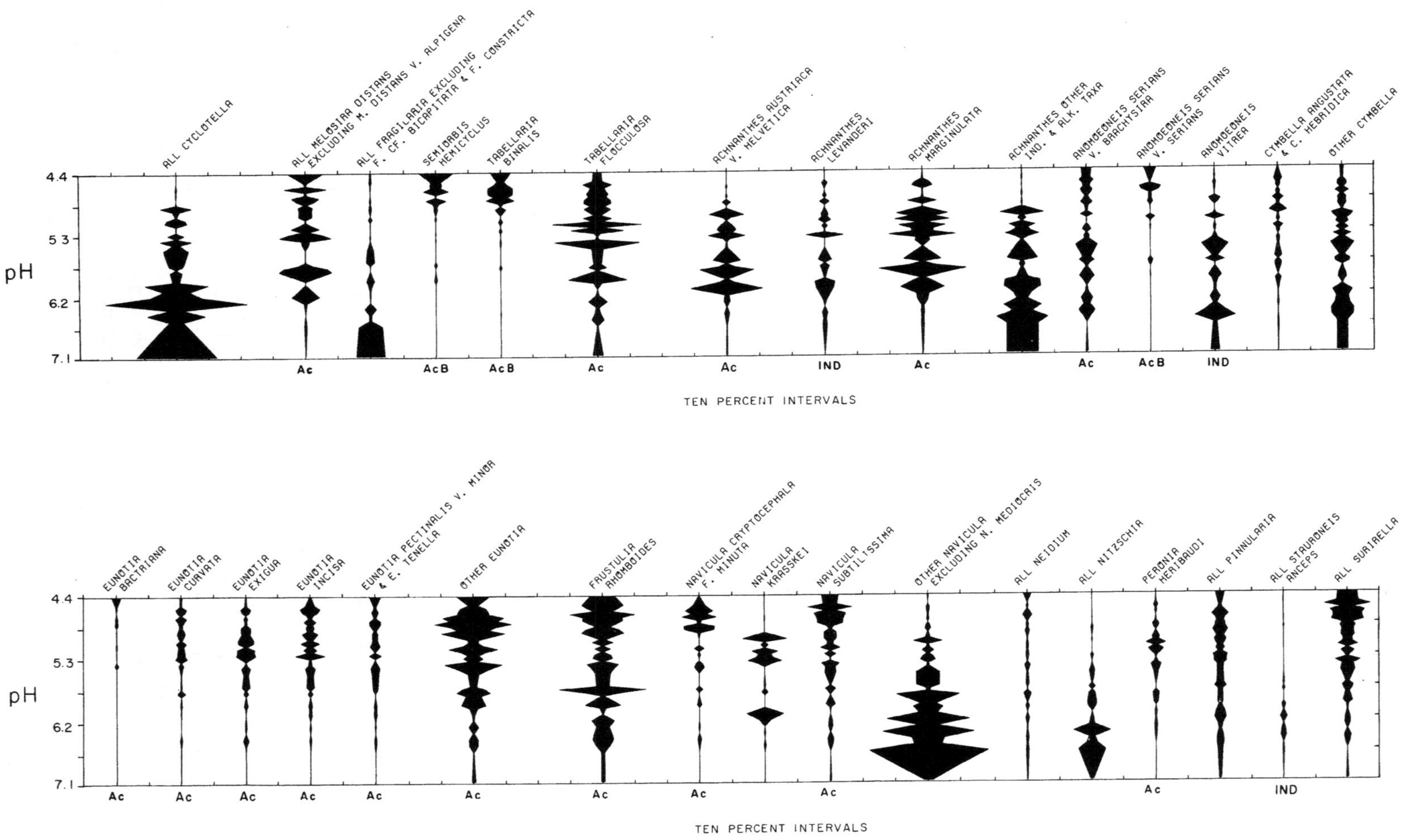

Fig. 4. Diatoms in surface-sediments from 36 Norwegian lakes ranging in pH from 4.4 to 7.1. Only the more abundant taxa are given.

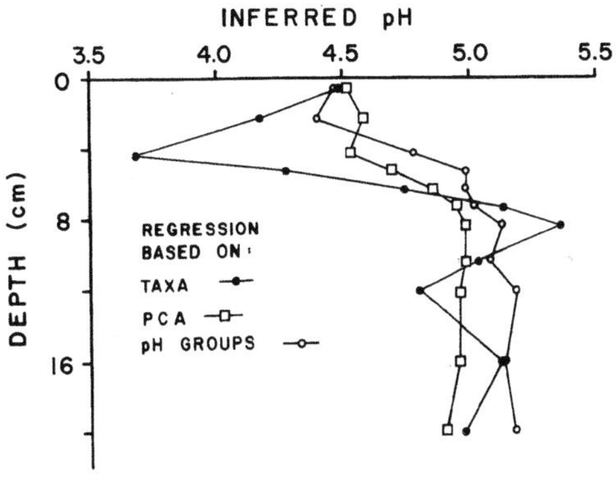

Fig. 5. Inferred pH down the core from Holmvatn.

sions gave inferences of pH >5.7, then the results of the full range regressions are given.

Holmvatn pH reconstruction. Omitted from the inferred pH's for Holmvatn (Fig. 5) are inferences based on the alpha index because of infinity values of the index (no alkaliphilic taxa) in the core. The inferred pH values of 4.46 to 4.51 for the 0.0–0.5 cm level approximate the present water pH of 4.7. Except for the outlier pH 3.67 at 4.25 cm inferred from the taxon regression, the results of the three regressions are in good agreement. Characteristically (as in the other lakes), inferences based on the taxon regression are more erratic due to the behavior of individual taxa. These results indicate that Holmvatn was quite acidic (pH 4.80–5.36, mode ca. 5.1) prior to the acidification which set in at 7.25 cm (ca. 1927). Fishes did not disappear until the 1960's (Johannessen & Wright 1980).

The total decrease in pH of ca. 0.6 units at Holmvatn may be typical of such lakes in the region, as suggested by the sporadic pH data compiled by Wright (1977) who compared early pH readings (1925–1945) to readings in the 1970's from the same lakes. For lakes with such data which are on the same precipitation-pH isopleth as Holmvatn (pH 4.3 in 1974) and which are closest to Holmvatn (six lakes in Aust-Agder, ca. 50 km to the W and SW; 12 lakes in Telemark, ca. 50 km to the NE) pH decreased 0.2 to 1.1 units (geom. mean 0.72). In this region, the known dates of disappearance of fishes from lakes are mostly in the 1960's (Johannessen & Wright 1980).

pH reconstruction, changes in cladocera, and heavy metal accumulation at other lakes

In Norway (Fig. 6), pH decreases of 0.6–0.8 units are inferred for the three lakes which are now less than pH 5.0. The decreases began in ca. 1918 (Hovvatn), ca. 1890 (Nedre Malmesvatn), and ca. 1927 (Holmvatn). Fish disappeared from these lakes in the 1940's (Hovvatn) and 1960's (Johannessen & Wright 1980). For Hovvatn, the 1918 date is a correction for the erroneous 1964 date (Davis & Berge 1980) based on the Cs-137 profile. At the

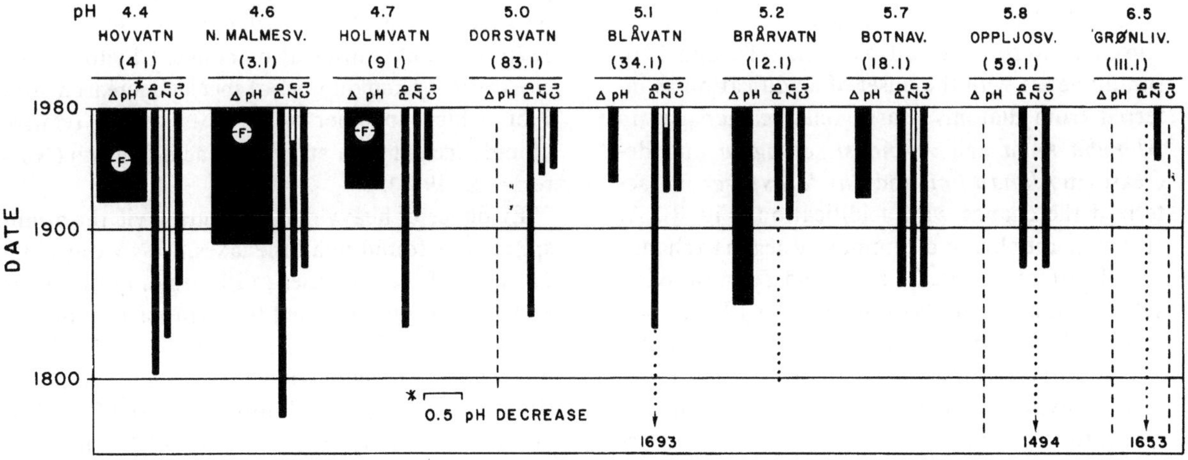

Fig. 6. Chronology of decrease in water pH and increase in Zn, Pb, and Cu concentration in sediment of Norwegian lakes. Above each lake name is the present pH; below the name is the SNSF-project lake number. The bars span the period of anthropogenic decrease in pH and increase in heavy metals. The upper, white section of the Zn bar is the period of recent decreases in concentration in the sediment. Dashed lines indicate no change; dotted lines minor or doubtful increases up-core. F = date of fish disappearance.

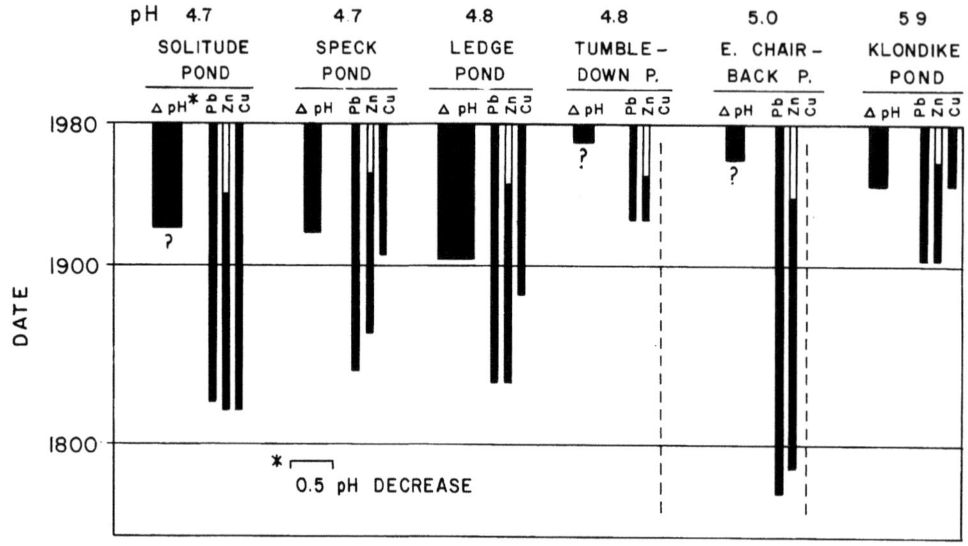

Fig. 7. Chronology of decrease in water pH and increase in Zn, Pb, and Cu concentration in sediment of New England lakes. Solitude Pond is in New Hampshire, the others in Maine. Question marks indicate that application of some regression types (see text) indicated no change in pH. All the lakes had fishes in 1980. See Fig. 6 caption for other details.

four lakes with present pH's 5.0 to 5.7, the pH decreased zero to 0.3 units, and no decrease was observed in the two higher pH lakes. The six New England lakes decreased 0.2–0.4 units (Fig. 7). In at least three cases (Ledge, Tumbledown and E. Chairback), the decrease may be due in part or entirely to reforestation and soil development following mild eutrophication associated with lumbering or other disturbance.

At Nedre Målmesvatn, a major change in the chydorid Cladocera assemblage as indicated by cluster analysis occurred between 1890 and 1910 simultaneous with the onset of acidification as inferred from diatoms. Large relative increases in *Alonella nana* and *Alonopsis elongata* and decreases in *Alona affinis* and *Chydorus piger* characterized the change with acidification (Fig. 8). At Hovvatn, the chydorid community began to change shortly after the increase in Pb and Zn in the early 1800's, but major shifts occurred after 1900 coincident with the decrease in lake water pH (dates altered from Brakke 1980). In both lakes, responses of the chydorid community to acidification preceded the elimination of fishes and included declines of species (e.g. *Alona costata, Chydorus piger*) characteristic of surface sediments of higher pH lakes (Brakke 1980). In the New England surface sediments (Norton *et al.* 1981), species diversity and the number of chydorid species decrease along the gradient of decreasing pH of the lakes; ratios of the regressions are highly significant. At elevations greater than 500 to 750 masl diversity does not increase with pH. At Speck Pond and Tumbledown Pond our analyses indicate that the most important changes in chydorids occurred at the same sediment depths where diatoms indicate acidification (Norton *et al.* 1981). At Ledge Pond, diatom stratigraphy suggests pH change in 1920 and again in the past decade. Another disturbance occurred at about 1820, when chydorids also change. Chydorid species diversity and number of species decreased after 1950 which together with shifts in the relative abundances of taxa suggest an acidification (Norton *et al.* 1981).

Evidence of heavy metal pollution via the atmosphere was found in all the lakes. In Norway (Fig. 6), some of the increases in Pb began in the 1600s and 1700s (approximated by extrapolation of Pb-210 CIC model), but mostly in the 1800s. Lead increases in the sediment 5.5 to 23 times pre-pollution concentrations. Pollution by Zn and Cu is first detected in individual lakes in the 1800s and early 1900s, although earlier and later first-detection dates occur (Fig. 6 & 7). Concentrations reach 5–6 times pre-pollution values in some lakes. In most of the lakes, Zn concentrations in the sediment de-

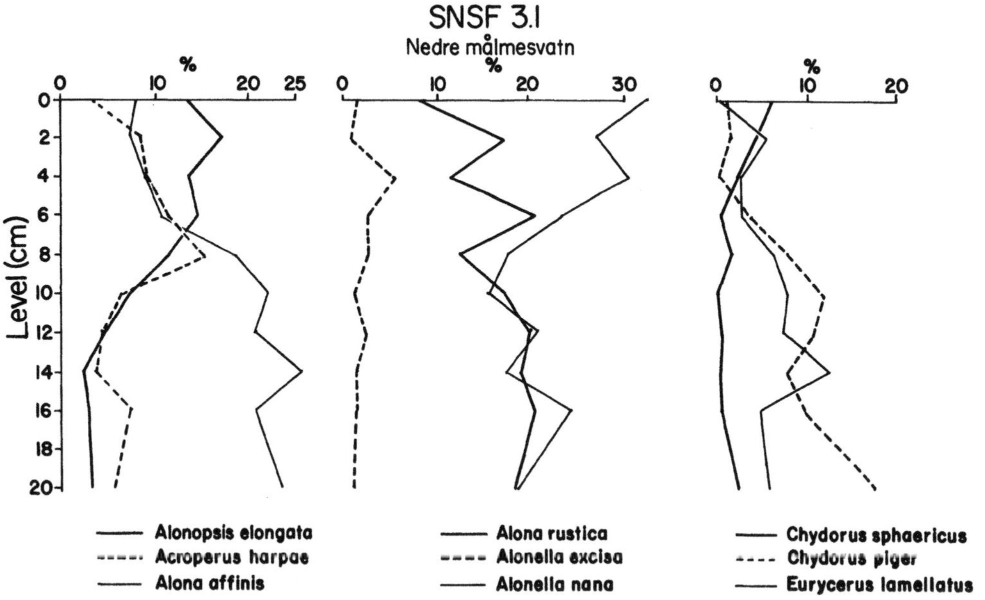

Fig. 8. Changes in selected chydorid Cladocera of Nedre Målmesvatn (SNSF 3.1), Norway. Chronology of pH change and heavy metal deposition is given in Fig. 6. 2 cm = 1965; 6 cm = 1917; 10 cm = 1877.

crease in the most recent 20–40 years, a phenomenon which we have suggested (Norton *et al.* 1980) is due to leaching of Zn from sediment precursors by increasingly acidic solutions. This process is probably accompanied by increasing concentrations of Zn in lake water (Norton *et al.* 1980). In many of the lakes, Zn and Cu increases in the sediment are correlated with up-core increases in organic matter and may be due in part to partitioning into the organic matter. However, we believe that the increasing Zn and Cu are in part (some lakes) or entirely (other lakes) due to pollution inputs, as evidenced by (1) greater relative increases than the organic matter; (2) recent up-core decreases in Zn while organics continue to increase; and (3) in some lakes, increases not associated with up-core increases in organic matter.

The lag of many decades to over a century in the first diatom indications of lake acidification following the early appearances of heavy metal pollutants (Figs. 6, 7) may be due to some combination of: (1) a reduction in recent decades of particulate emissions (viz. fly ash) which previously neutralized acid emissions, the reduction resulting from air pollution controls and the transition from coal to oil (Schriescheim & Kirshenbaum 1981); (2) insensitivity or slowness of diatom response; and (3) buffering of acid inputs by lakes and their watersheds. We believe the third factor to be most important. Our results indicate that lake acidification began well before the abovementioned fuel changes and any significant controls of emissions. It seems unlikely that slowness of diatom response is a significant factor, given the short generation time of diatoms and their rapid response in experiments (Patrick 1977). The initial anthropogenic acidification of precipitation may have accompanied the early pollution by heavy metals; the sources of the acid precursors (SO_X, NO_X) and metals are largely the same (sulfide ore smelting; fossil fuel combustion).

The geographic distributions of the acidification and heavy metal pollution of these lakes follows the air pollution gradients in Norway (Henriksen & Wright 1978) and New England (Semonin *et al.* 1981), with greatest lake pollution in the south and southwest, respectively.

Concluding comments

Based on the ^{210}Pb chronology in lake watershed systems with young, granitic sediment precursors and acidic waters the Cs-137 profile fails as a dating tool due to elevated concentrations at the sediment surface and significant concentrations in pre-fallout sediment. Longmore *et al.* (1981) have observed '... an exponential increase of ^{137}Cs towards the sediment surface,' in a lake surrounded by siliceous sand, and have postulated recycling of ^{137}Cs in organic materials.

It is very difficult to find lakes whose sediments are completely free of the influences of human and non-human disturbances to the watersheds. At Holmvatn and certain of the other lakes it was not possible to distinguish with full confidence the relative influences on lake pH of acidic precipitation versus watershed changes unrelated to atmospheric inputs.

Acknowledgements

Persons who assisted in the gathering and analyses of the data include D. S. Anderson, R. S. Anderson, D. R. Beeson, F. Berge, J. A. Gilotti, E. M. Hall, D. W. Hanson, K. D. Hoagland, M. L. Hoskins, S. T. Jagels, and P. R. Sweets. Funding was provided from NSF DEB7810641, USFWS 1416000979040, USDA Hatch ME08465, and USOWRT A048ME.

References

Appleby, P. G. & Oldfield, F., 1978. The calculation of lead-210 dates assuming a constant date of supply of unsupported Pb-210 to the sediment. Catena 5: 1-8.

Battarbee, R. W., 1973. A new method for the estimation of absolute microfossil numbers, with reference especially to diatoms. Limnol. Oceanogr. 18: 647-653.

Beck, K. C., Reuter, J. H. & Perdue, E. M., 1974. Organic and inorganic geochemistry of some coastal plain rivers of the southeastern United States. Geochim. Cosmochim. Acta 38: 341-364.

Brakke, D. E., 1980. Atmospheric deposition in Norway during the last 300 years as recorded in SNSF lake sediments. III. Cladoceran community structure and stratigraphy. In: Drabløs, D. & Tollan, A. (Eds.), Ecological Impact of Acid Precipitation. Proc. int. Conf. Ecol. Impact Acid Precip., SNSF Project. Oslo, Norway: 272-273.

Buckley, D. E. & Cranston, R. E., 1971. Atomic absorption analyses of 18 elements from a single decomposition of aluminosilicate. Chem. Geol. 7: 273-284.

Covell, D. F., 1956. Determination of gamma-ray abundance directly from the total absorption peak. Anal. Chem. 31: 1785-1790.

Davis, R. B. & Berge, F., 1980. Atmospheric deposition in Norway during the last 300 years as recorded in SNSF lake sediments. II. Diatom stratigraphy and inferred pH. In: Drabløs, D. & Tollan, A. (Eds.). Ecological Impact of Acid Precipitation. Proc. int. Conf. Ecol. Impact Acid Precip., SNSF Project. Oslo, Norway: 270-271.

Davis, R. B. & Doyle, R. W., 1969. A piston corer for upper sediment in lakes. Limnol. Oceanogr. 14: 643-648.

Drabløs, D., Sevalrud, I. & Timberlid, J. A., 1980. Historical land-use changes related to fish status development in different areas in southern Norway. In: Drabløs, D. & Tollan, A. (Eds.). Ecological Impact of Acid Precipitation. Proc. int. Conf. Ecol. Impact Acid Precip., SNSF Project. Oslo, Norway: 367-369.

Eakins, J. D. & Morrison, R. T., 1978. A new procedure for the determination of lead-210 in lake and marine sediments. Int. J. appl. Radiat. Isot. 29: 531-536.

Hendrey, G. R., Galloway, J. N., Norton, S. A., Schofield, C. L., Shaffer, P. W. & Burns, D. A., 1980. Geological and hydrochemical sensitivity of the eastern United States to acid precipitation. U.S. Envir. Prot. Ag., E.P.A. 600/3-80-024: 1-100.

Henriksen, A. & Wright, R. F., 1978. Concentrations of heavy metals in small Norwegian lakes. Wat. Res. 12: 101-112.

Hustedt, F., 1939. Systematische und okologische Untersuchungen uber die Diatomeen - Flora von Java, Bali, und Sumatra. III. Die okologischen Faktoren und ihren Einfluss auf die Diatomeen-flora. Arch. Hydrobiol. Suppl. 16: 274-394.

Johannessen, M. & Wright, R. F., 1980. Regional surveys of lakes in southern Norway. SNSF Project, TN 53/80. Oslo, Norway. 28 pp.

Krishnaswami, S. D. L., Martin, J. M. & Maybeck, M., 1971. Geochronology of lake sediments. Earth planet. Sci. Lett. 11: 407-414.

Longmore, M. E., O'Leary, B. M. & Rose, C. W., 1981. Caesium-137 profiles in a near meromictic lake on Great Sandy Island (Fraser Island), Queensland, Australia. III. int. Symp. Paleolimnol., Joensuu, Finland: 40 (abstract).

Norton, S. A., 1973. Laterite and bauxite formation. Econ. Geol. 68: 353-361.

Norton, S. A., in press. The effects of acidification on the chemistry of ground and surface waters. U.S. Fish Wildl. Serv., Proc. Symp. Acid Rain Impact U.S. Fish. am. Fish. Soc. 20 pp.

Norton, S. A., Davis, R. B. & Brakke, D. F., 1981. Responses of northern New England lakes to atmospheric inputs of acids and heavy metals. Completion Report, U.S. Dep. Interior Project A-048-ME. Washington, D.C. 90 pp.

Norton, S. A., Hanson, D. W. & Campana, R. J., 1980. The impact of acidic precipitation and heavy metals on soils in relation to forest ecosystems. In: Miller, P. R. (Ed.). Effects of Air Pollutants on Mediterranean and Temperate Forest Ecosystems. U.S. Dep. agric. Gen. Tech. Rep. PSW-43: 152-157.

Nygaard, G., 1956. Ancient and recent flora of diatoms and Chrysophyceae in Lake Grisbo. Folia limnol. scand. 8: 32-99, 254-262.

Overrein, L. N., Seip, H M. & Tollan, A., 1980. Acid precipitation - effects on forest and fish. Final Report of the SNSF Project 1972-1980. Oslo, Norway. 175 pp.

Patrick, R., 1977. Ecology of freshwater diatoms and diatom communities. In: Werner, D. (Ed.). The Biology of Diatoms. Bot. Monogr. 13. Univ. Cal. Press, Berkeley: 284-332.

Rosenqvist, I. Th., Jørgensen, P. & Rueslatten, H., 1980. The importance of natural H^+ production for acidity in soil and water. In: Drabløs, D. & Tollan, A. (Eds.). Ecological Impact of Acid Precipitation. Proc. int. Conf. Ecol. Impact Acid Precip., SNSF Project. Oslo, Norway: 240–241.

Schreisheim, A. & Kirshenbaum, I., 1981. The chemistry and technology of synthetic fuels. Am. Scient. 69: 536–542.

Semonin, R. G., Bowersor, V. C., Gatz, D. F., Pedan, M. E. & Stensland, G. J., 1981. Study of atmospheric pollution scavenging. III. Inst. nat. Res. COO-1199-63. 126 pp.

Toonkel, L. E., 1980. Environmental measurements laboratory. Envir. Quart., U.S. Dep. Energy. Washington, D.C. 221 pp.

Wright, R. F., 1977. Historical changes in the pH of 128 lakes in southern Norway and 130 lakes in southern Sweden over the period 1923–1976. SNSF Project TN 34/77. Oslo, Norway. 71 pp.

Wright, R. F., Torstein, D., Henriksen, A., Hendry, G. R., Gjessing, E. T., Johannessen, M., Lysholm, C. & Støren, E., 1977. Regional surveys of small Norwegian lakes. SNSF Project, IR 33/77. Oslo, Norway. 153 pp.

Cultural disturbances and trophic history of a small meromictic lake from central Canada

John P. Smol[1], S. R. Brown[1] & R. N. McNeely[2]
[1] *Department of Biology, Queen's University, Kingston, Ontario, Canada K7L 3N6*
[2] *Geological Survey of Canada, 601 Booth St., Ottawa, Ontario, Canada K1A OE8*

Keywords: paleolimnology, pollen, diatoms, Synuraceae, Chrysophyta, meromixis, eutrophication

Abstract

The recent history of Little Round Lake, a small meromictic lake in southeastern Ontario, is considered. Pollen analyses were used to identify past changes in terrestrial vegetation, whilst limnological conditions were interpreted on the basis of diatom and chrysophyte microfossils. Contemporaneous with the arrival of European settlers (ca. A.D. 1850), the predisturbance assemblage of oligotrophic *Cyclotella* diatoms was replaced by *Synedra* spp., which then succeeded to a eutrophic flora dominated by *Stephanodiscus hantzschii*. Meanwhile, synuracean algae were almost completely excluded. Over the last 30 years, the algal microfossils indicate that the lake underwent a marked return to oligotrophy. Evidence is presented which suggests that this shift was related to the cultural enhancement of meromixis by the seepage of road salt.

Introduction

Patterns of trophic change related to human activity have been well established in studies of sedimentary cores from holomictic lakes. The effects of cultural disturbances on the developmental histories of meromictic lakes are less well known, even though their sediments are potentially freer from artifacts introduced by post-sedimentation processes.

Munch (1980) discussed changes in fossil diatom and synuracean communities in an acid lake in which meromixis was destroyed by road construction. In contrast, this study considers the recent history of an alkaline lake in which human activity may have enhanced the meromictic stability. Pollen profiles and algal microfossils are considered.

Little Round Lake (44°47′30″ N, 76°41′30″ W) occupies a small, deep, cup-shaped basin in southeastern Ontario (Fig. 1). Its drainage basin is underlain by metamorphosed Precambrian (Grenville) bedrock composed of granite, granite gneiss, iron-rich schists, as well as marble. Consequently, it has both the characteristics of Canadian Shield lakes in that its waters are slightly colored and has cation ratios similar to igneous drainage waters, but has anions more typical of hard bicarbonate waters (McNeely 1973). The lake is presently oligo-mesotrophic and is chemically stratified with a distinct chemocline near the 11 m level.

The watershed has been subjected to moderate cultural disturbances over the last century; first by loggers and the building of a railroad adjacent to the lake, and later by the construction of two highways (Fig. 1). Although poorly endowed with soils, portions of the drainage basin were farmed in the past. Presently this is reverting to the original forest cover consisting of a mixed deciduous forest of the Great Lakes – St. Lawrence type, with both Deciduous and Boreal affinities (Rowe 1959).

The lake has a legacy of sedimentological studies (e.g. Whittaker & Vallentyne 1957; Vallentyne 1960, 1963; Tippett 1964; Brown *et al.* 1977; Daley *et al.* 1977). The contemporary limnology and a complete site description were given by McNeely (1973). Munro (1978) described the phytoplankton dynamics of the lake.

Hydrobiologia 103, 125–130 (1983).

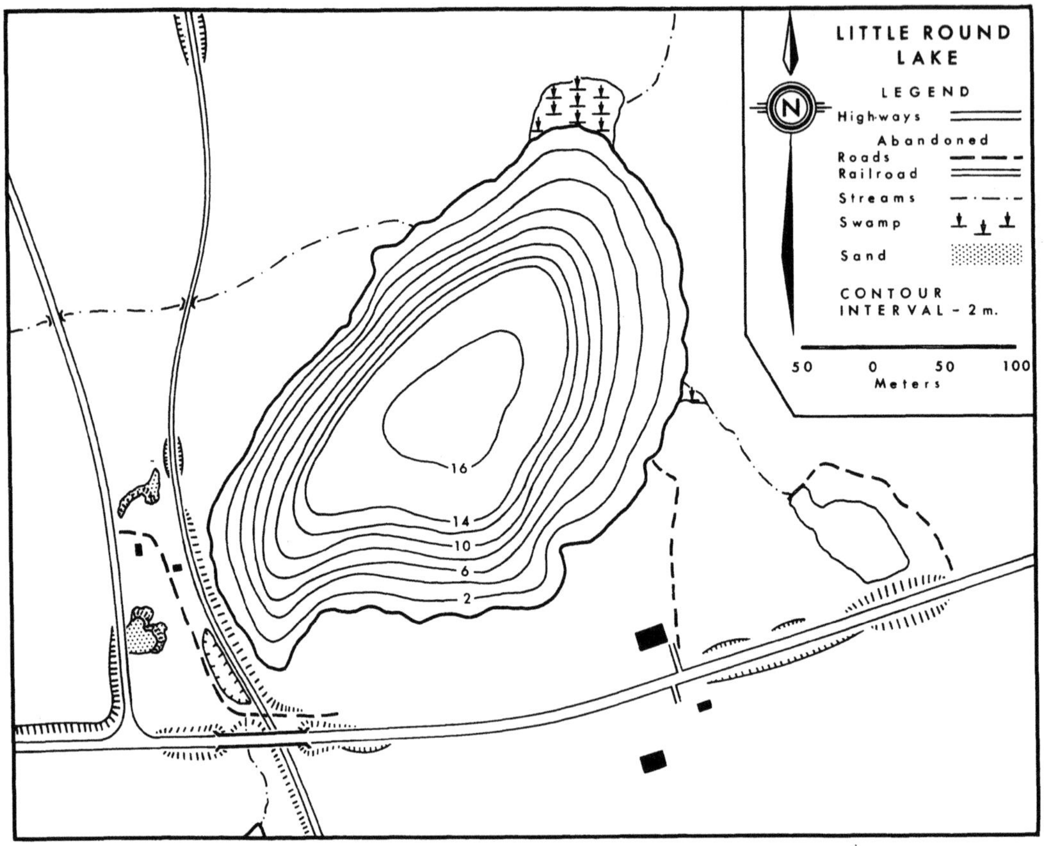

Fig. 1. Morphometric map of Little Round Lake (drainage basin area = 1.09 km², surface area = 7.4 ha, maximum depth = 16.8 m, mean depth = 8.3 m; from McNeely 1973).

Materials and methods

In May 1981, a sediment core was obtained from the central plateau of the basin with a Brown (1956) corer. In the laboratory the core was immediately sectioned at 1 cm intervals using a hydraulic extrusion system, similar in design to the fractionator described by Fast & Wetzel (1974). The sediment sections were then stored in the dark at 4 °C. This study reports analyses of the upper 20 cm.

Sedimentary organic matter was estimated by weight loss on ignition at 550 °C (Dean 1974) and expressed as percentage of the dry weight.

Pollen preparation followed Faegri & Iversen (1975). A minimum of 450 pollen grains were identified and counted at each sediment level using a Nikon microscope (400 × magnification).

Evaporation trays, similar to those described by Battarbee (1973), were used to prepare quantitative slides of the algal microfossils. Sediment preparations followed Smol & Dickman (1981) for the diatoms and Smol (1980) for the chrysophytes. Usually, at least 650 diatom valves and chrysophyte scales were counted from each slide; however, at certain sediment levels synuracean scales were so sparse that counting was discontinued once 650 fields were enumerated. Diatom taxonomy was based on Hustedt (1930–1966), Huber-Pestalozii (1942), Cleve-Euler (1951–1955), Patrick & Reimer (1966, 1975), Stoermer & Yang (1969), and Lowe (1975), whereas synuracean scales were identified following Asmund & Hilliard (1961), Takahashi (1978), and Cronberg & Kristiansen (1980). Chrysophycean statospores were also counted, but no attempt was made to classify their morphotypes (cf. Nygaard 1956). Counting variance was estimated at two points in the core by enumerating four replicate slides. Because of problems of interpretation (cf. Smol 1981) diversity indices were not calculated.

Results and discussion

Changes in the recent vegetational history of Little Round Lake (Fig. 2) are similar to those observed in other culturally disturbed areas in northeastern North America (e.g. Brugam 1978). The increase in non-arboreal pollen such as ragweed (*Ambrosia*) and grasses (Gramineae) at the 16 cm level indicates the arrival of European settlers (Bassett & Terasmae 1962). The decline in hemlock (*Tsuga*) and later pine (*Pinus*) probably reflects lumbering activity. These trees were then partially replaced by successional species such as alder (*Alnus*) and birch (*Betula*).

Throughout the same time-frame, four changes in the lake's fossil algal communities (Figs. 3, 4) can be identified: 1) an oligotrophic assemblage of *Cyclotella* diatoms and synuracean chrysophytes dominated by *Mallomonas pseudocoronata/crassisquama*; 2) a period of increasing nutrient enrichment in which a *Synedra* flora and *M. tonsurata* Teiling emend Krieger increases; 3) a eutrophic phase in which *Stephanodiscus hantzschii* Grun. dominates and the Synuraceae are reduced to trace levels; and 4) a return to oligo-mesotrophic conditions indicated by the reoccurrence of *Cyclotella* spp. and *Mallomonas*, most notably *M. fastigata* Zacharias (Syn. *M. caudata* Iwanoff emend Krieger).

The succession from the predisturbance assemblage of oligotrophic *Cyclotella* diatoms to the *Synedra* flora and later to *Stephanodiscus hantzschii* (Fig. 3) was contemporaneous with the rise in non-arboreal pollen (Fig. 2). In contrast, the marked return of oligotrophic algae in the lake's most recent sediments do not coincide with any changes recorded in the pollen profile.

The diatom succession from *Cyclotella* spp. to a *Synedra* assemblage (Fig. 3) is very similar to the one recorded by Smol & Dickman (1981) following road construction in the watershed of Found Lake. It was postulated in that study that thinly silicified diatoms would be more competitive than faster sinking taxa should light penetration decrease due to turbidity and/or a higher standing crop of algae. Although the dominant *Synedra* species were different in the two lakes (i.e. *S. delicatissima* var. *angustissima* Grun. in Found Lake and *S. filiformis* var. *exilis* Cl.-Eul. in Little Round Lake) the morphologies of the two taxa are very similar, and an analogous argument may be applied in this study. The synuracean data confirm that some enrichment had occurred since scales of *Mallomonas tonsurata*, a taxon characteristic of eutrophic waters

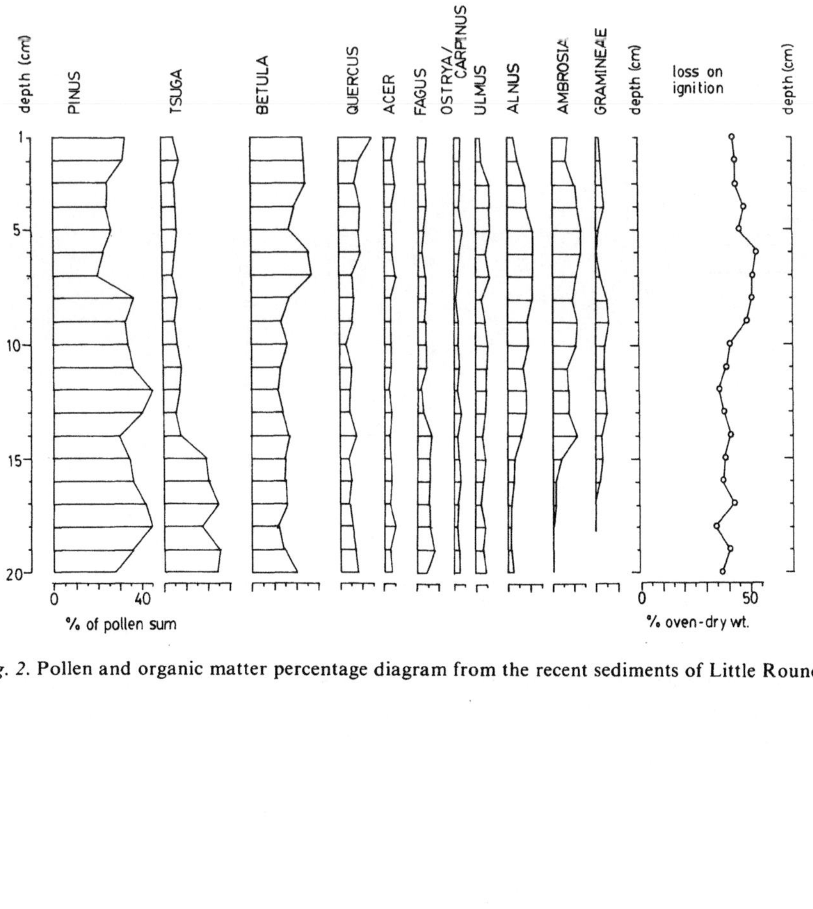

Fig. 2. Pollen and organic matter percentage diagram from the recent sediments of Little Round Lake.

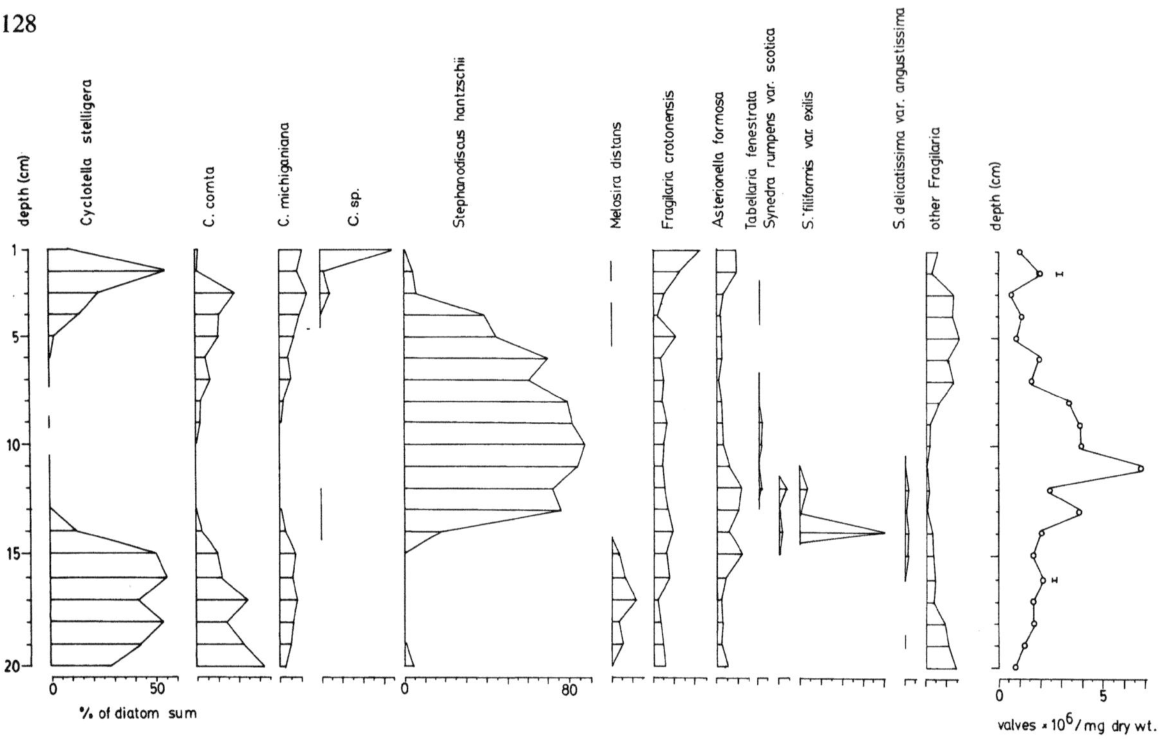

Fig. 3. Profiles of the relative frequency of diatom species and concentration of diatom valves in the recent sediments of Little Round Lake. Although 99 diatom taxa were identified, only 12 were sufficiently abundant (i.e. greater than 2% of the total diatom sum at any stratigraphic level) to warrant their inclusion in this diagram. Error bars represent 95% confidence limits.

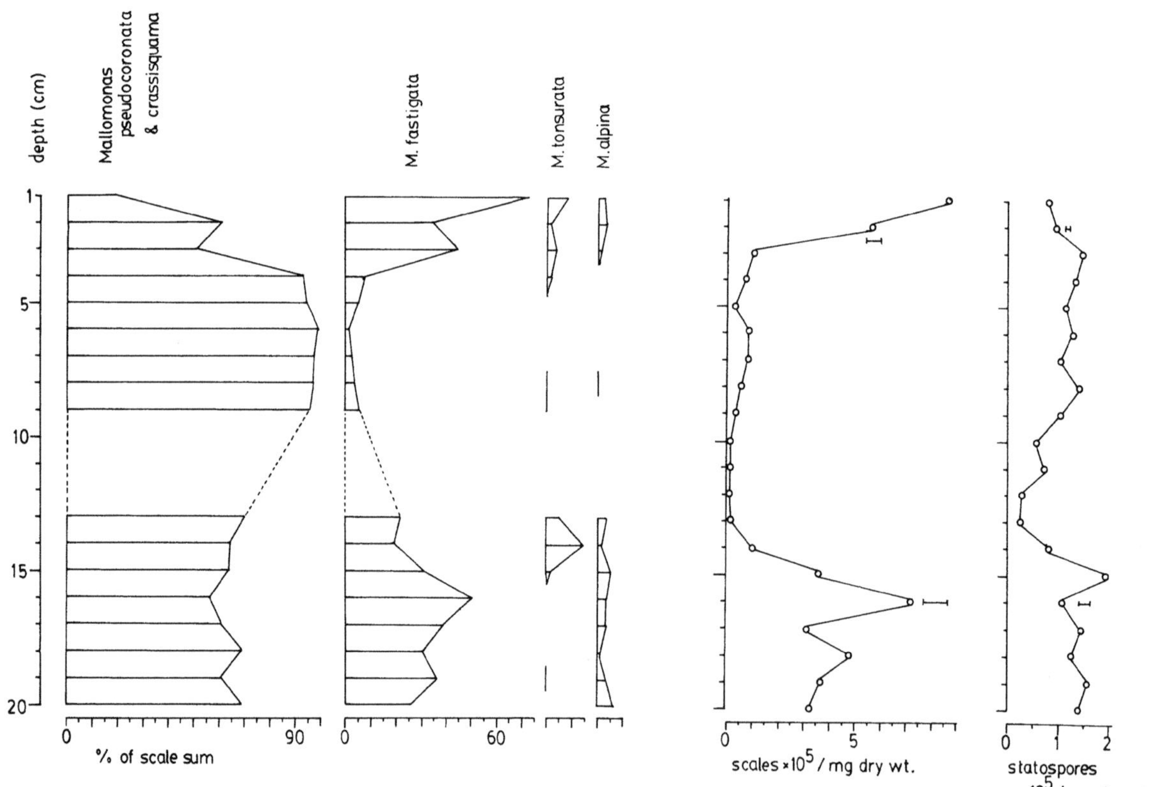

Fig. 4. Profiles of the relative frequency of synuracean species and concentrations of chrysophycean scales and statospores in the recent sediments of Little Round Lake. Synuracean scales were so sparse between 13 and 9 cm that it was considered that an analysis of species composition was not justified. Error bars represent 95% confidence limits.

(Tarapchak 1973), peaked abruptly at this time.

These *Synedra* populations were soon replaced by a much more eutrophic species – *Stephanodiscus hantzschii* (Fig. 3). Brugam (1979), following an analysis of 48 Minnesota lakes, concluded that *S. hantzschii* was the most reliable indicator of eutrophication in his alkaline lakes. However, even in Brugam's most enriched lakes (i.e. total phosphorus > 150 $\mu g \cdot l^{-1}$), *S. hantzschii* never exceeded 70% of the diatom sum. This suggests that Little Round Lake was very eutrophic near the turn of the century since this taxon comprised close to 90% of the diatoms.

Coincident with this eutrophic diatom assemblage, synuracean algae were almost completely excluded (Fig. 4). Since 'scaled chrysophytes' tend to be less competitive in eutrophic waters, their near disappearance is not surprising. Furthermore, in a contemporary study of the lake's phytoplankton, Munro (1978) found that *Mallomonas* spp. bloomed in deep water populations below the thermocline, and it is probable that they would be shaded out during periods of high productivity. The continued presence of chrysophycean statospores in the fossil record during this eutrophic phase (Fig. 4) may be explained by Munro's observations that 'non-scaled chrysophytes' do not form discreet, deep water populations in this lake.

Both the diatom and chrysophyte microfossils recorded a marked return to oligotrophy over the last 30 or so years. For example, *Stephanodiscus hantzschii* was reduced to trace levels and the diatom assemblage was dominated by oligotrophic taxa such as *Cyclotella stelligera* (Fig. 3). In addition, synuracean scales were common (Fig. 4).

To our knowledge, no other lake has ever recorded such a striking return to oligotrophic conditions, and consequently an explanation suggesting that watershed soils had simply reconsolidated over the last 30 years appears inadequate. An alternate hypothesis, one that is related to the low meromictic stability of the lake (ca. 80 g-cm cm^{-2}), stems from McNeely's (1973) limnological studies. McNeely calculated that, should the lake undergo a complete circulation, phosphorus levels in the epilimnion would triple and nitrogen concentrations would increase tenfold. McNeely also found that salt concentrations in the monimolimnion were far in excess of what could be explained on geological grounds (i.e. sodium, 58.4 $mg \cdot l^{-1}$; chloride, 103.7 $mg \cdot l^{-1}$). Since NaCl has commonly been used as a de-icing agent on Ontario roads since 1945, probable sources for these elevated monimolimnetic concentrations are the two highways adjacent to the lake, as well as seepage from a salt storage shed situated in its drainage basin (Fig. 1). McNeely further proposed that one of the reasons Little Round Lake was presently meromictic (and oligotrophic) was that meromixis was being maintained by this ectogenic source of salt.

Although an increase in meromictic stability appears to have greatly reduced algal productivity, the species composition in the surface sediments (Figs. 3, 4) suggests that the lake is still slightly more eutrophic than it was prior to man's arrival. For example, the shift from *Mallomonas pseudocoronata/crassisquama* to *M. fastigata* suggests enrichment (Smol 1980), as does the increase in *Fragilaria crotonensis* Kitton (Bradbury 1975).

Conclusions

Contrary to trends observed in holomictic lakes, the history of Little Round Lake under the influence of man is not one of ever increasing eutrophy. Near the turn of the century, human disturbances in the watershed coincided with marked changes in algal communities indicating nutrient enrichment. More recently, however, cultural enhancement of meromixis may have been a major factor influencing the rapid return to oligotrophic conditions. This does not preclude the occurrence of meromixis during the lake's eutrophic phase, but suggests that less entrainment of nutrient rich water from the monimolimnion has occurred in recent years.

Acknowledgements

This paper represents a portion of the senior author's Ph.D. thesis, and was supported by the Natural Sciences and Engineering Research Council of Canada in the form of an operating grant (A-808) to S. R. B. as well as a postgraduate scholarship to J. P. S. & H. J. McIntosh assisted with the collection of the core. We thank B. Asmund for information concerning *Mallomonas pseudocoronata*.

References

Asmund, B. & Hilliard, D. K., 1961. Studies on Chrysophyceae from some ponds and lakes in Alaska. I. Mallomonas species examined with the electron microscope. Hydrobiologia 17: 237–258.

Bassett, I. J. & Terasmae, J., 1962. Ragweeds, Ambrosia species in Canada and their history in post-glacial time. Can. J. Bot. 40: 141–150.

Battarbee, R. W., 1973. A new method for the estimation of absolute microfossil numbers, with reference especially to diatoms. Limnol. Oceanogr. 18: 647–653.

Bradbury, J. P., 1975. Diatom stratigraphy and human settlement in Minnesota. Geol. Soc. Spec. Pap. 171: 1–74.

Brown, S. R., 1956. A piston sampler for surface sediments of lake deposits. Ecology 37: 611–613.

Brown, S. R., Daley, R. J. & McNeely, R. N., 1977. Composition and stratigraphy of the fossil phorbin derivatives of Little Round Lake, Ontario. Limnol. Oceanogr. 22: 336–348.

Brugam, R. B., 1978. Pollen indicators of land-use change in southern Connecticut. Quat. Res. 9: 349–362.

Brugam, R. B., 1979. A re-evaluation of the Araphidineae/-Centrales index as an indicator of lake trophic status. Freshwat. Biol. 9: 451–460.

Cleve-Euler, A., 1951–1955. Die Diatomeen von Schweden und Finnland. K. svenska Vetensk. Akad. Avh. 5: 1–232.

Cronberg, G. & Kristiansen, J., 1980. Synuraceae and other Chrysophyceae from Central Småland, Sweden. Bot. Notiser 133: 595–618.

Daley, R. J., Brown, S. R. & McNeely, R. N. 1977. Chromatographic and SCDP measurements of fossil phorbins and the post glacial history of Little Round Lake, Ontario. Limnol. Oceanogr. 22: 349–360.

Dean, W. E. Jr., 1974. Determination of carbonate and organic matter in calcareous sediments and sedimentary rocks by loss of ignition: Comparison with other methods. J. Sediment. Petrol. 44: 242–248.

Faegri, K. & Iversen, J., 1975. Textbook of pollen analysis. Blackwell scientific Publications, Oxford. 295 pp.

Fast, A. W., & Wetzel, R. G., 1974. A close interval fractionator for sediment cores. Ecology 55: 202–204.

Huber-Pestalozzi, G., 1942. Das Phytoplankton des Süsswassers. Binnengewasser 16: 367–599.

Hustedt, F., 1930–1966. Die Kieselalgen Deutschlands, Österreichs und der Schweiz. In: Rabenhorst, L., (Ed.), Kryptogamen-Flora, Verlag von J. Cramer, Leipzig. 556 pp.

Lowe, R. L., 1975. Comparative ultrastructure of the valves of some Cyclotella species (Bacillariophyceae). J. Phycol. 11: 415–424.

McNeely, R. N., 1973. Limnological investigations of a small meromictic lake, Little Round Lake, Ontario. Ph.D. thesis, Queen's University, Kingston. 185 pp.

Munch, C. S., 1980. Fossil diatoms and scales of Chrysophyceae in the recent history of Hall Lake, Washington. Freshwat. Biol. 10: 61–66.

Munro, K. A., 1978. Phytoplankton dynamics in meromictic Little Round Lake, southeastern Ontario, with emphasis on growth of Cryptomonas rostratiformis in nature and in culture. M.Sc. thesis, Queen's University, Kingston. 154 pp.

Nygaard, G., 1956. Ancient and recent flora of diatoms and Chrysophyceae in Lake Gribsø. Folia limnol. scand. 8: 32–262.

Patrick, R. & Reimer, C. W., 1966. The diatoms of the United States. 1. Fragilariaceae, Eunotiaceae, Achnanthaceae, Naviculaceae. Acad. nat. Sci., Philadelphia. 688 pp.

Patrick, R. & Reimer, C. W., 1975. The diatoms of the United States. 2 (1). Entomoeidaceae, Cymbellaceae, Gomphonemaceae, Epithemiaceae. Acad. nat. Sci., Philadelphia. 213 pp.

Rowe, J. S., 1959. Forest regions of Canada. Dep. northern Affairs nat. Res., Ottawa. 172 pp.

Smol, J. P., 1980. Fossil synuracean (Chrysophyceae) scales in lake sediments: a new group of paleoindicators. Can. J. Bot. 58: 458–465.

Smol, J. P., 1981. Problems associated with the use of 'species diversity' in paleolimnological studies. Quat. Res. 15: 209–212.

Smol, J. P. & Dickman, M. D., 1981. The recent histories of three Canadian Shield lakes: A paleolimnological experiment. Arch. Hydrobiol. 93: 83–108.

Stoermer, E. F. & Yang, J. J., 1969. Plankton diatom assemblages in Lake Michigan. Univ. Mich. Great Lakes Res. Div. Spec. Pap. 47, Ann Arbor. 268 pp.

Takahashi, E., 1978. Electron microscopical studies of the Synuraceae (Chrysophyceae) in Japan. Tokai University Press, Tokyo. 194 pp.

Tarapchak, S. J., 1973. Studies of phytoplankton distribution and indicators of trophic state in Minnesota lakes. Ph.D. thesis, Univ. Minn. Minneapolis. 227 pp.

Tippett, R., 1964. An investigation into the nature of the layering of deep water sediments in two eastern Ontario lakes. Can. J. Bot. 42: 1693–1709.

Vallentyne, J. R., 1960. On fish remains in lacustrine sediments. Am. J. Sci. 258: 344–349.

Vallentyne, J. R., 1963. Isolation of pyrite spherules from recent sediment. Limnol. Oceanogr. 8: 16.

Whittaker, J. R. & Vallentyne, J. R., 1957. On the occurrence of free sugars in lake sediment extracts. Limnol. Oceanogr. 2: 98–110.

Diatom and chrysophyte relict assemblages in the sediments of Blelham Tarn in the English Lake District

Elizabeth Y. Haworth
Freshwater Biological Associations, Ambleside, Cumbria LA22 0LP, England

Keywords: paleolimnology, recent sediments, diatoms, chrysophytes, ^{210}lead

Abstract

Diatom and chrysophyte remains in recent sediments have been examined. The upper stratigraphy is dated by reference to phytoplankton records whilst the lower depends on dates from ^{210}Pb analysis, which shows some unexplained variation. Recent changes, due to farming and domestic drainage, form part of the long term response of the lake to events upon the catchment.

Introduction

Algal populations studies have been mainly concerned with planktonic algae and the present lake environment. In order to relate these data to the study of sediment assemblages I have concentrated on assessing the numbers of planktonic diatom valves and chrysophyte scales in the top 1 m of the sediments of Blelham Tarn. A previous study (Haworth 1980) established the comparative nature of the two sets of data and indicated that algal changes also ocurred prior to the period for which planktonic data was available.

Methods

Acid cleaned valves and scales were counted as numbers per slide, which was equivalent to 0.0001 cm^3 wet sediment collected at 0.5 cm intervals from a Mackereth wide diameter minicore (Haworth 1980). Percentage analyses of the diatom assemblages correlated the study core with one used for radio-isotope and chemical studies by Pennington *et al.* (1976, 1977). Direct comparison of diatom assemblages and ^{210}Pb analyses has provided a timescale before the period of plankton monitoring.

Results

The diatom assemblage (Fig. 1) in the early nineteenth century was dominated by *Cyclotella comensis* Grun., *C. praetermissa* Lund, *Achnanthes minutissima* Kütz. and *A. microcephala* (Kütz.) Grun. A series of changes in assemblages at horizons A–D (Fig. 1) has resulted in the present domination of *Asterionella formosa* Hassall and *Melosira italica* subsp. *subarctica* O. Müller, with other indicators of nutrient enrichment also present, ie. *Fragilaria crotonensis* Kitton and *Stephanodiscus astraea* var. *minutula* (Kütz.) Grun. Small peaks in numbers of siliceous chrysophyte scales of *Synura spinosa* Kors., *Mallomonas crassisquama* (Asmund) Fott and *M. reginae* Teiling occurred at the decline of *Cyclotella* spp. and the increase in *Asterionella formosa* ca. 1940 A.D. (Haworth, in press). Post-1947 sediments were identified by comparison with plankton records, and algal changes during that period indicate the environmental responses to changes in farming practice and domestic water supplies (Lund 1979). It is evident from pollen and chemical data that man-induced changes in sediment deposition have occurred at various times during the last 5 000 years. A diatom profile by Evans (1969) shows that there were sharp changes

Hydrobiologia 103, 131–134 (1983).

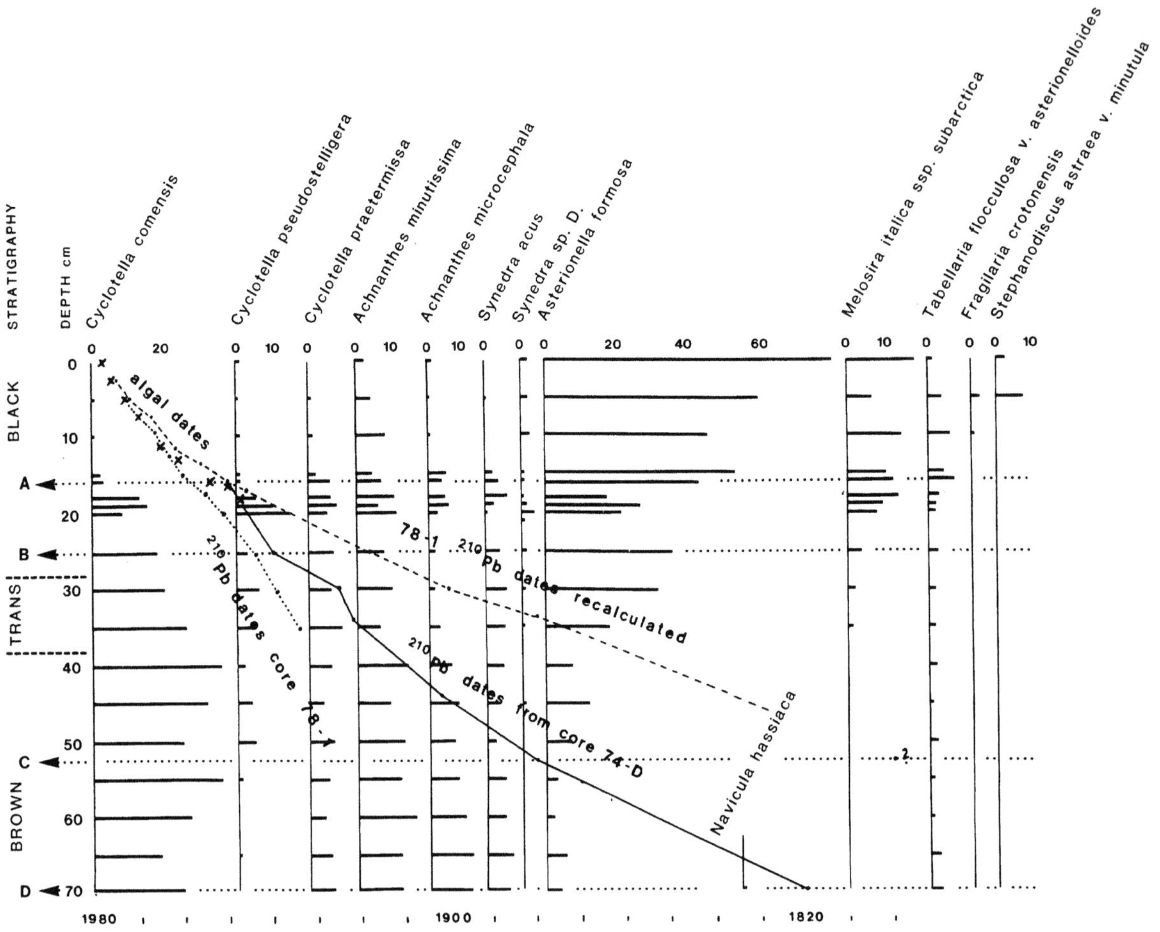

Fig. 1. Recent diatom assemblages in Blelham Tarn sediments. Percentages of major taxa in core 78–1, with the timescale provided by algal records and the variation in ^{210}Pb dates.

in *Cyclotella* spp. and in the presence or absence of *Melosira* and *Stephanodiscus* spp. throughout the Postglacial.

Correlation between minicores shows that the early ploughing, indicated by geochemistry, caused little change in the diatom assemblage. Old maps show that land drainage took place between about 1830 and 1880 A.D. and included a slight lowering of the tarn water level. An increase in sediment accumulation rate at about 35–25 cm in core 74 D (Pennington, pers. comm.) coincides with the initial increase in *Asterionella* and the reappearance of *Cyclotella pseudostelligera* Hustedt / *C. glomerata* Bachman (a variable population that included specimens typical of both, according to Lowe & Collins, unpublished key). The steepest increase in *Asterionella formosa* and appearance of *Melosira italica* subsp. *subarctica* coincides with the transition to black mud and the appearance of large numbers of *Cyclotella pseudostelligera* and *C. comensis*. This was followed by an increase in various chrysophyte scales, *Asterionella, Tabellaria flocculosa* var. *asterionelloides* (Grun.) Knudson and *Melosira*, and the decline of the *Cyclotella* spp.

Timescale

Dating these diatom and chrysophyte changes, so that they could be related to historical records, has not proved easy (Fig. 1). Although ^{137}Cs analysis and diatom indicators agreed closely over the 1963 horizon the ^{210}Pb dates for core WD78-1 appeared increasingly younger below 1962 and originally ascribed the 1950 *Cyclotella praetermissa* decline to 1956. The recorded disappearance of *Mallomonas reginae* in 1949 (Lund unpublished) confirmed the discrepancy as, on the ^{210}Pb timescale, no scales of this taxon were deposited in the sediments until after its disappearance. Recalculation of ^{210}Pb results, by Dr P. G. Appleby, has made post-1950 dates agree with other indicators but earlier diatom horizons now appear significantly older in core 78-1 than in 74-D (Haworth, in press).

References

Evans, G. H., 1969. Pollen and diatom analyses of late-quaternary deposits in the Blelham basin, north Lancashire. New Phytol. 69: 821–874.

Haworth, E. Y., 1980. Comparison of continuous phytoplankton records with diatom stratigraphy in the recent sediments of Blelham Tarn. Limnol. Oceanogr. 25: 1093–1103.

Haworth, E. Y., 1983. Stratigraphic changes in algal remains (diatoms and chrysophytes) in the recent sediments of Blelham Tarn, English Lake District. In: Lake Sediments and Environmental History, Haworth, E. Y. & Lund, J. W. G. (eds). Leicester University Press, in press.

Lund, J. W. G., 1979. Changes in the phytoplankton of an English lake, 1947–1977. Hydrobiologia J. 14: 6–21.

Pennington, W., Cambray, R. S., Eakins, J. D. & Harkness, D. D., 1976. Radionuclide dating of the recent sediments of Blelham Tarn. Freshwater Biol. 6: 317–331.

Pennington, W., Cranwell, P. A., Haworth, E. Y., Bonny, A. P. & Lishman, J. P., 1977. Interpreting the environmental record in the sediments of Blelham Tarn. F.B.A. Ann. Rep. 45: 37–47.

Changes of the benthic fauna of the profundal zone of Traunsee (Austria) due to salt mining activities

H. Löffler
Limnological Department, University of Vienna, Berggasse 18, A-1090 Vienna, Austria

Keywords: paleolimnology, Traunsee, artificial meromixis, ostracods

Abstract

Traunsee is the deepest (189 m) lake in Austria and has been holomictic until the beginning of industrial discharges from the alkali works in Ebensee. The input of soluble compounds such as $CaCl_2$ and NaCl have produced meromictic conditions. The relatively high oxygen content of the monimolimnion is due to water movements promoted by the large inflow and wind-action.

The industrial discharges have caused both meromictic conditions and the deposition of highly alkaline sediment in the southern portion of the profundal zone. The effect of the discharges on the benthic fauna was evaluated by comparing the subfossil benthic fauna of four cores taken inside and outside the affected area.

The ostracod fauna resembles that of other large lakes in the Salzkammergut, but also includes groundwater species such as *Kovalevskiella* and *Cypridopsis subterranea. Leucocythere mirabilis* has been found for the first time in an Austrian lake.

Ostracods and other benthic organisms such as Testacea were not apparently influenced by the meromictic condition. They are, however, missing in all areas covered by the alkaline sludge.

Introduction

The majority of meromictic lakes in Austria became meromictic at the same time as the climate change towards the end of the Pleistocene and the onset of the Holocene (Löffler 1975 & 1978). A few of the small meromictic lakes returned to holomictic conditions when the monimolimnion filled with sediment (Löffler 1977 & 1979). Beside these naturally meromictic lakes, artificial meromixis has been caused by salt mining and the alkali works in the Saltzkammergut (Upper Austria). The most remarkable example of such an artificially induced meromictic condition is Traunsee, the deepest lake and the lake with the largest catchment area in Austria (Table 1).

In contrast to naturally meromictic lakes, which lack any important inflow and therefore are inclined to form a chemical gradient, Traunsee is supplied by a large river, the Traun, and therefore has a rather short retention time. One of the important consequences of this short retention time and the dynamic power of the large river discharge is the permanent supply to the monimolimnion with oxygen, which in naturally meromictic lakes is absent or at least considerably decreased. According to observations available, values of oxygen saturation in the hypolimnion of Traunsee never drop below 60%.

Table 1. Background data, Traunsee.

Altitude (m)	422
Catchment area (km²)	1417
Area (km²)	24.4
Volume (10^6 m³)	2200
Max. depth (m)	189
Mean depth (m)	91
Retention time (months)	9

Hydrobiologia 103, 135-139 (1983).

Since the turn of the century the alkali works at Ebensee, founded in 1883, release daily up to 30 tons of largely insoluble substances (mainly $CaCO_3$, $MgCO_3$, $CaSO_4$) and about 200 tons of soluble compounds $CaCl_2$ and NaCl). The latter have resulted in a permanent chemical gradient and thus in meromictic conditions. Values of chloride content above the chemocline vary between 65 and 127 mg l^{-1} (1956–1975), those below the thermocline vary between 94 and 163 mg l^{-1} (Dobesh & Neuwirth 1980).

There is no indication that the chlorinity has influenced the species composition. Ruttner (1937) stressed the similarity of the Traunsee phytoplankton with that of the uninfluenced Wolfgangsee and it is clear that no major change has been observed since then.

In contrast to the planktonic and upper benthic communities, the profundal zone of Traunsee is strongly influenced by the insoluble substances which have been accumulating for more than seven decades. Therefore it seemed desirable to explore whether any change in the profundal fauna, especially ostracods, had occurred in connection with this industrial impact.

Material and methods

During 1980 28 cores of 4 cm diameter were taken with a heavyweight corer (Meischner & Rumohr) and analysed with respect to pH and Eh (Schneider 1980). It was found that the sediment originating from the alkali works has a range of pH between 10.2 and 11.8 and a range of Eh between −100 and −400 mV. Even mixtures of natural sediment and mud from the alkali works were easily recognized by high pH-values and a strong decrease of redox-value. In natural sediment, pH values between 6 and 8, and Eh values mainly above zero were observed. Because of the striking differences between the natural sediment and the sludge from the alkali works it was possible to map the distribution of industrial sediment in Traunsee (Fig. 1). The map needs improvement, but the general features of the distribution have been identified. Close to the deepest part of Traunsee the sediment from the alkali works exceeds 1 m in depth.

10 additional cores were obtained with a 6 cm diameter piston-corer (EAWAG, Switzerland). Figure 1 shows the location of the sampling sites for benthic organisms obtained with both corers. In order to remove the fine turbid material which impedes the inspection of the samples, 5 cm³ from selected depths of the cores were carefully washed in a 50 μm plankton filter.

Results

Diatoms were not well represented in the samples. The most abundant species was *Campylodiscus sp.*, but since diatoms at these depths are allochthonous they were not studied in detail. Moreover, it is most likely that diatoms as well as Testacea with siliceous cases are dissolved in the

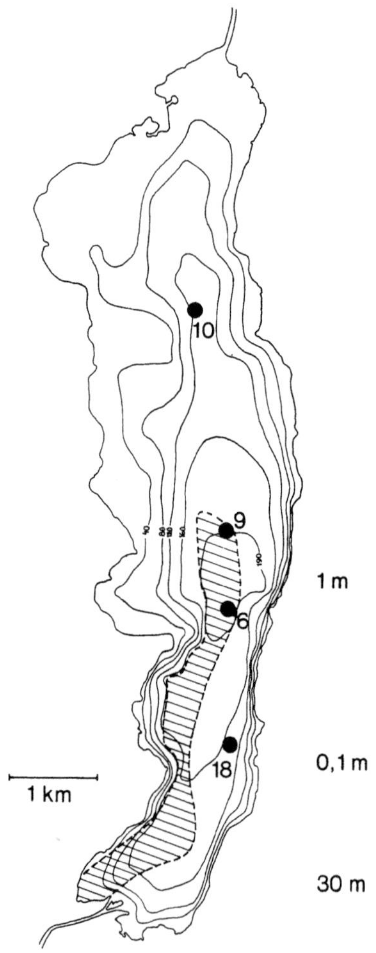

Fig. 1. Sampling sites. Cores (4 cm ∅) 80/10, 141 m, and 80/18, 162 m; EAWAG-cores (6 cm ∅) 80/9, 190 m, and 80/6, 190 m.

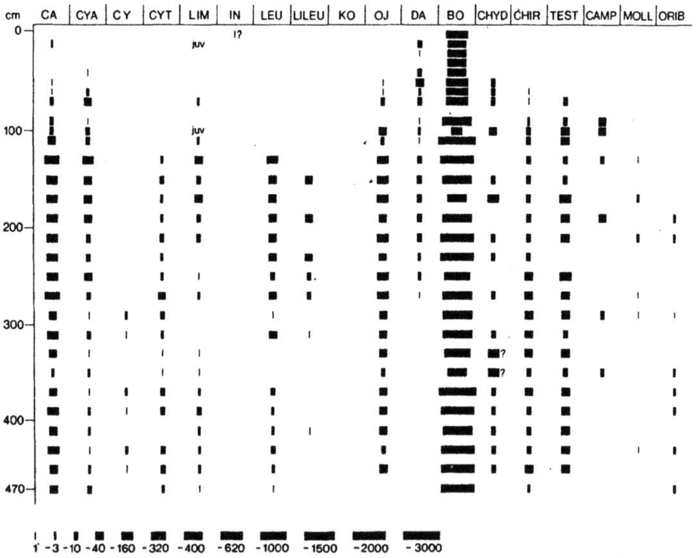

Fig. 2. Core 80/6. Ca: *Candona,* mainly *neglecta;* Cya: *Cypria ophthalmica;* Cy: *Cyclocypris ssp.;* Cyt: *Cytherissa lacustris;* Lim: *Limnocythere sanctipatricii* In: *Limnocythere inopinata;* Leu: *Leucocythere mirabilis;* Lileu: *Limnocythere* and *Leucocythere* juv.; Ko: *Kovalevskiella* sp.; Oj: ostracods juv.; Da: ephippia *Daphnia;* Bo: *Bosmina sp.;* Chyd: chydorids; Chir: chironomids; Test: Testacea; Cam: *Campylodiscus sp.;* Moll: molluscs; Orib: *Oribatei.* Individuals or shells/5 cm³.

highly alkaline sediment from the soda works. The latter were well represented in core 80/6 and resemble the distribution of ostracods. They deserve more attention.

Because of the small size of the samples, molluscs *(Pisidium sp.* and gastropods) were found only occasionally and therefore disregarded. Chironomids (head capsules) were well represented only in cores 80/6 and 80/18. Their distribution follows the pattern of ostracods. Like the Testacea they deserve more attention in future. To a large extent chydorids and Oribatei are allochthonous at the depths surveyed. They will therefore not reflect the change in the profundal zone due to the alkali works.

In this paper most attention is given to the ostracods which live even at the maximum depth of the large Alpine lakes. So far, no detailed information exists on the contemporary ostracod fauna of Traunsee. However, deep holomitic lakes in the Eastern Alps have a very similar ostracod fauna. From lakes close to Traunsee such as Attersee, Mondsee and Wolfgangsee (Graf 1938; Löffler 1972) species such as *Candona candida* (O.F.M.), *Candona neglecta* Sars, *Cypria ophthalmica* (Jur.), *Cyclocypris laevis* (O.F.M.), *Limnocythere sanctipatricii* Br. & Rob. and *Cytherissa lacustris* Sars have been reported and most of these were also found in Traunsee sediments (Figs. 2 & 3). Only *Candona candida,* obviously restricted to the littoral (Graf 1938) was missing or very infrequent in the cores. Particularly interesting is the discovery of *Leucocythere mirabilis* Kaufm. in all cores except 80/18. This species has been reported from deep Swiss lakes, northern Europe (including parts of the Baltic Sea) and Corfu (Stephanides 1948). The examination of a core from very deep water in Attersee has shown that this species also occurs there and has probably been overlooked or confused with *Limnocythere.* More puzzling is the presence of *Kovalevskiella sp.* in core 80/18 only. This genus is known from groundwater (Balkan, Austria: Danube region, Turkey) and its presence in Traunsee must be related to karstic phenomena such as sublacustrine springs which are abundant in the region. This is supported by the occurrence of *Cypridopsis subterranea* Wolf, another groundwater species in the same core. Finally one specimen of *Psychodromus fontinalis* (Wolf) from core 80/10 should be mentioned.

The vertical distribution of ostracods in the core far outside the zone of industrial influence (80/10) shows clearly (Fig. 3) that no major change can be observed, and it is most likely that living speciemens of all species with subfossil records occur in this area of the profundal. In core 80/18 (with *Kovalevskiella)* at least the upper 30 cm show a

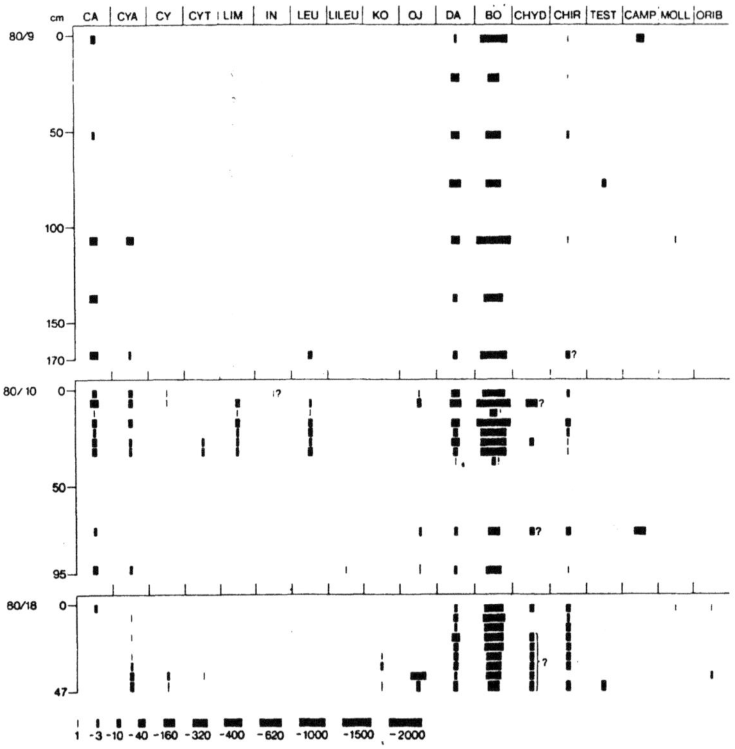

Fig. 3. Cores 80/9, 80/10, 80/18. Abbreviations as for Fig. 2.

decrease of ostracod shells and some species typical of the benthic zone, such as *Limnocythere* and *Leucocythere,* are absent. However, since only 47 cm were inspected, the full extent of any influence cannot be estimated. By far the best evidence for the industrial impact is indicated by core 80/6 (Fig. 2). Here the upper 120 cm lack *Cytherissa* and *Leucocythere,* and *Candona* and *Cypria* disappear from about 60–70 cm onward (the occurrence of single shells towards the surface may be explained by drifting). So far only *Limnocythere inopinata* is known to tolerate strongly alkaline conditions. It should be of interest to learn whether this species has replaced *L. sanctipatricii* there. However, the single shell found does not support such a possibility. In core 80/6 the chironomids and Testacea show a similar pattern of vertical distribution, but for a better understanding more systematic information is necessary.

Core 80/9, which was taken at the northern fringe of the influenced area, leaves one with the impression of a much more dramatic impact, but since the number of shells recovered was rather low no interpretation is possible. The finding of *Leucocythere* at 170 cm depth suggests that much longer cores from this location will be needed.

Discussion

Because of the absence or very low concentration of oxygen in the monimolimnia, metazoan fauna are usually missing in meromictic lakes. In Traunsee, however, because of the large influent, the supply of oxygen even in the deepest zones is sufficient to ensure the continued existence of the benthic fauna at least until the beginning of industrial discharges. From subfossil remains there is no indication of any major change occurring in cores not influenced by the sludge from the alkali works but well within the monimolimnion. Pollen analyses of the sediment at the maximum depth (4.7 m) of core 80/6 suggest an age of approximately 2 000–4 500 B.P. which allows sufficient time to reach back to the holomictic period.

The sediment from the alkali works which has spread over a large area within the southern portion of the profundal zone has apparently resulted in the extinction of the benthic fauna. It is most likely that the strong alkaline character of the sediment is the limiting factor.

In cores taken in the area influenced the limnocytherids generally disappear before *Candona* and *Cypria*. Similar observations have been made at the onset of eutrophication in Alpine lakes, when *Cytherissa* and *Limnocythere* always start to diminish first. Apparently *Leucocythere* behaves like other limnocytherids.

With core sampling only, living specimens of the benthic fauna have not yet been collected. Therefore further studies should concentrate on material from the surface sediment.

Acknowledgements

The author is very much indebted to Dr. J. Schneider, Göttingen, for the core samples and the information presented in Fig. 1 and to Dr. R. Schmidt for preparing the samples.

References

Colin, J. P. & Danielopol, D. L., 1980. Sur la morphologie, la systematique, la biogeographie et l'evolution des ostracodes Timiriasevinae (Limnocytheridae). Paleobiol. Continent. 11, 1: 1–52.

Dobesch, H. & Neuwirth, F., 1980. Die Stabilitätsverhältnisse des Traunsees. Limnologische Untersuchung Traunsee – Traun. Oberrösterr. Landesreg. 1: 1–59.

Findenegg, I., 1959. Die Gewässer Osterreichs. Ein limnologischer Überblick. 14. Intern. Limnologenkongress 1959, Biologische Station Lunz, 1–68.

Graf. H., 1938. Beitrag zur Kenntnis der Muschelkrebse des Ostalpengebietes. Arch. Hydrobiol. 33: 401–502.

Löffler, H., 1963. Die Ostracodenfauna Österreichs. Sitz. Ber. Akad. Wiss., math.-nat. Kl. I, 172: 193–211.

Löffler, H., 1972. The distribution of subfossil ostracods and diatoms in pre-alpine lakes. Verh. Internat. Verein. Limnol. 18: 1039–1050.

Löffler, H., 1975. The Onset of Meromictic Conditions in Alpine Lakes. Quaternary Studies, Roy. Soc. New Zealand, 211–214.

Löffler, H., 1977. 'Fossil' meromixis in Kleinsee (Carinthia) indicated by ostracodes. In: H. Löffler & D. Danielopol (Eds.) Aspects of Ecology and zoogeography of recent and fossil Ostracoda, pp. 312–325, Junk, The Hague.

Löffler, H., 1978. The Paleolimnology of some Carinthian lakes with special reference to Wörthersee. Pol. Arch. Hydrobiol. 25: 227–232.

Löffler, H., 1979. Die Entwicklung des Jeserzer Sees (Saisser See). Carinthia II, 169./89: 367–370.

Ruttner, F., 1937. Limnologische Studien an einigen Seen der Ostalpen. (Seen des Saltzkammergutes, des Ötscher- und Hochschwabgebietes). Arch. Hydrobiol. 32: 167–319.

Schneider, J., 1980. Wasserwirtschaftliche Planung Traunsee. Bericht über die bisherigen Arbeiten.

Stephanides, Th., 1948. A survey of the freshwater biology of Corfu and certain other regions of Greece. Praktika Helenic Hydrobiol. Inst. 2: 1–263.

Cladoceran remains as evidence of change in trophic state in three Swiss lakes

Mary M. Boucherle[1] & Hans Züllig[2]
[1] *Department of Biology, Indiana University, Bloomington, IN 47405, U.S.A.*
[2] *CH-9424, Rheineck, Switzerland*

Keywords: paleolimnology, lakes, sediments, eutrophication, Cladocera, Switzerland

Abstract

Species composition and density of cladoceran populations changed in Lakes Zürich, St. Mortiz and Baldegg as human populations increased in these watersheds. Lake Zürich sediments became annually laminated in the 1890's as a result of increased organic input as the size of the cities surrounding the lake grew. At the same time, the *Bosmina* species changed from a oligotrophic form (*longispina*) to a eutrophic form (*longirostris*). An increase in *Daphnia* spp. populations also occurred at this time in the lake's history. *Bosmina longispina* reappeared in the lake in 1965 as the lake's trophic status changed from eutrophic to mesotrophic due to effective sewage treatment facilities. Annual laminations appear in the Lake St. Moritz sediments about 1910. Shortly thereafter, a shift from *B. longispina* to *B. longirostris* occurred. This change in trophic status is associated with increased tourism in the area. Lake Baldegg sediments also show annual laminations beginning in 1885 and a similar shift in the *Bosmina* species. Other cladoceran remains were too scarce to be useful in interpreting the histories of these lakes.

Introduction

Historical information on trophic status provides useful data for management and restoration of lakes. Lake histories can provide information on the trophic status of a lake prior to human influence and on the change in lake conditions as it is influenced by human habitation in its watershed. Insight into the events in a lake's past are normally gained by paleolimnological methods, because there are few lakes with long-term records of limnological data.

Many plants and animals leave biochemical or morphological remains in lacustrine sediments. Cladocera are particularly suitable for paleolimnological studies, because their remains are the most diverse and abundant of the animal remains in sediments (Frey 1979). Their remains are morphologically distinct and can be identified easily to genus and often to species. The community composition and species abundance of Cladocera reflect the trophic status of the lake. Thus, analysis of cladoceran remains is an excellent paleolimnological tool that is useful in reconstructing lake histories (Frey 1961, 1976, 1979).

Sediment cores from lakes Zürich, St. Mortitz, and Baldegg were examined for Cladoceran remains. Each of these lakes contains sediments that became annually laminated in response to increased organic inputs from increased human populations in their watersheds (Züllig, 1982), Nipkow (1920) has demonstrated that laminations in Lake Zürich are truly annual because of the seasonal changes in diatoms preserved in them. Annually laminated sediments facilitate our ability to detect changes in the cladoceran fauna over short periods of time, even from year to year. Hence, wherever possible in this study, each pair of laminae was analyzed separately so that yearly fluctuations in population size and species composition could be assessed.

Hydrobiologia 103, 141–146 (1983).

Members of the families Bosminidae and Daphniidae were found to be abundant in the sediments of these three lakes. In each lake the effect of progressive eutrophication caused by cultural influences was reflected in the shift of *Bosmina* from an oligotrophic species to a eutrophic species. Each lake also showed an increasing abundance of *Daphnia* spp. as the lakes became more eutrophic. The partial recovery of Lake Zürich resulting from the installation of waste water treatment facilities was reflected in the cladoceran stratigraphy. Similarly, in Lake St. Mortitz the abundance of Cladocra decreased, beginning at the time sewage was diverted from the lake.

Methods

Sediment cores were raised from the deepest area of the three lakes by the method described by Züllig (1981). Information on the location of the lakes and some of their physical parameters is given in Table 1. Züllig (1981) described how the cores were raised and how they were separated for analysis into small groups of varves or individual varves. Sediment samples were dried at 80 °C for 2 h and cooled in a desiccator before weighing. Each sediment sample was heated in 10% KOH for 15 min to deflocculate the organic material, rinsed with distilled water, sieved through a 37 μm mesh net and stored in 95% ethanol.

A known volume (0.05 ml coverslip^{-1}) of sediment was mounted in Karo mounting medium (Taft 1978) and examined for cladoceran remains at 120×. Cladoceran parts such as headshields, carapaces, postabdomens, claws, and ephippia are found in sediments rather than intact animals. Cladocera can easily be identified to genus on the morphological differences among the parts and in the Chydoridae usually to species. The counting procedure consisted of counting every identifiable cladoceran fragment, but when counts were tallied only the most abundant fragment of a given species was used to avoid overestimating the population size (Frey 1979). Counts of remains belonging to the Bosminidae and Daphniidae were converted to the number of animals cm^{-2} a^{-1}. Because these lakes have small littoral zones, remains of other Cladocera, including the Chydoridae, were too scarce to be useful in interpreting the histories of these lakes.

Table 1. Location and some physical parameters of Lakes Zürich, St. Moritz and Baldegg.

Lake	Zürich	St. Moritz	Baldegg
Longitude	8°45′	9°45′	8°15′
Latitude	47°15′	46°30′	47°15′
Elevation (m)	409	1771	466
Surface area (km^2)	680	0.8	5.2
Maximum depth (m)	143	44	67
Length (km)	29.0	1.5	4.8

Results

Bosmina longirostris, B. longispina, and the *Daphnia longispina* group were found in each of the three lakes. The presence of *B. longispina* is indicative of oligotrophic conditions in lakes, while *B. longirostris* is indicative of eutrophic conditions (Frey 1955; Kiefer 1960; Patalas & Patalas 1966). In each of the lakes *B. longispina* was present prior to cultural influences on the lake (Figs. 1, 2, 3). It reappeared in Lake Zürich after water conditions had been improved by installation of effective sewage treatment facilities. *B. longirostris* appeared in each lake after its sediment became laminated, indicating that the lake has undergone cultural eutrophication. Members of the *D. longispina* group became more abundant in all the lakes after cultural influences began and decreased when measures were taken to reduce organic inputs.

Discussion

History of Lake Zürich

Lake Zürich has been transformed from an oligotrophic to a eutrophic lake in the last 85 years (Nipkow 1920; Thomas & Märki 1949). This change in trophic conditions has been attributed to a watershed population that has expanded 240% since 1850 and presently stands at 300 000. Lake Zürich is used as a source of drinking water for the surrounding communities. The first sign of change in water quality was the adoption in 1871 of slow sand filters to process drinking water.

Lake Zürich sediments became annually laminated in 1895 (Nipkow 1920). In 1896, *Tabellaria*

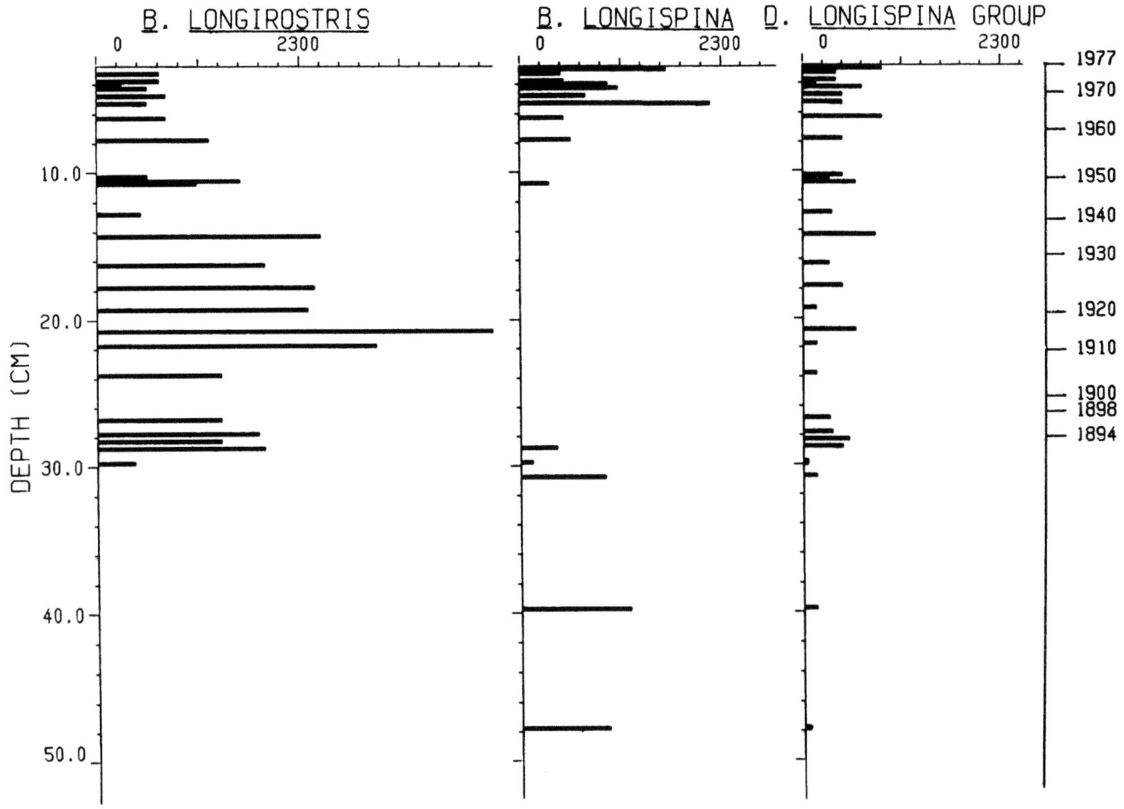

Fig. 1. Number of Cladocera remains ($cm^{-2} a^{-1}$) from Lake Zürich. The year of deposition is on the right.

fenestrata was very abundant in the lake and by 1899 it was displaced by *Oscillatoria rubescens*. The present study revealed that *B. longispina* was abundant in the lake until 1876, then declined rapidly to the point where its remains could not be found in the sediment deposited in 1894. Züllig (1981) found a large increase in algal pigments, especially oscillaxanthin, myxoxanthophyll, and raw carotenoids, indicating increased algae populations at this time. *B. longirostris* first appeared in the sediment of 1876, and by 1898 had become the only *Bosmina* in the sediment. The *Daphnia longispina* group was present in all the sediment samples, but it did not become abundant until 1899. The shift in *Bosmina* species and the increase in *Daphnia* accompany eutrophication of the lake and hence can be used to infer eutrophication.

Treatment of sewage entering Lake Zürich began in 1913 but the human population increased faster than the capacity of the treatment facilities. Lake conditions appear to have improved in the early years of World War II. A sediment sample includ ing varves for 1939–1941 contained *B. longispina* and showed a decrease in *Daphnia* abundance. By 1944, lake conditions had returned to those prior to 1939.

In the mid-1950's pipelines built to bypass some sewage to the Limmat River at the outflow of Lake Zürich (Dietlicher 1974) brought about some improvement as evidenced by the return of *B. longispina*. Removal of phosphate from sewage began in the early 1960's, and by 1967 there was an increase in hypolimnic oxygen (Züllig 1981). Concurrently, *B. longispina* increased in abundance and by 1977 no remains of *B. longirostris* were found in the sediment indicating that oligotrophic conditions had returned.

History of Lake St. Moritz

Changes in the trophic status of Lake St. Moritz have been reported by Borner (1922), Schmassman

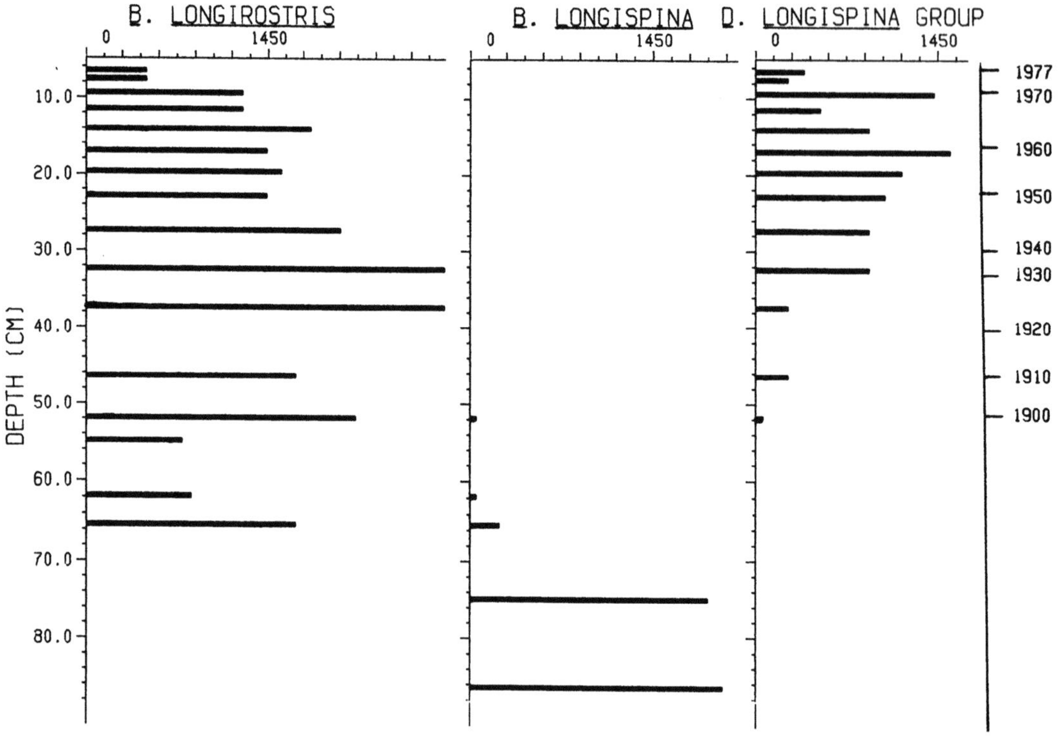

Fig. 2. Number of Cladocera remains (cm^{-2} a^{-1}) from Lake St. Moritz.

(1920), Schmassman and Schmassman (1948) and Messikomer (1965). St. Moritz was regarded as a resort as early as the 16th Century and was noted for its mineral baths. In 1850 the resident population of St. Moritz was 500, but by 1910 when the first hotel was built the population had increased to 2 500. The sedimentary record shows that *B. longispina* was the dominant zooplankter prior to 1875. *B. longirostris* first appeared in the sediments about 1880 and immediately became dominant, but *B. longispina* did not disappear entirely from the sedimentary record until 1920. At that time the *Daphnia* species became more abundant. Züllig (1982) has found a 10–20 fold increase in the abundance of diatoms and peridinians between 1920 and 1930.

The shift in *Bosmina* species and the large increase in algae are indicative of eutrophication. This change in the lake's condition can be attributed to increasing organic inputs including kitchen wastes received through a canal system. By 1970 the resident population of St. Moritz had stabilized at 5 000 but the overnight guest population fluctuated between a low of 120 000 during World War II to a high of 1 100 000 presently. Stepwise reduction in sewage flow to the lake began in 1965. In that year sewage was diverted from the lake, and by 1967 kitchen wastes were diverted, but the canals continued to carry wastes from dairy farms into the lake until 1973. *Daphnia* and *B. longirostris* populations continued to increase until 1975, two years after all the wastes had been diverted from the lake, and then decreased. Lake conditions appear to be improving although fall diatom blooms are still an annual occurrence.

History of Lake Baldegg

Prior to 1885 Lake Baldegg sediments were not laminated, and the algal pigments (Züllig, 1982) and the Cladocera records imply oligotrophic conditions. In 1885, when the lake sediments first became laminated, *B. longispina* decreased in abundance and *Daphnia* increased.

In Lake Baldegg these changes in the biota occurred more slowly than in the other two lakes because human habitation in its watershed increased more slowly. The shift in *Bosmina* species

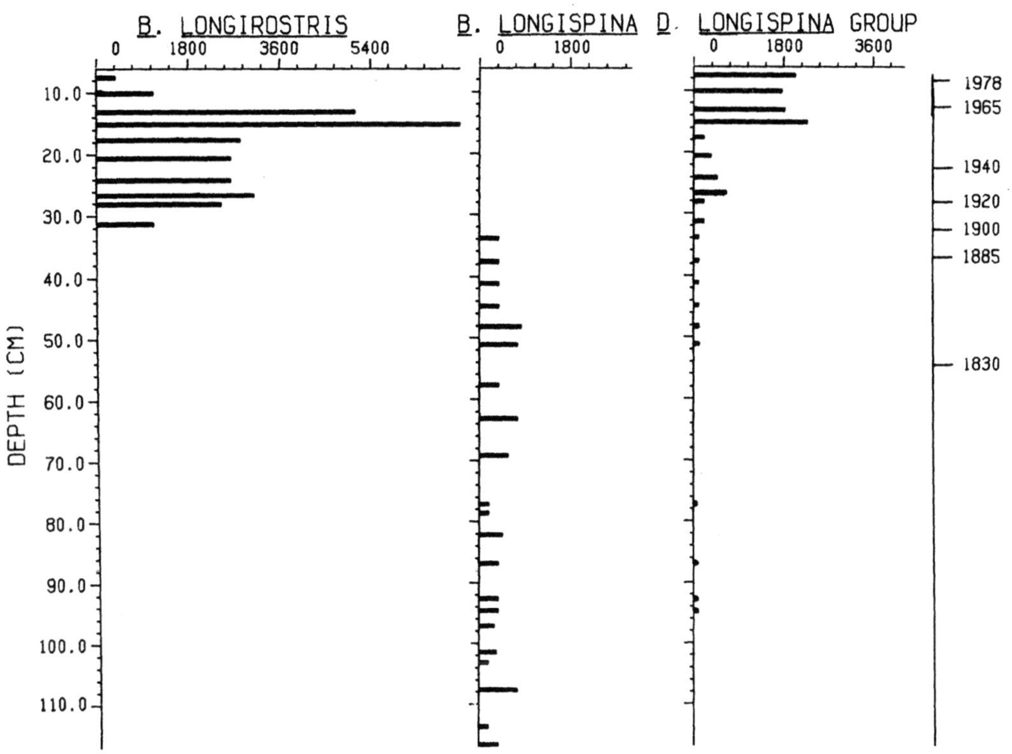

Fig. 3. Number of Cladocera remains ($cm^{-2}\ a^{-1}$) from Lake Baldegg.

and the increase in *Daphnia* abundance did not occur until 1920. An increase in the number of pennate diatoms and pigments of other algae in 1920 indicated eutrophication (Züllig, 1982). In 1970 efforst were initiated to divert sewage from the lake. The decrease in abundance of cladocera and algae in 1976 suggests that the lake is becoming less eutrophic as a result.

Summary

Cladocera are sensitive to changes in the trophic status of lakes and can be used in paleolimnological analysis to deduce past lake conditions. Each of the three lakes studied showed an oligotrophic phase followed by eutrophication as human influence increased. Similarly, each lake showed a return to mesotrophic or oligotrophic conditions after sewage had been diverted away from it. These changes in trophic state are reflected in a shift of *Bosmina* species as well as a change in cladoceran abundance.

Acknowledgements

We thank Dr. David G. Frey for his encouragement, help with identification of Cladocera, and criticism of the manuscript. We also appreciate the help of Dr. Brenda J. Hann with Cladocera identification and Dr. Dorothy B. Berner for comments on the manuscript.

References

Borner, L., 1972. Die Bodenfauna des St. Moritzer-See. Arch. Hydrobiol. 13: 1–91.

Dietlicher, K., 1974. The water quality of the lakes of Zürich and Walensee – an essay of description. Zürich Waterworks, Zürich. 40 pp.

Frey, D. G., 1955. Längsee: a history of meromixis. Mem. Ist. ital. Idrobiol. Suppl. 8: 141–164.

Frey, D. G., 1961. Developmental history of Schleinsee. Verh. int. Ver. Limnol. 10: 476–488.

Frey, D. G., 1976. Interpretation of Quaternary paleoecology from Cladocera and midges, and prognosis regarding usability of other organisms. J. Zool. 54: 2208–2226.

Frey, D. G., 1979. Cladocera analysis. In: Berglund, B. E. (Ed.). Project Guide to Paleohydrological Changes in the Temperate Zone in the last 15 000 years, Subproject B. Lake and Mire environments. IGCP Project 158 vol. II chapt. 10.10. Univ. Lund, Sweden: 227–257.

Kiefer, F., 1960. Der Bodensee in Gefahr! Neuere Ergebnisse aus Untersuchungen im Pelagial des Bodensees. Umschau 60: 65–68.

Messikommer, E., 1965. Materialien zur Algenkunde des Oberengadins Schweiz. Z. Hydrol. 27: 115–162.

Nipkow, F., 1920. Vorläufige Mitteilungen über Untersuchungen des Schlammabsatzes im Zürichsee. Schweiz. Z. Hydrol. 1: 100–122.

Patalas, J. & Patalas, K., 1966. The crustacean plankton communities in Polish lakes. Verh. int. Ver. Limnol. 16: 204–215.

Schmassman, W., 1920. Die Bodenfauna hochalpiner Seen. Arch. Hydrobiol. Suppl. 3: 1–105.

Schmassman, W. & Schmassman, H. J., 1948. Chemische Untersuchungen im St. Moritzersee. Schweiz. Z. Hydrol. 10: 21–35.

Taft, C. E., 1978. A mounting medium for fresh-water plankton. Trans. am. micros. Soc. 97: 263.

Thomas, E. A. & Märki, E., 1949. Der heutige Zustand des Zürichersees. Verh. int. Ver. Limnol. 10: 476–488.

Züllig, H., 1981. On the use of carotinoid stratigraphy in lake sediments for detecting past developments of phytoplankton. Limnol. Oceanogr. 26: 970–976.

Züllig, H., 1982. Stratigraphische Carotinoidbestimmungen an Bohrkernen aus zehn Schweizerseen zur Erkundung früherer Planktonentfaltungen. Schweiz. Z. Hydrol. 44: 1–98.

Changes in diatom flora deposited annually during the freshening of the impounded, sea-bay of Gennarbyviken in southern Finland

Matti Räsänen[1] & Kimmo Tolonen[2]
[1]*Department of Quaternary Geology, University of Turku, 20500 Turku 50, Finland*
[2]*Department of Botany, University of Helsinki, Unioninkatu 44, 00170 Helsinki 17, Finland*

Keywords: paleolimnology, diatoms, salinity ecology, annual lamination, sulphide mud, meromixis, Baltic

Abstract

Diatoms have been studied in an annually laminated sulphide mud which was deposited after the impoundment of the sea-bay of Gennarbyviken. It was possible to follow how the measured decrease in salinity in the basin was reflected by the diatom flora of the sediment which had been deposited concurrently.

Introduction

One way of investigating the salinity ecology of diatoms is to study changes in basins artificially isolated from the sea, cf. Kolbe (1927), Johansen *et al.* (1933–36) and Beaufort (1954). We followed the changes in diatom communities during the freshening of the dammed sea-bay of Gennarbyviken using annually laminated sediments. Of the authors K. T. initiated the study and was responsible for the field work. Both authors contributed to the writing, but all other parts of the study were carried out by M. R.

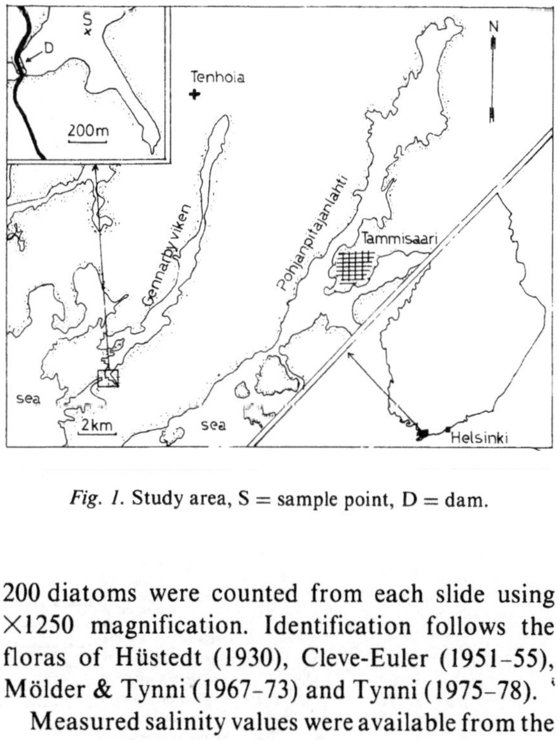

Fig. 1. Study area, S = sample point, D = dam.

Site and methods

Gennarbyviken is a 17 km-long sea-bay on the northern coast of the Gulf of Finland, situated mainly in the commune of Tenhola. It was dammed in 1957–58 to form a fresh-water supply. For this study a sediment sample was taken from a 32 m-deep depression (Fig. 1).

The sediment sample was taken in March 1979. A box freezer was used (Huttunen & Meriläinen 1978). The frozen sample was preserved in a deep-freeze until slides were made.

Qualitative diatom analyses were carried out from sediments treated with 10% H_2O_2. At least 200 diatoms were counted from each slide using ×1250 magnification. Identification follows the floras of Hüstedt (1930), Cleve-Euler (1951–55), Mölder & Tynni (1967–73) and Tynni (1975–78).

Measured salinity values were available from the same site from which the sediment sample was taken (Purasjoki, unpubl.). Salinity was measured at 2.5 m intervals from surface to bottom several times a year until the water was regarded as fresh ($AgNO_3$ titration and conversion to ‰ values by the formu-

Hydrobiologia 103, 147–152 (1983).

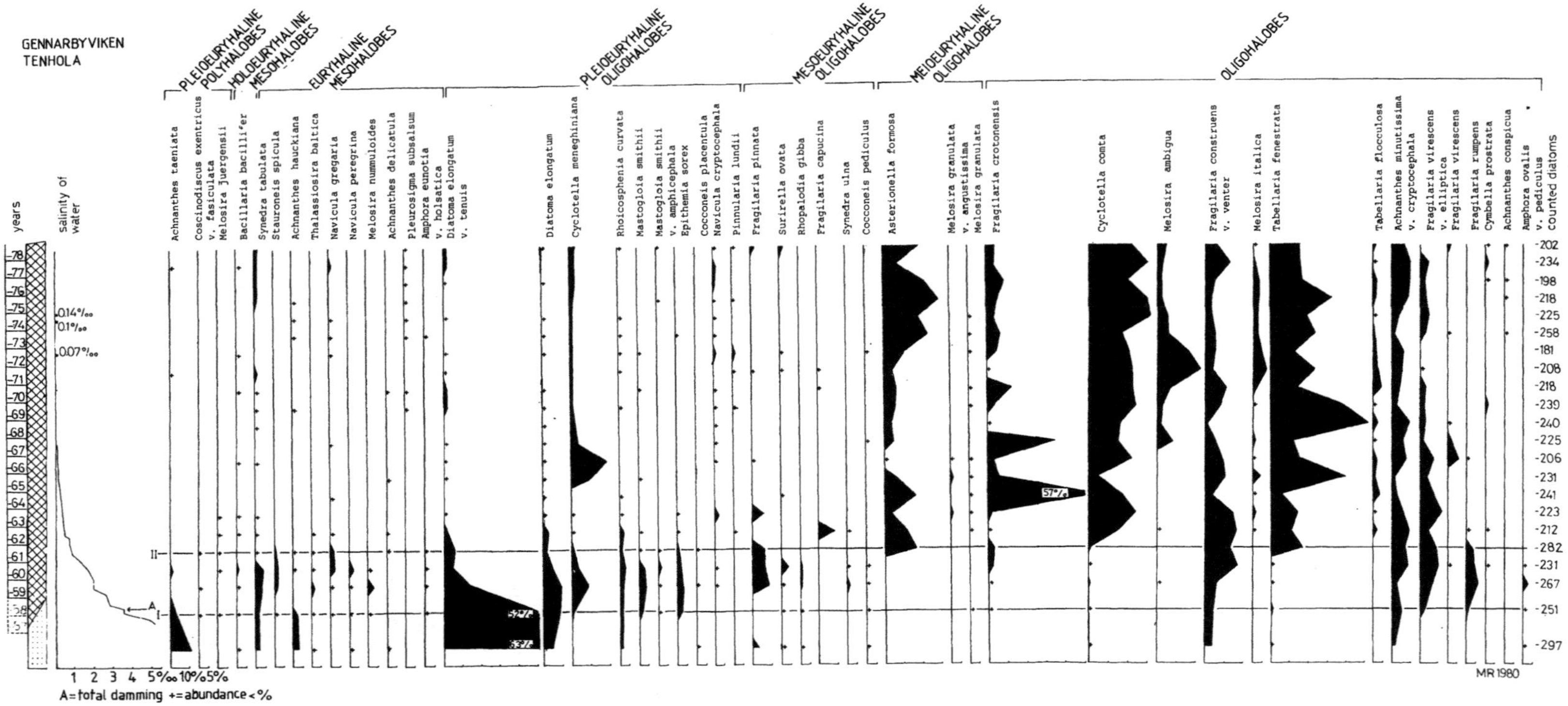

Fig. 2. Diatom diagram from Gennarbyviken. The most rapid change in salinity occurs between lines I and II.

la of Zarins & Ozolin (1935). In the diagram (Fig. 2) the average value of the salinity above the halocline was used to represent the decreasing salinity of the water.

Description of sediments and origin of varves

The sediment of the post impoundment stage is sulphide mud in which a strong laminated structure can be seen (Fig. 3). The mud is underlain by coarse sand and gravel with interbedded finer sand and clay. In the sand there are several shells of edible mussel (*Mytilus edulis*) apparently *in situ*.

Sand and gravel are part of the minerogenic bottom sediment. Some of the finest minerogenic material probably derives from the time when the dam was built, but most of the gravel, sand and clay was deposited before damming.

Finer sediments were not deposited at this time because of water currents. A distinct halocline formed when water began to freshen. The salinity of the surface water decreased strongly during the construction, which lasted from February 1957 to April 1958 (Fig. 2). In October 1957 the halocline was situated between 17.5 and 20 m and the salinity had already decreased by 3‰. Meromictic anoxic conditions may have developed relatively quickly resulting in the precipitation of sulphide mud in the depression.

The 18–20 cm varved sulphide mud was deposited between the time the dam was finished in April 1958 and March 1979 when the sediment sample was taken (21 years). From the sediment we can distinguish 20 visually homologous varves, separated by dark bands (Fig. 2) and representing annual periods of corresponding length.

The formation of varved sediment in meromictic

Fig. 3. The laminated sulphide mud of Gennarbyviken.

Ontario lakes was related to the sedimentation of photosynthetic anaerobic bacteria (Dickman 1979). In this model bacteria increase during summer in the lower part of the mixolimnion. During the late fall anaerobic bacteria die and sink to the bottom causing a dark layer in the sediment. Willén (1962) observed sulphur bacteria during meromictic conditions in a sea-bay, and Lindholm (1976) described a meromictic lake in which photosynthetic purple bacteria and chemosynthetic bacteria occurred on the boundary of the monimo- and mixolimnion. The same factors which Dickman described may be responsible for the origin of the dark layers in the sediment of Gennarbyviken. In the Ontario lakes bacteria accounted for 60% of the annual sedimentation, although the contribution of bacteria in Gennarbyviken is probably less.

Our sediment profile can be precisely dated by the very striking varve of 1973 (Figs. 3 & 4). The winter of 1972–73 was exceptionally mild. The sea was frozen outside Tammisaari (Fig. 1) on 20.1.1973 and melted on 22.3.1973 (Institute of Marine Research, 1978). During the same winter Lindholm (1975a, b) monitored the limnology of meromictic lakes at Ahvenanmaa, and refers to a special 'winter fall'. In spring the plankton flourished under the ice, because nutrients had moved from the monimolimnion during winter and early fall. The white layer in the lower part of the varve formed in 1973 and consists mainly of *Asterionella formosa*, which both Willén (1962) and Simola (1977) have described as abundant during fall.

On this basis we judge the two-layered white band, consisting of diatoms, to have been formed in 1973. Simola (1977) also found a two-layered white band consisting of diatoms in the 1973 varve for Lovojärvi.

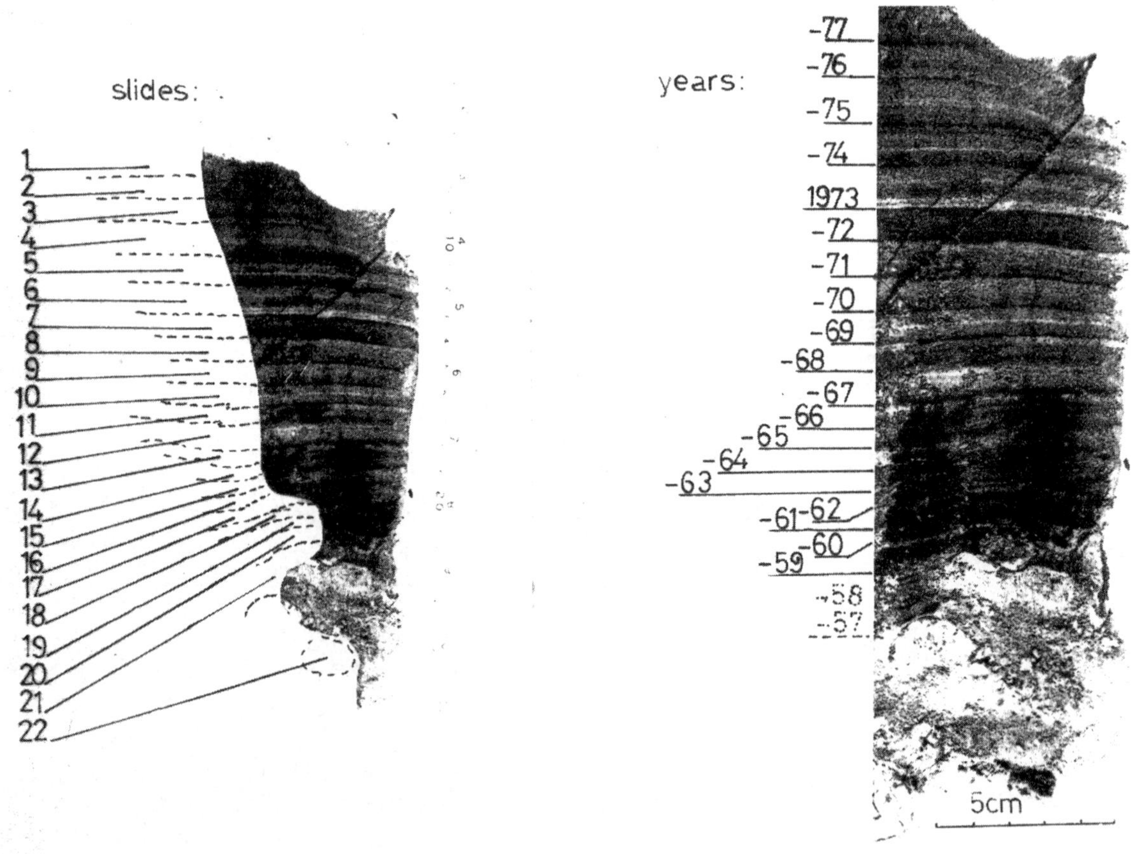

Fig. 4. Places where slides were made and the varves dated.

Results

In the diagram (Fig. 2) changes in the salinity have been compared with diatom changes on a calendar time scale. Because each slide was made from a single varve the diagram shows the relative annual changes among the diatom species. The sediment material of slide no. 21, however, represents the time between 1959 and period before the dam construction and slide no. 22 represents an undetermined period before 1959. The species are grouped according to their tolerance to salinity (Simonsen 1962).

Some of the most abundant planktonic species, such as *Fragilaria crotonensis* and *Tabellaria fenestrata*, vary considerably from year to year.

The diatoms reacted immediately to the decreasing salinity. During 1961 the salinity decreased from 1.4‰ to 0.7‰. During this year the most striking changes in diatom communities took place. Polyhalobous and halophilous species are still relatively abundant but oligohalobous forms are, for the first time, the most abundant group of species. More than ten species have their maximum abundance when the salinity is between 5 and 1‰. These species are mainly mesohalobous and halophilous forms. However, it should be remembered that the lowermost slide represents the sea-bay stage. Sedimentation conditions were different and there may have been differences in preservation compared with conditions afterwards. The abundance of *Asterionella formosa* and *Melosira ambigua* in the period following the dam construction suggests that the basin became eutrophic.

The data allow comments to be made about the ecology of certain species. *Synedra tabulata*, which is commonly regarded as mesohalobous appears relatively abundantly in the freshwater zone. *Pleurosigma subsalsum* appears only in small amounts and only in fresh water. It has also been regarded as a mesohalobe. Tynni (pers. comm.) has found it in eutrophic harbour pools and its appearance here could be explained in the same way. Otherwise the species occur as expected from their known ecology, although *Fragilaria rumpens* and *Amphora ovalis* v. *pediculus* seem to be more halophilous.

Because only one freshening episode is considered the changes which take place during it cannot be used to make generalizations about the ecology of the various taxa concerned. The results, however, can be compared with data from a study nearby which deals with salinity ecology of diatoms. Halme & Mölder (1958) studied recent diatoms from a continuous series of water samples in Pohjanpitäjänlahti (Fig. 1) representing the whole salinity gradient in Pohjanpitäjänlahti, from about 0.0‰ to about 6.5‰ (Fig. 5). Their data are also affected by changes of other ecological factors as well as salinity. This can be seen particularly when the occurrence of oligohalobe species in Pohjanpitäjänlahti and Gennarbyviken, for example *Cyclotella comta*, *Asterionella formosa* and *Fragilaria crotonensis*, are compared. The appearance of oligohalobe species in the freshest part of Pohjanpitäjänlahti has been restricted by other factors than salinity.

Conclusions

At the beginning of the freshening period (6–8 years), decreasing salinity was the most important factor which caused changes in the diatom communities. Diatoms reacted immediately to the decreasing salinity. The largest changes occurred between 1.4 and 0.7‰ salinity. Afterwards oligohalobes became the dominant species group. Certain species are characteristic of salinities between 5 and 1‰. Results were compared with another study

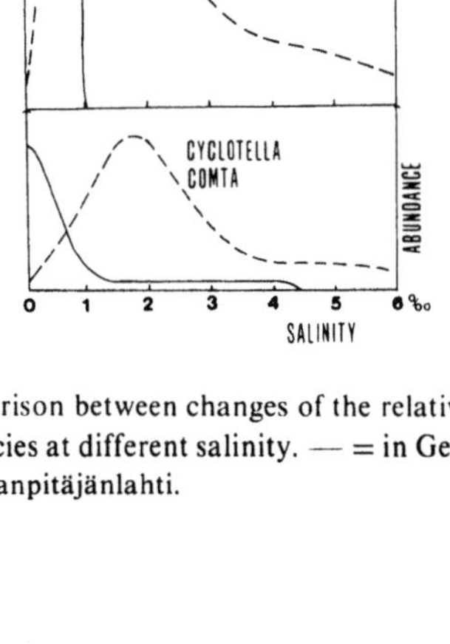

Fig. 5. Comparison between changes of the relative abundance of diatom species at different salinity. — = in Gennarbyviken, --- = in Pohjanpitäjänlahti.

and showed that, when changes are monitored through time, more precise information about the salinity ecology of diatoms can be acquired than when differences in space, where the sources of error are bigger, are studied.

Acknowledgements

We wish to express our thanks to Dr. K. Purasjoki for allowing us to use his unpublished results, to Associate Prof. G. Glückert and Dr. M. Eronen for valuable advice given in the course of the study. We are also grateful to Mrs. K. Porola Phil. Cand. who kindly revised the English manuscript.

References

Beaufort, L. F. De (Ed.), 1954. Veranderingen in de flora en fauna van de Zuiderzee na de afsluiting in 1932. Nederl. Dierk. Ver., Leiden.

Cleve-Euler, A., 1951–1955. Die Diatomeen von Sweden und Finnland. I–V kungl. svenska Vet.akad. Handl. 4. Ser., 2:1, 4:1, 5, 5:4, 3:3.

Dickman, M. D., 1979. A Possible Varving Mechanism for Meromictic Lakes. Quater. 11: 113–124.

Halme, E. & Mölder, K., 1958. Planktologische untersuchungen in der Pojo-Bucht und angrenzenden Gewässern. III Phytoplankton. Ann. bot. Soc. 'Vanamo'. 30: 1–71.

Hüstedt, F., 1930. Die Süsswasser flora Mitteleuropas. Heft 10: 1–466. Jena.

Huttunen, P. & Meriläinen, J., 1978. New freezing device providing large unmixed sediment samples from lakes. Ann. bot. fenn. 15: 128–130.

Institute of Marine Research, 1978. Jäätalvet 1971–1975 Suomen merialueella. Finn. mar. Res. 245: 1–63.

Johansen, A. G., Blegvad, H. & Spärck, R., (Eds.), 1933–1936. Ringköping Fjords Naturhistorie i Brackvattenperioden 1915–1931. Copenhagen.

Kolbe, R. W., 1927. Zur Ökologie, Morphologie und Systematik der Brackwasser-Diatomeen. Die Kiselalgen des Sperenberger Saltsgebiets. Pflanzenforschung 7: 1–149.

Lindholm, T., 1975a. Långsjön och Markusbölefjärden före och efter isoleringen från havet. Husö Biologiska Station, Meddelande 16: 28–49.

Lindholm, T., 1975b. Meromiktiska sjöar på Åland – en av Åländska sjöar med bräck vatten. Husö Biologiska Station Meddelande 17: 16–41.

Lindholm, T., 1976. Mekanismer i en meromiktisk kustsjö. Zool. Revy 38: 3–10.

Mölder, K. & Tynni, R., 1967–1973. Uber Finnlands rezente und subfossile Diatomeen I–VII Bull. Soc. Finl. 39–45.

Simola, H., 1977. Diatom succession in the formation of annually laminated sediment in Lovojärvi, a small eutrophicated lake. Ann. bot. fenn. 14: 143–148.

Simonsen, R., 1962. Untersuchungen zur Systematik und Ökologie der Bodentiatomeen der westlichen Ostsee. Intern. Rev. Hydrobiol. Syst. Beih. 1: 1–144.

Tynni, R., 1975. Uber Finnlands rezente und subfossile Diatomeen VIII. Geol. Surv. Finl. 174: 1–37.

Tynni, R., 1976. Uber Finnlands rezente und subfossile Diatomeen IX. Geol. Surv. Finl. 284: 1–37.

Tynni, R., 1978. Uber Finnlands rezente und subfossile Diatomeen X. Geol. Surv. Finl. 296: 1–55.

Willén, T., 1962. The Utål Lake Chain, Central Sweden, and its phytoplankton. Oikos, Suppl. 5: 1–164.

Zarins, E. & Ozolin, S., 1935. Untersuchungen über die Zusammensetzung des Meerwassers im Rigaschen Meerbusen und an der lettländischen Küste des Baltischen Meeres. J. Cons. int. Explor. Mer. 10 Copenhagen.

Eutrophication processes in Spanish reservoirs as revealed by biological records in profundal sediments

Narcis Prat & M. Victoria Daroca
Departament d'Ecologia, Facultat de Biologia, Universitat de Barcelona, Avda. Diagonal, 645, Barcelona - 28, Spain

Keywords: paleolimnology, eutrophication, reservoirs, *Bosmina*, Chironomidae

Abstract

Eutrophication of reservoirs can be detected by changes in the abundance of insect and crustacean remains in the sediments. In recently constructed reservoirs, the time of impoundment can be determined through the presence of chironomid head capsules. The initial phase is characterised by highly enriched water as can be seen by the abundance of *Bosmina* remains. If further nutrient input into the reservoir remains low, the upper part of the sediment column is almost completely lacking in invertebrate remains.

Introduction

Paleolimnological studies in Spain have been few, and devoted only to paleo-lakes or estuarine bays (Margalef 1957, 1962, 1969). In other parts of the world studies of recent eutrophication revealed by changes in the biostratigraphic record, are common and a considerable amount of data exist (Carter 1977; Bengtsson & Persson 1978; Moss 1980). There are also more general accounts of the Quaternary history of lakes (Frey 1974, 1976). The association of different dating methods with historical records provides a good basis for studying the trophic status of lakes in the past, as exemplified by the fine study of the Bay of Quinte (Warwick 1980). But eutrophication processes in man-made reservoirs, although well documented in terms of contemporary plankton and benthos, have received little paleolimnological attention (Thomas & Soltero 1977).

As part of a study of more than 100 reservoirs in Spain (Margalef *et al.* 1976; Estrada 1975; Armengol 1980; Prat 1980), cores were taken in a number of these reservoirs and the results are presented in this paper.

Material and methods

Samples were taken with a simple gravity core of 1 m length, with a weight of 25 kg. Mud was retained in a plastic liner with an inside diameter of 2.4 cm, and a plastic valve underneath. The sampling method was not very effective and the maximum length of sediment collected was 50 cm.

Cores were frozen *in situ* with dry ice and preserved until examination, when they were cut at 2 cm intervals except for the core from the Yesa reservoir which was cut at 1, 2, and 4 cm intervals.

One gram of sediment was treated with 10% KOH according to Frey (1964) and filtered through two nets of 40 and 150 μm mesh. The material retained was diluted with water and examined with a stereoscopic microscope at 10 × and 100 ×. The mesh size was sufficient to retain chironomid and crustacean remains (Warwick 1980). Some diatom, dinoflagellate, and green algal remains were also found.

Hydrobiologia 103, 153–158 (1983).

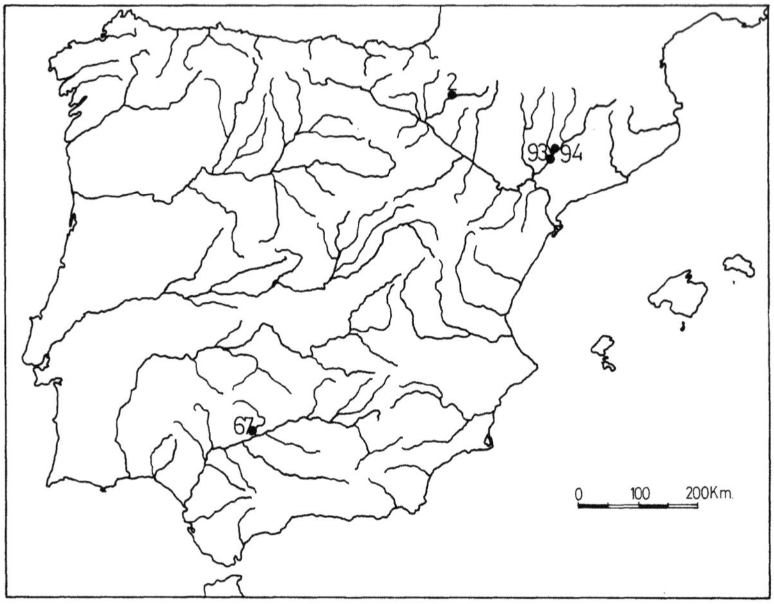

Fig. 1. Geographical situation of the four reservoirs sampled. 2 - Yesa, 67 - La Breña, 93 - Santa Anna, 94 - Camarasa.

Reservoir characteristics

Only a few cores contained more than 20 cm of sediment and a number of them were selected for detailed analysis by V. Daroca. The four reservoirs selected were Yesa (2), La Breña (67), Santa Anna (93), and Camarasa (94). The La Breña reservoir is the only one situated in the south of the Iberian peninsula (Fig. 1). The number in brackets is the identification code of the reservoirs as used in the general survey carried out on Spanish reservoirs (Margalef *et al.* 1976). The most important morphological and physicochemical features of the reservoirs, in addition to core data and profundal fauna taken with a grab sample are provided in Table 1.

Table 1. Geographical, morphometrical, chemical and core data for four Spanish reservoirs.

	Santa Anna	Yesa	La Breña	Camarasa
River	Noguera Ribagorzana	Aragón	Guadiato	Noguera Pallaresa
Impoundment year	1961	1960	1935	1920
Altitude (m)	375	490	121	333
Maximum depth (m)	80	60	52	91
Surface area (ha)	768	1900	600	624
Volume (10^6 m^3)	241	470	115	180
Flow (10^6 m^3/year)	946	1216	180	2181
Alkalinity (meq/l)	1'9	2'6	1'9	1'9
Phosphorous (μg-at PO_4-P l^{-1})	0'09	0'12	2'24	0'87
Chlorophyl (mg chl a m^{-2})	19	15	64	88
Core sampling date	30.4.1974	2.5.1974	14.3.1974	29.4.1974
Reservoir depth	54	48	23	60
Sediment in the core (cm)	18	32	28	30
Profundal fauna taken with grab	Tubificidae	Tubificidae *Procladius* Tanytarsinii	Tubificidae *Procladius* *Chironomus* *Microchironomus*	Tubificidae *Procladius*

The four reservoirs considered are large, deep, man-made lakes that differ in their morphology and flow conditions. Camarasa is deep and narrow but Yesa and La Breña are shallower and broader (Table 1).

All the reservoirs are alkaline, but not very eutrophic (Table 1) as can be seen from the phosphorous content of the waters and the mean chlorophyll *a* content per m^2. La Breña and Camarasa were included in the group of moderately eutrophic reservoirs (group V) and Yesa and Santa Anna were in the oligotrophic group (group VI) in the typological classification made by Margalef *et al.* (1976).

As can be seen from Table 1, tubificidae were frequent but chironomids scarce (Prat 1980). Larvae of *Chironomus* were collected only in La Breña. The most frequent Chironomidae were the larvae of the genus *Procladius*. However, *Procladius* was absent from our sediment data, perhaps because of the small size of the ligula that was not retained in the analysed samples.

Results

In Camarasa no remains of Chironomidae were found (Fig. 2). Only planktonic crustaceans and some Chydoridae were present in the core. Well-preserved remains of pennate diatoms and *Pediastrum* were also found. There are clear fluctuations in the number of individuals in the core but interpretation is difficult because the sediment accumu-

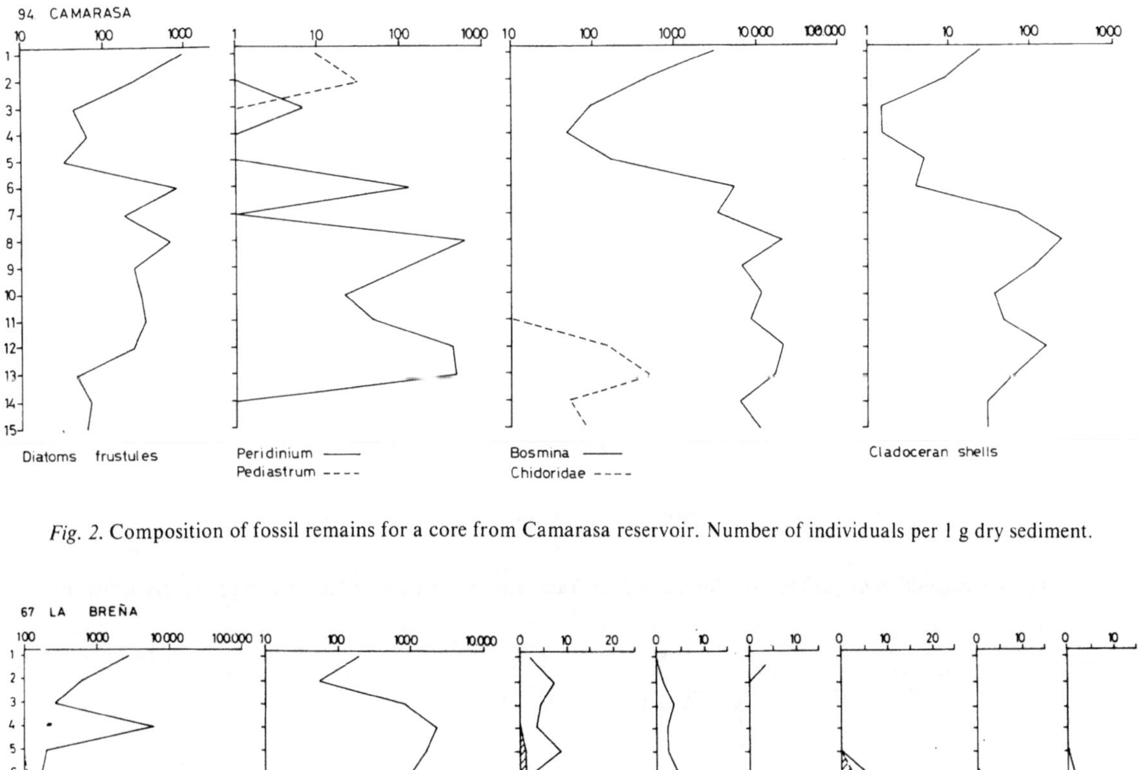

Fig. 2. Composition of fossil remains for a core from Camarasa reservoir. Number of individuals per 1 g dry sediment.

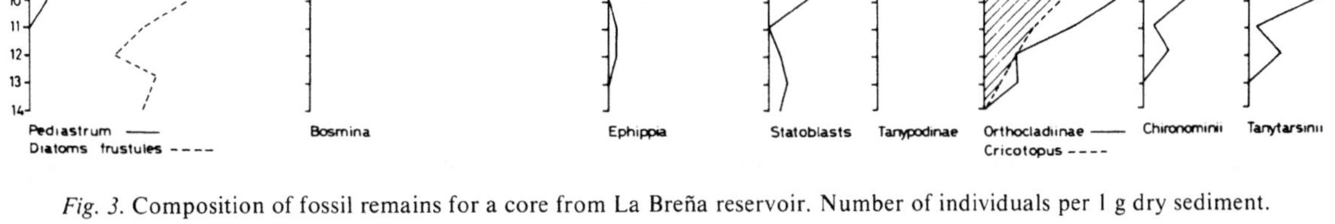

Fig. 3. Composition of fossil remains for a core from La Breña reservoir. Number of individuals per 1 g dry sediment.

lation rate is not known. It may be that differences in the retention time of the water in the reservoir is responsible for these differences. Rainy years with a rapid throughflow may correspond to lower accumulation of planktonic animals in the sediment.

In La Breña, (Fig. 3) the Chironomidae appear in the middle levels of the core and dominance by diatoms in the profundal layers is replaced by an increase in *Pediastrum* in the upper layers. *Bosmina* appears at level 8 and rises and remains high in the upper layers. Ephippia of *Daphnia* and Ceriodaphnia were also more abundant in the upper layers. In layers 9 to 11 the number and diversity of Chironomidae was very high. Using Hofmann (1971) we have found representatives of *Procladius, Cricotopus, Psectrocladius, Chironomus, Parachironomus, Dicrotendipes, Paratendipes, Stictochironomus, Paracladopelma*, and unidentified Tanytarsinii. But in the upper layers the rich chironomid fauna disappears.

In the more recently built reservoirs a clear difference exists between base and top of the core. In the upper layers (1 to 6) of Yesa reservoir, few remains of chironomids were found (Fig. 4). In the deepest layers (11–12) Chironominii were abundant (*Chironomus, Paracladopelma*, and *Polypedilum*) and Tanytarsinii were found more abundantly in the middle layers, between levels 6 to 9. At the base of the core Orthocladiinae were also abundant, especially *Psectrocladius, Cricotopus*, and *Parakiefferiella*. *Bosmina* appears at the base of the core, but its number is not as high as in the other reservoirs.

The most recent reservoir, Santa Anna, had a

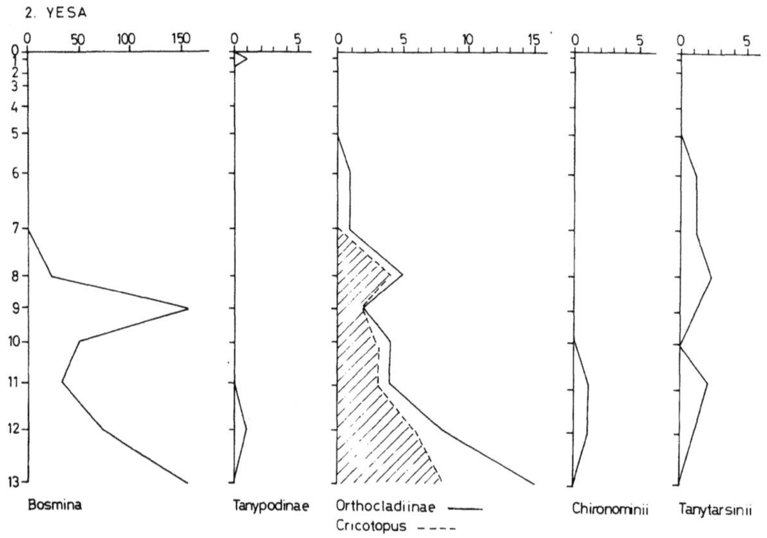

Fig. 4. Composition of fossil remains for a core from Yasa reservoir. Number of individuals per 1 g dry sediment.

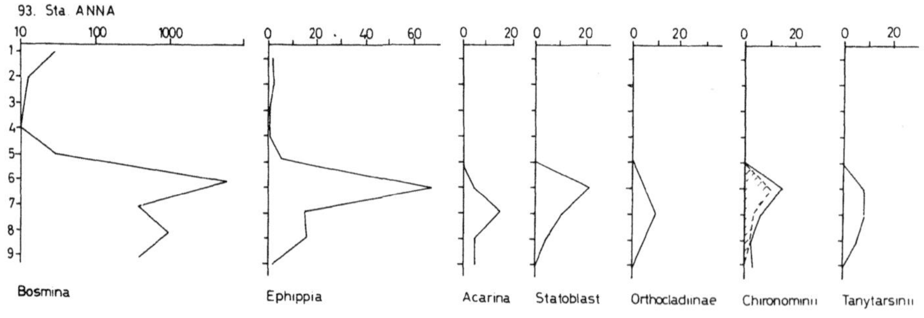

Fig. 5. Composition of fossil remains for a core from Santa Anna reservoir. Number of individuals per 1 g dry sediment.

clear lithostratigraphic change in the core taken in deep water. The lower sediment was dark and compact, similar to inundated soil; the more extensive upper zone was light brown, typical of mud from deep oligotrophic lakes. The upper layer had few fossil remains, while the lower one was very rich in such remains as *Bosmina, Daphnia* ephippia, statoblasts, and chironomids (*Chironomus, Polypedilum, Stictochironomus, Cryptochironomus, Cricotopus,* and *Psectrocladius*), and the concentration of organisms was especially high at layer 6 (Fig. 5), with abundant remains of *Chironomus* and *Bosmina* suggesting a eutrophication phase. The fauna disappears in the upper layers, and the reservoir was very oligotrophic at the time of sampling.

Discussion

The rate of sediment accumulation in reservoirs is very variable. Thomas & Soltero (1977) found a rate of 2–4 cm a^{-1} and 10 cm a^{-1} may be accumulating in some reservoirs. The rate can be very variable depending whether the reservoir is situated in a headstream, or located in a lower part of a series of man-made lakes. The rate will be higher in the upper basins. Camarasa and Santa Anna are the third basins in a reservoir series, while La Beña and Yesa are located in zones without other reservoirs upstream. In the absence of laminated sediments and radio-isotope dating, it is very difficult to assess the filling rate of the reservoirs. Undoubtedly 20 to 40 cm of sediment taken by our corer (Table 1) only provides data for the more recent events.

In Yesa and Santa Anna, recently completed dams, the base of the core could represent the time of reservoir impoundment. This may especially be the case in Santa Anna, where the mud structure and fossil remains are very different from the upper sediments. In Yesa, although the core went deeply into the mud it is unlikely that initial sediments were sampled since the reservoir is situated in a head water and probably has a higher accumulation rate than Santa Anna.

The variety of chironomid fauna when present, with especially *Psectrocladius, Cricotopus* and Chironominii other than *Chironomus*, suggests a rich development of the littoral fauna (Stahl 1969). The absence of littoral organisms can be due to morphometric and flow conditions of the reservoirs. Deep steep-sided reservoirs with a low water retention time (as Camarasa) restrict the development of littoral fauna and only planktonic elements were found in the sediments (Fig. 2).

In reservoirs with high retention times, and situated in broader valleys (as La Breña) fluctuations in water level may be greater, but when water level is low, and little water enters the reservoir, the temporary stability can allow a diverse littoral fauna, (with *Cricotopus, Psectrocladius, Glyptotendipes,* or *Paracladopelma*) to develop. Eutrophic conditions can occur at low water level situations and the replacement of diatoms by *Pediastrum* and the presence of *Chironomus* in the intermediate levels of the sediments in La Breña reservoir suggest this process. Later, the disappearance of chironomids and the dominance of *Pediastrum* and *Bosmina* remains, suggests a fluctuation in water level and eutrophic water.

As we have suggested the Santa Anna sedimentary record shows a complete history of reservoir development. The basal mud was very different in structure and some typical riverine insects (like *Euorthocladius*) were found. Many typical littoral chironomids were also found, suggesting a stable, low water level.

At level 6 of Santa Anna (Fig. 5), many *Bosmina* and *Chironomus* remains suggest an enrichment phase, in which the terrestrial vegetation decomposed and released nutrients to the water. After this phase, as the water entering the reservoir is nutrient poor, plankton and benthos declined, and *Bosmina*, for instance, became very scarce. Grab samples taken in Santa Anna demonstrated the absence of Chironomidae in the profundal zone and only some tubificids were found (Table 1). This oligotrophic phase explains the absence of fossil remains in the upper part of the core taken in Santa Anna (Fig. 5).

We can conclude that if reservoirs are built in a zone without incoming nutrients, after an initial eutrophic phase, reservoirs become oligotrophic leaving few fossil remains in the sediments. But water level fluctuations are very important for the trophic status of reservoirs. A period of low water level may result in reservoir eutrophication and the development of a rich littoral fauna. More investigation is needed to understand how the morphometric pattern, flow and water level fluctuations affect to trophic conditions of the reservoirs and the remains preserved in the sediments.

Acknowledgements

Many thanks are due to Prof. R. Margalef for remarks and comments. Anna Mª Domingo made the drawings. Financial support has been provided by the Ministerio de Obras Públicas.

References

Armengol, J., 1980. Colonización de los embalses españoles por crustáceos planctónicos y evolución de la estructura de sus comunidades. Oecol. aquat. 4: 45–70.

Bengtsson, L. & Persson, T., 1978. Sediment changes in a lake used for sewage reception. Pol. Arch. Hydrobiol. 25: 17–33.

Carter, C. E., 1977. The recent history of the chironomid fauna of Lough Neagh from the analysis of remains in sediment cores. Freshwat. Biol. 7: 415–423.

Estrada, M., 1975. Statistical considerations of some limnological parameters in Spanish reservoirs. Verh. int. Ver. Limnol. 19: 1849–1859.

Frey, D. G., 1964. Remains of animals in Quaternary lake and bog sediments and their interpretation. Arch. Hydrobiol. Ergeb. Limnol. 2: 1–116.

Frey, D. G., 1974. Paleolimnology. Mitt. int. Ver. Limnol. 20: 95–123.

Frey, D. G., 1976. Interpretation of Quaternary paleoecology from Cladocera midges and prognosis regarding usability of others organism. Can. J. Zool. 54: 2208–2226.

Hofmann, W., 1971. Zur taxonomie und Palokölogie subfossiler Chironomiden (Diptera) in Seesedimenten. Arch. Hydrobiol. Ergeb. Limnol. 6: 1–50.

Margalef, R., 1957. Paleoecologia del lago de la Cerdanya. P. Inst. Biol. Apl. 25: 131–137.

Margalef, R., 1962. Registro fósil de fluctuaciones climáticas de corto periodo en el area pirenaica y en época miocénica. Actas 3er congreso Internac. Est. Pirenaicos, Girona, 1958. 217–226.

Margalef, R., 1969. Size of centric diatoms as an ecological indicator. Mitt. int. Ver. Limnol. 17: 202–210.

Margalef, R., Planas, D., Armengol, J., Vidal, A., Prat, N, Guiset, A., Toja, J. & Estrada, M., 1976. Limnologia de los embalses españoles. Publicaciones del Ministerio de Obras Públicas. 123, Madrid. 454 p.

Moss, B., 1980. Further studies on the paleolimnology and changes in the phosphorous budget of Barton Broad, Norfolk. Freshwat. Biol. 10: 261–279.

Prat, N., 1980. Bentos de los embalses españoles. Oecol. aquat. 4: 3–43.

Stahl, J. B., 1969. The uses of chironomids and other midges in interpreting lake histories. Mitt. int. Ver. Limnol. 17: 111–125.

Thomas, J. R. & Soltero, R. A., 1977. Recent sedimentary histories of an eutrophic reservoir, Long Lake, Washington. J. Fish. Res. B. Can. 34: 669–676.

Warwick, W. F., 1980. Paleolimnology of the Bay of Quinte, Lake Ontario: 2800 years of cultural influence. Can. Bull. Fish. aquat. Sci. 206: 1–117.

Seasonal eutrophication by wildfowl in basins isolating from the sea

A working hypothesis

Gunnel Linnman
Public Department, Swedish Museum of Natural History, Box 50007, S-104 05, Stockholm, Sweden

Keywords: paleolimnology, marine/brackish-freshwater, transition basin, wildfowl, gregarious water fowl, allochthonous nutrients, seasonal eutrophication

Abstract

A new and very special course of events starts in a freshwater basin, just isolated from the sea. In NW Europe this is mainly caused by land uplift in Holocene times or by regression of the sea. It has been thoroughly documented by the gradual change in the composition of the fossil diatom floras. Attention is drawn to the possible impact of migrating birds, especially wildfowl, breeding, resting and moulting in those often shallow, sheltered lakes along the Scandinavian coastlines. The gregarious wildfowl species might seasonally increase the allochthonous input of nutrients in the lake ecosystems and cause marked changes in biotic interactions. A working hypothesis is suggested, that the wildfowl species in question were appreciably more numerous before the arrival of man. Consequently, their effect on seasonal eutrophication cannot be neglected in the paleolimnological context.

Introduction

In Scandinavia the transition from marine or brackish to freshwater environments is caused mainly by land uplift or by regression of the sea. Eutrophication in connection with isolation has been recorded in several Scandinavian lake basins, sometimes even directly preceding the final isolation (Böyum 1976). Florin (1977) investigated the pollen, spore and diatom flora of Långa Getsjön in Southern Central Sweden. Due to the land uplift the Getsjö archipelago of the Yoldia Sea became an area of more sheltered bays on the sunny side of the coast. Around 10 000 years B.P. these passed the transition stage to freshwater. Peaks of *Pediastrum* spp. and *Fragilaria* spp. 'indicate a transitory improvement of the trophic status of the basins. The higher water temperature increases the productivity of the water . . .'. The author notably points out the expansion of the nettle, *Urtica* (p. 22), which 'indicates that the nitrogen status of the soil was good during this period, perhaps improved by the presence of sea-birds'.

Digerfeldt *et al.* (1980) described the development of Långsjön, a shallow basin not far away from Stockholm, Central Sweden. It was isolated more than 5 000 years B.P. The diatom flora in the sediments of lake Långsjön indicates highly eutrophic conditions immediately after the transition to freshwater.

In Norway Kjemperud (1980) investigated the diatom changes in sediments of five lake basins, possessing marine-lacustrine transitions, in Frosta, Nord Tröndelag. The freshwater stratigraphical zone always starts with a maximum in the frequency of both diatoms and other algae (*Pediastrum* and *Botryococcus*). The first period of isolation is characterized by a pronounced increase in productivity with the dominance of alkaliphilous forms. Kjemperud stresses that the nutrient-rich environment existing just after the isolation might be due to a combination of several factors.

Hydrobiologia 103, 159–163 (1983).

The course of transition

It is well-established that there can be a great difference in the rate of shore displacement during the time of transition, due to the eustatic as well as isostatic factors in different regions of Scandinavia.

Ingmar (1973) has thoroughly examined the special course of events following the transition from marine/brackish to freshwater conditions in lake basins, by studying a lake chain (Utål) in Uppland, North of Stockholm. Ingmar points out that the isolation will sometimes be very diffusely registered in the sediment strata, sometimes more distinctly, and he wishes to distinguish between the level of *biological* isolation of a freshwater basin, and the *hydrological* isolation. They are not always coincident.

This article will be restricted to freshwater basins in the cold and also in the temperate regions of north-western Europe, which were isolated in early Holocene times and before any recognizable impact of *Homo sapiens*. The turnover of a newly isolated waterbody has been documented in many investigations and from many points of view (Wetzel 1975). It seems important, however, to examine closely the ecological interrelationships as a whole, including macroflora and macrofauna in former times, under natural conditions and before the arrival of man.

Documented former wildlife

The fossil remains of large animals are deplorably few, compared with what is left of the former microfauna. In fact we have to utilize the information gathered from documented historical time. Animal refuse, 'kökkenmöddings' (originally a Danish term), from permanent human dwellings dates back to the Neolithic age in Scandinavia. Some conception, though rather diffuse, of the wildlife in Sweden of those days, is given to us by the faunal ossiferous remains of the Iron age settlements as Eketorp, Öland, close to the south-western Baltic shoreline (Boessneck *et al.* 1979). For example, in the case of the shelduck, *Tadorna tadorna* (page 261) they state: 'bereits ein häufiger Brutvogel nahe Eketorp, wie die zahlreichen gefundenen Knochen erkennen lassen (Tab. 90)' and for the Whooper Swan, *Cygnus cygnus* (page 274): 'Durchzügler und Wintergast in Eketorp . . .'

The father of diatom research in the Baltic, Herman Juhlin-Dannfelt in Uppsala, Sweden, published his paper in 1882 'On the Diatoms of the Baltic sea'. His interest was concentrated on the shallow waters, 'open to the sunshine, but protected from winds and waves'. My intention is also to focus on those environments, and to draw attention to the last link in many of the food chains of the lake basins in the stage of isolation in the clay sediment regions of north western Europe. I am concerned especially with birds, and particularly wildfowl.

The gregarious water fowl

The most interesting group are the gregarious bird species, resting and breeding in great numbers in fairly shallow basins. The different groups of wildfowl, Anatidae, are ducks, swans and true geese, but even other bird groups such as waders, herons and storks are characteristic members of that society. Wildfowl sometimes have just a short stop of a few days in that environment, when migrating from southern continents to the western Palaearctic, and back again. Sometimes wildfowl stay for weeks or the whole spring and summer season, breeding, roosting and/or moulting. A few species are permanent inhabitants of the cold and temperate zones. They move only enough to have access to open water throughout the winter and often sleep in big flocks on the ice of the lakes or estuaries. The vast majority of the migrating wildfowl arrives early in the spring, in March or April. The return to the South is more irregular, lasting from June to October.

Seasonal eutrophication

We cannot overlook the fact that wildfowl and other birds are seasonally important links in food-chains and affect the interplay of organisms in lake basins, and even in an open bay or a fjord of restricted dimensions. They increase fertility by their droppings and by other organic matter, such as feathers and down, although they also take nutrients away from the water they visit. Birds such as swans and ducks also disturb the ordinary sedimentation of organic matter on the bottom by dipping and diving, and by trampling in the littoral zone. It is obvious that the wildfowl take part in the processes of cycling and recycling nutrients.

Input of allochthonous nutrients

Among the Anatidae, swans, ducks and geese, many species graze both in the water and on land. Some of them graze mainly in the littoral zone, others prefer terrestrial surroundings. Some of them feed on small creatures on the shore, like insects, frogs and molluscs. Others spend much time sleeping on the shore or in the water, but the vast majority spend hours swimmming or nesting in the sheltered waters (Cramp *et al.* 1978). It is striking that these bird species prefer the coastal waters and especially those in the transition stage from marine/brackish to freshwater. As we know and are able to see in a human life span, these kinds of basin have a relatively short existence. Often the high sediment load coming from streams and rivers, and generally shallow conditions, where macrophytic vegetation can grow, alters the productivity of the lake system (Wetzel 1975). However, as long as wildfowl are present, the input of allochthonous nutrients into the water from the surrounding land and from the organic matter in rain and meltwater will increase. It could be called a 'seasonal ornithological eutrophication', repeated annually and lasting from a few decades to some hundreds of years.

Migrating wildfowl species

Four wildfowl species have an important value in this context: the Shelduck, *Tadorna tadorna,* the Whooper Swan, *Cygnus cygnus,* the Bean Goose, *Anser fabilis* and the Greylag Goose, *Anser anser* (Svensson 1978). Maps of their recent distribution (Figs. 1 & 2) derive from Cramp *et al.* (1978). In addition, Svensson (1979) gives a list of the highly ranked resting and breeding sites for wildfowl in Sweden at present (Fig. 3). Those sites are situated in sheltered bays, and in estuaries, but mainly in newly isolated freshwater lakes. Some localities are also situated inland, along very old Pleistocene or Holocene coast lines, in sheltered areas within the two largest Swedish lakes, Lake Mälaren (M) and Lake Vänern (V).

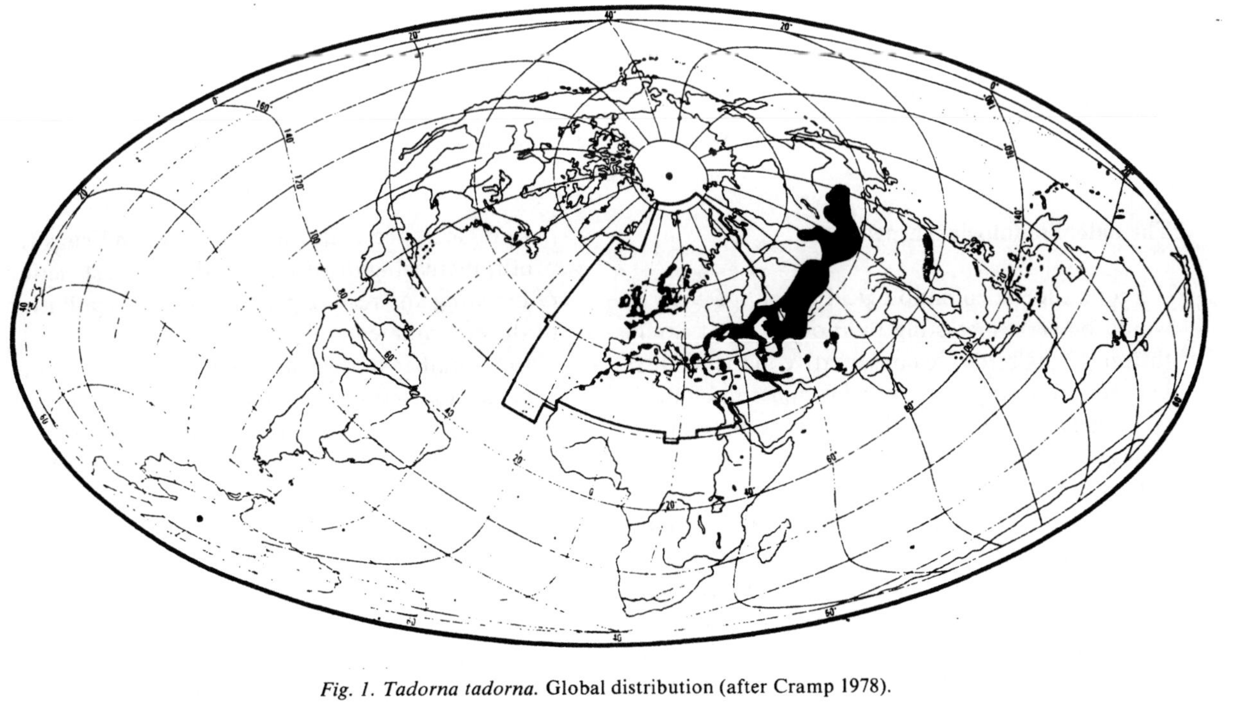

Fig. 1. Tadorna tadorna. Global distribution (after Cramp 1978).

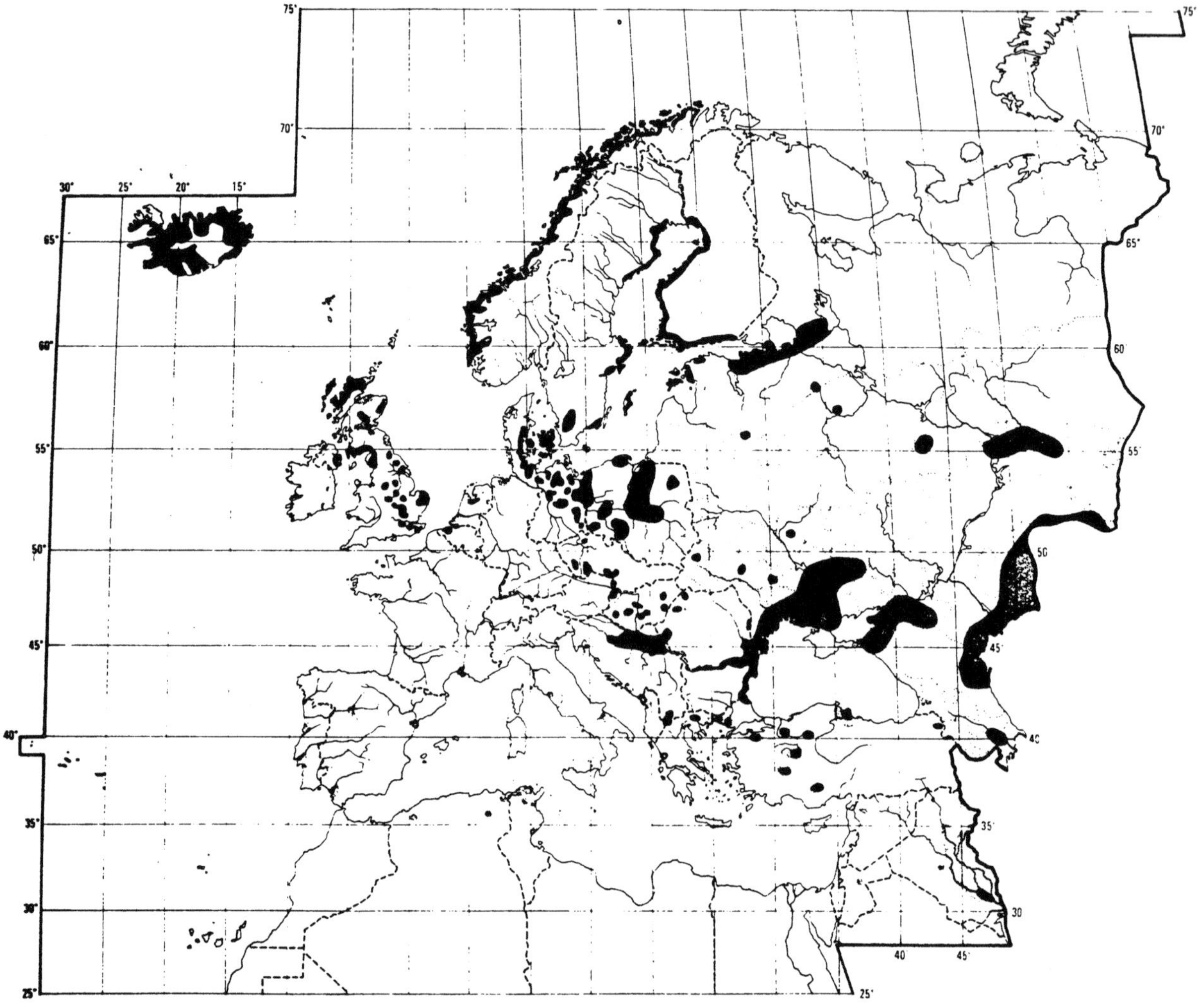

Fig. 2. Anser anser. Western Palaearctic distribution (after Cramp 1978).

The paleolimnological aspect

From the paleolimnological point of view it would be very interesting to know to what extent the migrating birds are conservative in their routes between wintering regions and summer habitats. It is already well-known that long term climatic changes can be of influence.

Ekman (1922) stressed that the ability of birds to fly is itself a conservative power among the migrators. In our time, however, we only see what is left of the wildfowl fauna, patches of a formerly very large distribution or the last refuges of formerly abundant species, some of them at present near extinction. Historical data records enormous hunting pressure, the collecting of young and eggs for profit, increasing disturbances and the destruction of breeding and resting places, and man's pollution of the environment.

In a number of coastal water basins it might be expected that a seasonally increased input of organic matter into lake ecosystems and changes in biotic interactions in lakes could be due to wildfowl feeding and moulting.

The impact of water fowl, particularly some species of *Anatidae,* must have been of great importance in the past, especially during the stages of transition from a marine/brackish to a freshwater environment and it cannot be neglected when considering the lake system as a whole.

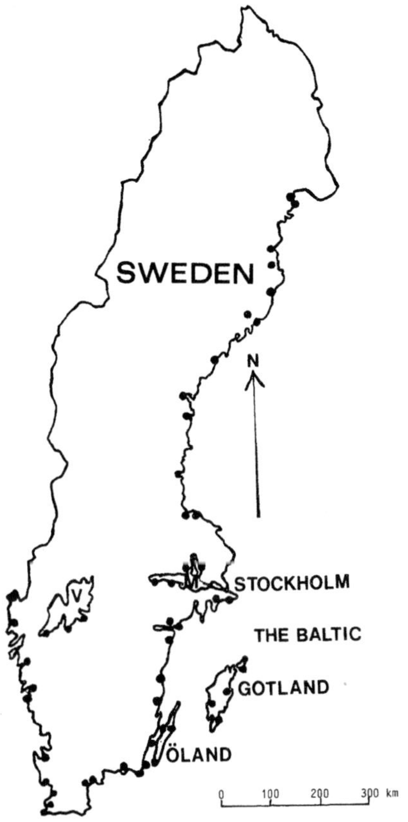

Fig. 3. Wild geese and ducks in Sweden: stop and breeding sites (comp. from Svensson 1979).

One of the conceivable consequences of great numbers of birds concentrating in freshwater lakes and along early coastlines might have been the arrival of man, hunters and fishermen attracted to those areas.

Acknowledgements

I wish to thank Gunnar Digerfeldt (Lund) for kind support, Maj-Britt Florin (Uppsala), Urve Miller (Stockholm) and my husband, Nils Linnman, for valuable comments on the manuscript and Mrs Joanne Rowley, who checked my English.

References

Boessneck, J., von den Driesch, A. & Stenberger, L., (Eds.), 1979. Eketorp. Befestigung und Siedlung auf Öland/Schweden. Die Fauna. Königl.Akademie der Literatur, Geschichte und Altertumsforschung. Almqvist & Wiksell, Stockholm.

Böyum, A., 1976. Limnology and Paleolimnology of Lake Nordvann, South-Eastern Norway. Arch. Hydrobiol. 77: 277–329.

Cramp, S., (Ed.), 1978. Handbook of the Birds of Europe, the Middle East and North Africa. The Birds of the Western Palearctic. Vol. 1. Oxford Univ. Press. London.

Digerfeldt, G., Håkansson, H. & Persson, T., 1980. Paleoecological studies of the recent development of the Stockholm lakes Långsjön, Lillsjön and Laduviken. Univ. of Lund, Dept of Quatern. Geol. Rep. 20.

Ekman, S., 1922. Djurvärdens utbredningshistoria på skandinaviska halvön. Albert Bonniers, Stockholm.

Florin, M.-B., 1977. Late-Glacial and Pre-boreal Vegetation in Southern Central Sweden. II. Pollen, Spore and Diatom Analysis. Striae. 5: 1–22.

Ingmar, T., 1973. Sjöavsnörningar från aktualgeologiska synpunkter. En översikt. Univ. of Lund, Dept. of Quater. Geol. Rep. 3: 48–90.

Juhlin-Dannfelt, H., 1882. On the Diatoms of the Baltic sea. (thesis), Uppsala.

Kjemperud, A., 1980. Diatom changes in sediments of basins possessing marine/lacustrine transitions in Frosta, Nord-Tröndelag, Norway. Boreas 10: 27–38.

Svensson, L., (Ed.), 1978. Checklist of Swedish birds. Swedish Ornithological Association. Stockholm.

Svensson, L., (Ed.), 1979. Fågellokaler i Sverige. Bonniers. Stockholm.

Wetzel, R. G., 1975. Limnology. W. B. Saunders Company, Eastbourne. 743 pp.

Eutrophication and the disappearance of lakes in the Brodnica Lake District, Northern Poland as a result of human interference

Zygmunt Churski
Copernicus University, Institute of Geography, Department of Hydrology, Toruń, Poland

Keywords: paleolimnology, lakes, eutrophication

Abstract

Problems relating to the disappearance and eutrophication of lakes as a result of human activity are presented. A study area was selected located in an agricultural/forest region where human influence on lakes used to be minimal.

In the past 25 years a number of lakes have completely disappeared and others have become rapidly enriched. Human interference has included: the pulling down of water mills, drainage and river canalisation, construction of holiday resorts, development of intensive farming, increased mineral fertilization, and the extension of roads and road traffic.

Introduction

Ever since their formation after the retreat of the Weichselian ice sheet the lakes in Northern Poland have been filling in or overgrowing, mainly by natural processes. Many of them have been drained dry by rivers. Since the Last Glaciation the total lake area has been reduced by 2/3 (Kalinowska 1961). Many of the remaining lakes are in a vestigial condition with terraces and exposed lacustrine deposits, identifiable on topographic maps, providing evidence of their former extent. Detailed field research also shows that there were many more lakes which disappeared soon after the ice retreat, many leaving little obvious field evidence.

While carrying out research in the proglacial area of Vatnajökull, Iceland, Churski (1973) examined the bottoms of lakes which had existed for only several years. These lake basins were changed by glacial waters and ice melting processes to such an extent that the range of these lakes could only be established on the basis of aerial photographs taken in 1945. Therefore, it is likely that it will not be possible to trace the presence of similar lakes in Poland.

The history of lakes which still exist or which have left clear evidence of their former existence are better known (Stasiak 1971). Moreover, contemporary dating methods allow the period in which a lake came into existence and the stages of its subsequent development to be determined in a relatively precise manner.

In the present paper I deal with the most recent history of lakes, the postwar period, when, due to changes in economic conditions great changes have taken place, especially as a result of eutrophication (Vollenweider 1970). In Poland this problem calls for particular attention requiring simultaneous biological, hydrological and chemical research.

Methods

The research area included about 100 lakes in Brodnica lake district in the Drwęca river basin (Fig. 1). Fifty of the lakes examined have an area of

Hydrobiologia 103, 165–168 (1983).

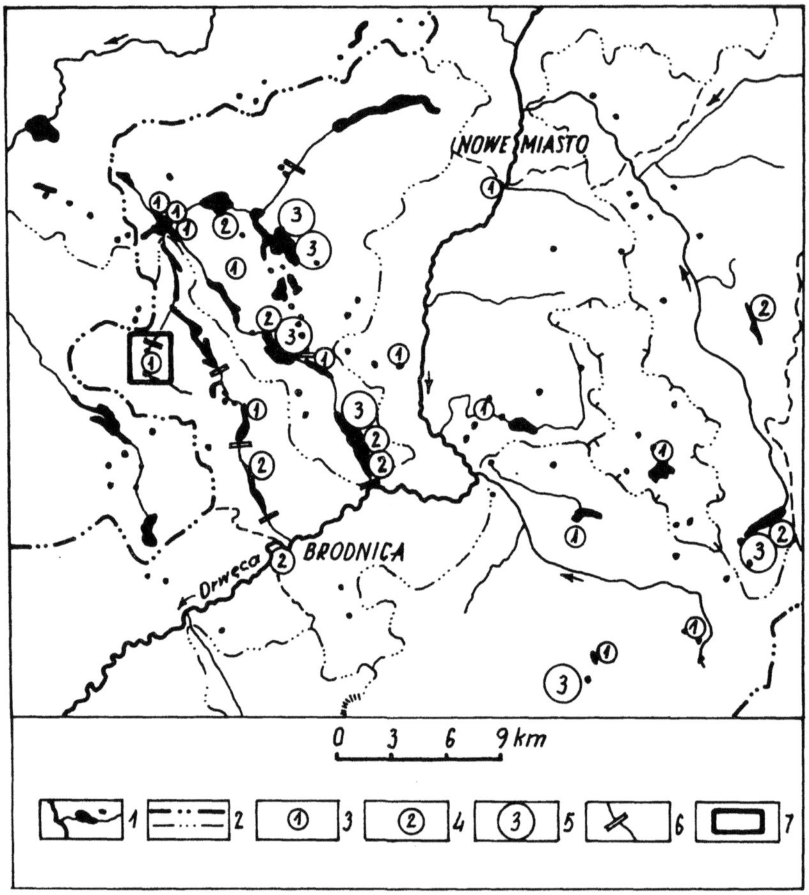

Fig. 1. Map of research area: 1. rivers and lakes over 1 hectare, 2. watershed, 3. holiday resorts < 100 people, 4. holiday resorts 100–300 people, 5. holiday resorts > 300 people, 6. location of partly destroyed dams, 7. area shown in Fig. 2.

less than 1 ha and a depth less than 5 m. The remaining 50 lakes have areas ranging from 1–300 ha and depths from 3–40 m. Data collection began in 1950 and was intensified in 1960. Field work was carried out mainly in summer and included observations of hydrological conditions (checks on water level changes), physical and chemical analysis, and the degree of overgrowth. In addition, economic and industrial activities in the area and their impact on the lakes were examined.

Results and discussion

Of the 50 smaller lakes located in kettleholes, 29 have disappeared completely and 21 are almost dry. Of the large lakes, 6 have disappeared completely and 3 are in a stage of advanded eutrophication and overgrowing. An example of land where lakes have been lost is shown in Fig. 2. The remaining large lakes have all been polluted and have undergone partial eutrophication.

The factors that brought about such rapid changes can be classified into two groups. In the first group are factors that have caused a lowering of water level and these include the removal of water driven mills, the straightening of river channels, the drainage of agricultural land, the introduction of mechanised agriculture, forest clearance and the intensification of tourist traffic. In the second group are factors that increase pollution: the construction of lakeside holiday resorts, intensification of mineral fertilization, intensive cattle raising, the development of small industrial plants and the introduction of rural piped-water systems.

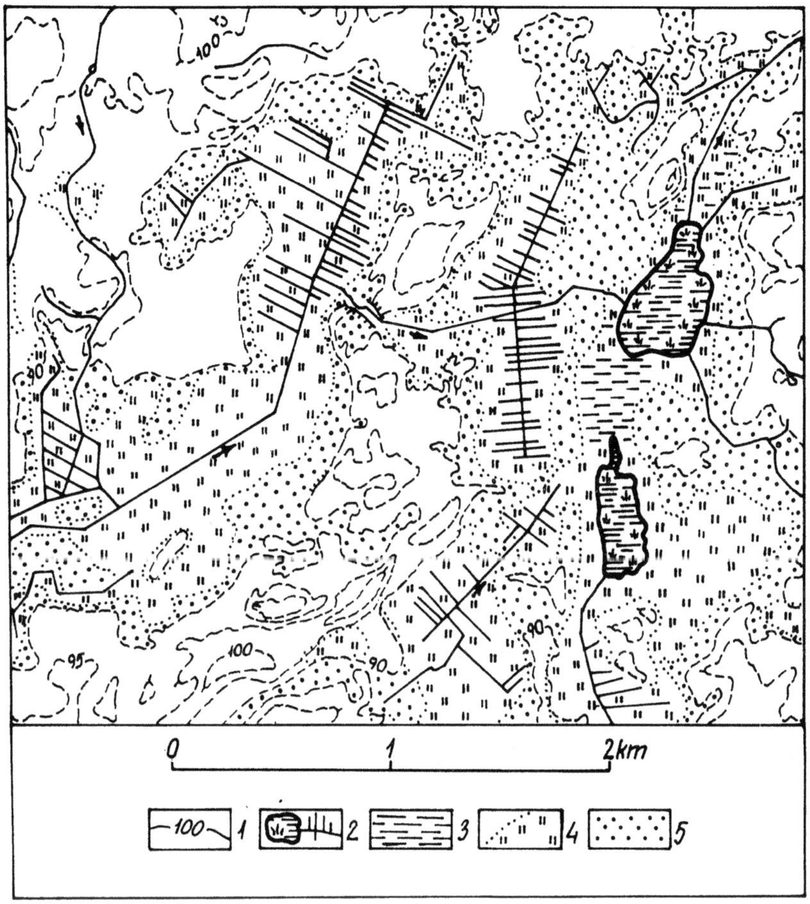

Fig. 2. Lakes which have been lost as a result of drainage and overgrowing: 1. contour line, 2. lakes and ditches, 3. marshes, 4. meadows in place of lakes, 5. lake terraces.

The effect of these factors has been especially dramatic since almost all the changes occurred simultaneously as a result of a change in the economic system.

The function of small local water mills has been taken over by large state plants. The disused water dams deteriorated quickly resulting in their disappearance or in a decrease of water level leading to an increase of river bed erosion and the filling up of lakes downstream.

Land drainage schemes were introduced to drain or decrease the level of shallow lakes and gain land for meadows and pastures. As a result, even large lakes have been drained completely and converted into pastures (Fig. 2). In general, drainage in agricultural areas has caused a lowering of the water table and the disappearance of small kettlehole lakes.

The mechanization of agriculture and the clearance of forested areas has led to a considerable increase in soil erosion. The use of tractors and heavy machines causes soil loosening, and, especially after ploughing, in the wet season the fine particles are washed directly into rivers and lakes. In addition, on the slopes where trees have been removed, soil erosion is also accelerated and some lakes are filled with alluvial cones. In the forest areas bank erosion occurs where holiday resorts have been constructed. The forest undergrowth is destroyed both from footpath pressure and by vehicles. In a part of the area examined, on three lakes totalling 10 km^2, hotels have been built for 3 000 people. After each season lake banks are worn out and macrophyte beds are destroyed.

The presence of holiday resorts is the most frequent cause of lake eutrophication because of the lack of proper purification plants. In farming areas the cause of water pollution is the intensive use of mineral fertilisers, the setting up of intensive rearing farms, and small plants such as creameries, distilleries and sugar mills. Records show the increase of fertilizer application from 35 kg ha^{-1} in 1960 to 200–250 kg ha^{-1} in 1980. In addition manure from large cattle raising farms is often washed off the land into lakes and causes pollution of the ground water. Lack of purification plants in creameries, distilleries and other factories also causes pollution of lakes and rivers.

Lakes that in 1960 had a total dissolved solid content of 200 mg l^{-1} and a PO_4 content of 0.4 mg l^{-1} now have 370 mg l^{-1} TDS and 0.6 mg l^{-1} PO_4.

Pollution varies in intensity throughout the year. In regard to tourism pollution is most intensive during summer, whereas in farming areas spring and autumn are the worst seasons.

These processes and problems continue to increase and the situation requires urgent attention. These studies are continuing with the further aid of aerial photographs. The present studies indicate the necessity to change the cultivation and development methods in the lake districts. In order to restore the lakes it is imperative to:

1. reconstruct dams and restore water levels
2. control the drainage system
3. control fertilizer use
4. limit tourism
5. set up sewage purification plants in all tourist resorts and factories in the lake district
6. afforest lake zones
7. introduce landscape park status wherever lake ecosystems are endangered.

References

Churski, Z., 1973. Hydrographic features of the proglacial area of Skeidarárjökull.Geogr. Pol. 26: PWN Warszawa.

Kalinowska, K., 1961. Disappearance of the post-glacial lakes on the territory of Poland. Przegląd Geogr. 33: 3.

Stasiak, J., 1971. Holocen Polski Potnocno-Wschodniej PWN Warszawa.

Vollenweider, R. A., 1970. Scientific fundamentals of the eutrophication of lakes and flowing waters, with particular reference to nitrogen and phosphorus as factors in eutrophication. OECD, Paris.

An attempt to trace eutrophication in a shallow lake (Balaton, Hungary) using chironomids

György Dévai & Judit Moldován
Department of Ecology, L. Kossuth University, Debrecen, Hungary

Keywords: paleolimnology, *Chironomidae*, subfossil remains, shallow lake, eutrophication, Lake Balaton

Abstract

Lake Balaton, the largest shallow lake in Central Europe, is about 20 000 years old. An enormous increase in tourism and the disproportionate building development of the last few decades has resulted in the acceleration of eutrophication in the lake. Widespread research to reveal the causes of water-quality deterioration and possible ways of protection against it have recently started. The investigation of the larvae of non-biting midges *(Diptera: Chironomidae)* in the sediment of the open-water zone has also begun. The contemporary faunal composition strongly correlates with the trophic gradient along the longitudinal axis of the lake. We therefore supposed that the eutrophication process should be identifiable from the analysis of subfossil chironomid head capsules from the upper (15 cm thick) layer of the sediment. We found that quantitative results could only be obtained when fragments as well as relatively intact head capsules are considered. Our data verify that the originally oligo-mesotrophic community has been gradually replaced by eutrophic species in a west to east direction. Large-bodied larvae belonging to the *Chironomus plumosus* group mix the sediment down to 15 cm as they build their tubes and consequently alter the original proportions of head capsules at the different levels. So the sequence of communities through the sediment-layers is not quite reliable.

Introduction

Chironomid data have been used by several authors for the classification of lake types since the beginning of this century (Thienemann 1913) and subfossil remains have also proved to be useful for lake-history studies. It was Sebestyén (1965) who first noticed that sediment samples from the 20 000 year old Lake Balaton contained many chironomid head capsules. The same phenomenon was observed during our investigations of the recent chironomid fauna of the lake. We supposed that these subfossil remains might enable us to trace eutrophication in the largest shallow lake in Central Europe.

Lake Balaton lies in the western part of Hungary at an altitude of 104 m; its area is 589 km^2 and its average depth is 3 m. The lake is dissected by three natural promontories into four basins, Keszthely, Szigliget, Central and Northeastern basins (Fig. 2), which differ in their water quality. The average annual temperature of the water is 12 °C (maxima 25-27 °C) and no true thermal stratification can be observed in the lake. The transparency measured with a Secchi-disc is 30–70 cm. The low mean depth in conjunction with the alignment of the longitudinal axis with the prevailing winds results in significant physical mixing. This is one of the reasons for the favourable O_2 supply of the water characterized by 20–200% oxygen saturation values. No permanent oxygen depletion has been noticed in the lake.

Acceleration of the eutrophication process has been observed over the last two decades. Industrialization, agricultural activity and urbanization have increased progressively, resulting in enormous

Hydrobiologia 103, 169–175 (1983).

quantities of waste-waters and feedlot wastes flowing into the lake in a relatively unaltered state. The high tourist season population in the surrounding area (647 000) exceeds the estimated capacity of the lake (600 000). The stands of reeds which function as natural filters along the shore are perishing. The proportion of undisturbed shoreline is also decreasing, now being only 60% of the total circumference. Three quarters of the catchment area belongs to the two western basins which constitute 26% of the total lake volume and 75.8% of the total discharge of rivers and streams. Thus, in those basins eutrophication is in an advanced state. There is an increase in phosphorus-loading from the east to the west, also due to differing basin volumes (environmental condition data from Plósz, unpublished). Biologically available phosphorus, phytoplankton biomass and primary production values per unit volume of water correspond to the trend of loading along the longitudinal axis of the lake. Besides this spatial trend a significant increase in primary production can also be observed. Its values are 2.5 times higher at Tihany, 3.5 times at Szigliget and 8.2 times at Keszthely than they were at the beginning of the 1960's (Herodek 1979). The average throughout the lake is 14.1 t C ha^{-1} a^{-1}. Bacterial production is also high (4.7 t C ha^{-1} a^{-1}) while the total organic carbon content of the water (5.36–6.42 mg dm^{-3}), the COD (3.8–4.9) and BOD (4.5–6.2) values are low. The total organic carbon content of the sediment is 1.5–1.8%. (Water-quality data after Biró 1977).

Recent chironomid fauna

The above-mentioned differences between the lake basins can also be detected in the recent chironomid fauna of the open water zone. In 1978 sediment samples were taken at nine sampling points along the longitudinal axis (H_1-H_9) and bottom-dwelling chironomid larvae were examined in them (Fig. 1). Although the counts of larvae decreased from the central sampling points to the west, mean biomass values showed an increase from the eastern points to the western points. Thus mean dry weight values per larvae increased in the same way. The explanation of this discrepancy can be seen in Fig. 2, which shows characteristic genera from two neighbouring basins having different water and sediment qualities. In Szigliget basin, and even more characteristically in Keszthely basin, large-bodied larvae belonging to the *Chironomus plumosus* group dominate while in the two eastern basins *Tanypodinae* dominance mainly with *Procladius* and with a few *Tanypus* exemplars can be observed.

Examining the results from five sampling points in Keszthely basin we can conclude that the qualitative and quantitative presence of chironomids is

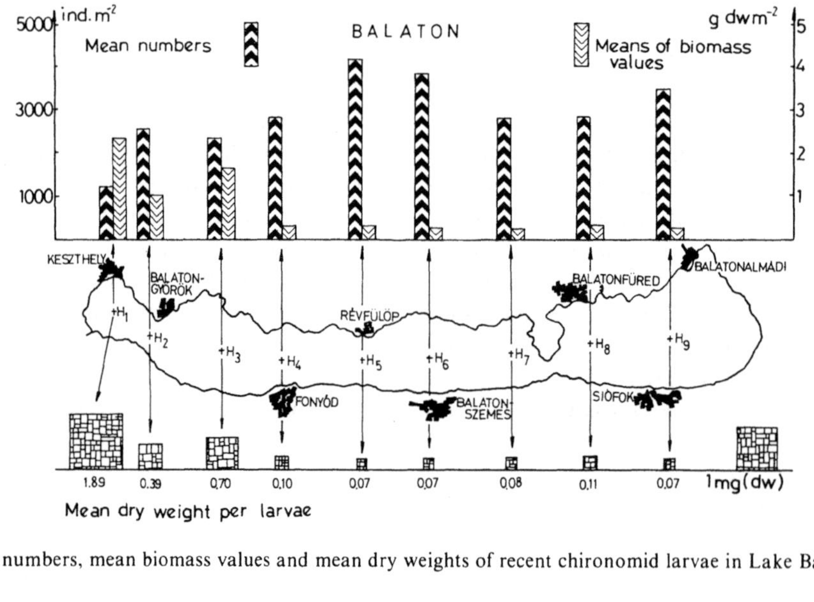

Fig. 1. Mean numbers, mean biomass values and mean dry weights of recent chironomid larvae in Lake Balaton in 1978.

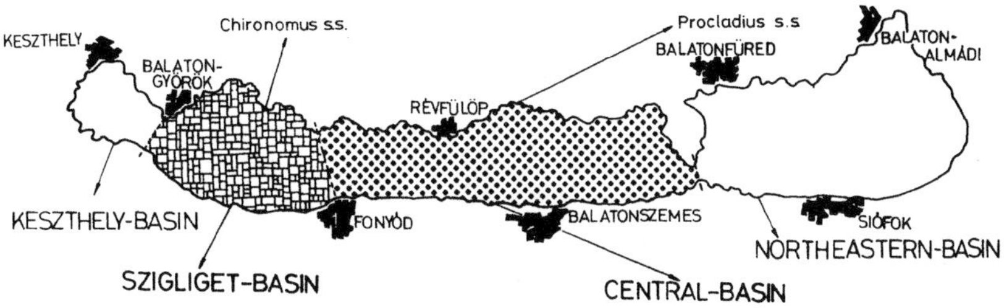

Fig. 2. The dominant chironomid genera from the recent fauna of Lake Balaton.

much more uniform within basins than between them.

Hence, the differences in the water quality of the basins noted above, can also be detected in the sediment. These results verify our basic assumption that spatial differences of the chironomid fauna in Lake Balaton are most likely due to the water quality changes which have taken place over the last few decades. The analysis of samples along the longitudinal axis reveals a distinct change of species in which the oligo-mesotrophic assemblage is gradually replaced by species characteristic of water bodies with high organic content in the sediment. This process can be considered as more or less complete not only in Keszthely basin but also in Szigliget basin. These changes are also under way in other basins. Earlier studies, though less systematic and taxonomically less reliable than present data, suggest that the original fauna of the lake is similar in species composition to the Northeastern basin of the present.

Methods

For the analysis of chironomid remains sediment samples were taken at two points in the Keszthely and Northeastern basins. Sampling was carried out using a Hargrave sampler (Hargrave 1969) with a surface of 225 cm^2 which worked without leakage. We took special care that the mud–water interface was horizontal and the upper loose sediment layer was undisturbed.

Subsamples were taken from the sampled mud with plastic tubes, 15 cm long and 3.6 cm in diameter, and sealed with rubber stoppers. During transportation to the laboratory the tubes were held in a vertical position and protected from agitation.

In one set of samples the whole core was processed as a single unit. Any material adhering to the wall of the tube was washed out with the help of a wash-bottle. Another set of samples was frozen at $-18\ ^\circ C$ in a refrigerator for 12 h and before further processing the cores were cut into three pieces, each of them 5 cm long. We observed no apparent damage to the animal remains due to the freezing.

The samples were washed under a moderate water jet through a 56 μm mesh truncated net made of silk sieve cloth. The residue was preserved in 4% formaldehyde solution and stored in patent-sealed vials. All of the head capsules and fragments were picked from other debris at 25× magnification under a Zeiss Technival stereo-microscope. Identification was carried out mainly with the same microscope at an interval of 10 to 100× magnification, while in some critical instances the capsules were mounted in euparale and examined under a Zeiss NU microscope. The larvae were identified using the keys of Fittkau (1962) and Hofmann (1971).

Head capsules in this shallow lake corrode and fragment relatively quickly compared with those in deep lakes due to the increased importance of wind action and water currents. To obtain a valid picture for quantitative evaluations and comparisons we had to take the fragments into account (Fig. 3). To avoid counting any head capsules more than once we considered only those fragments which, in the case of *Chironominae* had more than half of the labial plate, and in the case of *Tanypodinae* possessed the whole ligula or at least both of the paralabial combs. Head capsules which were obviously bisected were counted as one unit. The relatively well-preserved capsules which did not possess any characteristics which enabled positive identification were grouped together as 'unidentifiables'.

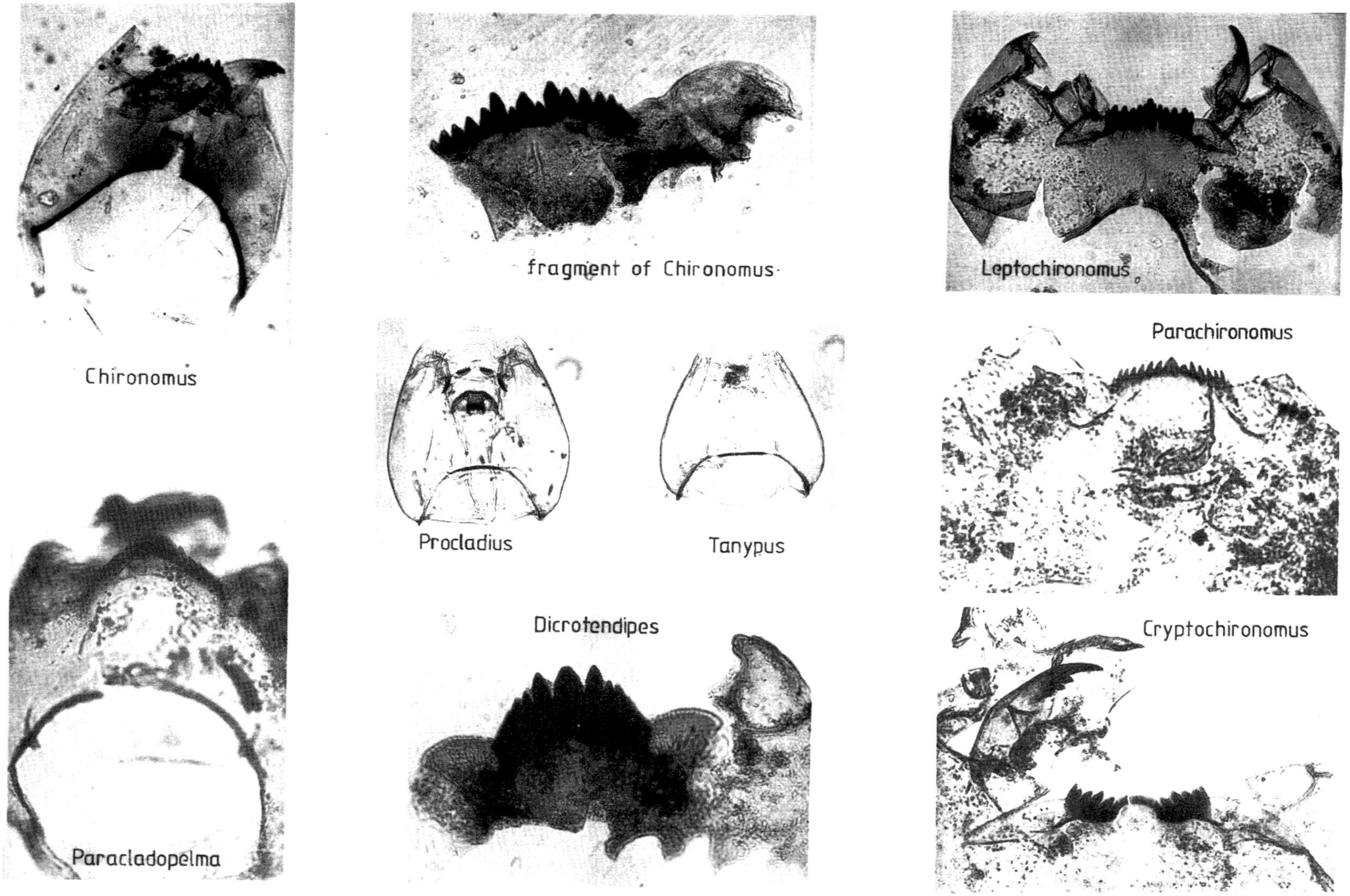

Fig. 3. The main types of subfossil chironomid head capsules in the sediment of Lake Balaton.

Tanypodinae capsules probably constitute the major portion of it.

In the analysis of species composition only 4th instar head capsules were counted but in the cases of the most important genera, *Procladius* and *Chironomus*, the ratio of 3rd to 4th instar capsules was determined.

No isotope-dating of the sediment has been carried out in Lake Balaton so far and the results of palynological investigations published by Zólyomi (1953) concern only the deeper sediment layers. So in the determination of the age of the upper sediment layer we could rely only on sedimentation rate data. The annual sedimentation in the lake is about 0.8 mm, so the examined layers represent the past 150–250 years of the lake's history. (The upper limit refers to the Northeastern basin and the lower one to the Keszthely basin.)

Results and discussion

The proportions of different chironomid head capsules in the two basins are shown in Fig. 4. Percentage distribution values confirm the results obtained in the investigation of the recent fauna. In Keszthely basin *Chironomus* and *Procladius* remains dominant, having nearly the same proportions. They are followed by *Chironominae* other than *Chironomus* (hereinafter called 'other *Chironominae*') and *Tanypus* head capsules are present in low numbers (Fig. 3). On the other hand an expressed *Procladius* dominance can be observed in Northeastern basin with the 'other *Chironominae*' group also being significant. Here *Chironomus* and *Tanypus* head capsules are present at nearly equal but rather low proportions. Although the proportion of *Tanypus* is higher here than in Keszthely basin, it remains under 5% in both basins. The proportions of 'unidentifiables' are similar in both cases.

As reported above, in spite of the advanced eutrophication process oxygen availability is good throughout Lake Balaton, due to its high surface area to volume ratio. Hence our data support Warwick (1975) who stated that the presence of the different chironomid groups was primarily determined by food availability and not by oxygen supply. Although neither Brundin (1958) nor Saether (1979) mentioned the genus *Procladius* in their lake-type systems, several authors (e.g. Carter 1976, 1977; Wiederholm & Eriksson 1979) considered associations dominated by the genera *Chironomus* and *Procladius* as indicators of moderate eutrophy. Hence the chironomid community of Keszthely basin indicates these conditions.

In the Northeastern basin *Procladius* capsules dominated the samples. This seems to be quite a new phenomenon, since in the literature this genus was found to be present only at low rates throughout oligotrophic phases. Kajak (1980) considered *Procladius* larvae to be more characteristic of oligotrophic than of eutrophic waters. Hence the *Procladius*-dominated community in the Northeastern

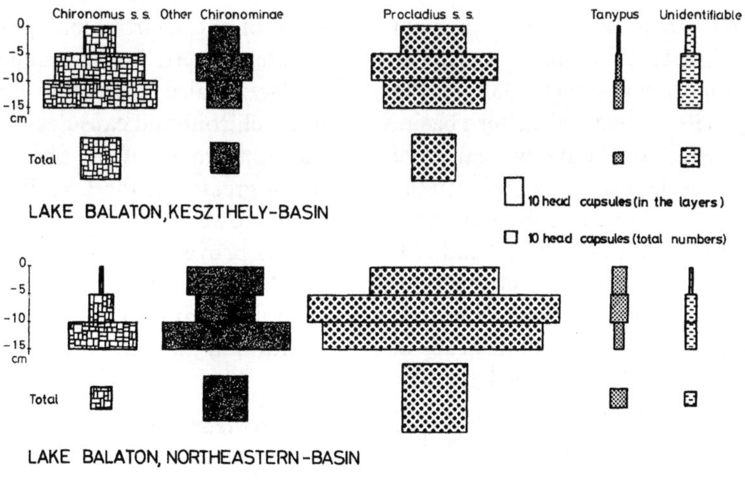

Fig. 4. Numbers and proportions of chironomid remains in two total samples.

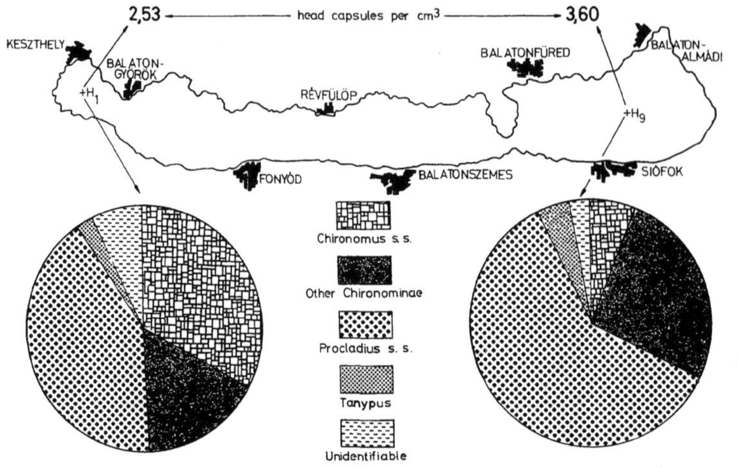

Fig. 5. Numbers of chironomid remains in three sediment layers of two basins.

basin probably indicates oligo-mesotrophic conditions and the low diversity of the lake's fauna is revealed here.

The differences in the fossilization of head capsules in the two basins are also remarkable. While 3rd instar *Procladius* capsules occur in great numbers in the Northeastern basin (165% of the 4th instar head capsules) smaller capsules break and vanish sooner in Keszthely basin due to higher energy currents, smaller depth and coarser deposits. The ratio of 3rd to 4th instar head capsules is about 20%. Furthermore a considerable part of 4th instar *Chironomus* capsules is fragmented in the Keszthely basin.

To trace the eutrophication process through time we examined samples from both basins in 5 cm thick layers (Fig. 5). The total number of 4th instar head capsules in the samples were 387 in the Keszthely basin and 550 in the Northeastern basin. Using the sedimentation rate data we calculate 0.20–0.29 head capsules deposited $cm^{-2} a^{-1}$. In the three layers the numbers of capsules increase from the upper to the bottom level (75, 160, 152 and 133, 198 and 219 in the two basins, respectively). Examining the numbers for different chironomid groups we find the same picture for *Chironomus* heads in both basins and for *Tanypus* heads in the Keszthely basin. In the Northeastern basin the *Tanypus* heads show a slight decrease in number downwards. The numbers of *Procladius* and 'other *Chironominae*' capsules do not vary in a definite way. The proportion of the different groups in all three layers do not differ from those in the total samples.

These results seem to be somewhat different from what would be expected. There are reasons for this other than resuspension and redeposition of the sediment. During our field work we noticed that large-bodied larvae of the *Chironomus plumosus* group live in their tubes 5 to 15 cm deep in the sediment. They make an entire network, aerating the mud and perhaps protecting it from oxygen depletion. This phenomenon was also observed in laboratory experiments (Tátrai, unpublished). A great proportion of the larvae moults and pupates in the lower layers causing the apparent contradiction in the numbers and proportions of head capsules of the *Chironomus plumosus* group in the different layers. The rearrangement of the sediment by larvae also alters the vertical distribution of other chironomid capsules. This causes the erroneous appearance of the abundance of chironomids being greater in the lake 100–200 years ago than now. However, recent differences between the two basins prove the contrary. So the effect of tube-building modifies stratigraphic distribution of capsules and reduces the reliability of the faunal composition in the different layers.

Acknowledgements

This study was undertaken while the authors were in receipt of a grant from the Hungarian

Academy of Sciences. We wish to thank Dr. Ferenc Máté for his helpful advice and valuable criticism. We also wish to thank Mr. Albert Bagyó, István Báthori and Imre Molnár who assisted with field work.

References

Biró, P., 1977. Effects of exploitation, introductions and eutrophication on percids in Lake Balaton. J. Fish. Res. Bd Can. 34: 1678–1683.

Brundin, L., 1958. The bottom faunistical lake type system and its application to the southern hemisphere. Moreover a theory of glacial erosion as a factor of productivity in lakes and oceans. Verh. int. Ver. Limnol. 13: 288–297.

Carter, C. E., 1976. A population study of the Chironomidae (Diptera) of Lough Neagh. Oikos 27: 346–354.

Carter, C. E., 1977. The recent history of the chironomid fauna of Lough Neagh, from the analysis of remains in sediment cores. Freshwat. Biol. 7: 415–423.

Fittkau, E. J., 1962. Die Tanypodinae (Diptera: Chironomidae) Akamie-Verlag, Berlin, 453 pp.

Hargrave, B. T., 1969. Epibenthic algal production and community respiration in the sediment of Marion lake. J. Fish. Res. Bd Can. 26: 2003–2026.

Herodek, S., 1979. Eutrofizálódás, a Balatont fenyegető közvetlen veszély. (Eutrophication, the immediate danger for Lake Balaton.) MTA Biol. Oszt. Közl. 22: 323–336.

Hofmann, W., 1971. Zur Taxonomie und Palökologie subfossiler Chironomiden (Dipt.) in Seesedimenten. Arch. Hydrobiol. Beih. Ergebn. Limnol. 6: 1–50.

Kajak, Z., 1980. Role of invertebrate predators (mainly Procladius sp.) in benthos. In: Murray, D. A. (ed.) Chironomidae. Proc. 7th int. Symp. on Chir., Dublin, 1979. Pergamon Press, Oxford: 339–348.

Saether, O. A., 1979. Chironomid communities as water-quality indicators. Holarct. Ecol. 2: 65–74.

Sebestyén, O., 1968. Pediastrum kawraisky Schmidle maradványok balatoni üledékekben. Tótörténeti tanulmány. (Pediastrum kawraisky Schmidle remains in the sediments of Lake Balaton. A lake-history study.) Hidrológiai Közlöny 48: 1–6.

Thienemann, A., 1913. Der Zusammenhang zwischen dem Sauerstoffgehalt des tiefen Wassers und der tiefen Fauna unserer Seen. Int. Revue Hydrobiol. 6: 243–249.

Warwick, W. F., 1975. The impact of man on the Bay of Quinte, Lake Ontario, as shown by the subfossil chironomid succession (Chironomidae, Diptera). Verh. int. Ver. Limnol. 19: 3134–3141.

Wiederholm, T. & Eriksson, L., 1979. Subfossil chironomids as evidence of eutrophication in Ekoln Bay, Sweden. Hydrobiologia 62: 195–208.

Zólyomi, B., 1953. Die Entwicklungsgeschichte der Vegetation Ungarns seit dem letzten Interglacial. Acta Biol. 4: 307–430.

Zn, P and ATP in the sediments of two lakes in South Finland

L. Koivo[1] & R. Oravainen[2]
[1] *Tampere University of Technology, P.O. Box 527, 33101 Tampere 10, Finland*
[2] *Water Protection Agency, Teiskontie 1, 33500 Tampere 50, Finland*

Keywords: paleolimnology, sediment chemistry, heavy metals, ATP

Abstract

This investigation is concerned with the impact of industrial and municipal waste discharge on lakes near the city of Tampere, Finland. The record of P, Zn and ATP in the recent sediments of Vanajavesi I, a polluted lake, and Mallasvesi, a largely unpolluted lake, are compared.

Introduction

The aim of the present study is to trace man's impact on some Finnish lakes on the basis of analyses of water and sediment samples. The study lakes are situated in the Kokemäenjoki watercourse (A = 27 040 km^2), South Finland (Fig. 1). Sources of pollution in the area are pulp and paper mills, textile factories, municipal sewage plants and fishfarms. In this paper we present preliminary evidence of such pollution by comparing the recent records of P, Zn and ATP in the sediments of two lakes.

Material and methods

The situation of the two lakes, Vanajavesi and Mallasvesi and the sampling sites for water and sediments are shown in Fig. 1. The water samples for chemical analyses were collected in March and August of 1980 with a Ruttner sampler. The total amount of phosphorus in the water was determined according to the Finnish standard method (INSTA-VH 23). The chlorophyll a determination followed a method reported by Iwamura *et al.* (1970), in which 2 h methanol extraction and spectrophotometric measurement is used. Primary production was determined in constant light and temperature according to the ^{14}C method (SFS-3049).

The sediment samples were taken through late winter ice, in March–May 1980, using a freezing technique (Huttunen & Meriläinen 1978). The water depth in the sampling sites varied from 3 to 60 m. Physical and chemical analyses of sediments followed the IGCP Project 158B guidebook (Bengtsson 1979). The final analytical stage for zinc determinations was made using AAS (Instrumentation Laboratory Inc.) at the Tampere University of Technology. Adenosine triphosphate (ATP) was measured using the firefly bioluminescent reaction adapted from Karl (1979).

Results and discussion

Table 1 shows data concerning the trophic state of the two lakes in 1980, and Tables 2 & 3 present the results of phosphorus (P), zinc (Zn), and adenosine triphosphate (ATP) analyses in the recent sediments of the lakes.

Mallasvesi is a clear-water lake with a low nutrient and electrolyte status although oxygen deficiency occurs in depressions in the lake bed during stratification. It is an almost unpolluted lake and as

Hydrobiologia 103, 177–179 (1983).

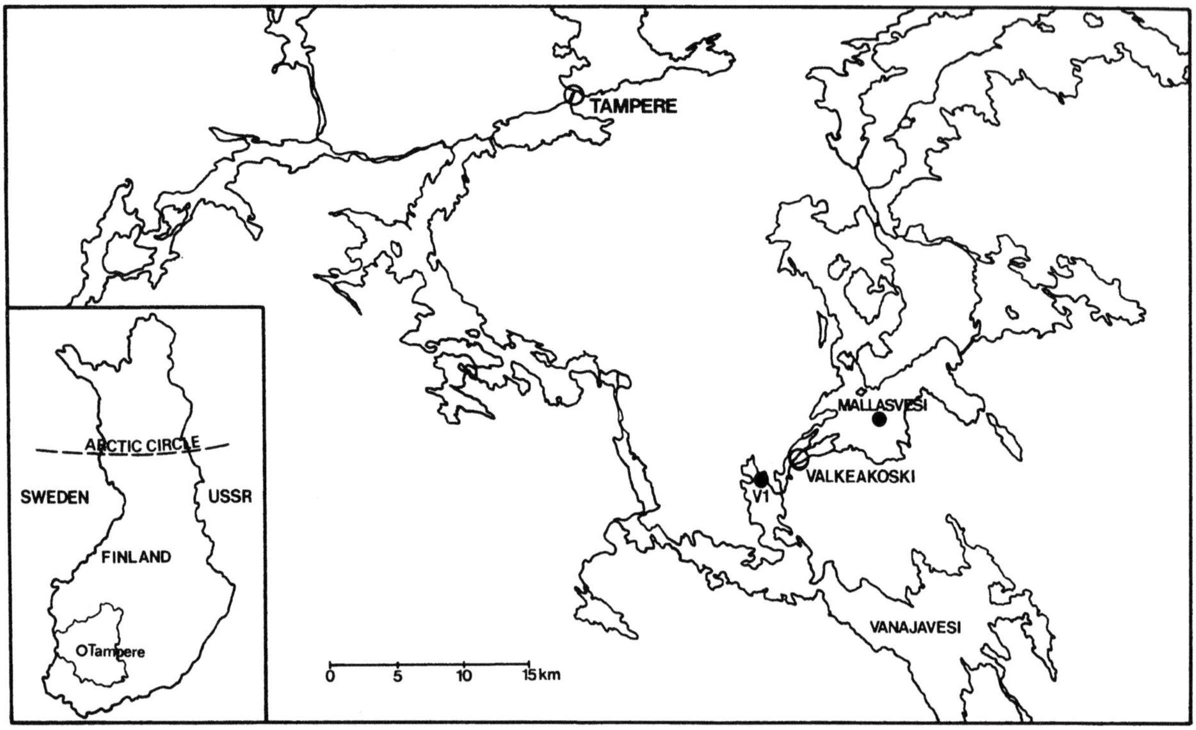

Fig. 1. Locations of sampling points for Mallasvesi and Vanajavesi (V1).

Table 1. Productivity characteristics of the study lakes.

Lake	Year	Secchi disk transparency (dm)	Total P ($\mu g\ l^{-1}$)	Chlorophyll a ($mg\ m^{-3}$)	Primary production – 5000 lux, 20 °C ($mg\ C\ m^{-3}\ d^{-1}$)
Mallasvesi	1980	20	16	2.5	80
Vanajavesi I	1980	15	34	10	420

Table 2. Dry weight (DW), organic matter (OM), total P, Zn and ATP in core samples from Lake Mallasvesi.

Depth (cm)	DW ($g\ kg^{-1}$ FW)	OM ($g\ kg^{-1}$ DW)	P ($g\ kg^{-1}$ DW)	Zn ($g\ kg^{-1}$ DW)	ATP ($\mu g\ g^{-1}$ FW)
0– 1	16.3	4.1	2.7	0.2	0.25
1– 2	36.1	5.0	2.3	0.2	0.25
2– 3	56.2	7.9	2.7	0.3	0.32
5– 6	83.2	8.9	2.0	0.4	0.43
7– 8	100.4	10.8	1.9	0.4	0.13
10–11	134.1	14.7	1.9	0.4	0.13
13–14	134.2	14.5	1.8	0.4	0.10
15–16	158.0	15.4	1.6	0.4	0.12
16–17	164.1	17.1	1.5	0.5	0.12
18–19	182.7	19.0	1.4	0.5	0.09
20–21	194.9	19.7	1.5	0.5	–

Table 3. Dry weight (DW), organic matter (OM), total P, Zn and ATP in core samples from Lake Vanajavesi I.

Depth (cm)	DW ($g\ kg^{-1}$ FW)	OM ($g\ kg^{-1}$ DW)	P ($g\ kg^{-1}$ DW)	Zn ($g\ kg^{-1}$ DW)	ATP ($\mu g\ g^{-1}$ FW)
0– 2	109.9	33.4	2.2	13.2	0.52
2– 4	134.1	32.4	1.5	8.2	0.52
5– 7	150.5	25.9	1.2	3.7	0.50
7– 9	146.7	24.4	1.1	3.0	0.42
9–11	188.3	27.3	1.0	3.3	0.42
15–16	201.1	25.9	0.9	1.9	0.45
20–21	204.3	26.3	0.9	1.1	–
25–26	238.8	28.3	0.9	0.5	–
30–31	286.3	33.1	0.9	–	
33–35	312.2	35.9	0.8	0.7	–

such it is being used as a control site in the present project. It can be seen from Table 2 that, although there is a slight increase in P values over the upper 30 cm, Zn values are consistently low throughout.

Vanajavesi I, on the other hand, is heavily polluted. The main pollution comes from pulp and paper mills, but the lake also receives the discharge of industrial and municipal effluent. In the beginning of the 1970s the water column was completely anoxic during the late winter and summer. The situation has improved after a wastewater treatment plant commenced operation. The zinc load (1 000 kg d^{-1}) from a textile factory has also decreased and the load is now 100 kg d^{-1}.

Evidence for this pollution can be clearly seen in the sediment record (Table 3). When compared with data from Mallasvesi (Table 2) it is clear that the upper 20 cm shows elevated Zn levels.

The ATP data show no trends at either site. However, there have been few measurements of ATP in lake sediments and a number of technical problems need to be overcome if the results are to be interpreted with confidence. Eventually ATP analysis could become a useful palaeolimnological technique.

More data on these and a range of other lakes in the Tampere region are being gathered and sediment cores are being dated by ^{137}Cs and ^{210}Pb techniques. When such results are available a more comprehensive appreciation of the impact of man on these lakes will be possible.

Acknowledgement

Financial support from the Academy of Finland (Natural Science Research Council) is greatly appreciated.

References

Bengtsson, L., 1979. Chemical analysis. In: Palaeohydrological changes in the temperate zone in the last 15 000 years. IGCP, Project 158 B. Lake and Mire Environments. Project Guide II: 113–132, (Ed.) Berglund, B. E. Lund.

Huttunen, P. & Meriläinen, J., 1978. New freezing device providing large, unmixed sediment samples from lakes. Ann. Bot. Fenn. 15: 128–130.

INSTA-VH 23, 1975. Veden kokonaisfosforin määritys.

Iwamura, T., Nagai, H. & Ichimura, S.-E., 1970. Improved methods for determining contents of chlorophyll, protein, ribonucleic acid, and deoxyribonucleic acid in planktonic populations. Int. Revue ges Hydrobiol. 55: 131–147.

Karl, D. M., 1979. Adenosine triphosphate and guanosine triphosphate determinations in intertidal sediments. In: Methodology for Biomass Determinations and Microbial Activities in Sediments. (Eds.) Litchfield/Seyfried, ASTM-STP 673: 5–20.

Clostridium perfringens (Holland) as an indicator of human effluent in the sediment of Lake Tuomiojärvi, central Finland

Kaj Granberg
University of Jyväskylä, Hydrobiological Research Centre, Seminaarinkatu 15, SF 40100 Jyväskylä 10, Finland

Keywords: paleolimnology, *Clostridium perfringens*, sewage pollution, eutrophication

Abstract

The horizontal and vertical distribution of the gram-positive, non-motile, spore-forming and rod-shaped bacterium *Clostridium perfringens* Holland was studied. The aim of the study was to estimate the quantity of *C. perfringens* at different depths of the sediment and evaluate the effect of human effluent which the lake received between 1940 and 1956. *C. perfringens* lives in the colon of man. Because it is spore forming and cannot multiply under a temperature of 20 °C and, according to the studies of Seppänen *et al.* (1979) it can be at least 300 years old, it may be a suitable paleolimnological indicator of pollution by human effluent.

The results showed that the amounts of *Clostridium* increased at the same level where redox potential decreased in the sediment due to the beginning of effluent disposal at a depth of 40 mm. The maximum number of *Clostridium* colonies occurred between 0–30 mm depth.

Introduction

Lake Tuomiojärvi is a mesotrophic lake situated near the town of Jyväskylä. The lake is used for water supply by the town of Jyväskylä and has been under intensive limnological study since 1968. Evidence for recent eutrophication from diatom and other paleolimnological analyses has earlier been presented (Granberg 1972; Granberg & Nyrönen 1978).

The surface area of the lake is 3.1 km^2, the water volume is 11.5×10^6 m^3, the theoretical retention time is 252 days and the drainage area is 56 km^2. Lake Tuomiojärvi received human effluent from 1940–1956 and the lake became enriched. The lake is still eutrophic due to diffuse nutrient inflow from the drainage area.

The aim of this paper is to describe the horizontal and vertical distribution of *Clostridium perfringens* Holland in the sediments of Lake Tuomiojärvi. *C. perfringens* is an anaerobic, spore-forming, gram-positive and rod-shaped bacterium. It lives in the colon of man. The bacterium does not reproduce under +20 °C (Bonde 1963), and thus it exists as spores in lake water or in sediments, and it is a suitable indicator of waste water pollution. *C. perfringens* survives well in sediments. Seppänen *et al.* (1979) have found the species in the sediments of eutrophic Lake Hiidenvesi at a depth of 15–20 cm, which, according to pollen and ^{14}C analyses, represents 170–300 years BP (Harjula 1972). It is possible, therefore that the bacterium can also be used to indicate past pollution.

Methods

Clostridium was cultivated with an anaerobic plate technique using Marshall agar (Merck 5264).

The pH of the medium at 37° is 7.0–7.4. The sediment samples were pasteurized before cultivation at 80 °C for 5 min. Anaerobic conditions in the medium were obtained with the use of Ott's anaerobic ring. The cultivation temperature was 47 °C

Hydrobiologia 103, 181–184 (1983).

and the cultivation time 16–18 h. The colonies were black. According to Seppänen *et al.* (1979) the selectivity of the medium is relatively good.

Samples were taken from the sediment surface at eleven sites (Fig. 1), and a core was taken from site T1, using a Kajak sampler. The core was 11 cm long, and was divided into 12 slices, 5 mm thick from 0–5 cm, and then 1 cm thick, with a 1 cm interval between slices. The sediment accumulation rate of Lake Tuomiojärvi is approximately 1.4–1.5 mm a^{-1} (Granberg 1972), thus the lowest sample (100–110 mm depth) is about 70 years old.

Dry weight and ignition losses were determined for all sediment samples and chlorophyll *a* and Cu:Zn-ratios were measured for the core samples. Chlorophyll *a* was measured from dried material after acetone extraction for 24 h. Extinctions were measured after centrifugation and the results were calculated according to Parsons & Strickland (1963). The results are expressed in μg chlorophyll g^{-1} dry weight (Palomäki 1978).

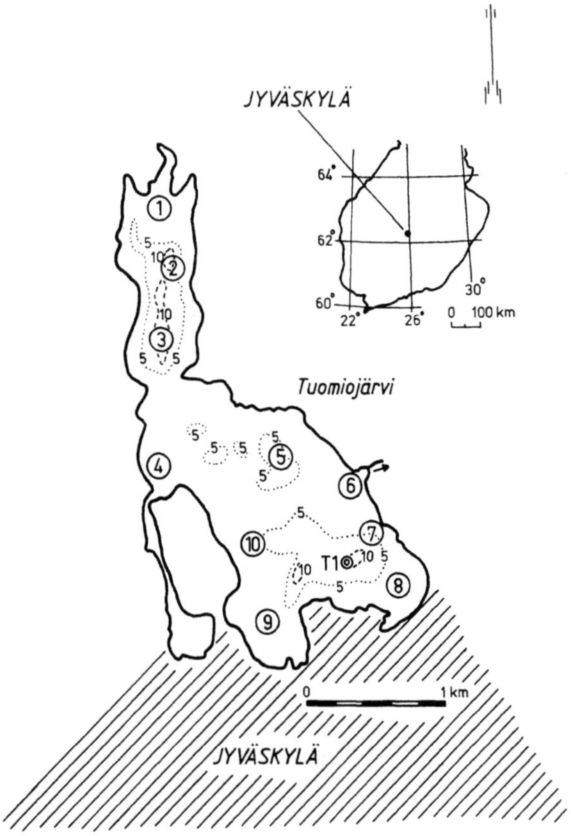

Fig. 1. Sampling stations (1–10) in Lake Tuomiojärvi. T1 = the deepest site (ca. 10 m). Dotted lines are depth curves m.

Copper and zinc were measured by AAS at the laboratory of Helsinki City waterworks. Cu:Zn-ratio indicates palaeoredox conditions (Hallberg 1974) and has been used by Vuorinen (1978) and Granberg & Nyrönen (1978).

Statistical treatment of Clostridium results

The error estimate for the number of colonies was solved according to the method of Niemelä (1978) by assuming that the bacteria are distributed in the sediment according to the negative binomial distribution. The standard deviation (s) in this case is:

$$(1) \quad s = \frac{V}{V_l} \sqrt{c + \frac{c^2}{k}}, \text{ where}$$

V_l = the volume of the original sample
V = the volume in which the number of bacteria is presented
c = the number of colonies counted
k = the empirical constant

With the aid of the coefficient $\frac{V}{V_l}$ the number of colonies is expressed either as the number of colonies per dry weight or colonies per fresh weight, and equation (1) gives the error limits for the counts. Because the empirical constant k varies in different lakes it must be calculated with the aid of the equation

$$(2) \quad k = \frac{\bar{x}^2}{(s^2_{\bar{x}} - \bar{x})}, \text{ where}$$

$\bar{x}$ is the mean number of *Clostridium* for the eleven sampling stations, and $s^2_{\bar{x}}$ is the variance. Means and variances were calculated from the original counts (dilution 10^{-1}), and the value of k was 84.789.

Results

The results are presented in Fig. 2. The Cu:Zn-results from the 1975 core (Granberg & Nyrönen 1978) are presented in parallel with the results of the

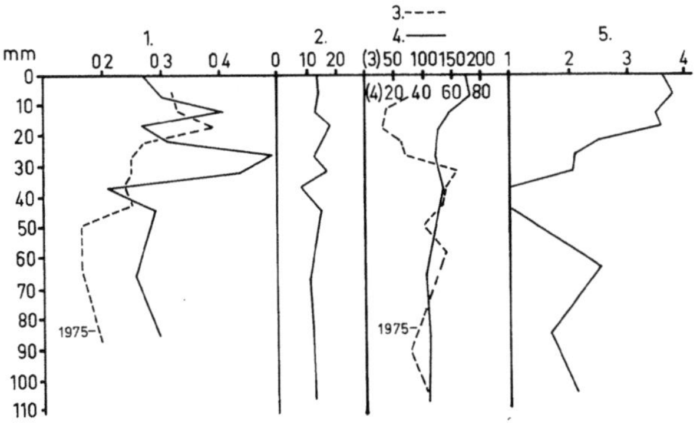

Fig. 2. The results of core analysis T1. (1) Cu:Zn-ratio, (broken line = results from 1975); (2) loss on ignition %; (3) the number of diatoms × 10^6 g^{-1} dry weight; (4) chlorophyll-*a* μg^{-1} dry weight; (5) number of *Clostridium perfringens* $\log_{10}$ colonies g^{-1} dry weight.

Table 1. Clostridium perfringens in the surface sediment of Lake Tuomiojärvi.

Station	Colonies g^{-1} fresh sediment	Colonies g^{-1} dry sediment
1	330	1 822 ± 374
2	360	2 927 ± 584
3	530	4 274 ± 755
4	160	1 082 ± 295
5	–	–
6	115	714 ± 225
7	210	1 743 ± 425
8	110	1 294 ± 415
9	110	738 ± 237
10	120	690 ± 213
T 1	290	4 866 ± 1 046

1979 core after correction for differences in sediment accumulation rate. The decrease in redox potential, indicated by the increase in the Cu:Zn-ratio begins at about 35–40 mm. The maximum Cu:Zn-ratio seems to have been in the 1950's. In the results of Vuorinen (1978) from Lake Hännisenlampi in North Karelia Cu:Zn-ratios of 0.3–0.4 represented redox values of -200 mV. Here Cu:Zn-ratios from 0.2–0.5 were obtained.

The slight increase in loss on ignition values (Fig. 2) may be related to enrichment processes in the lake. The amount of chlorophyll *a* in the sediment varied between 42 and 74 $\mu g\ g^{-1}$ dry weight. An increase can be seen in the top 20 mm. The comparison of chlorophyll *a* to the numbers of diatom valves (Fig. 2 and Granberg & Nyrönen 1978) do not show a clear correlation. At 35 mm the number of diatoms decrease although the amount of chlorophyll *a* increases. This may have been due to the increase of other algal groups than the diatoms.

The amounts of *Clostridium* in the sediment profile are shown in Fig. 2 (note $\log_{10}$ scale). They range from zero (35–40 mm, 40–50 mm) to 880 ± 134 colonies g^{-1} fresh wt (5–10 mm).

The main increase occurs at a depth of 25 mm (Fig. 2), and there are a much greater number of colonies in the upper part of the sediment. Seppänen *et al.* (1979) estimated the amount of *Clostridium* in the eutrophic and polluted Lake Hiidenvesi, South Finland. They found 1340–2380 colonies in the sediment ml^{-1} fresh weight in the polluted areas, and 640–690 colonies ml^{-1} in the less polluted areas. The numbers of *Clostridium* are lower in Lake Tuomiojärvi sediment but they are considerably higher than the amounts found in the surface sediment of Northern Päijänne of 146 ± 154 (n = 15) colonies g^{-1} fresh weight or 853 ± 934 colonies g^{-1} dry weight were found (Granberg 1976).

The quantity of *Clostridium* in the surface sediment of Lake Tuomiojärvi at sites 1–10 and T 1 are shown in Table 1. The surface sediment values represent the present situation in Lake Tuomiojärvi. If these are compared to the results of Seppänen *et al.* (1979) it can be seen that Lake Tuomiojärvi is at present cleaner than both the polluted areas and the 'less polluted' areas of Lake Hiidenvesi. It is evident that the pre-pollution level of *Clostridium* in Lake Tuomiojärvi is less than 300 colonies g^{-1} dry weight, (30 colonies g^{-1} fresh weight), judging from the values found in the sediment deeper than 30 mm.

The results presented here are preliminary, and more investigation is needed to fully evaluate the usefulness of *C. perfringens* as a paleolimnological indicator of sewage pollution.

References

Bonde, G. J., 1963. Bacterial indicators of water pollution. 2nd ed. Teknisk forlag. Copenhagen. 422 pp.

Granberg, K., 1972. Tuomiojärven tutkimus vv. 1970–1971. Summary: The limnological studies of Lake Tuomiojärvi in 1970–1971. Rep. Hydrobiol. Res Inst. 31: 1–61.

Granberg, K., 1976. Asumajätevesien leviäminen Pohjois-Päijänteelle *Clostridium perfringens*-bakteerin perusteella. Summary: The distribution of human effluents into Northern Päijänne according to the amounts of *Clostridium perfringens* in the sediments. Rep. Hydrobiol. Res. Inst. 77: 1–16.

Granberg, K. & Nyrönen, J., 1978. Tuomiojärven limnologiset tutkimukset vv. 1975–1976. Summary: The limnological studies of Lake Tuomiojärvi in 1975–1976. Rep. Hydrobiol. Res. Inst. 83: 1–41.

Hallberg, R. O., 1974. Paleoredox conditions in the Eastern Gotland Basin during the recent centuries. Merentutkimuslait. Julk. no 238: 3–16.

Harjula, H., 1972. Hiidenveden ja Tuusulanjärven rehevöitymisen kehitys pohjasedimenttien valossa. Institute of Limnology, University of Helsinki.

Niemelä, S., 1978. Vesianalyysin tulostaminen ja tulosten merkitsevyyden arviointi. B. Mikrobiologiset menetelmät. Ympäristö ja Terveys 7: 479–482.

Palomäki, A., 1978. Paleolimnologinen tutkimus luonnontilaisesta, rehevöityneestä ja puunjalostusteollisuuden muuttamasta järvestä. Department of Hydrobiology, University of Jyväskylä.

Parsons, T. R. & Strickland, J. D. H., 1963. Discussion of spectrophotometric determination of marine plant pigments with revised equations for ascertaining chlorophylls and carotenoids. J. Mar. Res. 21: 155–163.

Seppänen, H., Ojanen, T. & Zaiss, U., 1979. Die vertikale Verteilung des Fäkalindikators Clostridium perfringens in den Sedimenten des Sees Hiidenvesi. Aqua Fennica.

Vuorinen, J., 1978. The influence of prior land use on the sediments of a small lake. Pol. Arch. Hydrobiol. 25: 443–451.

Paleolimnological studies of annually-laminated sediments in Loe Pool, Cornwall, U.K.

M. A. Coard, S. M. Cousen, A. H. Cuttler, H. J. Dean, J. A. Dearing[1], T. I. Eglinton, A. M. Greaves, K. P. Lacey, P. E. O'Sullivan*, D. A. Pickering, M. M. Rhead, J. K. Rodwell & H. Simola[2]
Department of Environmental Sciences, Plymouth Polytechnic, Plymouth PL4 8AA, U.K.
[1] *Department of Geography, Coventry Polytechnic, Coventry, CV1 5FB, U.K.*
[2] *Ecology Section, Karelian Institute, University of Joensuu, SF 80100 Joensuu 10, Finland*
* *Address for correspondence*

Keywords: paleolimnology, annual laminations, magnetism, sediment influx, sediment chemistry, eutrophication

Abstract

The sediments of Loe Pool, a eutrophic coastal lake in south west England, consist largely of laminated clays and clay-gyttjas. Studies of the diatom microstratigraphy of frozen sediment cores from the Pool indicate that the laminations are annual, and that they contain pairs of light and dark bands formed by seasonal variations in the supply of sediment to the Pool from its catchment. Analysis of the magnetic properties of individual laminations demonstrates the presence of physical and mineralogical microstratigraphic variations, which may also be related to seasonality.

A varve chronology, which is confirmed by ^{137}Cs analysis and historical records, has been used to provide a timescale for the interpretation of data from other paleolimnological studies. A close agreement between variations in the abundance of sedimentary Sn, and the history of mining in the catchment, has been found. Similarly, analysis of total organic matter, total phosphorus, sedimentary chlorophyll *a*, sterols, diatoms and Cladocera in the uppermost sediments all indicate eutrophication of the Pool in the period AD 1940 to the present.

Introduction

The sediments of Loe Pool (Fig. 1), a small eutrophic coastal lake, 1 km south of the town of Helston, Cornwall (south-west England) are annually-laminated (Simola *et al.* 1981). Here we present a summary of studies undertaken so far on the paleolimnology of the Pool, which was originally formed by the damming of the mouth of the river Cober by a shingle bar. The morphometry of the present basin is shown in Fig. 1 and the major physical and hydrological characteristics of Loe Pool and its catchment are listed in Table 1.

Most of the northern half of the catchment is underlain by the Carnmenellis granite, with which several areas of former mining, especially for cassiterite (SnO_2), are associated. The major contemporary economic activity is farming and the main town in the catchment is Helston (population ~ 10 000). A further important centre of population is the military base of RNAS Culdrose, commissioned in 1947.

In a number of recent years, visible blooms of green (*Chlorella, Volvox* spp.) and blue-green (*Microcystis aeruginosa*) algae have appeared in Loe Pool. The lake receives from its catchment some 300 t N and 14 t P a^{-1}. About 75% of the N comes from agricultural run-off, and ~80% of the P from two sewage treatment works, serving Helston and Culdrose. The main taxa present in the phytoplankton are characteristic of eutrophic lakes. In August, total chlorphyll *a* levels in the water column may exceed 500 $\mu g\ l^{-1}$ in sheltered parts of the lake.

Hydrobiologia 103, 185–191 (1983).

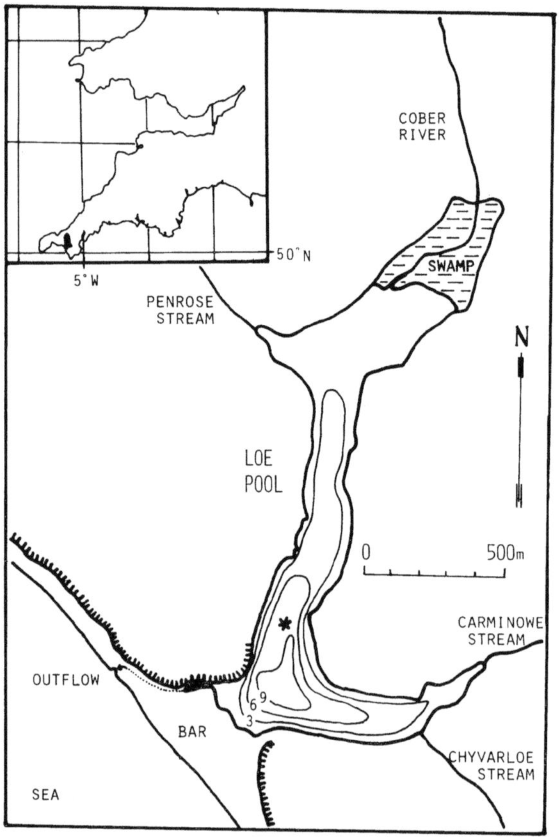

Fig. 1. Loe Pool, showing depth contours (m) and main coring site (asterisk). Inset map indicates the location of the lake's catchment area near the SW tip of Great Britain.

Table 1. Principal physical and hydrological characteristics of Loe Pool.

Latitude	50 ° 04′ N
Longitude	5° 17′ W
Altitude	4 m OD
Area (A)	55.6 ha
Length (L)	1.25 km
Breadth (B)	250 m
Maximum depth (Z_{max})	10.67 m
Mean depth ($\bar{Z}$)	3.47 m
Relative depth (Z_r)*	1.27
Volume (V)	1.93×10^6 m^3
Mean hydraulic residence time	20 days
Area of drainage basin (D)	50 km^2
D/A	98.9

* $Z_r = 50\ Z_{max} \sqrt{\pi}\,(\frac{1}{\sqrt{A}})$

Sediment stratigraphy and chronology

The location of the main coring site is shown in Fig. 1. Here occur sediments that are undisturbed either by the influence of the river Cober, or by marine incursions through Loe Bar. These appear to have taken place until the late nineteenth century.

The stratigraphy of the upper 3 m of Loe Pool sediment has been reported by Simola *et al.* (1981), and by O'Sullivan *et al.* (1982). At the top of the sediment are found 20–40 cm of dark brown clay-gyttja, underlain by a thick layer of red and grey

Fig. 2. Infra-red photograph of a frozen sample of the uppermost sediments of Loe Pool. Junction between brown clay-gyttja and red haematite-clay occurs in this section at 36 cm (Photograph: S. Johnson).

clay. In some cores, the uppermost, highly organic, section is laminated throughout (Fig. 2). In others, a bioturbated layer, corresponding approximately to the 1960's, occurs. The dark-brown layers represent the winter months and the light brown, which are more diatomaceous, the summer (Simola *et al.* 1981).

The red colour found in the irregularly laminated claycy section below is due to the presence of haematite (Fe_2O_3). At one of the former mines in the Cober catchment, that known as Bassett and Grylls/ Porkellis United, the cassiterite ore was embedded in a ground rock (or *mundic*) rich in this mineral. Several newspaper and other accounts dating from the 1920's and 1930's (Hamilton-Jenkins 1978) describe the Pool as being completely discoloured by mine waste material, rich in haematite, which was discharged into the River Cober. We therefore correlate haematite layers in the cores with periods of activity at this mine.

Basset and Grylls was closed in 1938, and since 1940, no mines have operated in the Cober catchment. The transition between the uppermost layer of haematite clay, which is a very prominent marker in Loe Pool sediments, and the overlying brown clay-gyttja can thus be dated to AD 1938.

The mean sediment accumulation rate of the clay-gyttja is therefore *ca.* 1 cm a^{-1}. The peak of ^{137}Cs activity (cf. Pennington *et al.* 1973) of AD 1963, lies at ~16 cm in this core, halfway between the top of the uppermost haematite clay, and the present sediment surface.

Deposition of the red and grey clays was very much more rapid. In all, only eight years are represented by *ca.* 80 cm of sediment. Below these are regularly laminated black/grey clays that accumulated at ~3 cm a^{-1}. In total therefore, a 3 m core from Loe Pool sediment may represent only the last 100–120 a.

Using the red clay as a marker, O'Sullivan *et al.* (1982) investigated the distribution of these three sediment types throughout Loe Pool. They found the average thickness of the clay-gyttja to be 35 cm, and that of the uppermost red haematite clay to be 20 cm. The depth of the black/grey clay is, as yet, unknown.

By calculating dry matter and ash content of each type of sediment, it is possible to show that the brown clay-gyttja represents the accumulation of some 18 t dry matter ha^{-1} a^{-1} over the basin of the Pool. This is the equivalent to the erosion of some 20 t km^{-2} a^{-1} of mineral matter from the Cober catchment as a whole.

During the 1930's, however, when mining was intensive, accumulation equalled 440 t dry matter ha^{-1} a^{-1}, and erosion rates some 450–550 t km^{-2} a^{-1}. The latter figure is however, probably a considerable overestimate of the general soil erosion level, as the main sources of matter during this period were very small mining areas. In the late nineteenth century the dry matter accumulation rate in the Pool was 163 t ha^{-1} a^{-1}, and the average erosion rate 236 t km^{-2} a^{-1} from the catchment.

Magnetic studies

Figure 3 illustrates the results of single-sample analysis of two cores of Loe Pool sediment for magnetic susceptibility (χ). On the right of the diagram is shown analysis of a 1 m Mackereth core, on the left analysis of a frozen sediment section taken using the 'box-freezer' sampler (Huttunen & Meriläinen 1978). In each case the position of the top

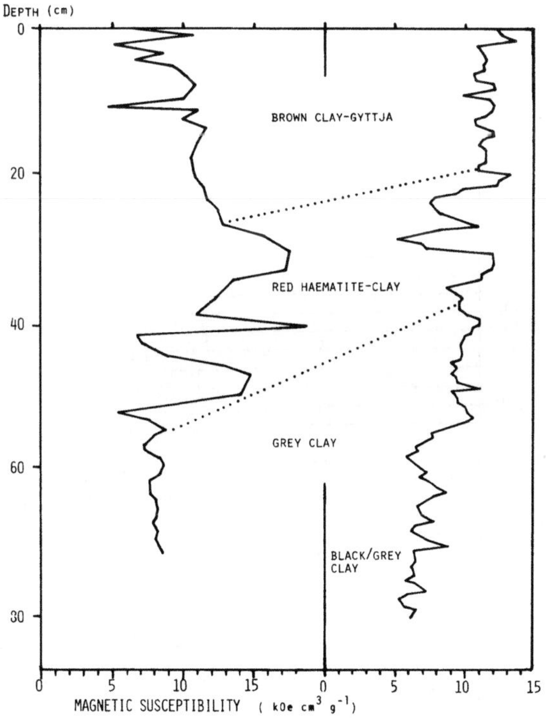

Fig. 3. Magnetic susceptibility (χ) of single samples from two cores of Loe Pool sediment.

and bottom of the uppermost red haematite clay is shown.

In both cores, χ rises to a maximum at the top of the red clay, but then falls sharply to a minimum, particularly in the Mackereth core. A pronounced peak in χ, associated with a black (summer) layer then occurs. At the base of the red clay, χ again reaches a maximum. These results suggest that χ may be used to identify precisely the location of the red clay layer in whole cores, and thus allow refinement of investigations such as that of sedimentation in the Pool described above.

At the top of the frozen section, a series of fluctuations in χ are associated with colour changes in the sediment. Peaks in χ are correlated with the paler (summer) layers, and minima with the darker (winter) laminations. Similarly, in the lower parts of both this section, and the Mackereth core, peaks in χ are associated with summer layers, and minima with winters. It may therefore be that magnetic parameters may be used to identify seasonal changes in sediment composition associated with lamination formation.

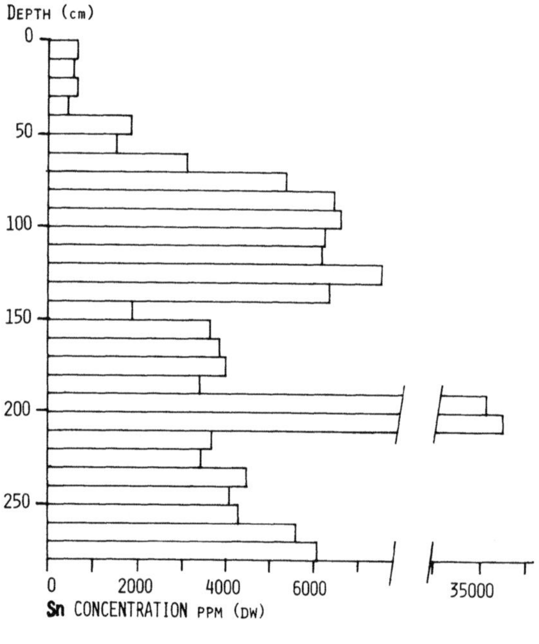

Fig. 4. XRF analysis of 10 cm segments of a 3 m Mackereth core from Loe Pool for Sn.

Tin (Sn) concentration

The Sn content of 10 cm sections of the same 3 m core as analysed by Simola *et al.* (1981) was measured using the technique of X-ray fluorescence (XRF). A ^{241}Am source, at 60 keV was employed. Results are shown in Fig. 4.

In the organic clay-gyttja (0–40 cm), Sn values are relatively low (<1 000 ppm), but in the red haematite-clay (80–140 cm) they rise to a peak of ~6 400 ppm. According to the varve chronology outlined above, this peak corresponds to AD 1925–1936, which coincides with the last documented phase of mining. Sn values then fall to ~3 600 ppm (below 140 cm), except for a peak of >35 000 ppm (3.5%!) at 200 ± 10 cm. This peak dates from the period AD 1900–1910, when mines were very active in the Cober catchment. Finally, at the base of the core, in the period AD 1870–1876, a peak of ~6 000 ppm occurs. This is the earliest period represented here, and one in which the greatest number of mines ever (28) were active in the Cober catchment.

There is thus a very close agreement between the results of XRF analysis for Sn, the varve chronology of the Loe Pool sediments, and the mining history of the catchment. The results confirm ideas about the origin of haematite-clays as mine wastes, and demonstrate the potential of XRF as a paleolimnological technique.

Organic geochemistry

Analysis of a 1 m Mackereth core for total organic matter (TOM), sedimentary chlorophyll *a*, and also total phosphorus (P_{tot}) are shown in Fig. 5. The results indicate that since AD 1938, TOM, sedimentary chlorophyll *a*, and P_{tot} have all significantly increased. We interpret these results as indicating higher internal P loading, and increased productivity of the Pool, in the period since AD 1940, as a result of greater nutrient inputs from its catchment.

At present, the precise cause of eutrophication of Loe Pool is uncertain. Inorganic fertilisers have been used in the area since the 1940's, and Helston sewage treatment works was first opened in 1930, extended in 1959, and again in 1974. Before that time, untreated sewage from the town was dis-

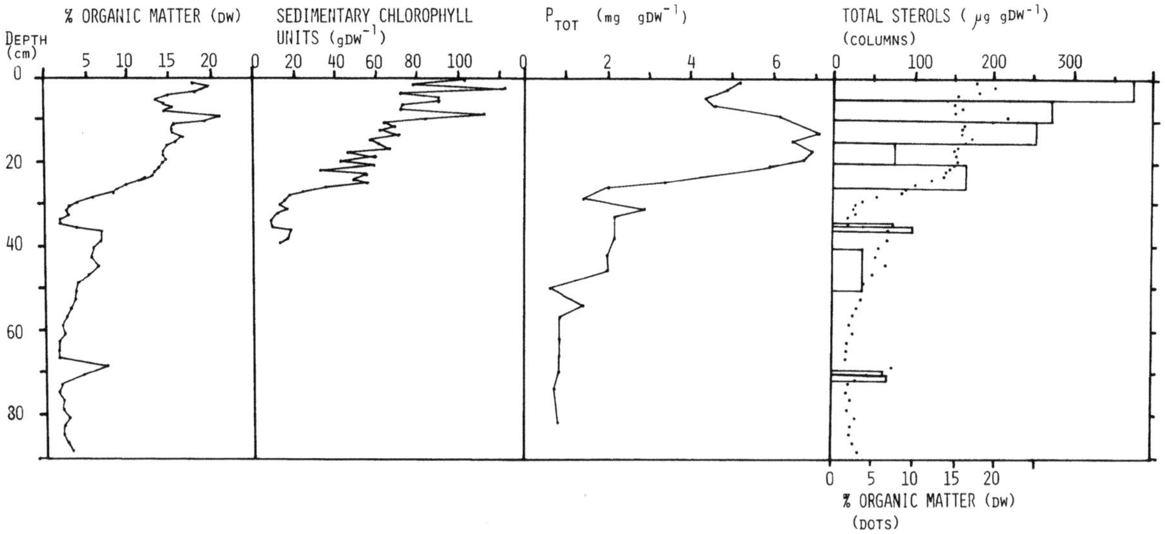

Fig. 5. Organic matter percentage, sedimentary chlorophyll *a*, total phosphorus (P_{tot}) and total sterol concentration in a 1 m Mackereth core from Loe Pool.

charged directly into the Cober, and even in AD 1900, was the cause of unpleasant odours (Vallentin 1903). RNAS Culdrose, which has a separate sewage works, was commissioned in 1947.

Eutrophication of Loe Pool over the last few decades is also suggested by analysis of sediment cores for sterols using both gas-liquid chromatography (GLC) and computerised gas chromatography mass-spectrometry (GCMS).

Different sterols or groups of sterols may be characteristic of various groups of organisms (Huang & Meinschein 1979). In particular, it has been shown that C_{27} sterols, which are produced by phytoplankton, are more abundant in the sediments of eutrophic lakes (Gaskell & Eglinton 1976) and that C_{29} sterols are more characteristic of higher plants.

In cores of both frozen and unfrozen sediments from Loe Pool, a total of 14 sterols were identified. The majority of these were C_{29} sterols, which suggests considerable allochthonous input of organic matter into the sediment. Total sterol abundance increases from <100 µg gdw^{-1} in the red and grey clays, to over 350 µg gdw^{-1} near the sediment surface, (Fig. 6). However, this distribution may be due as much to degradation of sterols below the sediment surface, as to any real increase in the original concentrations. More significant is the finding that the ratio of C_{27}:C_{29} sterols increases in the upper parts of the clay-gyttja (Fig. 7). This strongly suggests an increase in the internal productivity of the Pool in the period since AD 1940.

Figure 6 also shows the results of analysis of the sterol content of some individual laminations.

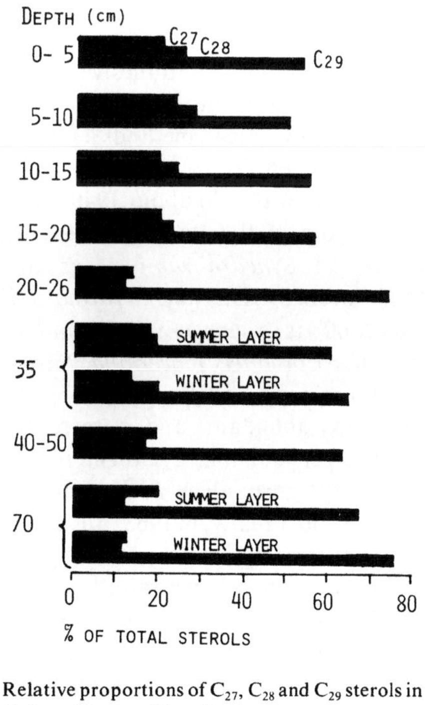

Fig. 6. Relative proportions of C_{27}, C_{28} and C_{29} sterols in various strata of the sediment of Loe Pool.

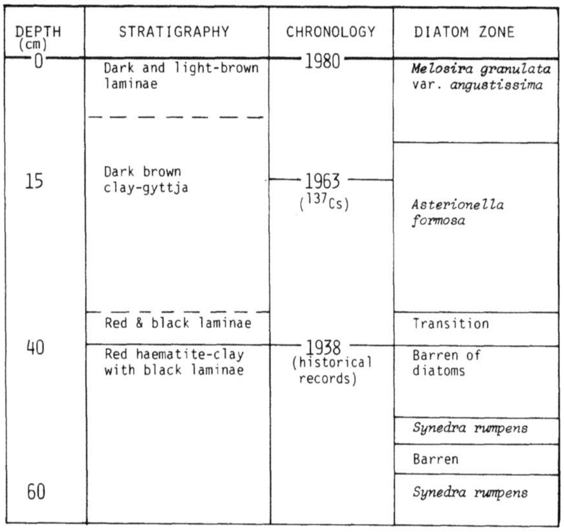

Fig. 7. Summary diagram of diatom stratigraphy of the uppermost sediment of Loe Pool.

Again, an increase in the proportion of C_{27} sterols is shown in the 'summer' laminations. This is consistent with the idea that this part of each lamination is deposited in the growth season for phytolankton (Simola *et al.* 1981).

Diatom analysis

A summary of results of an adhesive tape analysis of the uppermost sediments of Loe Pool are shown in Fig. 7. The analytical methods employed are identical to those of Simola (1977).

The results show that in about 1940 a change in the diatom flora of the Pool occurred, with the replacement of *Synedra rumpens* as the most abundant diatom by *Asterionella formosa.* During the 1950's & 60's taxa such as *Synedra pulchella, S. acus, S. ulna, Cocconeis placentula, Surirella capronii, Nitzschia* spp. and *Pinnularia* spp. gradually became less abundant, and were replaced by *Thalassiosira pseudonana, Cyclotella meneghiniana,* and also the green algae *Pediastrum* spp. and *Scenedesmus* spp. Finally, in 1968, *Melosira granulata* var. *angustissima* increased rapidly in abundance, *Scenedesmus* became rare, and *Cyclotella meneghiniana* and *Asterionella formosa* declined. We interpret these changes as indicating eutrophication of the Pool, in association with increased nutrient loadings.

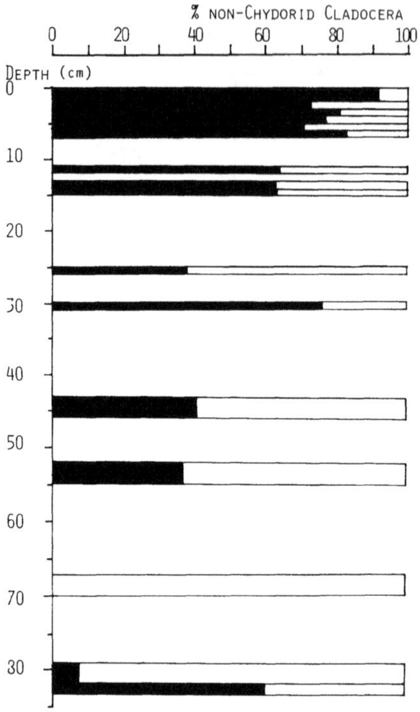

Fig. 8. Percentage of non-Chydorids (black column) of total Cladoceran remains at various levels in a 1 m Mackereth core from Loe Pool.

Cladoceran analysis

Results of analysis of Cladoceran remains are summarised in Fig. 8. The methods of extraction and counting employed were those described by Frey (1979). As in the case of the diatom analysis, a major change in the Cladoceran fauna of the Pool took place in the 1950's, when the proportion of non-Chydorid Cladocera present increased from <50% to over 60% of the total present. In the topmost sediments (1970–1980), non-Chydorids constitute 70–90% of the total. In the lower part of the core (below 40 cm), the main taxa encountered were *Chydorus sphaericus, Alona* sp. and *Daphnia* cf. *longispina.* In the upper sediments, however, *Bosmina* species, both *longirostris* and *coregoni*, are abundant. Again this result confirms the idea that the Pool has become more eutrophic in recent decades.

Acknowledgements

We would like to thank Lt. Cdr. J. P. Rogers, and the National Trust for England and Wales, for permission to work at Loe Pool, and for their continued interest in our research. We would also like to acknowledge the help of Professor Geoffrey Eglinton, who kindly allowed the use of GLC and GCMS facilities at the University of Bristol. Dr. Mary Boucherle, of the University of Indiana, USA gave helpful advice on Cladoceran analysis. Much of the work was carried out whilst M.A.C., K.P.L. and H.S. were supported by Devon Local Authority Research Assistantships, A.M.G. by a SERC studentship, and D.A.P. by a NERC studentship. Heikki Simola also received financial support from the Alfred Kordelin Foundation (Finland).

References

Frey, D. G., 1979. Cladocera analysis. In: B. E. Berglund (Ed.) Paleohydrological changes in the Temperate Zone in the last 15 000 years: Lake and Mire environments. Vol. 2, Specific methods. IUGS, UNESCO project No. 158B, Lund, Sweden.

Gaskell, S. J. & Eglinton, G., 1976. Sterols of a contemporary lacustrine sediment. Geochim. Cosmochim. Acta, 40: 1221–1228.

Hamilton-Jenkins, A. K., 1978. Wendron Tin. Wendron Forge, Wendron, Cornwall, U.K.

Huang, W. Y. & Meinschein, W. G., 1979. Sterols as ecological indicators. Geochim. Cosmochim. Acta 43: 739–745.

Huttunen, P. & Meriläinen, J., 1978. New freezing device providing large, unmixed sediment samples. Ann. Bot. Fenn. 15: 128–150.

O'Sullivan, P. E., Coard, M. A. & Pickering, D. A., 1982. The use of laminated lake sediments in the estimation and calibration of erosion rates. In: Recent Developments in the Explanation and Prediction of Erosion and Sediment Yield (Proceedings of the Exeter Symposium, July 1982) IAHS Publ. No. 137.

Pennington, W., Cambray, R. S. & Fisher, E. M., 1973. Levels of ^{137}Cs in British lake sediments. Nature 242: 824–826.

Simola, H., 1977. Diatom succession in the formation of annually laminated sediment in Lovojärvi, a small, eutrophicated lake. Ann. Bot. Fenn. 14: 143–148.

Simola, H., Coard, M. A. & O'Sullivan, P. E., 1981. Annual laminations in the sediments of Loe Pool, Cornwall. Nature 290 (5803), 238–241.

Vallentin, R., 1903. Observations on the plankton of Loe Pool. J. Roy. Inst. Cornwall 15: 328–345.

Recent geologic development of Lake Michigan (U.S.A.)

David L. Gross & Richard A. Cahill
Illinois Department of Energy and Natural Resources, State Geological Survey Division, Natural Resources Building, 615 East Peabody Drive, Champaign, IL 61820, U.S.A.

Keywords: paleolimnology, Lake Michigan, geochemistry, stratigraphy, Quaternary, trace elements

Abstract

The stresses placed on Lake Michigan since the advent of industrialization require knowledge of the sedimentology of the whole lake in order to make informed decisions for environmental planning. Sediment accumulation rates are low: areas of the lake receiving the most sediment average only 1 mm a^{-1}; deep-water basins average 0.1 to 0.5 mm a^{-1}; and large areas are not receiving any sediment. Sediment was deposited rapidly (typically 5 mm a^{-1}), in the form of rock flour, during the deglaciation of both Lake Michigan and Lake Superior Basins. Then the rate of accumulation decreased by 80–90% and has remained relatively constant since final deglaciation. Because active sedimentation occurs mostly in the deep water areas of the lake, the sediment remains undisturbed and contains a record of the chemical history of the lake.

Introduction

Lake Michigan is the only one of the five Great Lakes to lie entirely within the United States. The mean depth of the lake is 85 m; maximum depth is 281 m. The lake has a surface area of 58 000 km^2 and a drainage area of 176 000 km^2. The drainage basin is occupied by 14 milion people, and land use ranges from the forested, sparsely populated lands in the north to the heavily populated and industrialized areas in the south, which include both the City of Chicago and highly productive agricultural lands.

The basin of Lake Michigan was carved from Paleozoic bedrock by the action of southward-flowing glacial ice. The early history of the development of the basin is not known. The most recent episode of glaciation eroded all earlier glacial deposits and continued the deepening of the bedrock basin. All of the glacial tills known to exist in the lake are younger than 14 000 radiocarbon years B.P. Those tills and the postglacial sediment in the lake are the subjects of this paper.

Stratigraphy

The stratigraphy of Quaternary sediment in Lake Michigan has been described by a series of publications of the Illinois State Geological Survey. The stratigraphic classification for the lacustrine sediment was established by Lineback *et al.* (1970) and was extended to the whole lake basin by Wickham *et al.* (1978). The four underlying glacial tills of late Wisconsinan age were described by Lineback *et al.*(1974). The influence of the regional sedimentology on the fate of pollutants introduced to the lake was first noted by Gross *et al.* (1972) and Lineback and Gross (1972).

A segment of high-resolution, seismic reflection profile illustrates the stratigraphy (Fig. 1). Approximately the lower half of the Quaternary sequence in the lake is glacial till. The five tills in the lake represent small readvances of the waning Wisconsinan ice sheet, with each succeeding advance failing to reach as far south as the previous one. The contact between the two oldest tills (Fig. 1) illustrates the buried, sublake terminal moraine of the

Hydrobiologia 103, 193–198 (1983).

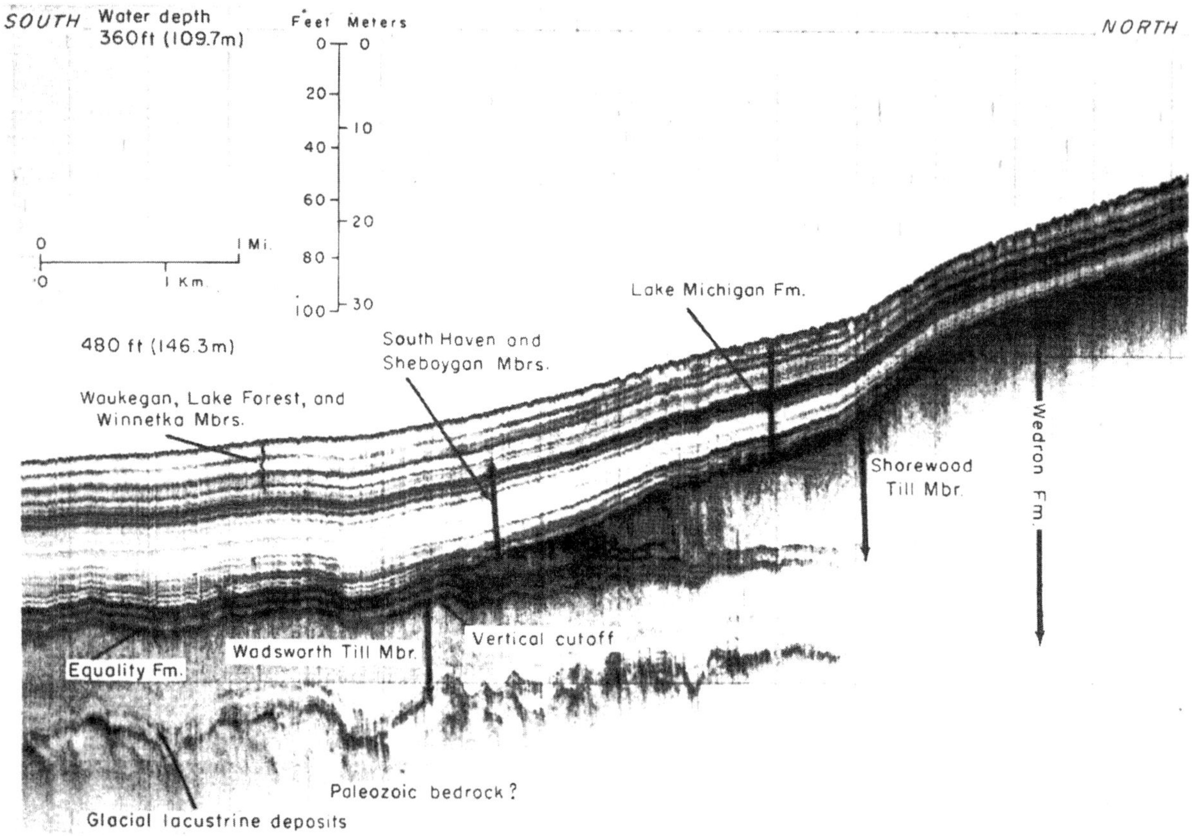

Fig. 1. Seismic reflection profile from the northern portion of the southern basin of Lake Michigan. (Reprinted from Lineback *et al.* 1974).

Shorewood Till. The Equality Formation is a thin unit, typically 1 or 2 m, consisting of silt, sand, and clay-pebble conglomerate laid down in the lake directly along the margins of the melting glaciers. The Equality is time transgressive, becoming younger to the north, and is the coarsest lacustrine sediment in the deep water portions of the lake. The absence of thicker outwash sediment is puzzling.

The upper half of the Quaternary sequence is the red, brown, and gray clay of the Lake Michigan Formation. The South Haven and Sheboygan Members are red, fine-grained clay (80% less than 2 μm), deposited as rock flour while active glacial ice was present in the Lake Michigan and Lake Superior drainage basins (Lineback *et al.* 1979). The Winnetka, Lake Forest, and Waukegan Members are brown and gray silty clay, derived from shoreline erosion and by postglacial stream erosion.

Sediment accumulation rates

The red clays, the glacially derived South Haven and Sheboygan Members, were mapped by Wickham *et al.* (1978). The clays range from 0 to 15 m thick and are found in approximately the deeper half, below 82 m of water, of the basin. Most of the clay was deposited as rock flour outwash of glaciers melting in Lake Superior (Lineback *et al.* 1979). Deposition began with the deglaciation of Lake Michigan about 11 500 radiocarbon years B.P. and ended soon after the connection between Lake Michigan and Superior was broken, at the time of or a little sooner than the end of glacial lacustrine deposition in Lake Superior about 9 000 radiocarbon years B.P. The deglaciation of these two Great Lakes is an extremely complex subject. The important point is that the red clays in Lake Michigan were deposited in less than 2 500 years. Typical thicknesses are 5 to 10 m. Using an average time of deposition of 2 000 years, accumulation rates during deglaciation are computed to be 2.5 to 5 mm a^{-1}. These rates are long-term averages. In order to correct for differences that are due to sediment compaction, accumulation rates ought to be reported in mass per unit area and time although it is easier to visualize thickness per unit time. Because the Lake Michigan Formation has not been compacted by overriding glacial ice and the density of sediment does not vary much with depth of burial (Silver & Lineback 1972), it is reasonable to report accumulation rates in thickness per unit time.

The gray and brown clays, the postglacial Winnetka, Lake Forest and Waukegan Members, were also mapped by Wickham *et al.* (1978) and are shown in Fig. 2a. The clays range from 0 to 18 thick and overlie the red clays. They are found in all but the shallow and nearshore areas of the lake. The time of their deposition is known much more accurately – beginning about 9 000 or 10 000 radiocarbon years B.P. and continuing to the present. Several independent methods can be used to compute accumulation rates for these sediments. Using total thicknes and total time of deposition to make the calculation, rates range from 0 to 1.8 mm a^{-1}. Radiocarbon dating has been used (Gross *et al.* 1973) to compute rates of 1 mm a^{-1} in the areas of thickest sediment and 0.02 mm a^{-1} in the deep water areas in the center of the southern basin. Contamination with geologically old carbon has caused some problems in the interpretation of the radiocarbon dates. Palynology, especially the periods of increase in ragweed pollen as a result of the large-scale forest clearance and development of intensive agriculture in the middle of the nineteenth century, was used (King *et al.* 1976) to compute rates of 1.0 and 0.8 mm a^{-1} in the areas of thick sediment and 0.5 mm a^{-1} in the center of the southern basin. Robbins and Edgington (1975) used ^{210}Pb to date eight cores and reported rates ranging from 0.1 to 4.2 mm a^{-1}. Christensen and Chien (1981) obtained accumulation rates of 0.01, 0.44, and less than 0.20 mm a^{-1}, using ^{210}Pb dating in the central portion of Lake Michigan.

The modern accumulation rates average less than 1 mm a^{-1} in the deep water areas and a few mm a^{-1} in the areas of thickest sediment. In contrast, accumulation rates for the older red clays were 2.5 to 5 mm a^{-1}.

Geochemistry of sediment

Shimp *et al.* (1971) examined the geochemistry of sediment cores from Lake Michigan and concluded that at least seven trace elements were accumulating in the uppermost, recently deposited portions of the organic-rich, fine-grained sediment. The concentrations of Br, Pb, Cr, As, Zn, Cu, and Hg gave good positive correlations with organic carbon and clay-sized material. For the accumulating elements, correlations were best for organic carbon. For the non-accumulating trace elements, correlations were best for clay-sized material.

Cahill (1981) utilized a systematic grid network of 286 sediment grab samples and for each sample examined the concentration of 48 chemical elements, Ph and Eh, and grain size. Figure 2b is one example of the many maps in that report. The highest levels of lead, mercury and zinc and other trace metals were found in the areas of thickest sediment accumulation (compare Fig. 2a to Fig. 2b). A similar pattern has been observed for the concentration of PCB (Frank *et al.* 1981). On the basis of the geologic mapping, Lake Michigan sediment can be described by a series of depositional and non-depositional areas. The concentrations of pollutant trace elements in the uppermost 3 cm of sediment differ significantly between the depositional and non-depositional areas (Table 1). Mean trace-element concentrations are similar in each of the five depositional sub-basins. Cahill (1981) concluded that the source of these elements were uniform around the lake, or that circulation patterns are effective, or both.

Table 1. Concentrations of selected elements in the uppermost 3 cm of sediment in Lake Michigan. Results are from 286 samples in a systematic sampling grid. (PPM unless otherwise noted.)

Element	All samples	Non-depositional areas	Depositional areas	Ratio dep./non.
As	10.5	9.4	12	1.8
Br	33	10	60	6.0
Cr	46	21	75	3.6
Cu	22	6	40	6.7
Hg (ppb)	107	57	176	3.1
Ni	24	13	37	2.9
Pb	40	11	77	7.0
Sb	1.1	.4	1.9	4.2
Zn	97	31	182	5.9
Organic	2.0	.6	3.6	6.3

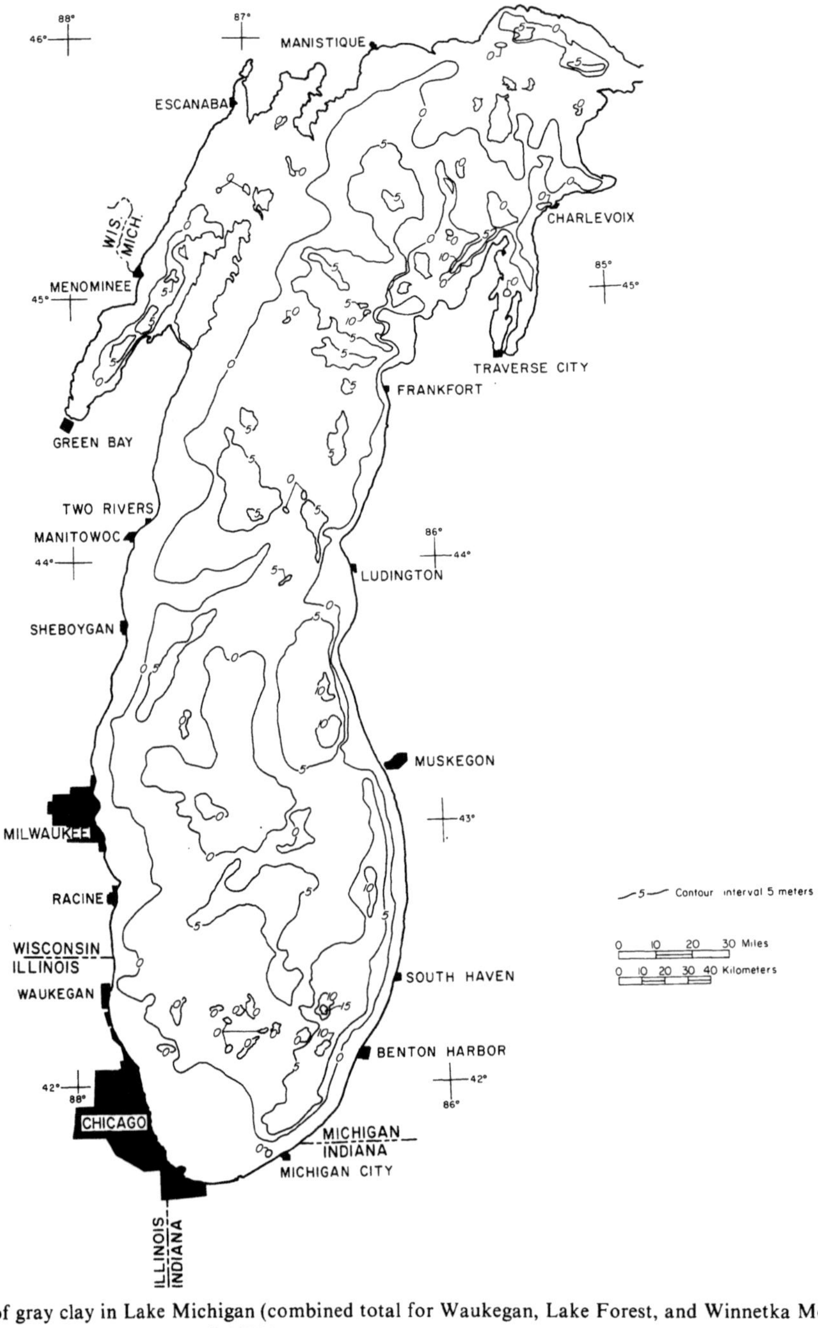

Fig. 2a. Thickness of gray clay in Lake Michigan (combined total for Waukegan, Lake Forest, and Winnetka Members of the Lake Michigan Formation). (From Wickham *et al.* 1978.)

Discussion

Sediment enters Lake Michigan from shoreline erosion, several small rivers, and by atmospheric transport. Except for the nearshore transport of sand by longshore drift, most of the sediment entering Lake Michigan is ultimately transported out to one of the depositional basins. The floors of those basins are below wave base, and sediment, especially in the south end of the lake, is not deposited uniformly over the basins. Instead it tends to be deposited on the margins of the basins where the

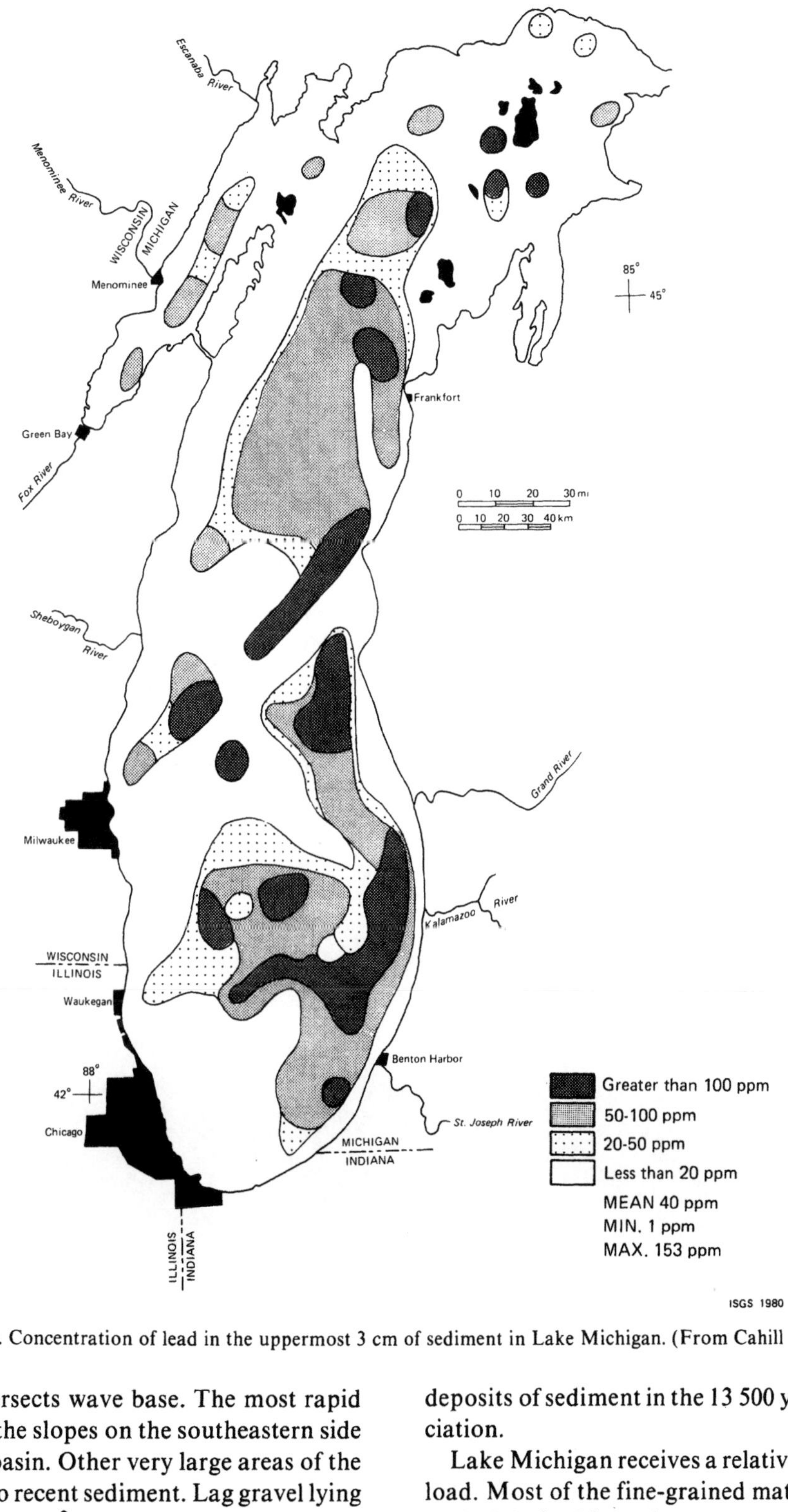

Fig. 2b. Concentration of lead in the uppermost 3 cm of sediment in Lake Michigan. (From Cahill 1981.)

lake bottom intersects wave base. The most rapid deposition is on the slopes on the southeastern side of the southern basin. Other very large areas of the lake floor have no recent sediment. Lag gravel lying on till covers 2 000 km^2 of the floor adjacent to the City of Chicago and has not received any significant deposits of sediment in the 13 500 years since deglaciation.

Lake Michigan receives a relatively low sediment load. Most of the fine-grained material introduced to the lake is ultimately transported to and deposited in one of the depositional basins. Clay-sized sediment is not being deposited in the shallow areas, and pollutants are ultimately deposited in association with organic-rich, fine-grained sediment. A close correlation between the ultimate resting place of pollutants in the lake and the location of point source pollutants to the lake would not be expected. A knowledge of the large-scale sedimentology of the lake is essential in evaluating long-term environmental impacts on the lake.

References

Cahill, R. C., 1981. Geochemistry of Recent Lake Michigan sediments. Ill. geol. Surv. Circ. 517: 96 pp.

Christensen, E. R. & Chien, N. K., 1981. Fluxes of arsenic, lead, zinc, and cadmium to Green Bay and Lake Michigan sediments. Envir. Sci. & Tech. 15: 553–558.

Frank, R., Thomas, R. L., Braun, H. E., Gross, D. L. & Davies, T. T., 1981. Organochloride insecticides and PCB in surficial sediments of Lake Michigan (1975). J. Great Lakes Res. 7: 42–50.

Gross, D. L., Coleman, D. D. & Lineback, J. A., 1973. Sedimentation rates in the deep water areas of southern Lake Michigan. Int. Ass. Great Lakes Res. 17th Conf.: 123–124.

Gross, D. L., Lineback, J. A., Shimp, N. F. & White, W. A., 1972. Composition of Pleistocene sediments in southern Lake Michigan, U.S.A. Int. geol. Congr. Montreal. Sect. 8: 215–222.

King, J. E., Lineback, J. A. & Gross, D. L., 1976. Palynology and sedimentology of Holocene deposits in southern Lake Michigan. Ill. geol. Surv. Circ. 496, 24 pp.

Lineback, J. A., Ayer, N. J. & Gross, D. L., 1970. Stratigraphy of unconsolidated sediments in the southern part of Lake Michigan. Ill. geol. Surv. envir. Geol. Note 35, 35 pp.

Lineback, J. A., Dell, C. I. & Gross, D. L., 1979. Clacial and postglacial sediments in Lakes Superior and Michigan. Geol. Soc. Am. 90: 781–791.

Lineback, J. A. & Gross, D. L., 1972. Depositional patterns, facies, and trace element accumulation in the Waukegan members of the Late Pleistocene Lake Michigan Formation in southern Lake Michigan. Ill. geol. Surv. envir. Geol. Note 58, 25 pp.

Lineback, J. A., Gross, D. L. & Meyer, R. P., 1974. Clacial tills under Lake Michigan. Ill. geol. Surv. envir. Geol. Note 69, 48 pp.

Robbins, J. A. & Edgington, D. N., 1975. Determination of recent sedimentation rates in Lake Michigan using Pb-210 and Cs-137. Geochim. Cosmochim. Acta. 39: 285–304.

Shimp, N. F., Schleicher, J. A., Ruch, R. R., Heck, D. B. & Leland, H. V., 1971. Trace elements and organic carbon accumulation in the most recent sediments of southern Lake Michigan. Ill. geol. Surv. envir. Geol. Note 41, 25 pp.

Silver, M. L. & Lineback, J. A., 1972. Velocity of sound in sediments cored from southern Lake Michigan. Ill. geol. Surv. Circ. 475, 20 pp.

Wickham, J. T., Gross, D. L., Lineback, J. A. & Thomas, R. L., 1978. Late Quaternary sediments of Lake Michigan. Ill. geol. Surv. envir. Geol. Note 84: 26 pp.

Paleolimnology of the Peten Lake district, Guatemala

I. Erosion and deposition of inorganic sediment as inferred from granulometry

Michael W. Binford
Department of Natural Sciences, Florida State Museum, University of Florida, Gainesville, FL 32611, U.S.A.

Keywords: paleolimnology, clay, cultural disturbance, erosion, Guatemala, Mayan archaeology, silt, particle-size analysis, tropical lakes

Abstract

Long-term changes of sedimentary particle-size distribution in two tropical lowland lakes were compared with changes of human population sizes, estimated archaeologically, in the drainage basins. Mean particle size of silt and clay fractions (<64 μm) varied between 3 and 15 μm. High positive skewness and kurtosis of the distributions were associated with smaller particle sizes; hence small mean size resulted from greater influx of small particles while influx of larger particles was probably constant. An inverse correlation between mean particle size and human population size is interpreted to mean that disturbance-induced erosion results in delivery of very fine inorganic particles at higher rates. Within any one basin, particle-size stratigraphy is more precisely related to archaeological time periods than is pollen stratigraphy. An absolute chronology still eludes us, owing to the failure of ^{14}C dating of calcareous, colluvial sediments, but our relative chronology is now more precise than before. If certain assumptions about past hydrologic relations can be met, particle-size analysis is a way of comparing the histories of geographically very different lakes, including lakes from tropical, temperate, and arctic regions.

Introduction

Studies of very long-term effects of human activities on ecosystems using standard ecological techniques are difficult because records of the causes and effects of disturbances are scanty. Paleoecological methods are likewise usually not adequate because baseline, or pre-human, conditions of past environments are not well-known or, except perhaps in Europe, records of human populations are non-existent. One area of the world where most of these objections are answered is the lake district of the Department of the Peten, Guatemala. The entire region is endorheic, so the only sink for materials, especially nutrients, is lake sediments. The region is the heartland of the ancient Maya, the only high civilization to develop in the lowland tropics. The Maya occupied the region from about 3 000 years ago with varying population size and stages of cultural development. Late in the 9th century A.D. the population declined precipitously, and, despite a brief Postclassic (? 13th to 17th centuries) resurgence, the Peten has become sparsely settled. As a result both pre- and post-disturbance conditions are found, and Deevey *et al.* (1979) have presented a quantitative model for per-capita nutrient flow to two of the lakes, Yaxha and Sacnab.

As the influence of Mayan activity on the lake-watershed systems has become better known, a problem in interpretation has emerged. Radiocarbon dating of Peten lake sediments is unreliable because of the hard-water lake error (Deevey & Stuiver 1964), so a well-controlled stratigraphic time scale is not available. Pollen diagrams from several lakes can be used to correlate archaeological events with sediment stratigraphy, but pollen stratigraphy has low resolution and expresses only a broad, regional pattern of vegetational disturban-

Hydrobiologia 103, 199-203 (1983).

ces. If settlement patterns around all the lakes were similar, correlation with pollen zones would be appropriate for all. However, recent unpublished work by D. S. and P. M. Rice (pers. comm.) indicates that settlement patterns in the western watersheds were rather different, particularly in the Early Classic, from those near the eastern lakes Yaxha and Sacnab. Given low stratigraphic resolution, this can mean that sedimentation rates calculated over pollen-stratigraphic zones may not track population sizes or material influxes within a basin (Brenner 1983). An independent and basin-specific means of correlating sediments with archaeology is desirable. The ideal index should be directly attributable to disturbances within the drainage basin of the lake in question.

For a long time paleolimnologists have noted an increased delivery of allochthonous inorganic material, especially silicates, to lake sediments during times of disturbances such as roadbuilding or agriculture (Hutchinson 1970; Warwick 1980; Deevey *et al.* 1979; Bradbury & Waddington 1973; Davis & Norton 1978). If the relationship between human activity and the influx of inorganic sediment could be expressed quantitatively, the index mentioned above would be found. In this paper I propose that particle-size analysis of the inorganic sediment of closed lakes provides such an index.

The objectives of this paper are threefold: 1) To develop the rationale of particle-size analysis, when used in conjunction with archaeological or historical evidence, as an index of human disturbance within drainage basins of single lakes; 2) To test several statistics of sedimentary particle-size distributions against known settlement histories in two Guatemalan lake basins; and 3) To use particle-size analysis to correlate periods of Mayan occupation with sedimentary stratigraphy.

Granulometry as an index of human disturbance

Human activities directly affect all the variables involved in soil erosion except the energy of rainfall (Wischmeier & Smith 1965). In all cases human activity leads to increased erosion of the total soil by disrupting the soil matrix and increasing the amount and velocity of runoff (Branson 1975). As water velocity and soil disturbance increases, both larger (>1.0 mm) and smaller (<0.1 mm) particles are moved (Hjulström 1935). The smaller particles, however, stay in suspension while large particles are deposited with only slight reduction of water velocity. In a lake, coarse particles are deposited very near

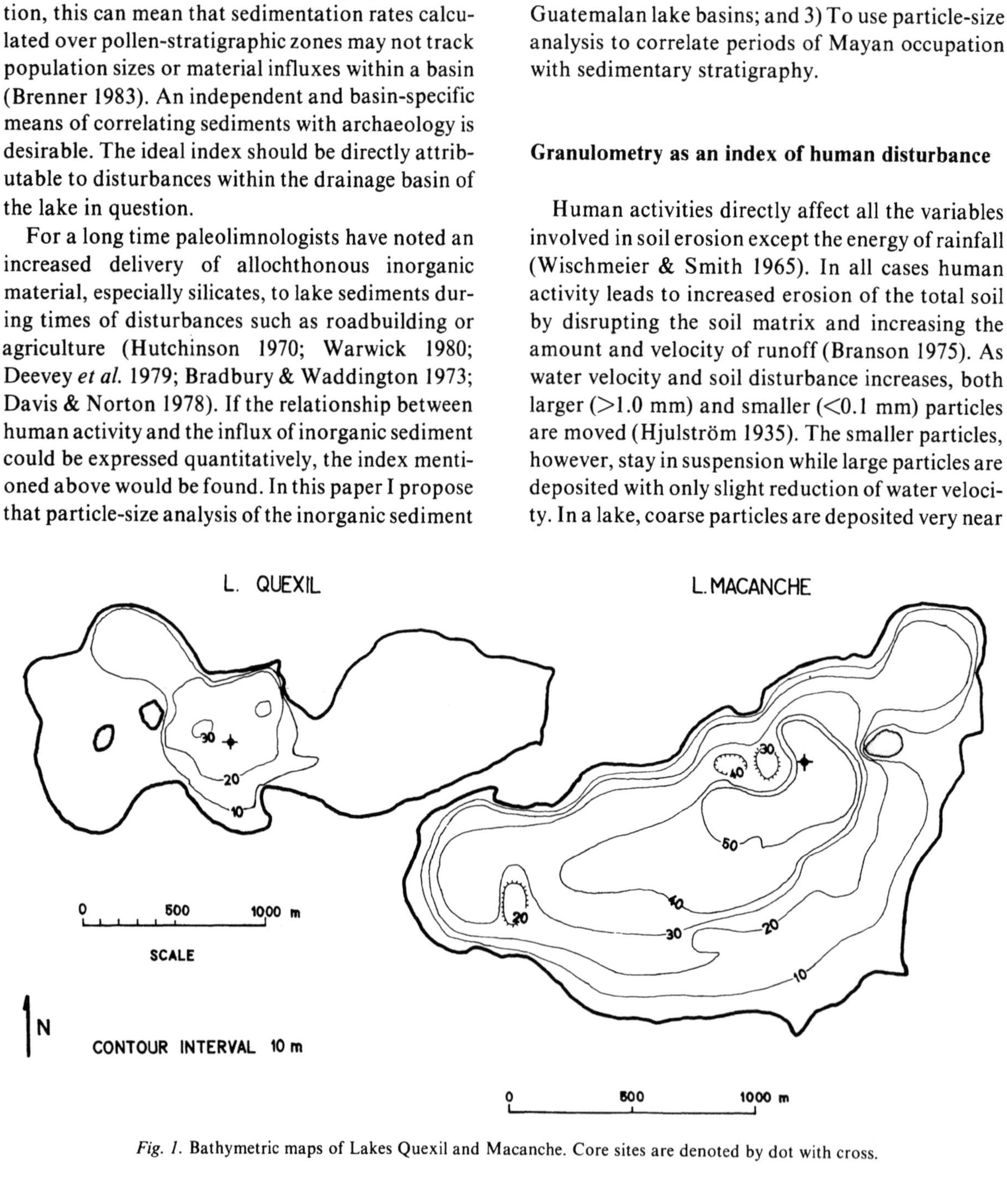

Fig. 1. Bathymetric maps of Lakes Quexil and Macanche. Core sites are denoted by dot with cross.

the shore, while the fines are transported throughout. Sedimentation at the center of a lake consists of the fine fraction of particles delivered to the lakeshore. Under the proper conditions, one would expect a sorting of inorganic particles from coarse, nearshore sediments to fine, offshore sediments. Such a sorting has been documented in both large and small lakes (Sly 1978; Davydova & Kurochina 1978). Two assumptions must hold in order to infer past disturbance patterns from lake sediments: 1) The lake level has not fluctuated significantly during the time in question; and 2) The effects of human disturbances in the drainage basin are much more important than other geological influences.

Materials and methods

Limnological characteristics and the geographical setting of the lakes of the Guatemalan lowlands have been reported by Deevey *et al.* (1980) and Brezonik & Fox (1974). Sediment cores were raised from approximately the deepest points in Lakes Quexil and Macanche (Fig. 1), with a modified Kullenberg coring apparatus. Sediment samples were dispersed ultrasonically in a 4% sodium hexametaphosphate solution, then filtered through a 75-μm sieve. The particle-size distribution of the suspension was then measured with a Model TA II Coulter Counter® (Coulter Electronics, Inc. 1975). I made no attempt to remove organic matter or carbonates from the routine samples. However, representative samples from all sediment types in Quexil that were treated with 30% H_2O_2 and 25% HCl to remove organics and carbonates, respectively, did not differ significantly from untreated samples.

D. S. and P. M. Rice estimated sizes of Mayan populations in the drainage basins of Lakes Quexil and Macanche according to methods published elsewhere (Rice 1976) and kindly supplied unpublished data collected in 1980.

Results

Frequency distributions of particle size can be described in terms of how well they resemble a Gaussian, or normal, distribution. The three distribution statistics of interest here are the mean particle size, $\bar{x}$, skewness, s, and kurtosis, k. A distribution with positive skewness has a longer tail to the right, with a greater proportion of the particles falling at lower sizes; while negative skewness indicates the opposite. Kurtosis is a measure of the relative flatness or peakedness of the curve. If $k > 3$ (leptokurtic) the curve is more peaked than a normal curve. If $k < 3$ (platykurtic), the curve is flatter. One would expect a small mean particle size, large positive skewness, and more leptokurtic curves at times of maximum disturbance within the drainage basin.

Mayan population estimates based on number of mounds occupied during a particular period for the drainage basins of Lakes Quexil and Macanche are shown in Figure 2a and b. The figure also shows the proposed correlation between population sizes and particle-size stratigraphy. Before and after Mayan occupation, mean particle size was relatively large and skewness and kurtosis relatively low and unchanging. During occupation, particle size decreased while skewness and kurtosis gradually increased in both lakes. The general trend was interrupted twice in Lake Quexil and possibly twice in Lake Macanche. Levels with larger mean particle size at 505–530 and 260–330 cm in lake Quexil are proposed to be contemporaneous with population decreases of the Terminal Preclassic and Early Classic, and the Terminal Classic, respectively. Likewise, the sediments of Lake Macanche indicate a slight decrease of population at 295 cm and a larger decrease at 110 cm, which I believe correlate with the Early Classic and Terminal Classic population declines. Sediments from Quexil seem to follow the population changes more accurately than those from Macanche, but the Quexil core was sampled at closer intervals and as a result has a better resolution of watershed disturbances.

In general, kurtosis and skewness vary inversely with mean particle size in these data, and therefore add little new information about Mayan disturbance. Both statistics are measures of sorting, however, and the long-term increases observed especially in the Maya zones of Lake Quexil may be related to sediment resuspension and focusing (Deevey *et al.* 1977). That is, they may imply that clay sediments injected by the first disturbance, about 3 000 B.P., were progressively better sorted in deep water as the long-time resultant of horizontal transport. However, no statistical significance can be assigned

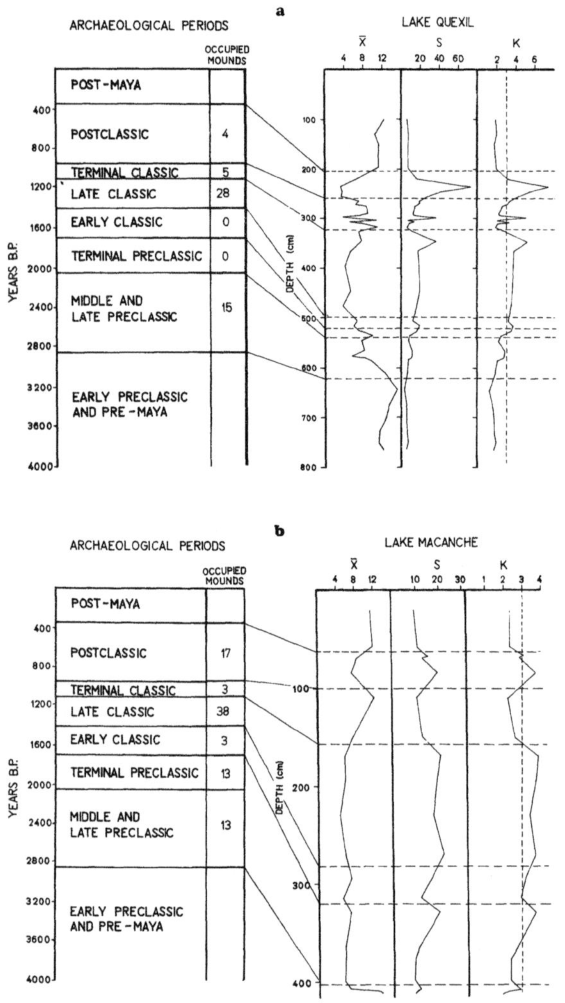

Fig. 2. Mean particle size ($\bar{X}$), coefficient of kurtosis (K), and coefficient of skewness (S) for a) Lake Quexil and b) Lake Macanche deepwater core H. The dotted line at $K = 3$ represents the kurtosis of a Gaussian, or normal curve. Occupied mounds are the total mounds occupied during a particular period, counted on three transects by the method of Rice (1976). Proposed correlation between archaeological periods and sediment stratigraphy is shown.

to the trend, nor does the coefficient of sorting as defined by Trask (1932) vary systematically through time.

Discussion

Brenner (1983) has shown that the resolution of pollen zones is not fine enough to accurately estimate influx rates of materials to sediments in Lake Quexil, and perhaps in other lakes. Particle-size analysis provides a finer resolution of disturbances and may provide a more precise correlation between Mayan periods and core stratigraphy than pollen-stratigraphic methods. Preliminary examination of Brenner's data from Lake Quexil does indicate that when sedimentation rates are calculated from dates assigned by correlation of particle-size analysis and Mayan population profiles the material influx closely tracks the changes of population size.

Particle-size analysis, especially of the silt and clay fraction, is not a commonly used method in paleolimnology. As far as I am aware, only Warwick (1980) has examined in detail the effects of disturbance on the sediments of a large lake, and used particle-size analysis for correlation between human activity and sediment stratigraphy. His study of the Bay of Quinte, Lake Ontario, prompted this study by suggesting that increased disturbance in a drainage basin results in increased delivery of very fine particles to the center of the lake. If lakes from other regions with long records of population changes are examined and show similar results, than the technique of particle-size analysis could be extended to areas that are less well-known archaeologically or historically.

Acknowledgements

I thank E. S. Deevey for extensive discussion and editorial assistance, Mark Brenner for discussions, and Brenner, Michael S. Flannery, and Sam Garrett-Jones for obtaining the sediment cores. Dr. Dan Haile, U.S. Department of Agriculture, Agriculture Research Service, generously allowed the use of the Coulter Counter. Drs. Don S. and Prudence M. Rice provided unpublished data of Mayan occupation for the two lakes. This research was supported by the U.S. National Science Foundation grants DEB-77-06629 and EAR-79-26330 to E. S. Deevey.

References

Bradbury, J. P. & Waddington, J. C. B., 1973. The impact of European development on Shagawa Lake, northeastern Minnesota. U.S.A. In: H. J. B. Birks & R. G. West (Eds.). Quaternary Plant Ecology. Blackwells, London: 289–307.

Branson, F. A., 1975. Natural and modified plant communities as related to runoff and sediment yields. In: A. D. Hasler (Ed.), Coupling of Land and Water Systems, Ecological Studies Analysis and Synthesis, 10. Springer-Verlag, New York: 157–172.

Brenner, M., 1983. Paleolimnology of the Peten lake district, Guatemala. II: Mayan population density and sediment and nutrient loading of Lake Quexil. Hydrobiol. 00: 000–000.

Brezonik, P. L. & Fox, J. L., 1974. The limnology of selected Guatemalan lakes. Hydrobiologia 25: 117–129.

Coulter Electronics, Inc., 1975. Operator's manual for the Coulter Counter Model TA II, 77 pp.

Davis, R. B. & Norton, S. A., 1978. Paleolimnological studies of human impact on lakes in the United States, with emphasis on recent research in New England. Pol. Arch. Hydrobiol. 25: 99–115.

Davydova, N. N. & Kurochina, A. A., 1978. Bottom sediments of large shallow lakes of the north-western part of the USSR in connection with hydrodynamic processes. Pol. Arch. Hydrobiol. 25: 87–92.

Deevey, E. S., Brenner, M., Flannery, M. S. & Yezdani, G. H., 1980. Lakes Yaxha and Sacnab, Peten, Guatemala: limnology and hydrology. Arch. Hydrobiol. Suppl. 57: 419–460.

Deevey, E. S., Rice, D. S., Rice, P. M., Vaughan, H. H., Brenner, M. & Flannery, M. S., 1979. Mayan Urbanism: Impact on a tropical Karst environment. Science 206: 298–306.

Deevey, E. S. & Stuiver, M., 1964. Distribution of natural isotopes of carbon in Linsley Pond and other New England lakes. Limnol. Oceanogr. 9: 1–11.

Deevey, E. S., Vaughan, H. H. & Deevey, G. B., 1977. Lakes Yaxha and Sacnab, Peten, Guatemala: Planktonic fossils and sediment focusing. In: H. L. Golterman (Ed.), Interactions between Sediments and Freshwater. Junk, The Hague: 189–196.

Hjulström, F., 1935. Studies of the morphological activity of rivers as illustrated by the River Fyris. Univ. Uppsala Geol. Inst. Bull. 25: 221–557.

Hutchinson, G. E. 1970. Ianula: an account of the history and development of the Lago di Monterosi, Latium, Italy. Trans. Am. Phil. Soc. N.S. 60: 1–178.

Rice, D. S., 1976. Middle Preclassic Maya settlement in the central Maya lowlands. J. Fld Archaeol. 3: 425–445.

Sly, P. G., 1978. Sedimentary processes in lakes. In: A. Lerman (Ed.), Lakes, Chemistry, Geology, Physics. Springer-Verlag, New York: 65–90.

Trask, P. D., 1932. Origin and Environment of Source Sediments of Petroleum. Houston, Texas. 67 pp.

Warwick, W. F., 1980. Paleolimnology of the Bay of Quinte, Lake Ontario: 2 800 years of cultural influence. Can. Bull. Fish. Aquat. Sci. 206: 1–117.

Wischmeier, W. G. & Smith, D. D., 1965. Predicting Rainfall Erosion Losses from Cropland East of the Rocky Mountains. U.S. Dep. Agric. Handbook 282. 47 pp.

Paleolimnology of the Petén Lake district, Guatemala

II. Mayan population density and sediment and nutrient loading of Lake Quexil

Mark Brenner
Florida State Museum, University of Florida, Gainesville, FL 32611, U.S.A.

Keywords: paleolimnology, colluvium, soil chemistry, clay, phosphorus loading, tropical lakes

Abstract

The long-term impact of Maya culture on a lowland tropical watershed is assessed, using data from a 9.2 m sediment core taken from deep water (28 m) in Lake Quexil. Human population growth, estimated by the 1980 archaeological survey, is associated with a shift in the composition of the sediment to a dominance by inorganic material, the Maya clay formation, beginning ca. 3500 B.P. Increasing settlement densities are correlated with accelerated influxes of phosphorus, carbonates, and siliceous sediment. However, chemical data do not track short-term population fluctuations closely. Because much of the sediment is delivered as colluvium, and not by running water, there is a lag between terrestrial disturbance and impact on the aquatic system. As an indication of this lag, contemporary high sedimentation rates are a residual of Maya activity that virtually ceased some 300–400 years B.P. Comparison of the deep-water core with a shallow-water (7 m) section, based on palynological correlation, reveals only minor differences in proximate chemical composition. Chemical influxes are much higher at the deep-water site, however, as a consequence of sediment focusing in this hyperconical basin. Chemical analyses of soil samples from 21 test pits in the Quexil basin support the principal conclusion that bulk soil movement was the mode of nutrient transfer to the lake, following forest clearance by the Maya.

Introduction

Northern Guatemala's Petén lake district (Fig. 1) provides ideal study sites for a paleolimnological investigation of the long-term interactions between human populations and aquatic systems of the tropical lowlands. The karsted, central Petén region covers some 20 000 km^2 and archaeological investigations there reveal that the area was once densely settled by the Maya. Ceramic evidence from the sites of Altar de Sacrificios and Ceibal dates the earliest human presence in Petén at 3000 B.P. (Rice 1976), although earlier occupation is known in Belize (Hammond *et al.* 1977). By classic times (A.D. 250–850), regional Maya population density exceeded 20 persons · km^{-2} (Sanders 1973), and Late Classic (A.D. 550–850) urban centres supported 200 persons · km^{-2}, with levels of 600–700 persons · km^{-2} plausible at some sites like Tikal (Culbert 1973).

Tsukada's (1966) study of Lake Petenxil sediments employed palynology to document a long history of human disturbance. Subsequent studies at Lakes Yaxhá, Sacnab and Quexil (Brenner 1978; Deevey *et al.* 1979; Vaughan 1979) have used sediment chemistry, pollen, and microfossils to assess Maya perturbations in the watersheds and consequent effects on the lakes. Archaeological examination of the Yaxhá and Sacnab watersheds revealed that populations in the basins increased exponentially from the Middle Preclassic (1000–250 B.C.) until the Late Classic (A.D. 550–850) and one environmental response to the population growth was an increase in the delivery rate of phosphorus to the

Hydrobiologia 103, 205–210 (1983).

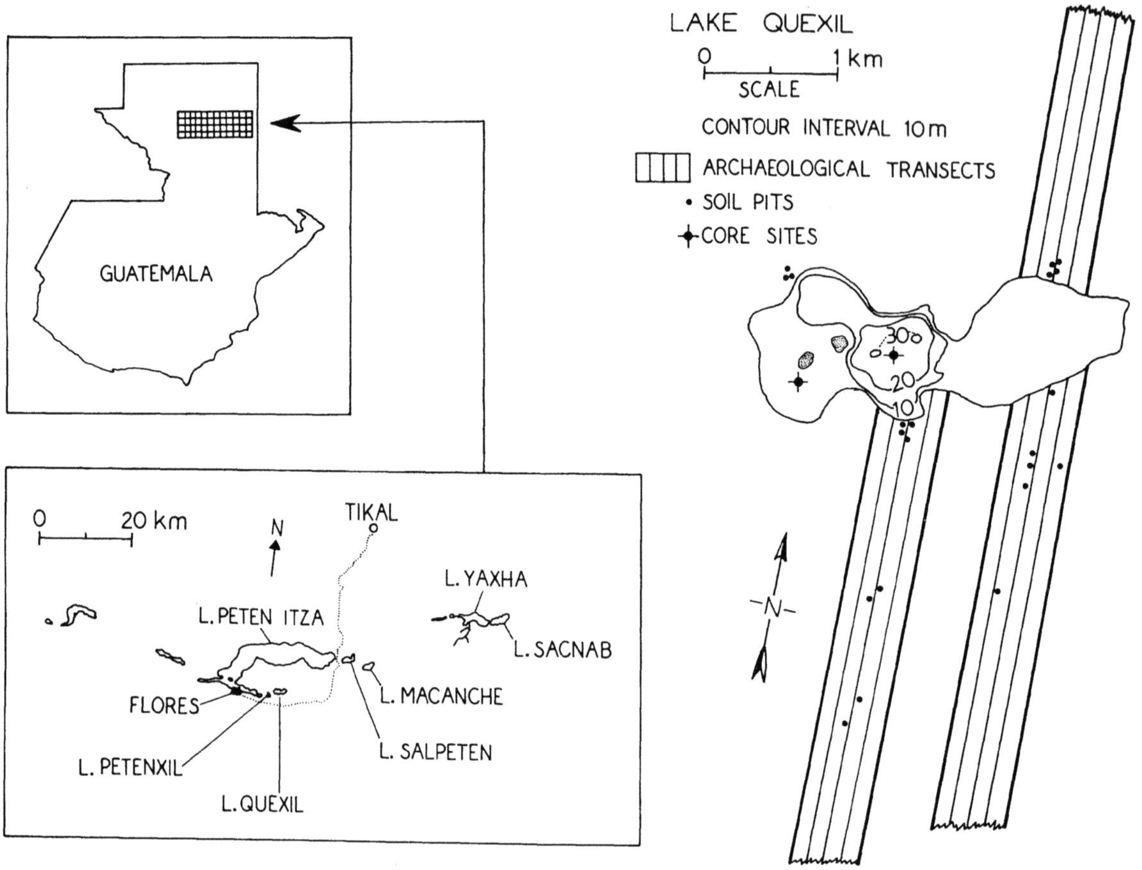

Fig. 1. Map of the Petén lake district and Lake Quexil showing locations of archaeological sampling transects, soil pits, and core sites.

lakes by colluviation (Deevey *et al.* 1979).

With the completion of the 1980 archaeological field season in the Quexil watershed, it is now possible to extend our research to a third lake, exploring correlations between Maya population fluctuations (Rice & Rice, unpublished) and the paleolimnological record. Quexil is particularly interesting because populations did not undergo the slow steady growth revealed at Yaxhá and Sacnab, but instead displayed a Terminal Preclassic-Early Classic (100 B.C.–A.D. 550) population decline before rising once again in the Late Classic (A.D. 550–850) prior to the as yet unexplained regional population collapse around A.D. 900.

This study employs both a shallow-water (7 m) and a deep-water (28 m) core (the latter designated Quexil H) to examine the sedimentary history of Lake Quexil. Both cores contain nearly complete Holocene profiles, document initial vegetation disturbance at about 5000 B.P., and demonstrate the profound effects of sediment focusing in this hyperconical basin (Lehman 1975; Deevey *et al.* 1977). Chemical analyses of basin soils substantiate our claim that bulk soil movement was the major mode of nutrient transport from watershed to lake following Maya forest clearance.

Results and discussion

Because radiocarbon dating fails to provide reliable results as a consequence of the hard-water lake effect (Deevey & Stuiver 1964) and colluviation, we are compelled to zone our cores by identifying sections consisting of discrete pollen assemblages and assigning archaeologically correlated dates to them (Vaughan & Deevey, 1981). Thus, as at Yaxhá and Sacnab, we rely on the changing regional pollen spectrum to date sections of the Quexil cores, estimates proximate sediment composition within zones, and calculate chemical influxes during the various archaeological periods.

The percentage chemical composition of equivalent zones in the shallow and deep-water cores is not dissimilar, though some differences are apparent (Table 1). The deep-water section is dominated by silica while the shallow core can be characterized as more organic. However, proceeding from the deepest, pre-Maya sediments to the modern muds, both cores display similar zone-to-zone shifts in the organic-inorganic ratio.

Human occupation of the watershed caused a shift toward inorganic domination of the sediments and depopulation has resulted in a partial return to more organic muds like those of the deepest, pre-Maya and low population zones. In view of this interpretation, Early Classic sediments are anomalous as peak inorganic percentages are associated with this low population period. However, modern sediments of the now-forested watershed contain less organic matter than pre-disturbance mud, and it appears that material transfer has not achieved equilibrium in the 400 years since the abandonment of the catchment. Apparently, equilibrium was not attained during the Terminal Preclassic–Early Classic population hiatus either.

Influx rates

Chemical delivery rates to Quexil are now examined as they may have controlled the trophic state of the lake. This chemical transfer implies removal of nutrients from the drainage basin, a process that perhaps reduced soil fertility.

Despite the problem of non-equivalent time correlations, total phosphorus influxes can be compared roughly to changing population densities as expressed by mound occupations (Fig. 2). In both cores it can be seen that human settlement of the basin was accompanied by higher phosphorus influxes. High Early Classic influxes (normal at Yaxhá and Sacnab) calculated for the Quexil cores are unexpected considering the low population of the period. With respect to this anomaly, it is noteworthy that post-Maya influxes at the shallow and deep sites are much higher than their respective baseline rates, indicating once again that long periods of time (>400 years) are necessary for equilibrium to be achieved.

Comparing equivalent zonal influx rates between the two cores (Fig. 2), it is apparent that more phosphorus is deposited per unit time in deep water. This stems from the fact that more bulk sediment is deposited in deep water and deep-site sediments contain more dry weight per unit volume. While some delivery rates calculated for the deep-water site exceed tolerable levels (Vollenweider 1968), it has been argued elsewhere (Brenner 1978; Deevey *et al.* 1979) that little of the sedimented phosphorus was available for eutrophication, thus the deep-water data are biased indicators of past trophic state and are simply presented as evidence of the strong sediment focusing that occurs in the hyperconical Quexil basin (Fig. 3).

Like phosphorus, other sediment components (C_{org}, carbonates, silicates) were transferred to the lake at higher rates as a consequence of human activity in the watershed (Table 2), but augmentation of carbonate and silicate influx rates exceeded changes in organic carbon deposition as expressed by the inorganic nature of disturbance-zone sediments. Diatoms and other microfossils are extreme-

Table 1. Percent composition of Lake Quexil shallow-water and deep-water core sediments.

Zone	Organic matter		$CaCO_3$		SiO_2		Fe_2O_3	
	Shallow	Deep	Shallow	Deep	Shallow	Deep	Shallow	Deep
Post-Maya	42	36	26	5	30	54	2	5
Late & Postclassic	18	9	30	5	48	81	4	5
Early Classic	14	8	6	6	75	81	5	5
Late Preclassic	66	40	5	5	27	51	2	4
Middle Preclassic	69	61	3	5	25	31	3	3
Early Preclassic	59		3		36		2	
Pre-Maya	69	48	3	6	24	41	4	5

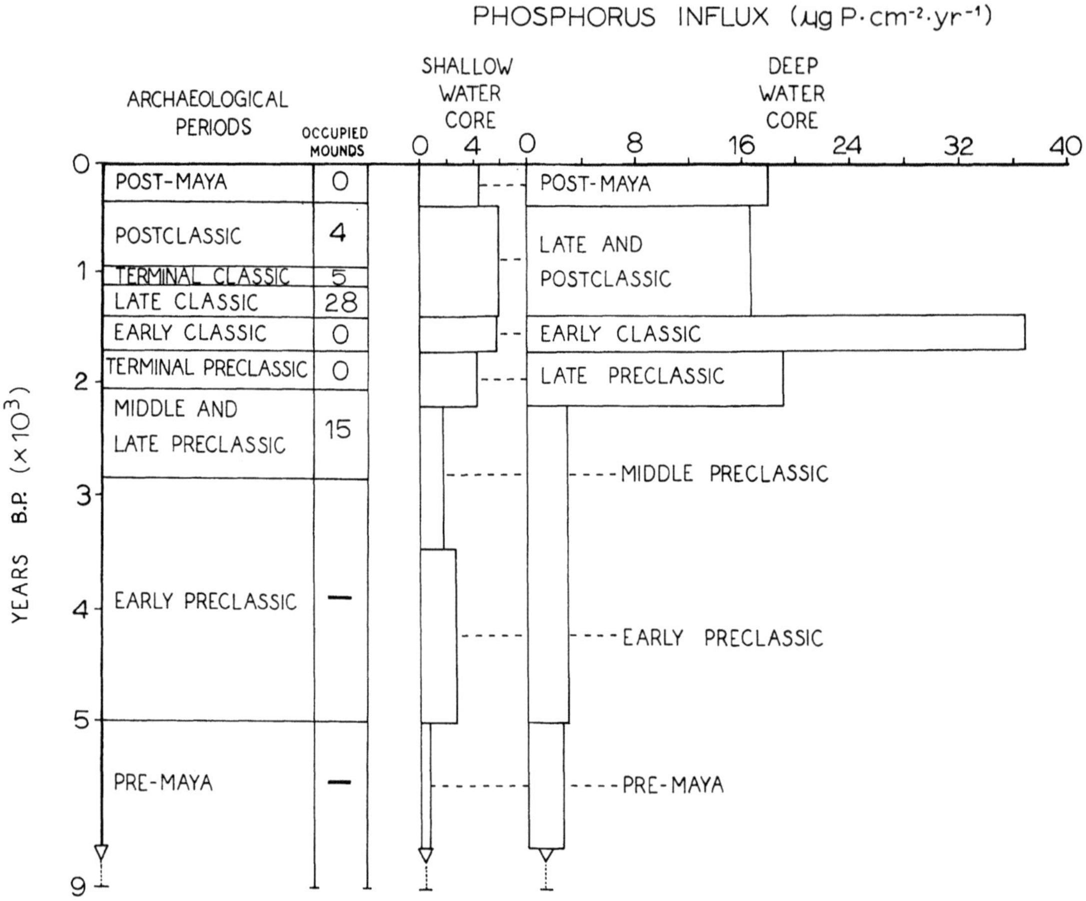

Fig. 2. Time correlation of Maya population levels and phosphorus influxes.

ly scarce in disturbance-level sediments and it is suspected that the bulk of the organic matter, carbonates and silicates deposited following vegetation removal was derived allochthonously. In fact, zone-to-zone phosphorus influx rates are positively correlated with deposition rates of carbonates and silicates in both cores, suggesting that a common delivery mechanism may be shared by all three chemical constituents.

Soils

Since Petén rainfall contributes only a small fraction of the annual phosphorus income to the lakes

Table 2. Influxes to the Lake Quexil sediments at the shallow-water and deep-water sites (amount · cm^{-2} · a^{-1}).

Zone	C_{org} (mg)		$CaCO_3$ (mg)		SiO_2 (mg)		P_{tot} (μg)	
	Shallow	Deep	Shallow	Deep	Shallow	Deep	Shallow	Deep
Post-Maya	2.2	4.8	3.1	1.8	4	20	4.4	18.0
Late & Postclassic	1.3	1.6	5.0	2.9	8	47	5.9	16.6
Early Classic	1.7	4.0	1.6	7.8	20	111	5.8	36.7
Late Preclassic	4.8	7.8	0.7	3.0	4	32	4.2	18.9
Middle Preclassic	1.8	1.8	0.2	0.4	1	2	1.7	2.9
Early Preclassic	3.3		0.3		4		2.7	
Pre-Maya	1.3	1.4	0.2	0.4	1	3	0.7	2.6

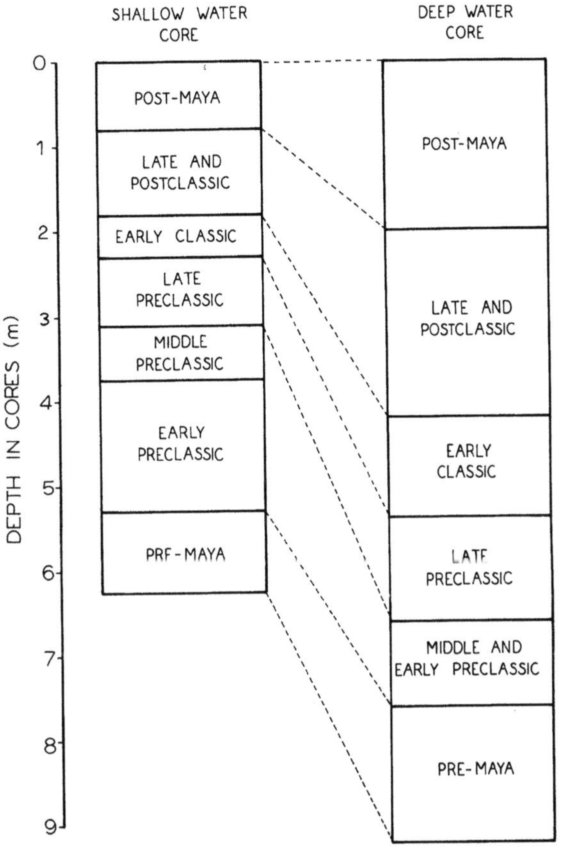

Fig. 3. Palynological zone correlation between the Lake Quexil shallow-water and deep-water cores.

(Deevey *et al.* 1979), it is likely that basin soils are the principal reservoir from which this nutrient, as well as carbonates and silicates were derived during Maya times. Leaching of phosphorus is held to be unlikely as surface soil (0–10 cm) concentrations from the 21 pits (Fig. 1) are about 2.5 times deep-sample values. Aluminum, calcium and sodium also display gradients in the profiles, but are richer in deep soils.

Though perhaps coincidental, total phosphorus concentrations in surface soils (0–10 cm) are surprisingly close to levels found in the lake sediments. In fact, Quexil surface soil concentrations ($278 \pm 156\ \mu g \cdot g^{-1}$) are not statistically different from values measured in the shallow core ($289 \pm 156\ \mu g \cdot g^{-1}$) and deep core ($358 \pm 122\ \mu g \cdot g^{-1}$) sediments. Likewise, surface soil iron concentrations are no different from sediment iron concentrations. These data confirm our original conclusions about phosphorus movement that were based on earlier soil sampling at Yaxhá and Sacnab: bulk soil movement was the principal means of nutrient transfer to the lake following Maya deforestation.

Conclusions

Human-induced deforestation of the Quexil watershed resulted in accelerated deliveries of phosphorus, carbonates, and silica to the lake, primarily via soil translocation. The environmental impact was sustained not only during the more than 3 millennia of Maya occupation, but continues today in the form of high modern influxes. Clearly, we can identify changes in sedimentary phenomena that resulted from human disturbance. However, we feel that using data from Quexil to refine our per capita phosphorus loading model developed at Yaxhá and Sacnab would be premature at this time. This is due to our inability to track short-term population fluctuations in the Lake Quexil cores. This difficulty results from several possible factors: 1) We are unable to zone the cores with the same archaeological time zone designations used by the social scientists. Hence, we present a single Late and Postclassic sediment zone that covers three identifiable ceramic periods possessing very different population levels, and in fact, includes the famous collapse. Likewise, our Late Preclassic zone overlaps two archaeological zones of very different population densities; 2) Long periods necessary to establish equilibrium for land-water nutrient transfers may in fact make it impossible to see responses to short-term population changes; 3) It may be that the environmental response we measure paleolimnologically reflects changes in land use rather than population numbers *per se;* 4) Finally, using a regional pollen stratigraphy to zone the Quexil core may produce archaeologically correlated palynological zonation which does not accurately reflect deforestation events in the Quexil watershed proper. Attempts at using magnetic susceptibility to establish interbasin stratigraphic correlations proved futile due to the lack of sedimented, magnetic particles. However, recent work with particle size analysis (Binford 1983) indicates that granulometry may be an appropriate technique for assessing basin-specific disturbance.

Acknowledgments

This study was supported by National Science Foundation grants nos. BMS-72-01859, DEB-77-06629, and EAR-79-26330 awarded to E. S. Deevey. Grateful acknowledgment is made to E. S. Deevey, M. S. Flannery, M. W. Binford, D. S. and P. M. Rice.

References

Binford, M. W., 1983. Paleolimnology of the Petén lake district, Guatemala. I. Erosion and deposition of inorganic sediments as inferred from granulometry. Dev. Hydrobiol. 00: 000–0000.

Brenner, M., 1978. Paleolimnological assessment of human disturbance in the drainage basins of three northern Guatemalan lakes. M.S. thesis, Univ. Florida. 128 pp.

Culbert, P. T., 1973. The Maya downfall at Tikal. In: P. T. Culbert (ed.). The Classic Maya Collapse. Univ. New Mexico Press, Albuquerque: 63–92.

Deevey, E. S. & Stuiver, M., 1964. Distribution of natural isotopes of carbon in Linsley Pond and other New England lakes. Limnol. Oceanogr. 9: 1–11.

Deevey, E. S., Vaughan, H. H. & Deevey, G. B., 1977. Lakes Yaxhá and Sacnab, Petén, Guatemala: Planktonic fossils and sediment focusing. In: H. L. Golterman (ed.). Interactions between sediments and fresh water. Proceedings of an international symposium held at Amsterdam, The Netherlands, Sept. 6–10, 1976. Junk Publishers, The Hague: 189–196.

Deevey, E. S., Rice, D. S., Rice, P. M., Vaughan, H. H., Brenner, M. & Flannery, M. S., 1979. Mayan urbanism: Impact on a tropical karst environment. Science 206: 298–306.

Hammond, N., Donaghey, S., Berger, R., de Atley, S., Switsur, V. R. & Ward, A. P., 1977. Maya Formative phase radiocarbon dates from Belize. Nature 267: 608–610.

Lehman, J. T., 1975. Reconstructing the rate of accumulation of lake sediment: The effect of sediment focusing. Quat. Res. 5: 541–550.

Rice, D. S., 1976. Middle Preclassic Maya settlement in the central Maya lowlands. J. Fld Archaeol. 3: 425–445.

Sanders, W. T., 1973. The Cultural Ecology of the Lowland Maya: a reevaluation. In: P. T. Culbert (ed.). The Classic Maya Collapse. Univ. New Mexico Press, Albuquerque: 325–365.

Tsukada, M., 1966. The pollen sequence. In: U. M. Cowgill, G. E. Hutchinson & others (eds.). The history of Laguna de Petenxil, a small lake in northern Guatemala. Conn. Acad. Sci. Mem. 17: 63–66.

Vaughan, H. H., 1979. Prehistoric disturbance of vegetation in the area of Lake Yaxhá, Petén, Guatemala, Ph.D. diss., Univ. Florida. 176 pp.

Vaughan, H. H. & Deevey, E. S., 1981. Pollen stratigraphy of two cores from the Petén lake district. Unpublished manuscript.

Vollenweider, R. A., 1968. Scientific fundamentals of the eutrophication of lakes and flowing waters, with particular reference to nitrogen and phosphorus as factors in eutrophication. OECD report DAS/CSI/68.27, Paris. 182 pp.

Paleolimnology of the Peten Lake District, Guatemala

III. Late Pleistocene and Gamblian environments of the Maya area

Edward S. Deevey, Jr., Mark Brenner & Michael W. Binford
Florida State Museum, University of Florida, Gainesville, FL 32611, U.S.A.

Keywords: paleolimnology, Pleistocene aridity, calcite, dolomite, gypsum, laminated gyttja

Abstract

The transition from an arid, glacial Late Pleistocene to an early Holocene (Gamblian 'pluvial') moist period has not been documented in Central America. Finding accessible volcanic lakes too youthful, and knowing that most Florida and Yucatan lakes were dry during glacial ages, we sought appropriate deposits lying deeper than 40 m in the deeper karst lakes of the Peten, in northern Guatemala. The bottom half of a 19.7 m core from Lake Quexil, and the bottom third of a 15 m core from Lake Salpeten, appear to be of Pleistocene age. The sediments contain lacustrine shells, sponge spicules, and *Pinus* pollen, and include several bands of humified gyttja with fragments of wood, but are dominantly montmorillonitic and mixed-layer clays and may be in part colluvial, like the later Holocene Maya clay. Calcite, gypsum, and (in presently saline Lake Salpeten only) dolomite indicate shallow, closed, moderately saline lakes 30–40 m lower than at present. In both cores a layer of inorganic sediment with gypsum dominant, perhaps recording the most arid phase of the glacial Late Pleistocene, overlies a similar clay layer with calcite dominant. As calcite and dolomite occur throughout the section(s), both minerals are believed to be detrital, but one source of calcite is algal crusts, formed and exposed today in the littoral and supralittoral zones. The early Holocene rise of lake levels formed several meters of fossiliferous gyttja with pollen of mesic tropical forest, now assigned to the pan-tropical Gamblian moist episode. Where deposited in oligomictic or meromictic lakes >30 m deep, Gamblian gyttja of pollen zone Pl is finely laminated, the dark layers being richer in P_{tot} and S_{tot} and poorer in Fe, Mn, Mg, and K than the light (clay) layers, but we cannot yet say that the laminae are annual.

Introduction

Though predicted by geophysical modelling (Kraus 1973; Gates 1976; Manabe & Hahn 1977; Kutzbach 1981), and verified paleolimnologically in the Old World tropics (Butzer *et al.* 1972; Livingstone 1975; Bowler *et al.* 1976; Street & Grove 1979), arid climates corresponding to high-latitude glacial ages have not been documented by dated, pollen-bearing deposits between Florida (Watts 1975; Watts & Stuiver 1980) and Guyana (Wijmstra & Van der Hammen 1966). Impressive new evidence from Lake Valencia, Venezuela (Bradbury *et al.* 1981) makes the absence of such evidence from Central America even more conspicuous.

Our search for Pleistocene deposits in the Central American highlands (Hutchinson *et al.* 1956; Deevey 1957; Tsukada & Deevey 1967) was defeated by the youthfulness of the volcanic lakes studied; anyway, deposits at altitudes above 1 500 m (Watts & Bradbury in press; Martin 1964; Van der Hammen & Gonzales 1960, 1965) tell us little about tropical climates near sea level. At low altitudes, tropical American lake deposits (Van der Hammen 1972, 1974; Bartlett & Barghoorn 1973; Covich & Stuiver 1974; Colinvaux & Schofield 1976) are either undated, deficient in pollen, or (in mangrove swamps) climatically ambiguous. Karsting of low-lying

Hydrobiologia 103, 211–216 (1983).

limestomes, in Yucatan as in Florida (Watts 1969, 1971) generally results in truncated Holocene sections, and these were also found in three of the deeper sinkhole lakes of the Peten lake district (Deevey 1978; Deevey *et al.* 1977, 1979, 1980; Deevey & Rice 1980; Vaughan & Deevey, in prep.). Despite formidable dating problems, caused chiefly by the colluvial nature of lake sediments of Maya age, the Holocene stratigraphy of the Peten is now fairly well understood. It has permitted our interdisciplinary team to consider quantitative relations between Mayan population densities and nutrient loading and siltation (Deevey *et al.* 1979; Rice *et al.*, in prep.; Deevey, in press).

In 1977 and 1978 we raised Kullenberg cores up to 10 m long in the deep basins of Lakes Macanche and Quexil. Pollen-analyzed by S. E. Garrett-Jones (Deevey *et al.* 1980), two of these cores are the subject of papers in this Symposium (Brenner 1983; Binford 1983). An important new finding is that pre-Maya sediments are finely laminated. At 631–635 cm in Quexil core H dark members of two laminar couplets are organic, richer in total P and total S and poorer in Fe, Mn, Mg, and K than two light members, which are dominantly montmorillonitic clay. No intact diatoms, and no other fossils suggestive of seasonal differences, have yet been seen in laminae. Radiocarbon measurements (Stuckenrath, in prep.) were only slightly more informative than those obtained earlier (Deevey 1978).

From the pollen stratigraphy it is clear that the Kullenberg cores failed to reach sediments older than the pre-Maya, high-forest pollen zone Pl, dated at ca. 8 400 ^{14}C years in core S. The 10 m length limitation, imposed by the stickiness of the Maya clay, left Pleistocene deposits tantalizingly out of reach. But an early Holocene stage of deep lakes and mesic forest is clearly correlative with the Gamblian stage of Leakey (1931), and Richardson & Richardson (1972).

Pleistocene deposits

Between March and May, 1980, operations by Daho Pozos, S. A., of Guatemala City, successfully raised long cores from the deep basins of Lakes Quexil, Salpeten, and Macanche. The Quexil and Salpeten sections include several meters of deposits older than the longest Kullenberg cores. A stratigraphic transect of Lake Quexil is shown in Fig. 1. Chemical, paleomagnetic, and microfossil studies are very incomplete; this preliminary report is based on lithologic description, assisted by x-ray analyses of minerals as reported by Professor Frank N. Blanchard, Department of Geology, University of Florida.

Lithologic sections of deposits of Lakes Quexil and Salpeten are shown in Fig. 2. We assign the lower, inorganic segments to the Late Pleistocene without direct proof of age, for we have pollen data only for Quexil core H and have no new ^{14}C results. In both lakes, however, abundance of gypsum and possible presence of kaolinite in a clay matrix dominated by montmorillonite and containing shells, sponge spicules, *Pinus* pollen, and humified organic layers or inclusions, indicate deposition in shallow, moderately saline lakes, 30–40 m shallower than at present, as expected during an arid phase. The overlying organic sediments of zone Pl were laid down in much deeper and probably meromictic lakes. In Lake Quexil, which is hard, but fresh today (TDS 103.9 mg · l^{-1}; Deevey *et al.* 1980), the Holocene lake level rose to within 15 m of the present depth by ca. 8400 B.P., and appears not to have fallen significantly thereafter. Lake Salpeten, which owes its present high salinity (TDS 4 758.1 mg · l^{-1}) to calcium and magnesium sulfate, was also shallower, but may not have been much more saline during its low-water stages than at present. Higher salinity might be inferred if the dolomite were authigenic; but this mineral occurs throughout the Salpeten section and is evidently detrital. Aragonite was not found in any sample, and is not expected here as an authigenic mineral (Bradbury *et al.* 1981).

The most abundant mineral in all samples examined is montmorillonite; as it is also the dominant constituent of soils in the lake district, we believe it to be the residue of long-continued dissolution of the regional limestone (Cowgill & Hutchinson 1963). Other clay minerals – palygorskite, kaolinite or dihydro-halloysite, various mixed-layer clays – are scarce or poorly ordered and give ambiguous x-ray traces. Some of these may be products of weathering of volcanic ash, but although a few volcanic shards have been detected in the Maya clay (Virginia Steen-McIntyre, pers. comm.), we have found no evidence of the massive tephra known

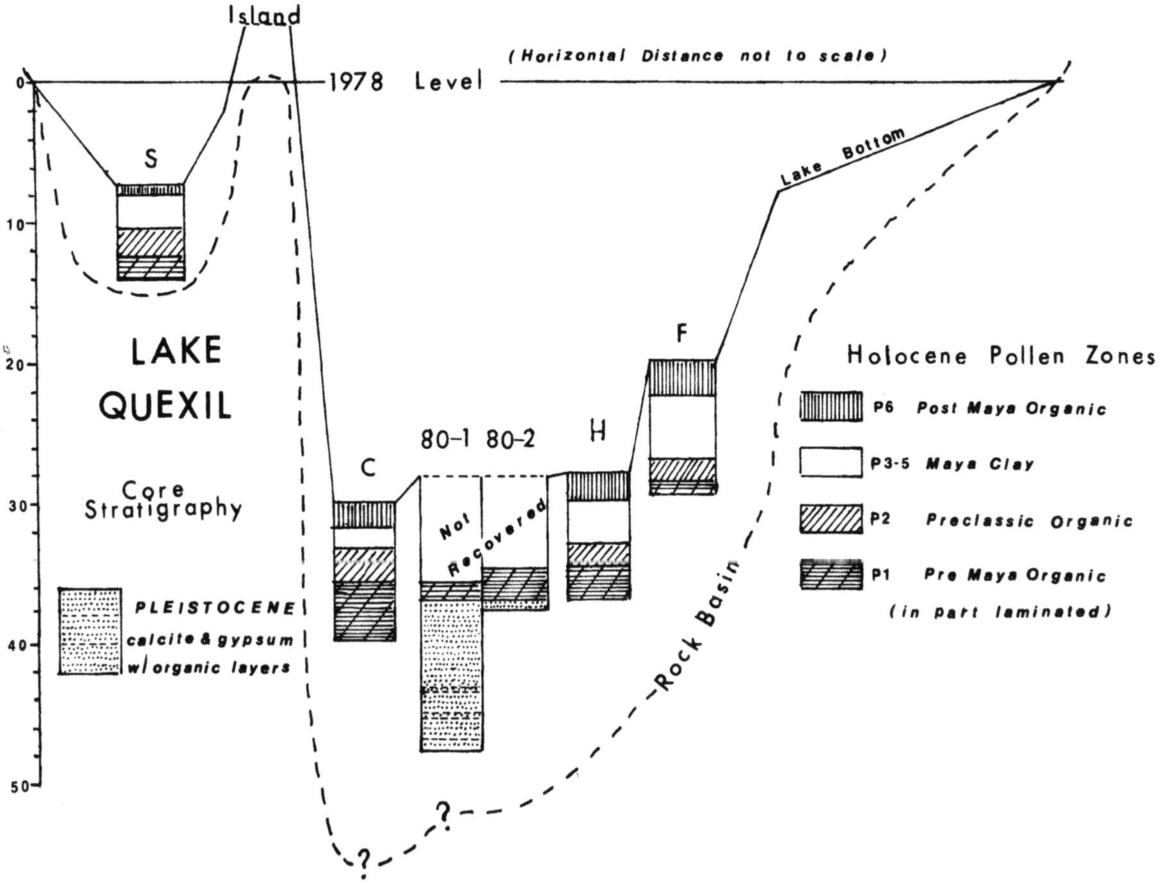

Fig. 1. Stratigraphic transect (horizontal distances not to scale) of Lake Quexil. Palynology of core S by Hague Vaughan (1979); of core H by S. E. Garrett-Jones (Deevey *et al.* 1980).

from volcanic districts farther south (Sheets 1931). Of non-clay minerals, calcite is abundant throughout both Quexil and Salpeten sections, as also in basin soils. Though clearly not formed *in situ* in deep water, calcite is the main constituent of algal crusts, which are extensively developed in the littoral zone of the Peten lakes (Deevey *et al.* 1980). Thus calcite, unlike montmorillonite (and quartz), is at least in part of lacustrine origin and need not have been delivered with colluvium.

Except for some calcite and for some traces of pyrite, detected in Quexil core H, the only mineral believed to be authigenic here is gypsum. Primary gypsum presumably accounts for the dominance of calcium sulfate in several Peten lake waters, but local exposures are not known. Recrystallized or secondary gypsum is common in Peten soils; but deep sediment samples composed almost entirely of gravel-sized gypsum crystals, seen especially in the Salpeten core, were surely precipitated from lake water. (Seen in the field, these crystals have not replaced calcareous fossils during storage, as suggested by Schnitker *et al.* 1980.) In the Pleistocene segment of the Quexil core, where gypsum occurs in all x-rayed samples except the uppermost, gypsum dominates over calcite in an intermediate layer, 38.3–42.7 m below the 1972 lake level, overlain and underlain by clayey layers in which calcite is relatively more abundant. A similar gypsum-dominated clay layer overlies a calcite-dominated layer in the Salpeten core. Scanty as the data are, they support our belief that Lake Quexil was substantially more saline during the Late Pleistocene than later, and suggest that the oldest (calcite-dominated) sediments recovered in both cores predate the period of maximum Late Pleistocene aridity. No other basis has yet been found for stratigraphic subvision of these rather unfossiliferous sediments.

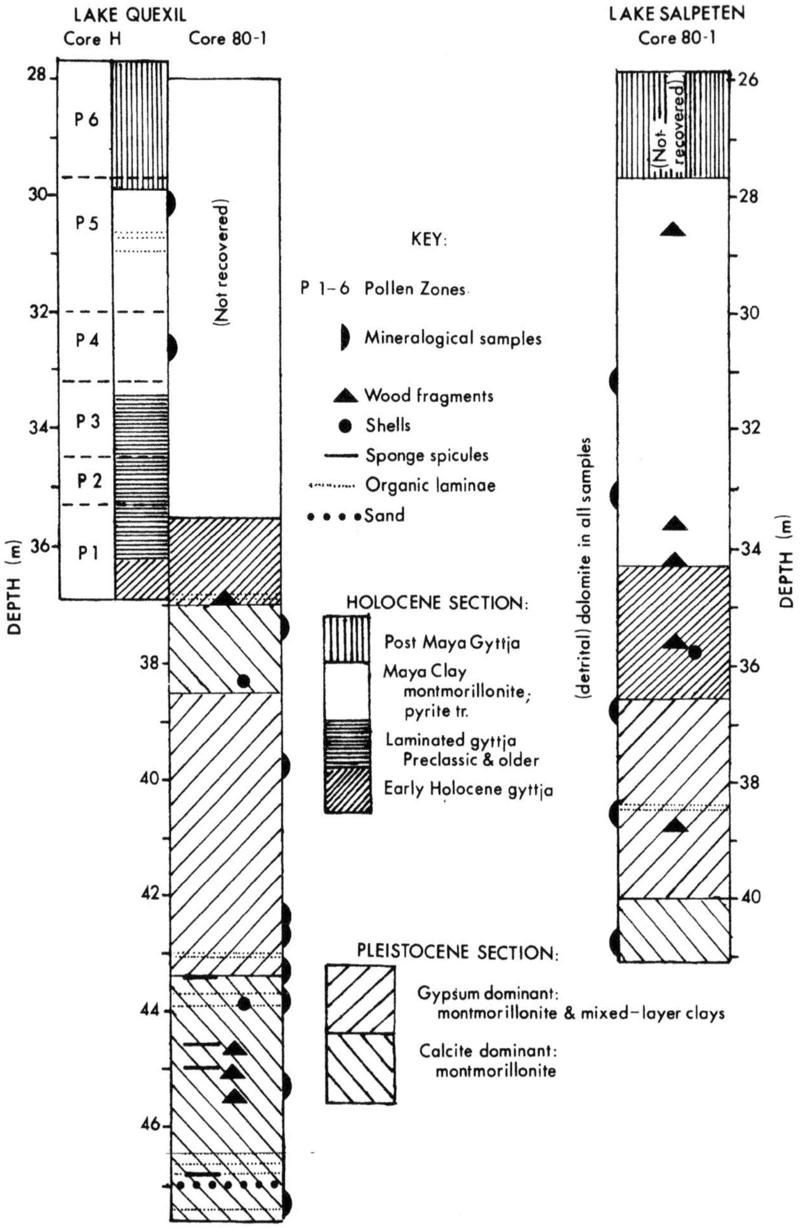

Fig. 2. Lithology of deep-water deposits in Lakes Quexil and Salpeten. Mineralogical assays by Frank N. Blanchard.

Discussion

These lacustrine histories – low-water stages during an arid Late Pleistocene, succeeded by deep, probably meromictic stages during an early Holocene moist interval – correspond in essential respects to those recorded elsewhere in the tropics. In the American tropics the resemblance to the history of Lake Valencia is particularly noteworthy. This lake, situated in an intermontane basin with a climate considerably more arid today than that of the Peten, was a closed swampy depression, ca. 67 m below its overflow level, prior to 10 500 B.P. (Bradbury *et al.* 1981). An early Holocene rise to the overflow level is recorded in a deep-water core by fossil diatoms, Cladocera, and ostracods, while pollen assemblages (Salgado-Labouriau 1980) record replacement of very arid savanna vegetation by mesic forest. Even when deepest and freshest, Lake Valencia was more saline and much richer in magnesium than any Peten Lake, for aragonite occurs throughout the core, and a band of dolomite lies at the top of the Pleistocene segment. Intermittent desiccation after ca. 8500 B.P. (a ^{14}C date that Bradbury *et al.* appear to accept without misgivings) has counterparts in African and Australian lacustrine histories, and is also suggested by ^{18}O data from Lake Chichancanab, Yucatan (Covich & Stuiver 1974). If comparable later Holocene climatic fluctuations affected the Peten, however, the evidence is masked by intensive human disturbance (Vaughan & Deevey, in prep.).

Recovery of Pleistocene lacustrine deposits culminates a forty-year search by the senior author, motivated by two kinds of questions. For climatology, the significant fact is that the Gamblian episode was pan-tropical: moister climate in the Old-World tropics may have resulted from postglacial intensification of the Asiatic monsoon, as suggested by the modelling experiments of Kutzbach (1981), but the weaker North American monsoon is unlikely to have generated comparable effects in Venezuela, Guyana, and Guatemala. For ecology, proof that Late Pleistocene aridity was also pan-tropical entails a biogeographic corollary, that the 'sempiternal' tropical forest, as deployed over thousands of square kilometers of the humid tropics, is probably no older than about ten millennia anywhere. The idea may not be accepted by all ecologists until the Amazonian rain forest is conclusively demonstrated to have supplanted xeric Pleistocene savanna, for the evidence adduced by Van der Hammen (1972) is not compelling. The Peten forest, with present rainfall between 1 500 and 2 500 mm, experiences both seasonal and edaphic aridity, and is not a rain forest. In the American tropics, however, no pollen assemblage more mesic than that of zone Pl has yet been shown to overlie saline lake deposits with vegetational evidence of aridity.

Acknowledgements

Our project has been supported by the National Science Foundation (BMS 72-01859, DEB 77-06629, BNS 78-137736) since 1972. Deep-water drilling and sedimentological work, including the drilling contract with Daho Pozos, S. A., have been supported by the current grant, EAR 79-26330.

References

Bartlett, A. S. & Barghoorn, E. S., 1973. Phytogeographic history of the Isthmus of Panama during the past 12 000 years (a history of vegetation, climate, and sealevel change). In: Graham, A. (Ed.). Vegetation and Vegetational History of Northern Latin America. Elsevier, New York, 203–299.

Bowler, J. M., Hope, G. S., Jennings, J. N., Singh, G. & Walker, D., 1976. Late Quaternary climates of Australia and New Guinea. Quat. Res. 6: 359–394.

Bradbury, J. P., Leyden, G., Salgado-Labouriau, M., Lewis, W. M., Schubert, C., Binford, M. W., Frey, D. G., Whitehead, D. R. & Weibezahn, F. H., 1981. Late Quaternary environmental history of a low elevation tropical lake: Lake Valencia, Venezuela. Science 214: 1299–1305.

Butzer, K. W., Isaac, G. L., Richardson, J. L. & Washbourn-Kamau, C., 1972. Radiocarbon dating of East African lake levels. Science 175: 1069–1076.

Colinvaux, P. A. & Schofield, E. K., 1976. Historical ecology in the Galápagos Islands, I. A Holocene pollen record from El Junco lake, Isla San Cristobal. J. Ecol. 64: 989–1012.

Covich, A. & Stuiver, M., 1974. Changes in oxygen 18 as a measure of long-term fluctuations in tropical lake levels and molluscan populations. Limnol. Oceanogr. 19: 682–691.

Cowgill, U. M. & Hutchinson, G. E., 1963. El Bajo de Santa Fe. Am. Phil. Soc. Trans. 53, 51 pp.

Deevey, E. S., 1957. Limnologic studies in Middle America, with a chapter on Aztec limnology. Connecticut Acad. Arts Sci. Trans. 39: 213–328.

Deevey, E. S., 1978. Holocene forests and Maya disturbance near Quexil Lake, Peten, Guatemala. Pol. Arch. Hydrobiol. 25: 117–129.

Deevey, E. S., in press. Stress, strain, and stability of lacustrine ecosystems. In: Haworth, E. Y. and Lund, J. W. G., (Eds.). Lake Sediments and Environmental history. Leicester University Press.

Deevey, E. S., Brenner, M., Flannery, M. S. & Yezdani, G. H., 1980. Lakes Yaxha and Sacnab, Peten, Guatemala: limnology and hydrology. Arch. Hydrobiol. Suppl. 57: 419–460.

Deevey, E. S., Garrett-Jones, S. E. & Vaughan, H. H., 1980. Impact of Mayan civilization on tropical lowland forest. 5th Int. Palynol. Congr. Cambridge, Abstr.: 107 [abstract, poster presentation].

Deevey, E. S. & Rice, D. S., 1980. Coluviatión y retención de nutrientes en el distrito lacustre del Petén central, Guatemala. Biotica 5: 129–144.

Deevey, E. S., Rice, D. S., Rice, P. M., Vaughan, H. H., Brenner, M. & Flannery, M. S., 1979. Mayan urbanism: impact on a tropical karst environment. Science 206: 298–306.

Deevey, E. S., Vaughan, H. H. & Deevey, G. B., 1977. Lakes Yaxha and Sacnab, Peten, Guatemala: planktonic fossils and sediment focusing. In: Golterman, H. L. (Ed.). Interactions between Sediments and Fresh Water. Junk, The Hague: 189–196.

Gates, W. L., 1976. Modeling the Ice-Age climate. Science 191: 1138–1144.

Hutchinson, G. E., Patrick, R. & Deevey, E. S., 1956. Sediments of Lake Patzcuaro, Michoacan, Mexico. Bull. Geol. Soc. Am. 67: 1491–1504.

Kraus, E. B., 1973. Comparison between Ice Age and present general circulation. Nature 245: 129–133.

Kutzbach, J. E., 1981. Monsoon climate of the early Holocene: climate experiment with the Earth's orbital parameters for 9 000 years ago. Science 214: 59–61.

Leakey, L. S. B., 1931. The Stone Age cultures of Kenya Colony. Cambridge Univ. Press, 283 pp.

Livingstone, D. A., 1975. Late Quaternary climatic change in Africa. Ann. Rev. Ecol. Syst. 6: 249–280.

Manabe, S. & Hahn, D. G., 1977. Simulation of the tropical climate of an ice age. J. Geophys. Res. 82: 3889–3911.

Martin, P. S., 1964. Paleoclimatology and a tropical pollen profile. 6th INQUA Congr. Warsaw 1961, Rep. 2: 319–323.

Rice, D. S., Rice, P. M. & Deevey, E. S., in prep. Classic Maya impact on a lacustrine environment.

Richardson, J. L. & Richardson, A. E., 1972. History of an African Rift lake and its climatic implications. Ecol. Monogr. 42: 499–534.

Salgado-Labouriau, M. L., 1980. A pollen diagram of the Pleistocene-Holocene boundary of lake Valencia, Venezuela. Rev. Palaeobot. Palynol. 30: 297–312.

Schnitker, D., Mayer, L. M. & Norton, S., 1980. Loss of calcareous microfossils from sediments through gypsum formation. Mar. Geol. 36: M35–M44.

Sheets, P. D., 1981. Volcanoes and the Maya. Nat. Hist. 90: 32–41.

Street, F. A. & Grove, A. T., 1979. Global maps of lake-level fluctuations since 30,000 yr B.P. Quat. Res. 12: 83–118.

Tsukada, M. & Deevey, E. S., 1967. Pollen analyses from four lakes in the southern Maya area of Guatemala and El Salvador. In: Cushing, E. J. & Wright, H. E. (Eds.). Quiternary Ecology. Yale Univ. Press, New Haven: 303–331.

Van der Hammen, T., 1972. Changes in vegetation and climate in the Amazon basin and surrounding areas during the Pleistocene. Geol. Mijnb. 51: 641–643.

Van der Hammen, T., 1974. The Pleistocene changes of vegetation and climate in tropical South America. J. Biogeogr. 1: 3–26.

Van der Hammen, T. & Gonzales, E., 1960. Upper Pleistocene and Holocene climate and vegetation of the 'Sabana de Bogota' (Colombia, South America). Leidse Geol. Meded. 25: 261–315.

Van der Hammen, T. & Gonzales, E., 1965. Late glacial and Holocene pollen diagram from 'Cienaga del Visitador' (Dept. Boyacá, Colompbia). Leidse Geol. Meded. 32: 193–201.

Vaughan, H. H., 1979. Prehistoric disturbance of vegetation in the area of Lake Yaxha, Peten, Guatemala. Ph.D. thesis, Zoology, Univ. Florida.

Vaughan, H. H. & Deevey, E. S., in prep. Pollen stratigraphy of two cores from the Peten lake district.

Watts, W. A., 1969. A pollen diagram from Mud Lake, Marion County, north-central Florida. Bull. Geol. Soc. Am. 80: 631–642.

Watts, W. A., 1971. Postglacial and interglacial vegetation history of southern Georgia and central Florida. Ecology 52: 676–689.

Watts, W. A., 1975. A Late Quaternary record of vegetation from Lake Annie, south-central Florida. Geology 3: 344–346.

Watts, W. A. & Bradbury, J. P., 1982. Late Pleistocene and Holocene paleoenvironments and human activity in the west-central Mexican Plateau: evidence from Lake Patzcuaro, Michoacan, and from the Cuenca de Mexico. Quat. Res. 17: 56–70.

Watts, W. A. & Stuiver, M., 1980. Late Wisconsin climate of northern Florida and the origin of species-rich deciduous forest. Science 210: 325–327.

Wijmstra, T. A. & Van der Hammen, T., 1966. Palynological data on the history of tropical savanna in northern South America. Leidse Geol. Meded. 38: 71–90.

The Holocene development of some low and high arctic Greenland lakes

Bent Fredskild
Greenland Botanical Survey, Botanical Museum, Gothersgade 130, DK-1123 Copenhagen K, Denmark

Keywords: paleolimnology, Greenland, Holocene, oligotrophication, pollen analysis, macrofossils, arctic

Abstract

During the Holocene most West Greenland lakes passed from an early eutrophic stage, rich in both flora and fauna, through a mesotrophic to an oligotrophic stage with very low productivity. Temperature conditions were limiting factors only in the very beginning, whereas chemical factors alone were decisive later on.

In the low artic area of Godthåbsfjord nine lakes were cored. Four of the cores have been analysed for pollen and macrofossils. In the high artic area of Melville Bugt cores from two lakes have been investigated (Fig. 1). Most lakes in these parts of Greenland were formed 9500–8500 years ago. As the analyses have spread over eight years, the diagrams differ in their information content. For example, the counting of chironomid labiae and the abdomens of *Alona* and *Acroperus* in the pollen slides has only recently started.

Godthåbsfjord

The highest July mean temperature (10.9° C) in Greenland has been registered at the head of Godthåbsfjord, but a short and fairly dry summer season with only five months of the year above 0° C prevents tree growth, and low, lichen-rich *Betula nana* heaths dominate. The cooler, moister outer coast area of Nordlandet is dominated by mossy *Empetrum* heaths.

The sediment cores usually start with clayey, vividly coloured and finely laminated, jelly-like algal gyttja, or even with lake marl or lime gyttja. The minerogenic content rapidly decreases. The jelly-like consistency changes, the sediment becoming more loose and watery, and when extruded, it often tends to be slightly liquid as a result of a water content of ca. 90%. The depth of the sediment ranges between 1.3 and 3.6 m, giving average sediment accumulation rates between 0.14 and 0.39 mm a^{-1}. In some of the lakes the early accumulation rate is relatively high, up to 0.7 mm a^{-1}, whereas in others the rate is fairly constant throughout the Holocene. In some of the lakes the rate decreases towards the top (Fredskild 1977).

The only two macrophytes occurring today all over Greenland (*Hippuris vulgaris* and *Ranunculus confervoides*) are among the first immigrants along with *Potamogeton filiformis* (N. limit 74°) (Fig. 2), whereas some of the most southerly species (*Isoëtes echinospora* and *Myriophyllum alterniflorum*) are late incomers. This may lead to the conclusion that the succession simply reflects the climatic amelioration. But in some lakes *Myriophyllum spicatum* (N. limit 70°) flourishes after only one century, and a fruit of *Potamogeton alpinus* (N. limit 67°) has been found in sediments a millennium later. In addition the big fruits of *Myriophyllum spicatum, Hippuris* and *Potamogeton filiformis* do not seem to convey more advantage than the small spores of a late immigrant like *Isoëtes*, and all lakes follow a Holocene succession starting with fairly eutrophic conditions, passing through a mesotrophic, and

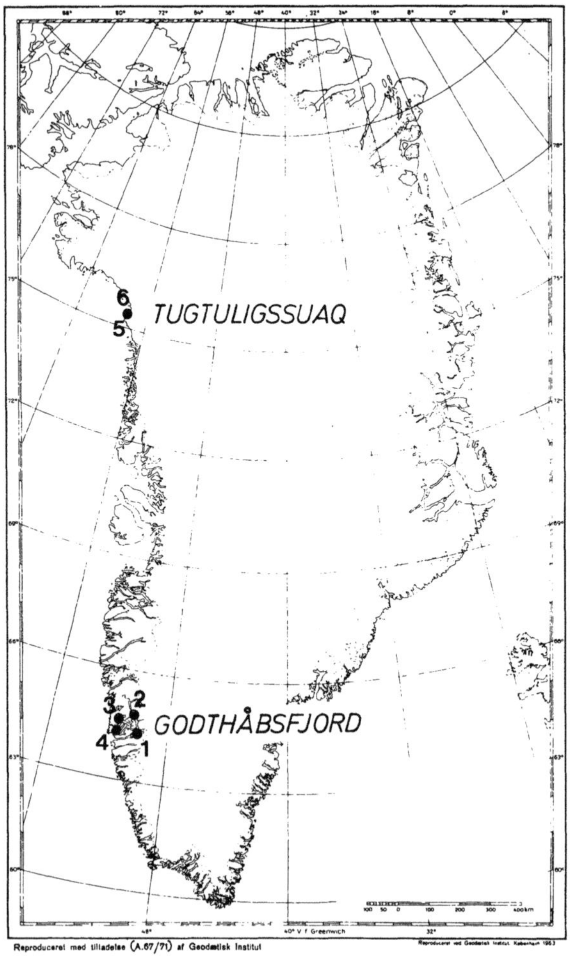

Fig. 1. Map of Greenland showing position of the lakes mentioned. 1 = Johs. Iversen Sø, 2 = Karra, 3 = Terte, Lake A, 4 = Sardloq, Sardlup qáqâ, 5 = Langesø, 6 = Rundesø.

ending in an oligotrophic stage. This leads to the conclusion that once the climatic threshold of all the macrophytes was passed around 8 000 B.P., their succession reflects the trophic stage of the lakes.

The length of each stage, its species diversity, and its productivity is highly dependent on the soil of the catchment area. Thus the two richest lakes, Johs. Iversen Sø in the interior, and Terte, Lake A on Nordlandet, are surrounded by till-covered, low hills, whereas Karra, at an elevation of 300 m in the interior, and Sardloq on Nordlandet are surrounded by barren rocks, and are, as they always were, comparatively poor and unproductive.

Characteristic algal and zoological remains

Pediastrum

By far the most important *Pediastrum* species in the Greenland lake sediments is *P. boryanum*, attaining its maximum number immediately or shortly after the formation of the lake (Fig. 3). The closely related *P. integrum* and *P. muticum*, included in the *P. boryanum* curve in some diagrams, usually attain a maximum later, but still in the early, eutrophic stage. In the beginning the total accumulation rate of these species ranges between 25–35,000 cm^{-2} a^{-1} in a rich lake (Johs. Iversen Sø) and 400–1 600 cm^{-2} a^{-1} in a poor lake (Karra), with an absolute maximum of ca. 85,000 cm^{-2} a^{-1}. *Pediastrum duplex* has been observed only once, viz. in Johs. Iversen Sø, where it follows the above species and characterizes the mesotrophic *Eupotamogeton-Isoëtes* stage. It is followed by *P. angulosum*, which together with *P. kawraiskyi* characterize the oligotrophic stage in the lakes.

Pediastrum tetras is rare, most often occurring sporadically in the eutrophic stage. *P. braunii* has only been observed in the two high arctic lakes. The total accumulation rate of *Pediastrum* in the upper samples of the four lakes range between 0.1 and 40 cm^{-2} a^{-1}.

The curves of *Botryococcus braunii* often parallel those of *Pediastrum boryanum*, but the rate of accumulation is lower, usually only a tenth to half that of *Pediastrum*.

Characeae

Oospores of *Chara* and *Nitella* are frequent in the lower part of the four low arctic lakes (Fig. 4), whereas they have not been found in the high arctic lakes.

Diatoms

Niels Foged has carried out diatom analysis from the four lakes. The species found have been grouped into alkaliphilous, acidophilous and indifferent types, and diagrams have been published (Foged 1977).

Zoological remains

The most prominent of the zoological remains in the macrofossil samples are mandible fragments of *Lepidurus arcticus,* larval cases of Tricoptera, most often of the circum-greenlandic *Apatania zo-*

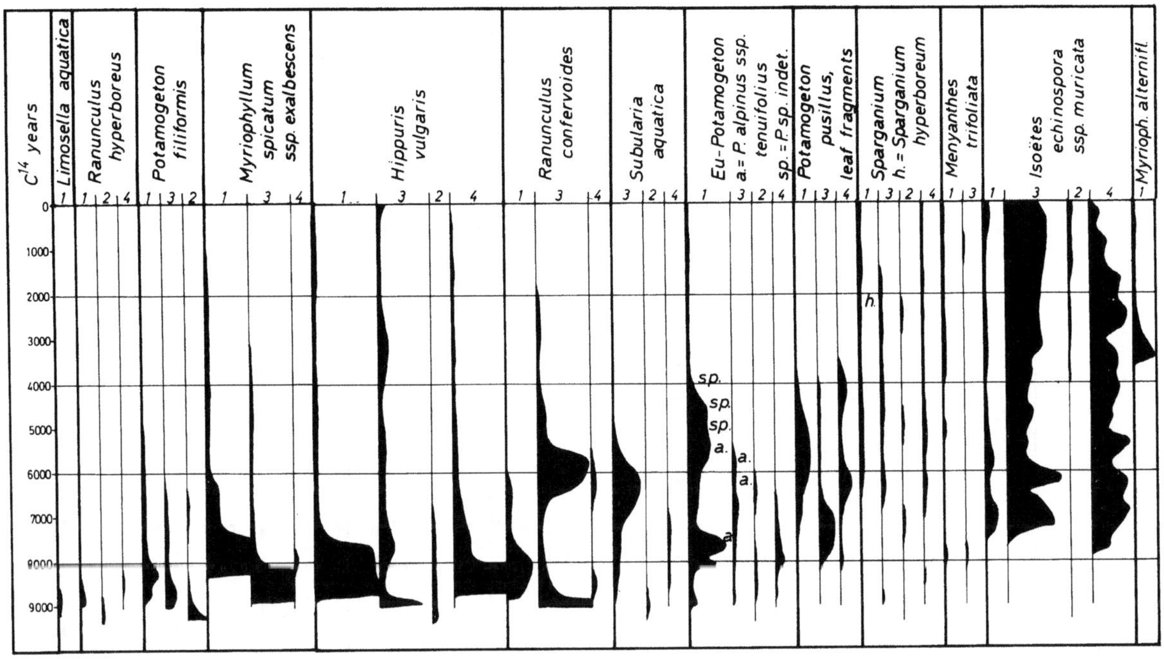

Fig. 2. The Holocene development of macrophyte vegetation of four lakes at Godthåbsfjord. The curves are based on the frequency of pollen, spores, seeds, fruits and leaf fragments. The sedimentation of organic material begins between 9 400 (lake 2) and 9 050 (lake 3) radiocarbon years ago.

Low Arctic	Johs. Iversen Sø	P. boryanum (incl. integr./mutic.)	P. duplex	P. angul.	P. tetras	(P. kawraiskyi)	
	Terte, Lake A	P. boryanum (incl. integr./mutic.)	P. tetras	P. braunii	(P. angulosum)	(P. kawraiskyi)	
	Karra	P. boryanum	P. integrum/muticum	(P. tetras)	P. kawraiskyi	(P. angulosum)	
	Sardloq	P. boryanum	P. integrum/muticum	P. angulosum	P. kawraiskyi		
High Arctic	Langesø	P. boryanum	P. integrum/muticum	(P. braunii)			
	Rundesø	P. boryanum	P. integrum/muticum	(P. braunii)			

Fig. 3. The Holocene succession of *Pediastrum* species. A name underlined by four lines denotes that during its maximum occurrence the content ml^{-1} gyttja exceeds 10^5, three lines 10^4, two lines 10^3 and one line 10^2, whereas a bracketed name denotes a content of less than 10^2 coenobia.

nella, epphipiae of *Daphnia pulex* and *Simocephalus vetulus*, and statoblasts of *Cristatella mucedo*.

Apart from a relict, high altitude occurrence in S. Greenland *Lepidurus arcticus* has its southern limit north of Godthåbsfjord, but it has also been found in the oldest sediments in some S. Greenland lowland lakes (Fredskild *et al.* 1975). Within its present distribution area it prefers rich ponds and lakes. Although its abundance in the deepest samples (Fig. 5) seems to be connected with the high productivity, its decrease and disappearance is most likely caused by an increase in the temperature of the water. In all four lakes the first statoblasts of *Cristatella* directly follow the last remnants of *Lepidurus* and points to a temperature control for *Cristatella* as well. So far *Cristatella mucedo* has not been found alive in Greenland, but as its statoblasts occur in the uppermost sediment in Johs. Iversen Sø, it may well occur in some lakes in this area.

The number of epphipiae of *Daphnia pulex* is high in the deeper part of the cores but eventually declines to zero. The species is distributed throughout Greenland, and its early abundance must be

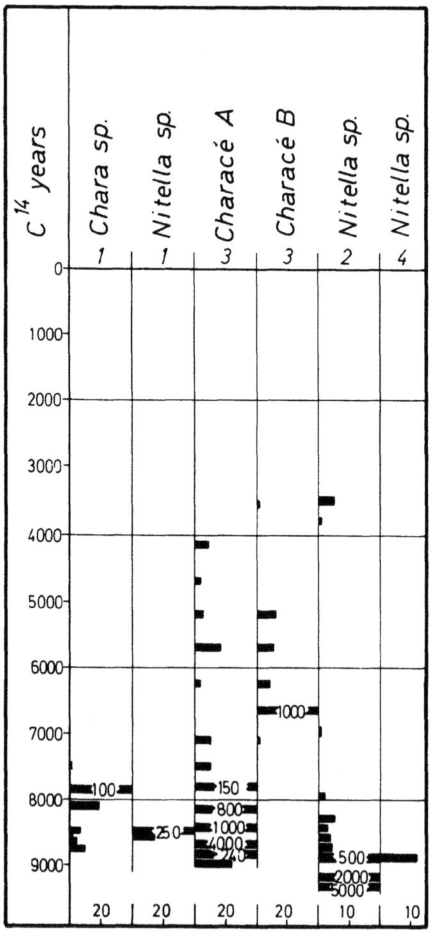

Fig. 4. The frequency of oospores of *Chara* and *Nitella* in four lakes at Godthåbsfjord.

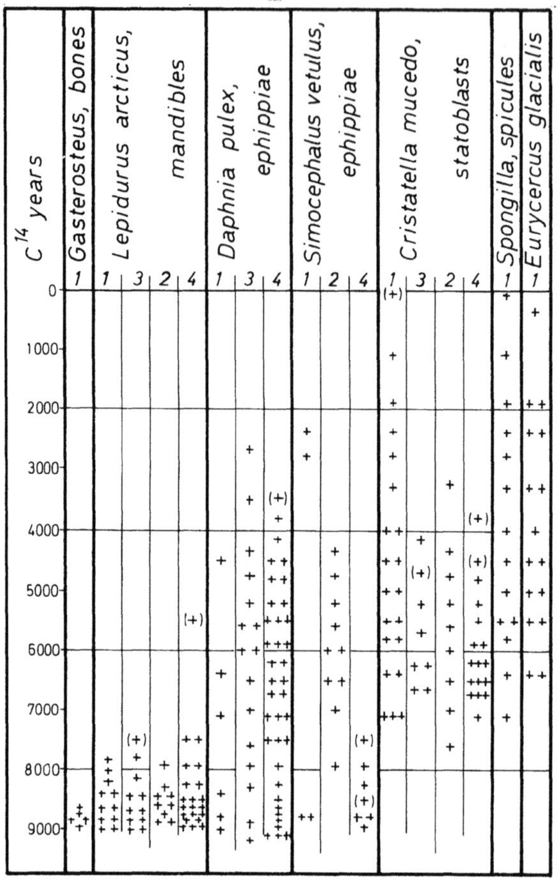

Fig. 5. The frequency of some zoological remains in four lakes at Godthåbsfjord.

associated with the richness of the water. This also seems to be the case with *Simocephalus vetulus* and the Tricoptera, yet single specimens of the latter may be found in the present benthos.

The spicules of *Spongilla lacustris* s.l. are dissolved in the standard preparation of pollen slides, but in one core (Johs. Iversen Sø) untreated samples have been searched for spicules. The species distribution, in accordance with its present demands, accords with that of *Isoëtes setacea* and *Cristatella*, and it is still found in this lake.

Johs. Iversen Sø, a rich lake

The development of a rich lake can be illustrated by the diagrams from Johs. Iversen Sø, named in honour of the late Danish ecologist, who was the first to publish pollen diagrams from Greenland lakes (Iversen 1954) when he presented a diagram from this lake, called by him 'Lake 100 m s.m.'. I cored it once again in 1973.

The first stage (Fig. 6) is dominated by pollen of *Hippuris* and, a little later, of *Myriophyllum spicatum*, and by huge amounts of *Pediastrum boryanum* s.l. and *Botryococcus*. The next stage is characterized by a decrease in these species, the expansion of *Eupotamogeton* and the emergence of *Isoëtes setacea, Cristatella* and *Sparganium*. A *Pediastrum duplex* stage follows with no indication of *Potamogeton filiformis* and *Characeae*, and with markedly decreasing frequencies of the early immigrants. A marked change in almost all curves appears at the next border. *Myriophyllum alterniflorum* pollen appear for the first time, and the pollen of *Menyanthes*, which immigrated at a very early

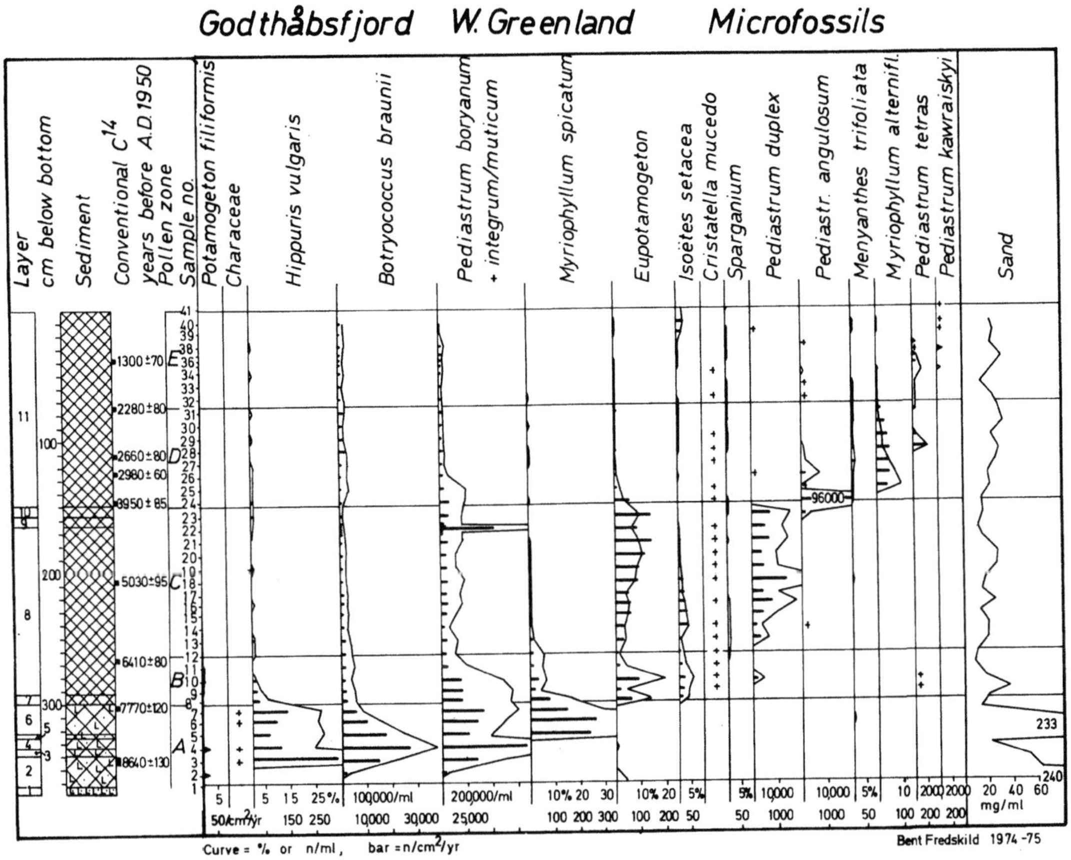

Fig. 6. Microfossil diagram from Johs. Iversen Sø (64°24′N, 50°12′W).

stage, appears in every sample. The final stage is oligotrophic with *Pediastrum kawraiskyi* as the last immigrant.

The macrofossil diagram (Fig. 7) confirms the picture of an early eutrophic stage with many species, a mesotrophic stage with different species and a final oligotrophic stage in which most species have disappeared.

The inferred history of the lake is also confirmed by the chemical analyses which show that the content of all ions decreases upwards, often with a marked drop at the sediment change at 300 cm.

Sardloq, a poor lake

The Sardloq lake has a rather mesotrophic beginning with *Nitella* spores in only one sample and with only a single pollen grain of *Myriophyllum spicatum* and no *Potamogeton filiformis* (Fig. 8). *Eupotamogeton* pollen are found only in the deeper part, but, from finds of leaf tips in the macrofossil samples, sterile *Potamogeton pusillus* persists during the following 3 000 years. In the final 3–4 000 years the lake is very oligotrophic with *Isoëtes* and *Pediastrum kawraiskyi* characterizing the flora.

Tugtuligssuaq, Melville Bugt

A 20 km long peninsula in Melville Bugt at 75° N is bordered by the inland ice and by the ice-filled sea. The climate is high arctic, maritime with a June average just above 0° C, a July approaching 5° C, and a winter starting at the beginning of September. The precipitation is very high, and even on Midsummer day everything is buried in snow (Jacobsen 1980).

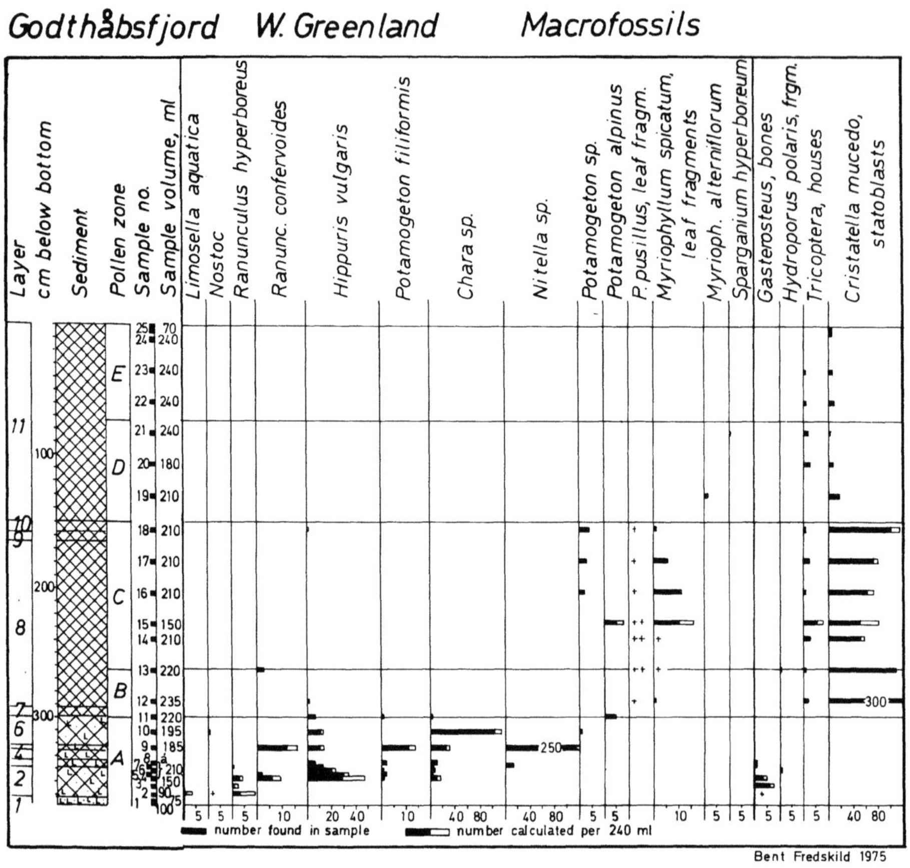

Fig. 7. Macrofossil diagram from Johs. Iversen Sø, (64°24′N, 50°12′W).

The usually very open vegetation is dominated by *Cassiope tetragona* and *Luzula confusa* with some *Salix arctica* and *S. herbacea*. Two lakes, connected by a streamlet, only 50 m long, were cored.

The catchment area consists of weathered gneiss rocks and blocks, resulting in very poor lakes, in which the only visible vegetation is some loose mats of the moss *Drepanocladus exannulatus*. However, the microfossil diagrams show that at the very beginning of their history their productivity was very high, starting with a maximum of *Botryococcus*, followed by maxima in *Pediastrum boryanum* and *P. integrum/muticum*. *Hippuris* grew at an early stage in one of the lakes (Fig. 9).

The most surprising fact is that the highest *Pediastrum* concentration ever found in a Greenland lake, viz. 5 10^6 ml^{-1} corresponding to an annual accumulation of the same order of magnitude as in the low arctic Johs. Iversen Sø, is found in the second lake. The high productivity is coincident with the content of clay in the sediment, and once the outwash of minerogenic matter stops, productivity decreases rapidly, and rarely more than 100 coenobia ml^{-1} are observed. In both lakes small amounts of *P. braunii* emerge at a late stage.

Tricoptera and chironomids are found throughout the cores, whereas preliminary macrofossil analyses have shown that *Daphnia pulex* and *Simocephalus vetulus* were common in the earliest stages. They do not live in the lakes today. *Lepidurus arcticus* was very common in the lowermost sediment, but from some 20 cm in the sediment upwards, no mandibles are found. This cannot be related to climate but must be an extinction as a result of too little prey, which itself is a result of the decreasing algal production. The contemporary increase in *Alona* and *Acroperus* may be related to the disappearance of one of their predators.

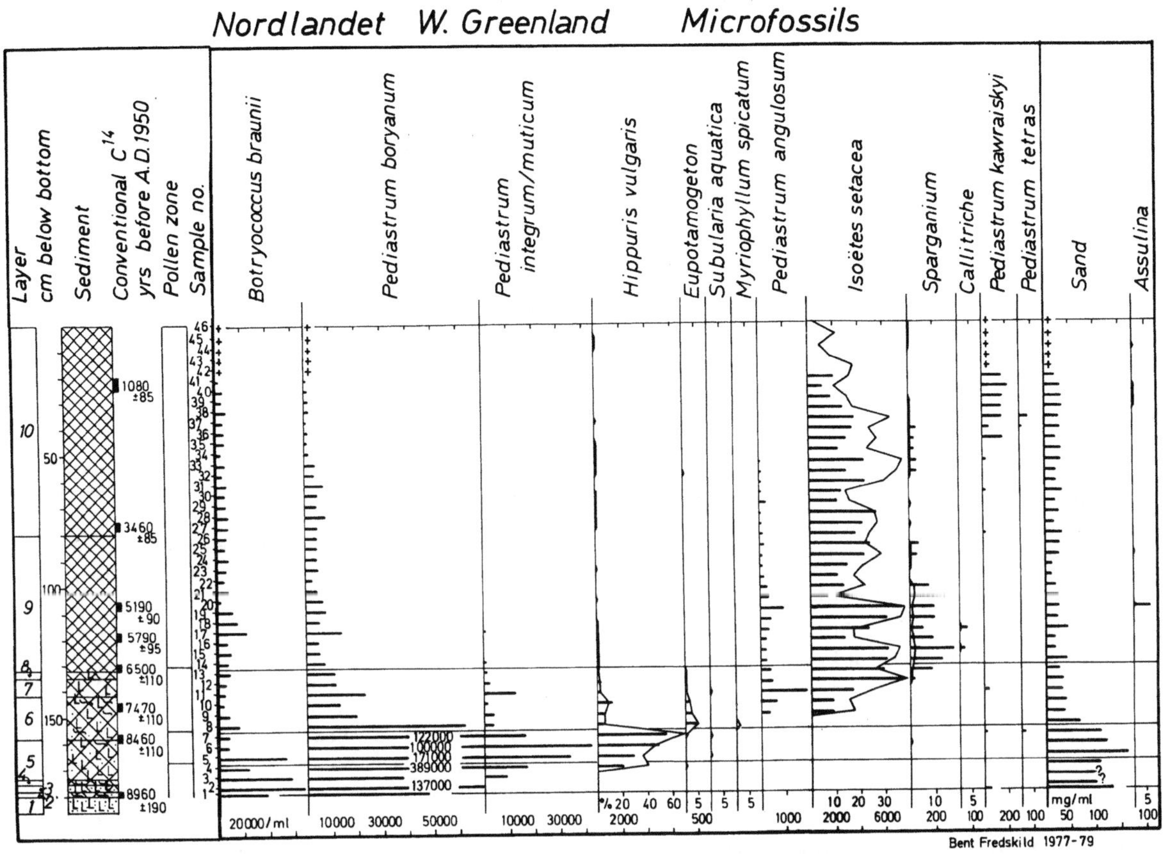

Fig. 8. Microfossil diagram from Sardloq, Sardlup qáqâ (64°24′N, 51°41′W).

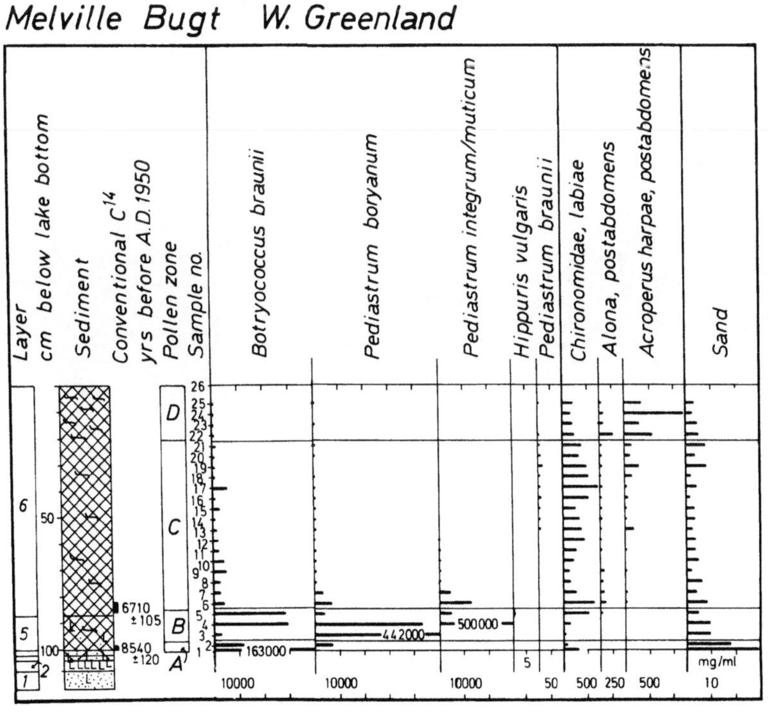

Fig. 9. Microfossil diagram from Tugtuligssuaq, Langesø.

Conclusions

All the lakes so far investigated in detail follow a series of stages. In the first stage(s) climatic amelioration was the dominant factor affecting life in the lakes, but very soon the chemistry and plant cover of the catchment area became the most important factors in determining the length and nature of the various stages. The only known exception to this general statement relates to saline lakes e.g. Store Saltsø in the highly continental Søndre Strømfjord area (Böcher 1949). The present day distribution in Greenland of the most nutrient-demanding macrophytes, *Myriophyllum spicatum* and *Potamogeton filiformis*, seems to be connected more with the occurrence in certain areas of lakes with high conductivity, yet usually with a skew composition of ions, than connected with climate. Thus *Myriophyllum spicatum* was present at an early stage in 8 of the 10 Godthåbsfjord lakes cored. It is now extinct in all of them.

Acknowledgements

Part of the field and laboratory work has been supported by the Danish Natural Science Research Council (Grant nos 511-1 847, 5 527 and 6 584). The Melville Bugt material was collected during my participation in the Knud Rasmussen Memorial Expedition in 1979, supported by the Knud Rasmussen Foundation. My thanks to both institutions. Dr. Sc. Ulrik Røen, Zoological Institute, University of Copenhagen, kindly identified the zoological macrofossils. The radiocarbon datings were carried out by the Carbon-14 Dating Laboratory, Geological Survey of Denmark and National Museum, Copenhagen.

References

Böcher, T. W., 1949. Climate, soil and lakes in continental West Greenland in relation to plant life. Meddel. Grønland 147: 1-63.

Foged, N., 1977. The diatoms in four postglacial deposits at Godthåbsfjord, West Greenland. Meddel. Grønland 199: 1-64.

Fredskild, B., 1973. Studies in the vegetational history of Greenland. Palaeobotanical investigations of some holocene lake and bog deposits. Meddel. Grønland 198: 1-245.

Fredskild, B., 1977. The development of the Greenland lakes since the last glaciation. Folia limnol. scand. 17: 101-106.

Fredskild, B., Jacobsen, N. & Røen, U., 1975. Remains of mosses and freshwater animals in some holocene lake and bog sediments from Greenland. Meddel. Grønland 198: 1-44.

Iversen, J., 1954. Origin of the flora of western Greenland in the light of pollen analysis. Oikos 4: 85-103.

Jacobsen, N. Kingo, 1980. The Knud Rasmussen Memorial Expedition. Geograf. Tidsskr. 80: 29-44.

Palynological investigations of the sediments of ten lakes in eastern Holstein, North Germany

Friz-Rudolf Averdieck
Universität Kiel, D 2300 Kiel 1, B.R.D.

Keywords: paleolimnology, pollen analysis, sediment accumulation rate, vegetation history, Holstein, Post-glacial

Abstract

Pollen analysis was carried out on the sediments of 10 lakes in eastern Holstein. Sediment accumulation rates varied considerably both from lake to lake and also between respective periods. Increases in sediment accumulation rate often coincide with the Neolithic settlement periods. Faunal remains have been examined from two of these lakes (cf. Günther 1983; Hofmann 1983).

Introduction

The Weichselian Glaciation of East Holstein covered over 5300 km². Within this region about 265 lakes ranging from 1 ha to 30 km² cover an area of more than 200 km². Additionally, there are quite a number of smaller lakes with a surface area less than 1 ha and many other former lakes which today are moorland or cultivated areas.

The big majority of the lakes are of glacial (dead ice) origin, but some (e.g. Lake Segeberg) have been formed as a result of salt tectonics. Of the modern lakes only a few have been the subject of sediment studies. Ten of them are presented here (Fig. 1), and they include lakes with a wide range of surface area and depths.

Sediment accumulation rates

In all lakes, even in Lake Segeberg, sedimentation starts in the Late-glacial mainly during the Younger Dryas (zone IV), more rarely during earlier periods. Figure 2 illustrates time-depth curves for the 10 lakes according to the thickness of different pollen zones (after Overbeck 1975, Nilsson 1964) and the ^{14}C datings of Averdieck *et al.* 1972. How-

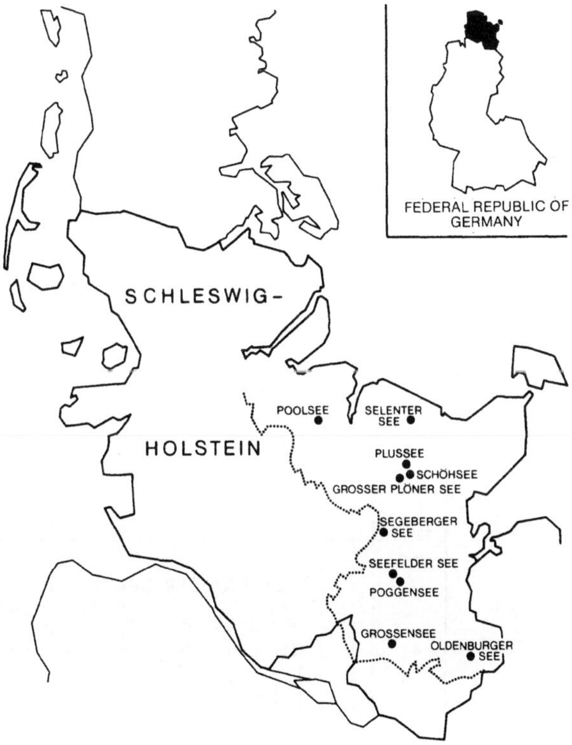

Fig. 1. Location of investigated sites in Schleswig-Holstein. Dotted line shows western border of Weichselian Glaciation.

Hydrobiologia 103, 225–230 (1983).

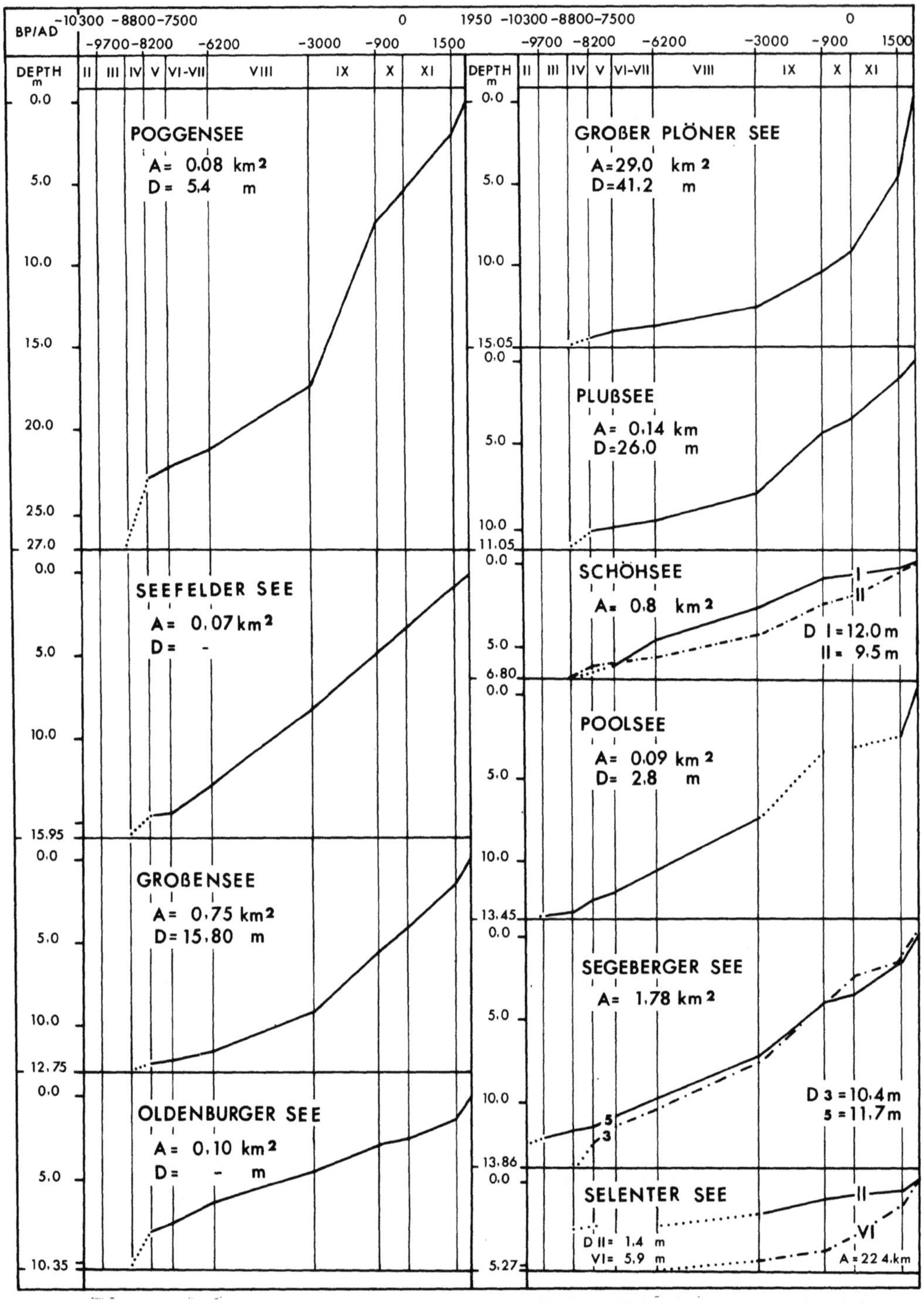

Fig. 2. Time-depth curves for 10 lakes (Schöhsee after Saad, 1970). Dotted lines show extrapolation.

ever, such dates are to be regarded cautiously, since there are sources of error in the ^{14}C dating of lake sediments (cf. Willkomm & Erlenkeuser 1972).

Late-Glacial, and often Pre-Boreal, sediments are minerogenic. Normally they are followed by chalky organogenic muds, often with visible seasonal laminations concentrated in small sections of the sediment cores. This phenomenon generally ceases at the Subboreal, although in Lake Segeberg the sediments are laminated between zone II and zone IX. During the later Postglacial lime-free humic muds are prevalent.

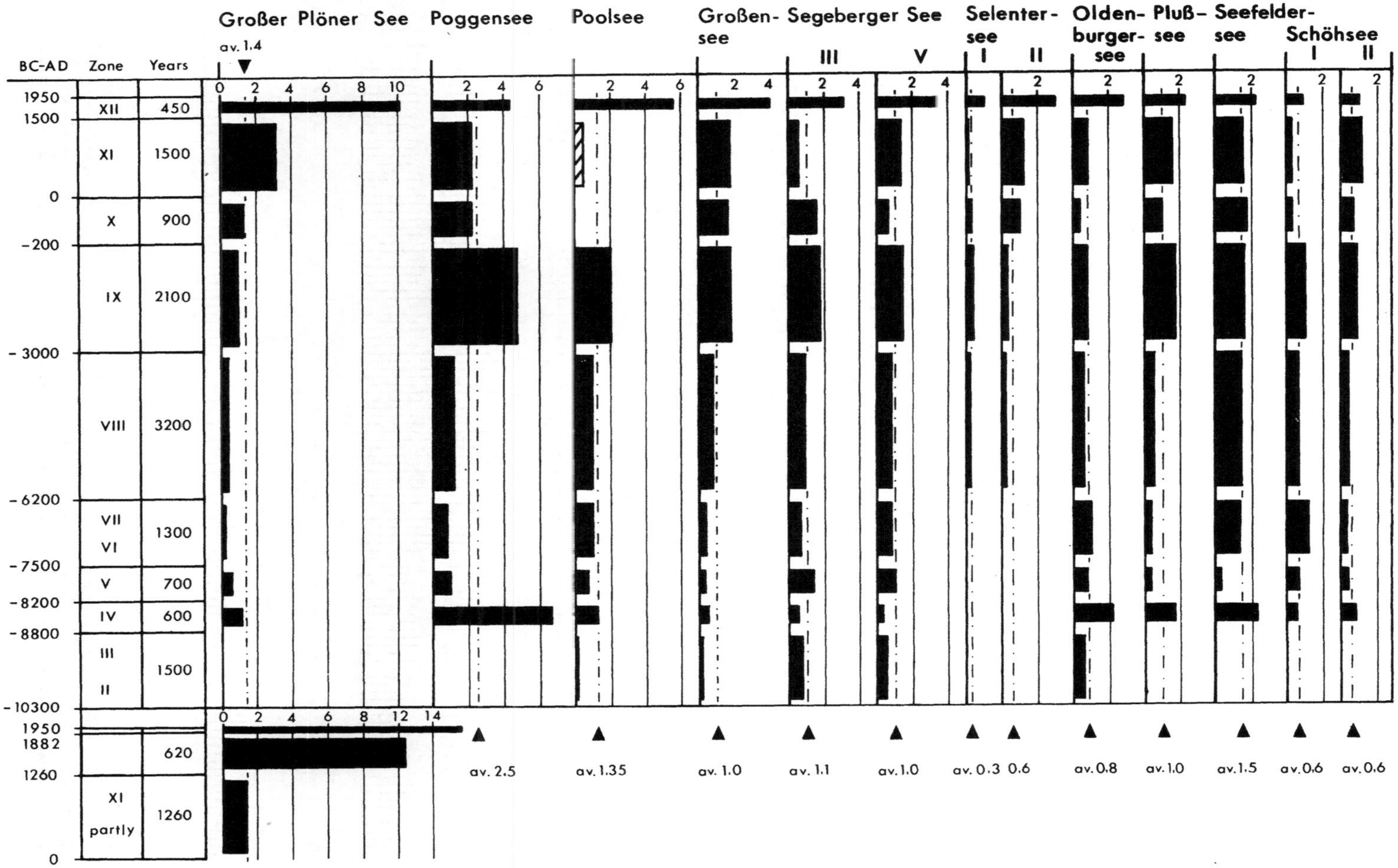

Fig. 3. Average sediment accumulation for each zone (mm a^{-1}). Hatched block = estimate.

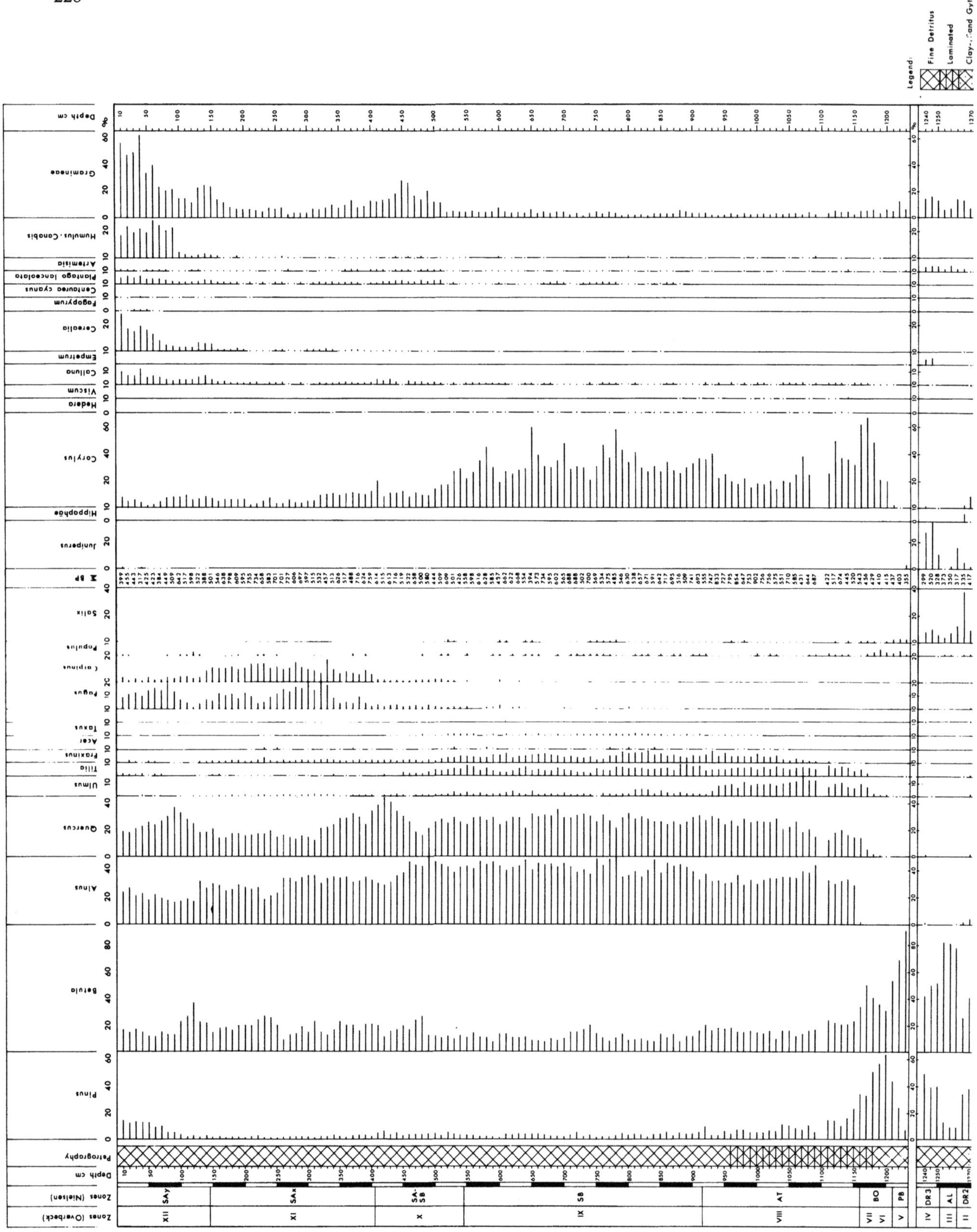

Fig. 4. Modified pollen diagram from Grossensee. Percentage of arboreal pollen excluding *Corylus*. Note scale expansion below 1 300 cm.

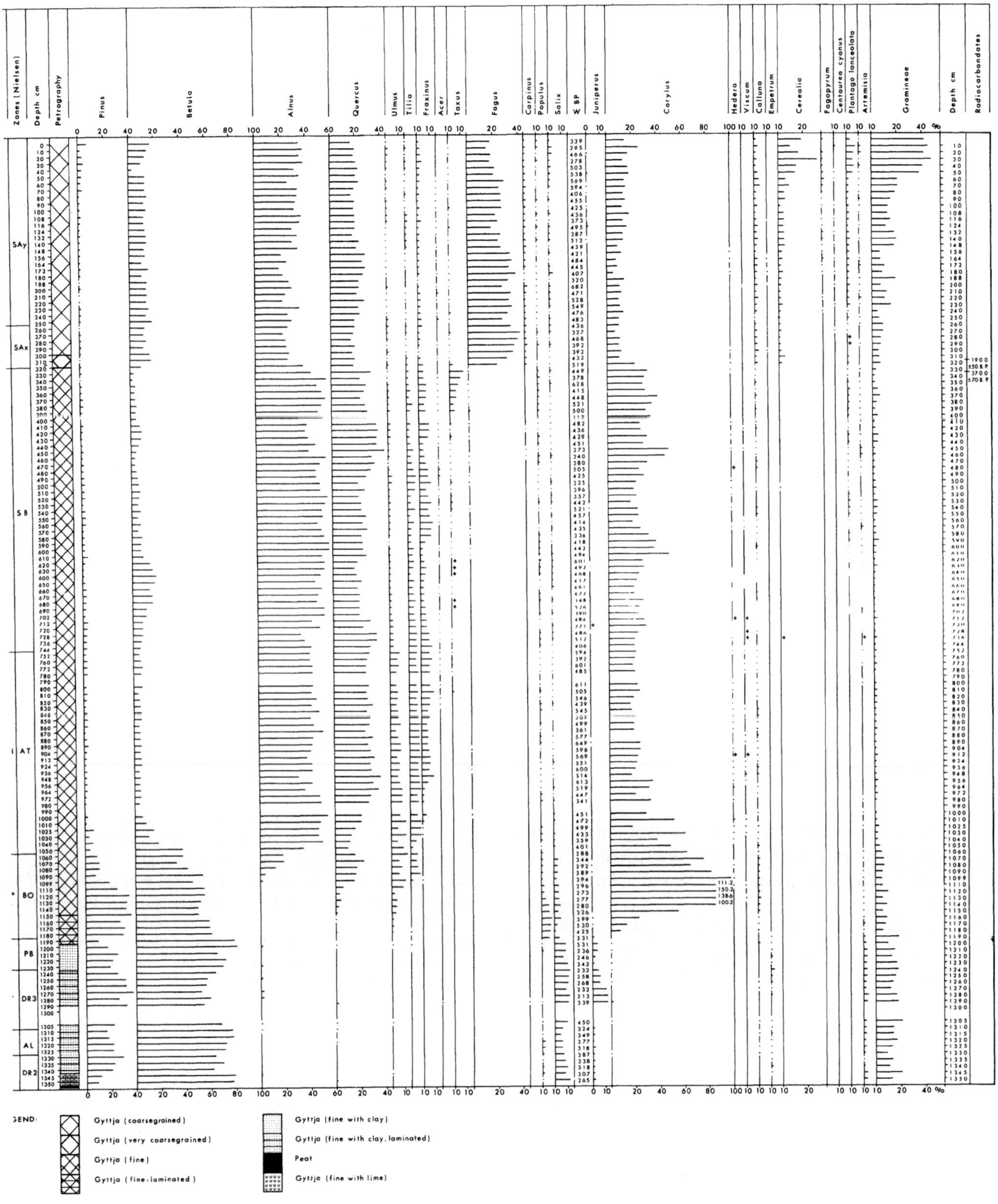

Fig. 5. Modified pollen diagram of Poolsee with hiatus and ^{14}C dates at 320 cm. Percentage of arboreal pollen excluding *Corylus*. Note scale expansion below 1 230 cm.

Sediment accumulation followed a constant rate only in rare cases, e.g. in Schöhsee and Lake Oldenburg. More often a large increase in accumulation rate occurs during the Subboreal or Subatlantic periods, especially in the small Poggensee and in the big Grosser Plöner See. In the Grosser Plöner See more than half of the sediments were deposited later than 1 260 AD. Because of the physiographic and environmental situations of the respective lakes this process will be strongly influenced by the morphometry of their basins. Sediments from different parts of the same lake will demonstrate different accumulation rates. The average annual accumulation rate for the various zones for each lake is shown in Fig. 3. Generally, accumulation rate increases with time but the Poggensee is an exception to this rule. Accumulation rates probably varied within zones, as demonstrated by dates from the Grosser Plöner See (Ohle 1972, Averdieck 1978, Erlenkeuser & Willkomm 1979). Increasing rates since the Subboreal might be caused by human impact, as forest clearance and agriculture are first documented during this zone.

Palynological investigations of Grossensee and Poolsee

The sediments of two lakes have been investigated using palynological methods. Microfaunal analyses have also been carried out and the results are presented in this symposium by Günther (1983) (Grossensee) and (Hofmann (1983) (Poolsee).

1. Grossensee is a medium sized oligotrophic lake with an area of 0.75 km^2 and a depth of 15.75 m. Sedimentation started before the Alleröd (Fig. 4), but about three quarters of the sediment has been deposited since the Subboreal where there is a change from chalky to lime-free mud. Several intensive settlement periods are indicated in the pollen diagram (Fig. 4). Exceptionally high percentages of *Humulus*/*Cannabis* pollen might be derived from the cultivation of one of these species. A series of ^{14}C dates is to be prepared.

2. Poolsee is a rather small lake with an area of almost 1 ha and a water depth of 2.8 m. Sedimentation in this lake also started before the Alleröd (Fig. 5). The most remarkable feature is the extremely high percentage of *Taxus* pollen (up to 11%, upper zone IX), infrequently found in the Postglacial.

The sediment of this lake is considerably disturbed at a depth of 3.20 m where a layer of coarse plant remains, trunks, twigs, and roots occurs. In the pollen diagram there is a hiatus between the last stage of the Subboreal and the late Subatlantic period and this can be seen by the sudden increase in the percentages of *Fagus, Carpinus, Taxus* and cereals, especially *Secale.*

^{14}C data (Willkomm, pers. comm.) indicate that the hiatus corresponds to a duration of at least 1 500 years. These uncorrected ^{14}C data might be too old, because the pollen analysis indicates that the upper horizon should be younger, presumably of medieval origin.

References

Averdieck, F.-R., 1978. Palynologischer Beitrag zur Entwicklungsgeschichte des Grossen Plöner Sees und der Vegetation seiner Umgebung. Arch. Hydrobiol. 85: 1–46.

Averdieck, F.-R., Erlenkeuser, H. & Willkomm, H., 1972. Altersbestimmungen an Sedimenten des Grossen Plöner Sees. Schr. Naturw. Ver. Schlesw.-Holst. 42: 47–57.

Erlenkeuser, H. & Willkomm, H., 1979. ^{13}C- und ^{14}C-Untersuchungen an Sedimenten des Grossen Plöner Sees. Arch. Hydrobiol. 85: 1–29.

Günther, J., 1983. Development of Lake Grossensee (Holstein, Germany): Variations in trophic status from the analysis of subfossil microfauna. Dev. Hydrobiol., this volume.

Hofmann, W., 1983. Stratigraphy of Cladocera and Chironomidae in a core from a shallow northern German lake. Dev. Hydrobiol., this volume.

Nilsson, T., 1964. Standardpollendiagramme und C^{14}-Datierungen aus dem Agerods Mosse im mittleren Schonen. Lunds Univers. Årsskr. N. F. Avd. 2, 59. nr. 7: 1–52.

Ohle, W., 1972. Die Sedimente des Grossen Plöner Sees als Dokumente der Zivilisation. Jb. Heimatkunde Kreis Plön. 2: 7–27.

Overbeck, F., 1975. Botanisch-geologische Moorkunde unter besonderer Berücksichtigung der Moore Nordwestdeutschlands als Quellen zur Vegetations-, Klima- und Siedlungsgeschichte. Wachholtz, Neumünster.

Saad, M. A. H., 1970. Entwicklungsgeschichte des Schöhsees aufgrund mikroskopischer und chemischer Untersuchungen. Arch. Hydrobiol. 67: 32–77.

Willkomm, H. & Erlenkeuser, H., 1972. C^{14} Measurements on Water, Plants and Sediments of Lakes. Proc. 8. Int. Conf. Radio Carbon Dating. Wellington, N.Z: 312–323.

Development of Grossensee (Holstein, Germany): variations in trophic status from the analysis of subfossil microfauna

Jutta Günther
Ordinariat für Bodenkunde, Universität Hamburg, Von-Melle-Park 10, D-2000 Hamburg 13, Federal Republic of Germany

Keywords: paleolimnology, lake development, trophic status, Chironomidae, Cladocera, Ostracoda

Abstract

Microfossil analysis was carried out on 90 samples of a sediment core from Grossensee. The diversity of subfossil Cladocera, Chironomidae, Chaoborus, and Ostracoda reflects at least five different stages of development. Three extended cycles correspond to different climatic periods in the first 10 000 years of the lake's history, while during the last 2 000 years, two shorter cycles can be identified which correlate with proofs or even historical events for human activity in the catchment area. During this time the profundal fauna changed from stenoxibiontic to euryoxibiontic species. In the planktonic fauna a succession of three *Eubosmina* types, *B. longispina, B. coregoni kessleri,* and *B. coregoni coregoni,* was observed. The chydorid and ostracod assembly of the littoral showed no clear change.

Introduction

Grossensee (Lake Grossen) is situated east of Hamburg (Fig. 1). It has an area of 75 ha, lies 38 m above sea level and has an average water depth of 12 m. The lake is wind-exposed and stretches over ca. 2.3 km from SSW to NNE. A small island marks the border line between the shallow southern and the deep northern part. According to pollen analysis (Averdieck 1983), sedimentation began after the last glaciation, about 12 000 BP.

Materials and methods

The investigated sediment core was taken from the deepest part of the lake, at 15.65 m. A modified Livingstone corer was used. The core was 12.93 m long and consisted of different types of gyttja. The lowest layers were rich in carbonates, clay and silt. From a depth of 9.50 m (ca. 5 000 BP) to the surface it consisted of lime-free gyttja. Coarse sands and calcareous material form the base of the core.

Microfossil analysis is based on the identification of the head capsules of chironomids, mandibles of *Chaoborus flavicans,* head shields or carapaces of cladocera, and valves of ostracods.

Chironomid analysis was performed on 52 samples of 1 ml volume and 38 samples of ca. 10 g sediment. For cladocera analysis 35 samples were selected, 1–7 for each climate period. Each sample was equivalent to 0.03 ml of sediment. Ostracods were collected from 38 larger samples.

The samples were prepared according to standard methods (Frey 1959; Hofman 1971; Löffler 1978). For identification papers of Diebel & Pietrzeniuk (1975), Flössner (1972), Hofmann (1971) and Klie (1938) were also used.

Results

I found 1778 head capsules of 44 different chironomid species. Each sample of 1 ml contained an average of 8 head capsules (2–25). From Alleröd (AL) until early Boreal (BO) period (12 000–9 000

Hydrobiologia 103, 231–234 (1983).

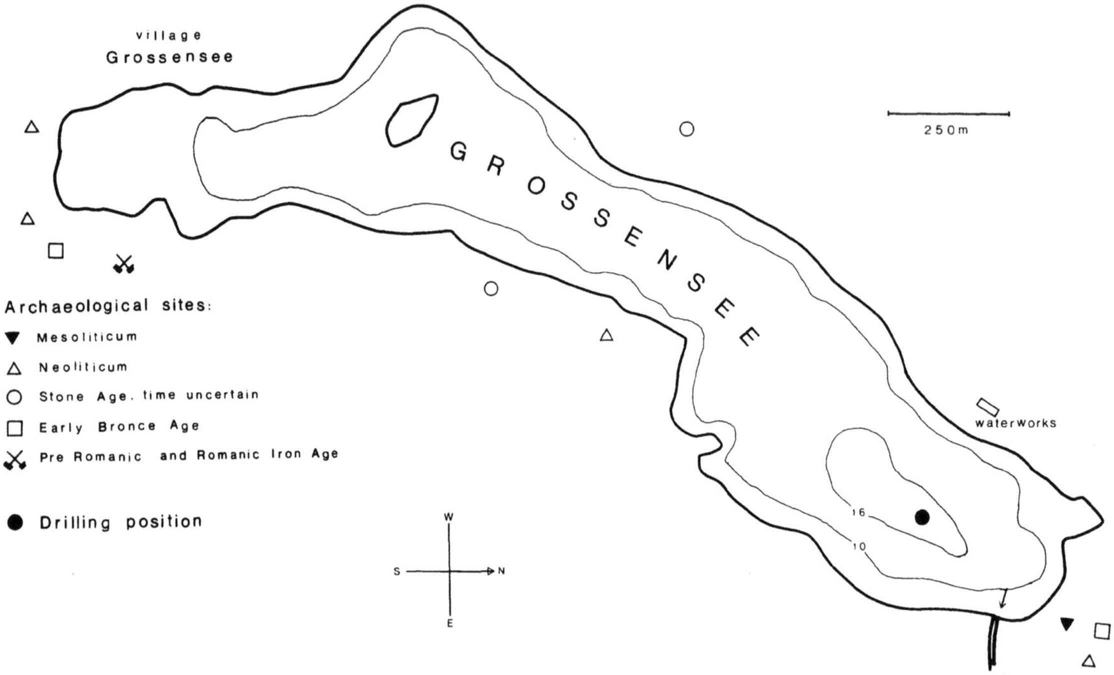

Fig. 1. Grossensee, core position and archaeological sites after Hingst (1959).

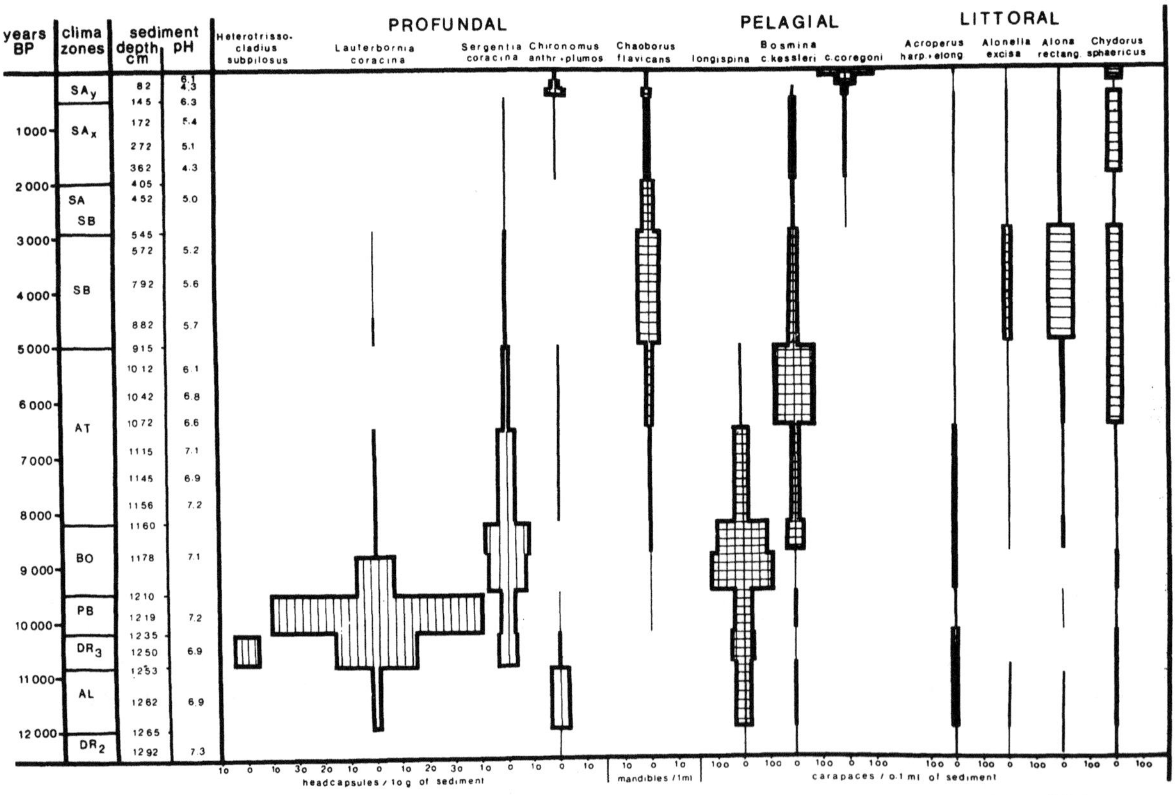

Fig. 2. Microfossils concentration: *Ch. anthr.* = *Ch. anthracinus, Ch. plumos.* = *Ch. plumosus, A. elong.* = *A. elongatus, A. harp.* = *A. harpae, B. c.* = *B. coregoni.*

BP) a higher number of specimens were found (72–158 head capsules). During AL shallow water dwellers: *Microtendipes, Dicrotendipes, Chironomus,* and Orthocladiae were most common. During Younger Dryas (DR_3) and Pre-Boreal (PB) these groups decreased in number, and high abundances of an oxibiont fauna, e.g. *Heterotrissocladius subpilosus* and *Lauterbornia coracina* were observed. *Sergentia coracina* is predominant from Boreal (BO) until early Atlantic (AT) (fig. 2). After its decrease in late Atlantic shallow water dwellers such as *Microtendipes, Dicrotendipes* and *Cladotanytarsus* reappear. High abundances of Tanypus species *(Alabesmyia* and *Procladius)* occurred through the whole Subboreal period (SB). A number of species of low frequency occurs in Subatlantic 1 (SA_x) and from the beginning of Subatlantic 2 (SA_y) *Chironomus anthracinus* increases sharply and becomes predominant (Fig. 2).

Of the 167 *Chaoborus flavicans* mandibles encountered the earliest occurrence was in Pre-Boreal (PB), with a peak in Subboreal and afterwards a slight decrease until recent times (Fig. 2).

5413 *Bosmina* shells were found, 75% belonging to *B. longirostris,* and 25% to *Eubosmina* species. There is a species shift from older to younger sediments as *B. longispina,* prevalent until Boreal (BO) becomes replaced by *B. c. kessleri* during Atlantic (AT). Since Subatlantic (SA_x), it is followed by *B. c. coregoni,* now the only *Eubosmina* species in the lake (Fig. 2).

I identified 1794 chydorid specimens of 23 species. They are most frequent in Alleröd, Subboreal and Subatlantic periods, with a smaller maximum occurring in recent times. The planktonic/littoral ratio (Fig. 3) shows negative correlations between Chydorid and Bosmina populations. Chydorus sphaericus (26%) is the most abundant chydorid species. Its maximum follows directly that of *Bosmina* sp. The maxima of *Alonella excisa* and *Alona rectangula* occur during the Subboreal (Fig. 2).

Seventy-five percent of the 779 ostracod valves encountered belong to the genera *Cypria* and *Cyclocypris.* The remainders belong to *Candona* and *Limnocythere.* Their valves are preserved only in the calcareous sediments. During Late-Glacial *Cyclocypris, Cypria* and *Candona* species can be found. *Limnocythere* species, especially *L. sancti patricii,* occur first in Pre-Boreal. Today, at least 18 species, including *L. sancti patricii,* occur in the lake (Keyser, pers. commun.).

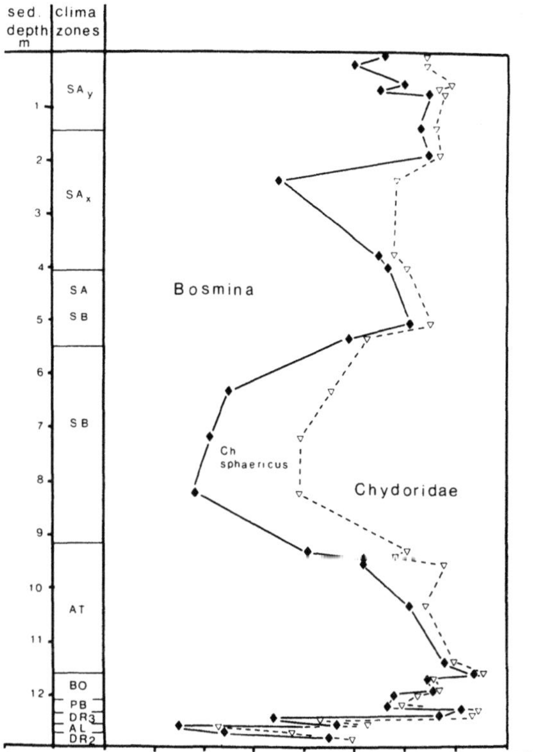

Fig. 3. Percentage of *Bosmina* and Chydorids (planktonic/littoral ratio) in the core.

Discussion

Changes in microfossil fauna for Grossensee can be compared with changes in Schöhsee, Grosser Plöner See, and Grosser Segeberger See (Hofmann 1971, 1978). Except for Grosser Segeberger See, a succession of the profundal fauna from species characteristic of oligotrophic lakes, to those found in mesotrophic waters and finally to inhabitants typical of eutrophic lakes is seen. Eubosmina populations switch from *B. longispina* through *B. c. kessleri* to *B. c. coregoni.*

A peculiarity of Lake Grossensee is a stepwise change in the subfossil assembly in periods of different length. High abundances of single species decrease slowly towards an equilibrium characterized by species diversity and high individual numbers. Except for Alleröd near the end of each period there is an increase in *Bosmina* abundances. Thereafter the population decreases rapidly and *Ch. sphaericus* spreads (Fig. 3), a precursor of a new biocoenosis and not necessarily indicative of more eutrophic status. Five cycles can be observed.

1. Initial status of water. Between Older Dryas and Alleröd an eutrophic shallow lake comes into existence.

2. Together with the decrease of temperature in Younger Dryas species diversity diminishes. A population of species characteristic for lakes (e.g. *L. sancti patricii*) has developed up to the end of Atlantic (Fig. 2). A high planktonic/littoral ratio is seen (Fig. 3). In the beginning the profundal fauna consists of highly oxibiontic species but it is followed successively by species with less oxygen dependence. In the end of Atlantic species diversity and high biomass of the whole microfauna reflect a climax state of a lake with high biological activity. The breakdown of this community at about 6 000 BP (Fig. 2) represents the end of the oligotrophic phase. The gyttja becomes lime-free at this point.

3. The Subboreal sediment accumulation rate is twice that of Atlantic, perhaps due to water level oscillations or catchment disturbance by human activity (Fig. 1). Only a few species present in Atlantic assembly remain. Of the species which emerge in Subboreal most thrive in more dystrophic conditions (e.g. *Chaoborus, Alonella excisa, Alona rectangula* (Fig. 2). The changes in chironomid population as well as the sharp decline of *Bosmina* in relation to the Chydorid assembly support the assumption that large lowland areas are influencing the metabolism of the lake (Ohle 1972). Ranking in a trophic scheme is difficult but increasing eutrophication up to early Subatlantic (SB/SA, about 3 000 BP) is clearly taking place.

4. Archaeological findings of early Iron Age (about 2 000 BP) indicate human activity. The planktonic/littoral ratio diminishes as does the species diversity of chironomids. The new cycle resembles the previous one, but due to human disturbance it comes to an end before climax. These observations correlate with a land clearing and settling period about 1 240 AD.

5. In the early part of Subatlantic 2 (SA_y) the number of chironomid species decreases to half of the previous value. In the newly emerging assembly *Chironomus anthracinus* becomes established and dominates in the profundal zone. *B. c. coregoni* is the only Eubosmina. The lake changes to a *Chironomus anthracinus* type (Thienemann 1954). Phases of stability and disturbance occur more often than in times before human impact. The species assembly, however, is stable and has not changed during SA_y.

Acknowledgements

I am indebted to Deutsche Forschungsgemeinschaft for financial support. I wish to thank Professor Dr. W. Ohle (Plön), who drew my attention to the problem and Professor Dr. W. Scharpenseel (Hamburg), who gave me the sediment samples and enabled me to carry out the study. I want to thank Dr. W. Hofmann for his scientific help, and also Vöge and Vöge who allowed me to use their data of water depth for Fig. 1. I should also like to thank my husband for his assistance.

References

Averdieck, F.-R., 1983. Palynological investigations of the sediments of ten lakes in Eastern Holstein, North Germany. Hydrobiologia, this volume.

Diebel, K. & Pietrzeniuk, E., 1975. Mittel- und jungpleistozäne Ostracodenfauna des Raumes Potsdam-Brandenburg-Rathenau in stratigraphischer und ökologischer Sicht. Z. geol. Wiss. Berlin 3: 1197–1233.

Flössner, D., 1972. Kiemen- und Blattfüsser, Branchipoda, Fischläuse Brachiura. In: Dahl, M. & Peus, F. (eds.). Die Tierwelt Deutschlands 60: 1–499.

Frey, D. G., 1959. The taxonomic and phylogenetic significance of the head pores of the Chydoridae (Cladocera). Int. Revue ges. Hydrobiologia 44: 27–50.

Hingst, H., 1959. Vorgeschichte des Kreises Storman. Wachholtz, Neumünster.

Hofmann, W., 1971. Die postglaziale Entwicklung der Chironomiden und Chaoborus Fauna (Dipt.) des Schöhsees. Arch. Hydrobiol. Suppl. 40: 1–74.

Hofmann, W., 1971. Zur Taxonomie und Palökologie subfossiler Chironomiden (Dipt.) in Seesedimenten. Arch. Hydrobiol. Beih. Ergebn. Limnol. 6: 1–50.

Hofmann, W., 1978. Analysis of animal microfossils from Grosser Segeberger See (F.R.G.). Arch. Hydrobiol. 82: 316–346.

Klie, W., 1938. Ostracoda, Muschelkrebse. In: Die Tierwelt Deutschlands. Dahl, Jena 34: 1–230.

Löffler, H., 1967. Ostracoda. In: Illies, J. (ed.). Limnofauna Europaea. Enke, Stuttgart: 162–172.

Ohle, W., 1972. Die Sedimente des Grossen Plöner Sees als Dokumente der Zivilisation. Jb. Heimatkunde Kreis Plön. 2: 7–27.

Thienemann, A., 1954. Chironomus. Die Binnengewässer 20: 1–834.

Stratigraphy of Cladocera and Chironomidae in a core from a shallow North German lake

Wolfgang Hofmann
Max-Planck-Institut für Limnologie, Abt. Allgemeine Limnologie, Postfach 165, D-2320 Plön, F.R.G.

Keywords: paleolimnology, cladoceran analysis, chironomid analysis, successions, lake development

Abstract

In the subfossil assemblages of chironomids and cladocerans significant changes were found during the late-glacial and postglacial development of a small, shallow North German lake. The successions observed were obviously not related to eutrophication but to factors such as climatic conditions and lake siltation. The absence of species of the subgenus *Eubosmina* during postglacial times as well as the predominance of *Alonella* species among the chydorids was due to the small dimensions of the water body.

Introduction

There is a general view concerning postglacial changes in Chironomidae (Diptera) and Bosminidae (Cladocera) faunas of deep, stratified lakes in North Germany. For chironomids a *Tanytarsus-lugens* community was replaced by *Chironomus* and in the case of the subgenus *Eubosmina* a succession from *Bosmina longispina* to *B. coregoni* f. *kessleri* and then to *B. coregoni* f. *coregoni* occurred (Hofmann 1971a, 1977, 1978a, 1978b).

The question that arises whether this sequence is also applicable to the development of shallow, unstratified lakes. Hence, a comparative study was carried out on a core from Lake Poolsee, a small (diameter ca. 100 m) and shallow (2 m) lake about 10 km west of Kiel (North Germany).

A chemical analysis of the lake water in August 1975 showed the following values: temperature 21.4 °C, pH 7.5–8.0, conductivity 405 $\mu S\ cm^{-1}$, total dissolved phosphorus 10 $\mu g\ P\ l^{-1}$, PO_4–P was not detectable.

Material and methods

Coring, using a Livingstone piston sampler, and pollen analysis was carried out by Averdieck (1983). The core had a total length of about 14 m. The sediments below 12 m were from late-glacial times. In the postglacial section there was a disturbance at a sediment depth of 3.20 m indicated by the absence of sediments from Subboreal 2.

Results

The content of organic matter, measured by loss on ignition, showed that the late-glacial layers were minerogenic (range: 2–6% ash weight) and the postglacial sediments were organic (range: 59–94% ash weight). The sediments in the transitional periods (PB, B1, B2) had intermediate values (range: 14–46% ash weight) (cf. Ungemach 1960).

The intention of the study was to identify major faunal changes on the basis of the succession of the predominant species (percentage > 10%).

Subfossil remains of chironomids and cladocerans were analysed in 36 samples; sampling distances were closer in the late-glacial and early postglacial sediments than during the later postglacial period (Figs. 1 & 2).

For chironomids the residue collected from siev-

Hydrobiologia 103, 235–239 (1983).

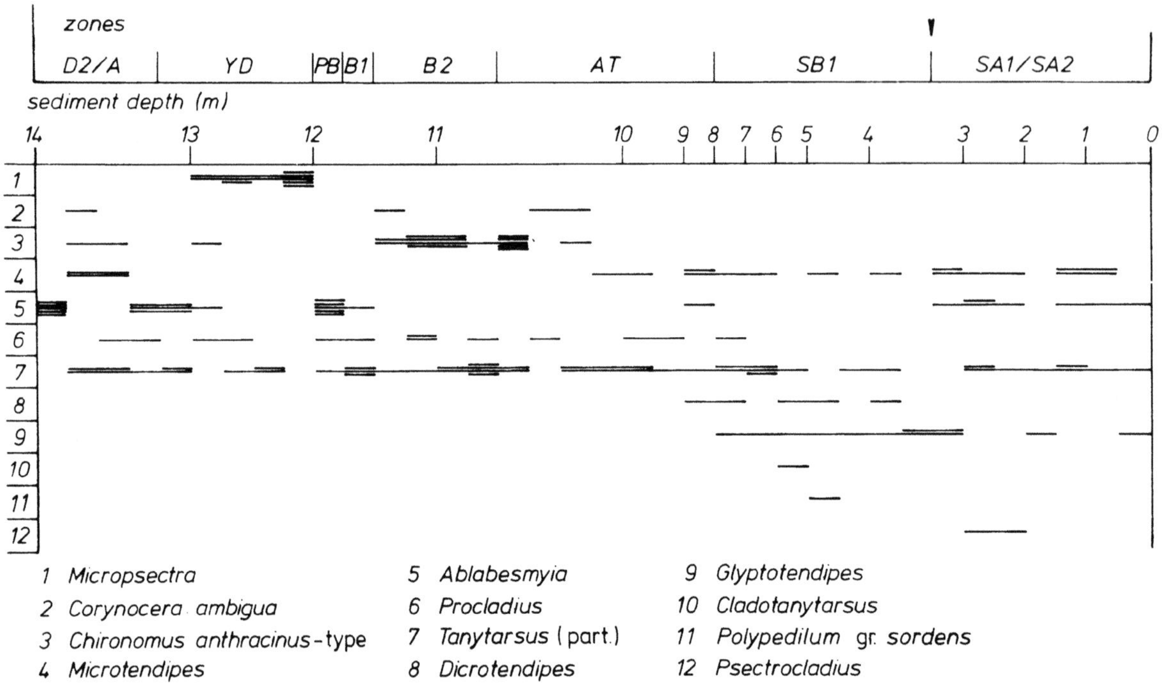

Fig. 1. Distribution of the predominant chironomids (>10% of the total); each line denotes a relative abundance ⩾10%; the arrow indicates a disturbance.

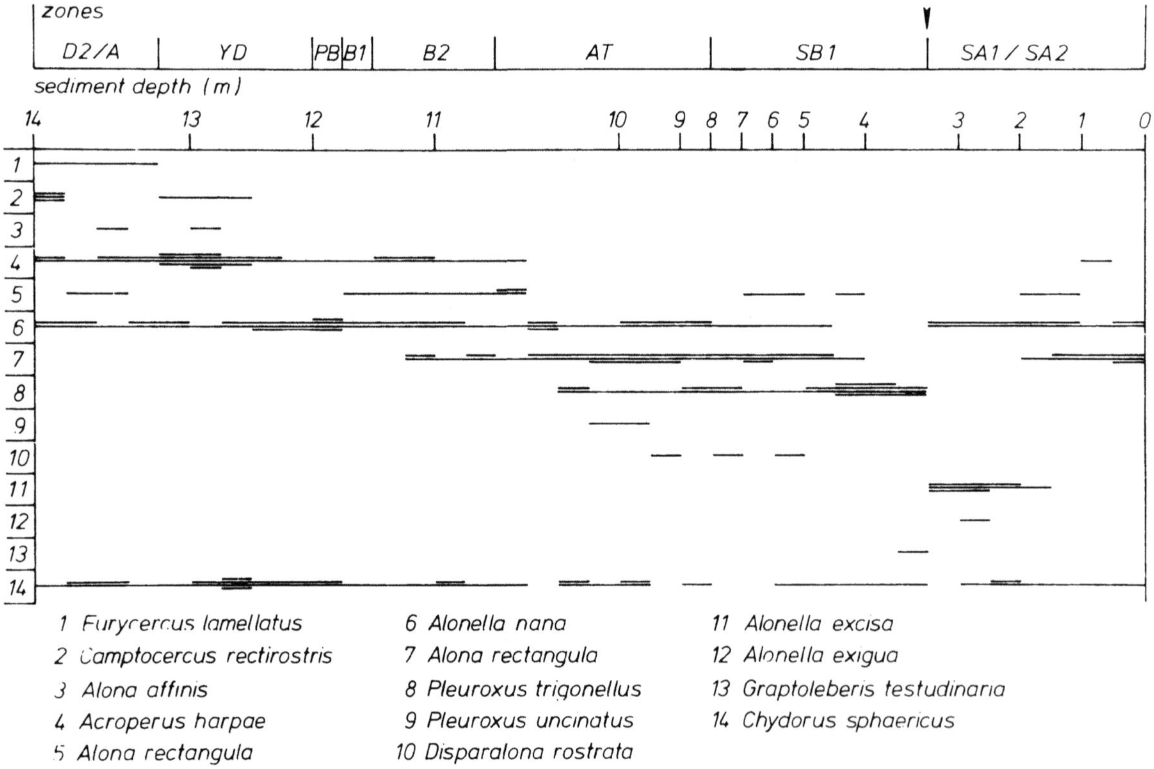

Fig. 2. Distribution of the predominant chydorids (>10% of the total); each line denotes a relative abundance ⩾10%; the arrow indicates a disturbance.

ing 10 g fresh sediment was examined (Hofmann 1978a). The number of head capsules varied from 14 to 500 specimens per 10 g of sediment.

Figure 1 shows the distribution of the chironomid taxa (Hofmann 1971b) accounting for more than 10% of the assemblage in any sample. Each line denotes a percentage of ≥10%. In the lowermost layers *Ablabesmyia* and *Corynocera ambigua* were dominant but were replaced by *Micropsectra* during Younger Dryas. *Ablabesmyia* was again the most important form during the Pre-Boreal and Boreal 1 periods. In Boreal 2 the *Chironomus anthracinus* type was most abundant.

Since the Atlantic period the chironomid assemblage was characterized by the predominance of sublittoral-littoral taxa: *Microtendipes, Ablabesmyia, Procladius, Dicrotendipes, Cladotanytarsus, Tanytarsus, Polypedilum* gr. *sordens,* and *Psectrocladius.*

The following taxa were confined to the late-glacial period: *Micropsectra, Protanypus, Monodiamesa, Paracladius, Sergentia coracina, Heterotrissocladius.*

For cladoceran analysis subsamples of the sieved fraction >100 μm equivalent to 0.05 to 1 g fresh sediment were examined (Hofmann 1978a).

The genus *Bosmina* was represented by two species only: *B. longirostris* occurred throughout the core, whereas *B. longispina* was found in the late-glacial sediment only. It was always less abundant than *B. longirostris.*

As Fig. 1, Fig. 2 which is based on the percentage of the most common chydorids, illustrates the changes in the assemblage during lake development.

Eurycercus lamellatus, Camptocercus rectirostris, Alona affinis had percentages (>10%) during the late-glacial period only. This period was also characterized by high abundances of *Acroperus harpae* and *Alonella nana. Acroperus harpae* decreased in the Atlantic period but *Alonella nana* occurred throughout and was the most abundant chydorid species in almost all of the samples.

The species which decreased in late-glacial or early postglacial times were replaced by species such as *Alona rectangula, Pleuroxus uncinatus, Disparalona rostrata, Alonella excisa, Alonella exigua,* and *Graptoleberis testudinaria,* which predominated for longer or shorter intervals during the late postglacial period. In the uppermost samples *Alona rectangula* and *Alonella nana* were most abundant.

With the exception of one sample from the Younger Dryas *Chydorus sphaericus* was not very abundant and its distribution was rather uniform.

The disturbance of the sediment at 3.20 m (see arrow in Fig. 2) is clearly visible in the chydorid diagram, indicated by an abrupt change in the assemblage.

Discussion

The chironomid taxa which are restricted to late-glacial times (*Micropsectra, Heterotrissocladius, Protanypus, Sergentia coracina, Paracladius* (probably *alpicola*)) and the predominating chydorid species of the same period seem to be mainly influenced by climate: the chironomids are cold stenothermal species (Brundin 1949) and the chydorids are known as pioneer species occurring under late-glacial conditions (Goulden 1964; Whiteside 1970). These authors did not mention *Eurycercus lamellatus* in this connection but this species has been frequently found in late-glacial sediments from Wallensen (Frey 1958) and Grosser Segeberger See (Hofmann 1978a).

Most of the chironomids typical of this early phase of lake development are members of the *Tanytarsus lugens* community (Brundin 1956) and are used as indicators of oligotrophic conditions (Saether 1975). However, in late-glacial times they might have occurred in the littoral zone and would thus not be indicative of the oxygen conditions in the hypolimnion (Brundin 1956; Hofmann 1979). They must not be used as indicators of trophic state in such cases.

The abrupt and highly significant changes in the chironomid assemblage during late-glacial and early postglacial times (*Ablabesmyia* – *Micropsectra*) cannot be discussed on the basis of this restricted material.

During postglacial times taxa of the *Tanytarsus lugens* community were not found. However, their disappearance might have been caused by rising temperatures rather than by eutrophication.

For an evaluation of the importance of temperature, information on former water levels is required. The *Bosmina*/chydorid ratio gives little information on whether the lake was ever deep and

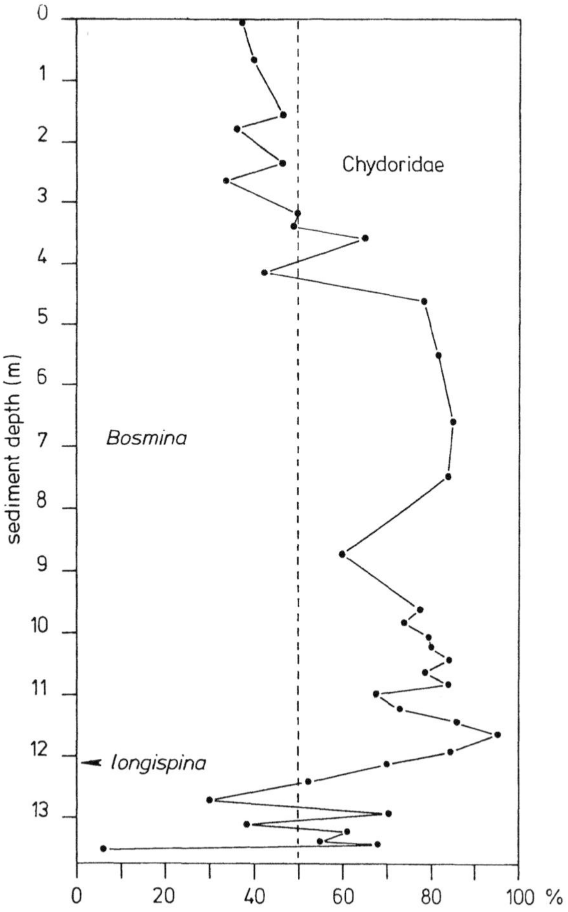

Fig. 3. Percentages of *Bosmina* and chydorids; the arrow indicates the upper limit of the occurrence of *Bosmina longispina.*

stratified but there is some significant variation in this ratio (Fig. 3). During postglacial times only *Bosmina longirostris* was involved. This species is very abundant in shallow waters and in the littoral zone (Flössner 1972) and is not a true indicator of the dimensions on the pelagic zone.

The predominance of the *Chironomus anthracinus* type during Boreal 2 is likely to indicate a deep, stratified *Chironomus*-lake with oxygen depletion in the hypolimnion. If the water level was as high as the present, the lake during the early postglacial periods would have been extremely deep relative to its surface area and a meromictic situation might have occurred. This is also suggested by the existence of varved sediments (Averdieck 1983).

Since the Atlantic period littoral chironomids have predominated. The interpretation of their distribution is difficult because the species involved are not known. *Tanytarsus* was abundant throughout the core, but its occurrence is not related to the trophic state of the lake (Hofmann 1979).

The low abundance of *Bosmina longispina* during late-glacial time and its disappearance at the end of this period as well as the complete absence of species of the subgenus *Eubosmina* during postglacial times is clearly related to the small dimensions of the water body.

The fossil assemblages of this shallow lake when compared with deeper lakes from the same region (Hofmann 1971a, 1977, 1978a, 1978b) are characterized by high percentages of littoral chironomids, the absence of species of the subgenus *Eubosmina* during postglacial times, and by the predominance of chydorid species typical of smaller water bodies such as *Alonella nana* and *Alonella excisa* (Whiteside 1970).

However, there were hardly any significant differences in the *Bosmina*/chydorid ratio between Poolsee and Grosser Segeberger See (178 ha, depth: 11.7 m). Only during the Subatlantic period was the ratio clearly lower in Poolsee (Hofmann 1978a).

The occurrence and high abundance of the *Alonella* species during late postglacial times indicate that pollution did not occur in Poolsee (Whiteside 1970) and this is in accordance with the present nutrient content of the water.

During lake development significant changes in chironomid and cladoceran assemblages were observed. These changes, however, do not reflect changing trophic conditions, but are related to factors such as climatic conditions and lake siltation. Recently, Whiteside (1983) has suggested that successions occurring during lake development are not only affected by nutrient conditions but by various other ecological factors, some of which were operative during the history of Poolsee.

References

Averdieck, F.-R., 1983. Palynological investigations on the sediments of ten lakes in Eastern Holstein, North Germany. Hydrobiologia, this volume.

Brundin, L., 1956. Die bodenfaunistischen Seetypen und ihre Anwendung auf die Südhalbkugel. Rep. Inst. Freshw. Res. Drottningholm 37: 186–235.

Flössner, D., 1972. Kiemen- und Blattfüsser, Branchiopoda, Fischläuse, Branchiura. In: Dahl, M. & Peus, F. (Eds.) Die Tierwelt Deutschlands 60: 1–499.

Frey, D. G., 1958. The late-glacial cladoceran fauna of a small lake. Arch. Hydrobiol. 54: 209–275.

Goulden, C. E., 1964. The history of the cladoceran fauna of Esthwaite Water (England) and its limnological significance. Arch. Hydrobiol. 60: 1–52.

Hofmann, W., 1971a. Die postglaziale Entwicklung der Chironomiden- und Chaoborus-Fauna (Dipt.) des Schöhsees. Arch. Hydrobiol./Suppl. 40: 1–74.

Hofmann, W., 1971b. Zur Taxonomie und Palökologie subfossiler Chironomiden (Dipt.) in Seesedimenten. Arch. Hydrobiol. Beih. Ergebn. Limnol. 6: 1–50.

Hofmann, W., 1977. Bosmina (Eubosmina) populations of the Grosser Segeberger See during late-glacial and postglacial times. Arch. Hydrobiol. 80: 349–359.

Hofmann, W., 1978a. Analysis of animal microfossils from the Grosser Segeberger See (F.R.G.). Arch. Hydrobiol. 82: 316–346.

Hofmann, W., 1978b. Bosmina (Eubosmina) populations of Grosser Plöner See and Schöhsee Lakes during late-glacial and postglacial times. Pol. Arch. Hydrobiol. 25: 167–176.

Hofmann, W., 1979. Chironomid analysis. In: Berglund, B. E. (Ed.) Palaeohydrological changes in the temperate zone in the last 15 000 years. Subproject B. Lake and mire environments. vol. 2. pp. 259–270. Lund.

Saether, O. A., 1975. Nearctic chironomids as indicators of lake typology. Verh. Int. Ver. Limnol. 19: 3127–3133.

Whiteside, M. C., 1970. Danish chydorid Cladocera: modern ecology and core studies. Ecol. Monogr. 40: 79–118.

Whiteside, M. C., 1983. The mythical concept of eutrophication. Hydrobiologia, this volume.

Ungemach, H., 1960. Sedimentchemismus und seine Beziehungen zum Stoffhaushalt in 40 europäischen Seen. Diss. Kiel.

Diatom succession in the post-glacial sediments of the Komořany Lake, North-West Bohemia, Czechoslovakia

Zdeňka Řeháková
Geological Survey of Czechoslovakia, 118 21 Praha 1 - Malá Strana, Malostranské nám. 19, Czechoslovakia

Keywords: paleolimnology, diatom succession, lake development

Abstract

The Late-glacial and Postglacial sediments of the former Komořany Lake in North-West Bohemia were studied by means of diatom analysis to trace the pattern of past environmental change in the lake. Several phases of lake development could be distinguished.

Originally the Komořany Lake was a eutrophic shallow water basin. The diatom succession in the early Postglacial sediment indicates alternating alkaliphilic, epiphytic and planktonic diatom communities associated with changes in water level in the basin. At the beginning of the Subatlantic period the water level began to rise and the development of planktonic diatoms, typical of small eutrophic lakes, occurred.

Later, the basin became shallow again and the number of indifferent and acidophilous diatoms increased. This is typical for the final phase of the infilling process of a lake before its transformation to peatland.

Introduction

The present paper summarizes the preliminary results of biostratigraphic and palaeoecological investigations of sediments in the former Komořany Lake obtained by the use of diatom analysis.

The Komořany Lake is one of the largest Quaternary lakes in Czechoslovakia. It is situated in north-western Bohemia, at the foot of the Krušné hory Mountains, about 2.5 km from the town of Most (Fig. 1). Its surface was accurately represented by Wolf (1880) in a detailed geological map of the North Bohemian brown-coal basin at a scale of 1:10,000. The surface extent of the lake was, according to this map, about 25 km^2; the elevation 230 m above sea level.

The Komořany Lake probably originated as early as the Pleistocene, as a Late-glacial barrier lake, a result of stone debris landslides blocking the valley of the River Bílina. The landslides occurred between the hills of Hněvìn (408 m) and Špičák (399 m) south of the town of Most (Hibsch 1929). Thus on the flat surface of the Neogene North Bohemian brown-coal basin, a large shallow water lake was created whose development can be traced throughout the whole Postglacial period.

In its early history the Komořany Lake had a larger extent than at present. Eventually the lake was overgrown and the area of its water surface reduced. As late as 1789, the surface area was 5.75 km^2, in 1831 only 2 km^2, but during the floods, it increased again to some 15 km^2. The filling in of the lake was accelerated by artificial draining in 1834. It occurred as a result of agricultural and continued brown-coal mining activity in this area.

At the present time, the sediments of the lake have been uncovered by open cast mining of two coal pits: one in the western part of the former lake, and the other in its eastern part. At these sites, a number of samples was taken which reflects the sequence of lake sediments in the littoral parts as well as in the centre of the lake.

Lithologically, the lake sediments are represented by basal gravel or coarse sand, followed by fine-

Hydrobiologia 103, 241–245 (1983).

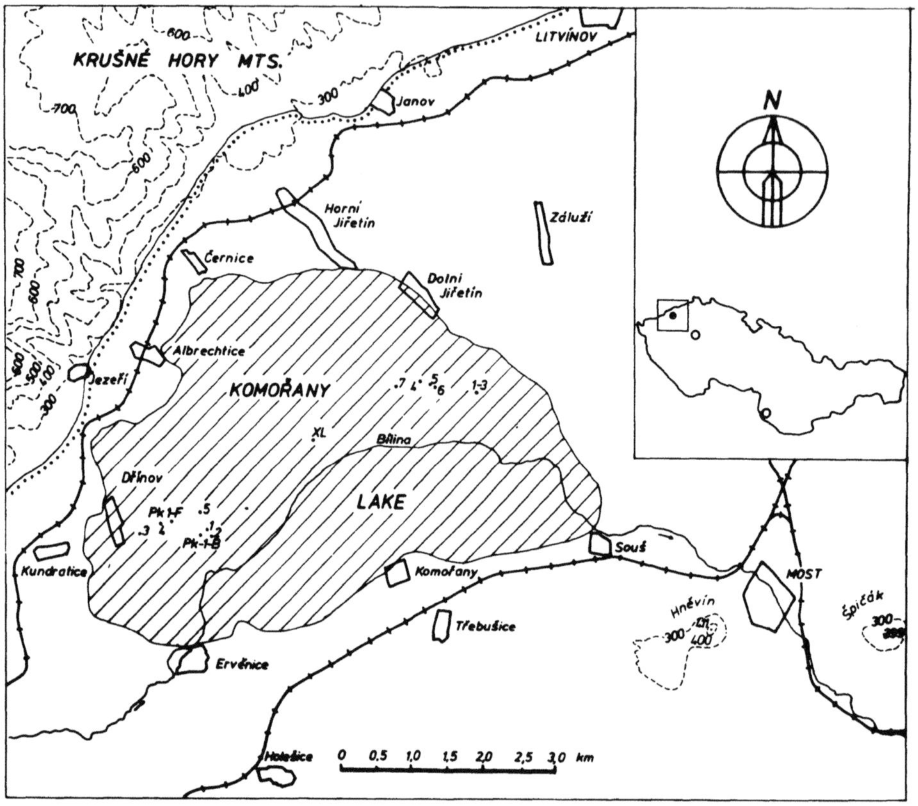

Fig. 1. Komořany Lake showing geographical position and location of sediment sections.

grained sand, sandy clay, gyttja and peat. The stratigraphic sequence has been influenced by local sedimentation conditions in various parts of the lake shown by the varying thickness of the sediments. For instance, in the western parts, at the foot of the Krušné hory Mountains, the sandy sediments are thicker than in the eastern part of the lake, and thick layers of fine sandy clay with a high mica content are deposited mainly in the centre of the lake. The gyttja facies represents undisturbed sedimentation, favourable to the accumulation of microorganisms, particularly diatoms and chlorococcal algae. The thickness of gyttja varies from 0.20 m to 1.50 m and thins towards the shore, so that the *Phragmites* or woody peat rests immediately on the mineral sediments. The thickness of the whole complex is not more than 6 m.

According to palynological analysis carried out by Rudolph (1926) and Losert (1940), the sediments in the Komořany Lake area range from Late-glacial to modern times.

Pollen analysis has been used to data the deposits. From the pollen sequence, the Postglacial sediments were divided into 6 zones which approximately correspond with the phases of the diatom succession. The diatom flora reflects the changing aquatic environment more sensitively, and the changes are consequently not parallel to the changes in the pollen spectra.

Material and methods

Samples for diatom analysis were taken from several profiles along a north-east to south-west transect. A profile (PK-1-B) taken by Jankovská was selected for combined micro-paleontological study. The pollen and diatom analyses were carried out at intervals of 1 cm on consecutive samples of equal size.

The diatom analysis was carried out according to the methods of Battarbee (1979). Diatoms were mounted in Pleurax, and identification was by means of a Leitz Orthoplan binocular light microscope using oil immersion planapochromatic objective 100:1 and x10 eye-pieces.

Results

The preliminary results of the diatom study described here are based on material from profile PK-1-B situated in the western part of the lake. So far 247 species, varieties and forms of diatoms have been identified. The flora comprises oligohalobous fresh-water taxa with the majority belonging to the alkaliphilous group. However in the diatom gyttja indifferent species of the temperate zone occur, and in the detritus gyttja and peat acidophilous forms predominate.

In the sandy sediments, probably deposited in the Pleistocene (Hurník 1969), no diatoms were found. It appears that the rapid deposition of mineral material was not favourable to the development of microphytes. High diatom concentrations occur only after the deposition of the sandy clay (Řeháková 1962, 1971).

The beginning of the diatom sequence has an alkaliphilous character. Littoral *Fragilaria* species are the dominant taxa. Only a small percentage belong to pH-indifferent diatoms, and planktonic taxa are very rare. The diversity of diatom communities is still small. Some *Navicula, Caloneis, Cymbella, Cocconeis* and *Ephithemia* species were established.

This diatom flora as well as the high percentages of *Pediastrum, Botryococcus* and other algae found here indicate that in this period the Komořany Lake was a shallow water eutrophic basin. The period extends form the Late-glacial to the Pre-Boreal. A division between the two periods is difficult because of the similarity of their diatom communities.

The diatom flora of the Boreal period is very similar to the previous one, though the diatom communities have a larger species diversity. Also the number of valves is much higher than at the end of the Pre-Boreal, particularly the littoral alkaliphilous *Fragilaria* species such as *Fragilaria construens* (Ehr.) Grun., *Fragilaria brevistriata* Grun., *Fragilaria pinnata* Ehr. Their varieties also have a high frequency. *Cyclotella comta* (Ehr.) Kütz. is also typical of this zone. At some levels, the maximum of *Fragilaria* species alternate with a maximum of *Melosira* species, especially *Melosira ambigua* (Grun.) O. Müll., a characteristic planktonic species of eutrophic basins. Of the other alkaliphilous or indifferent species the following are common: *Gomphonema acuminatum* var. *coronatum* (Ehr.) W. Sm., *Amphora ovalis* var. *libyca* (Ehr.) Cleve, *Caloneis silicula* (Ehr.) Cleve, *Neidium iridis* (Ehr.) Cleve, *Navicula schoenfeldii* Hust., *Navicula laterostrata* Hust., *Navicula radiosa* Kütz., *Cymatopleura solea* (Bréb.) W. Sm., *Cymbella aspera* (Ehr.) Cleve, *Cymbella cistula* (Hempr.) Grun. and *Epithemia zebra* (Ehr.) Kütz. and its varieties. The rising frequency of *Pediastrum* also points to eutrophication of the lake.

In the subsequent Atlantic period, which, according to the pollen spectrum, corresponds to a grey or brown-grey clayey gyttja two different diatom communities were established.

The diatom communities of the lower part of this period are characterized by the predominance of the littoral alkaliphilous *Fragilaria* species and the occurrence of *Melosira italica* (Ehr.) Kütz. a planktonic/littoral species. *Cyclotella stelligera* Cleve et Grun., *Navicula radiosa, Navicula laterostrata, Gomphonema acuminatum* Ehr., *Eunotia formica* Ehr., *Cymbella, Pinnularia* and *Epithemia* species are relatively common.

In the upper part of the Atlantic period, communities with a quite different composition were discovered. Besides *Navicula radiosa* and *Navicula laterostrata*, which culminate in the lower part of this zone, many epiphytic and epipelic species are present. *Tabellaria fenestrata* (Lyngb.) Kütz. and *Tabellaria flocculosa* (Roth.) Kütz. are common. Such a change in the composition of the diatom communities indicates a temporary change in the bathymetric and ecological conditions of the lake.

In the Subboreal, the littoral alkaliphilous *Fragilaria* species reach their second maximum. *Melosira ambigua* is also abundant. They quantitatively alternate throughout this period. In the upper part, the *Fragilaria* species for the first time are infrequent and the previously abundant *Melosira* reaches its first maximum. In some layers nearly pure communities of *Melosira ambigua* were found. *Cyclotella stelligera* is also frequent.

At the boundary between the Subboreal and Subatlantic periods, this assemblage declines suddenly and is replaced by communities in which epiphytic and epipelic forms prevail. The species diversity at this level is very high. The mainly littoral forms indicate a temporary overgrowing of the lake.

The diatom flora of the earlier Subatlantic consists largely of planktonic forms. *Melosira granulata* (Ehr.) Ralfs and its varieties are dominant, and together with *Melosira ambigua* reach a high percentage in the samples. In these communities, some species of pennate diatoms such as *Pinnularia nobilis* Ehr., *Pinnularia gibba* Ehr., *Pinnularia nodosa* Ehr., *Navicula americana* Ehr. were recorded. The plankton increase suggests a rise in water level. The presence of *Meridion circulare* Ag. and var. *constrictum* (Ralfs) V. H. indicates slightly moving water.

The diatom communities of the upper part of the early Subatlantic are not as uniform as in the lower part of this period. A great number of littoral forms are present. The alkaliphilous species again have a high frequency, but an increase of pH-indifferent species is also conspicuous. *Fragilaria construens* var. *exigua* (W. Sm.) Schulz appears in the assemblage for the first time. *Nitzschia scalaris* (Ehr.) W. Sm., *Navicula hungarica* var. *capitata* (Ehr.) Cleve, and many *Gomphonema, Eunotia, Cymbella, Achnanthes* and *Synedra* taxa are common.

At a higher level a mixture of oligohalobous, pH indifferent or alkaliphilous taxa with acidophilous forms appear. The water became more neutral and later, in the later Subatlantic, became slightly acidic.

The uppermost layers of the sediment studied belong to the period before artificial draining. The sediments consist of *Phragmites* peat and contains a predominance of acidophilous diatoms.

Discussion

According to the diatom analysis, several phases of succession in the Komořany Lake basin can be distinguished which characterize the typological development of the lake.

The correlation between pollen and diatom stratigraphy shows that organic sedimentation began near the end of the Late-glacial period. The species diversity of the diatom communities was low. The early postglacial period is characterized by high frequencies of *Fragilaria*. The species diversity slowly increases to reach its maximum at the beginning of the Atlantic period.

The early postglacial sediments were deposited in a shallow water eutrophic environment. The interchange of diatom communities from periphytic to planktonic forms can be explained by changes in productivity of individual species during the vegetation cycle, and later by oscillations of water-level depending on climatic changes in the Atlantic and Subboreal periods.

A significant change in the composition of the diatom communities occurs at the beginning of the Subatlantic period. The maximum of planktonic *Melosira* and *Cyclotella* species indicates the time of highest water-level in the lake. The species diversity is small, and diatoms typical of euthrophic water predominate. A high lake productivity during this period corresponds to the relatively constant conditions of the aquatic environment. This is also demonstrated by the great thickness of the diatom gyttja.

An increasing amount of *Achnanthes, Navicula, Nitzschia* and *Eunotia* species in the diatom communities at a later stage points to a change towards oligotrophic conditions. The appearance of indifferent epiphytic and epipelic forms, however, could have been associated with the shallowing of the lake. The low water level signals the beginning of the infilling process or the result of artificial draining.

A comprehensive change in the flora occurred in the later Subatlantic period with the expansion of indifferent and acidophilous diatoms. At the same time there is an increase in the content of the organic and mineral components in the sediments of the peaty gyttja. Much of the organic component can be related to the invasion of aquatic plants, offering a suitable habitat for the epiphytic and benthic diatoms.

The final stage of the development is characterized exclusively by acid water diatoms. The dominant taxa are *Eunotia, Pinnularia, Tabellaria* and *Frustulia* species. The species diversity is still quite high.

Similar diatom successions have been noticed in some lakes of N. Wales and in the Lake District in England (Round 1957; Alhonen 1968; Crabtree 1969) and in the Finnish lake region (Koivo 1976). There was also an analogous development in the Postglacial diatom flora of some small lakes in Sweden (Florin 1977), and in the Baltic regions of the U.S.S.R. (Kabailienė 1967).

References

Alhonen, P., 1968. On the Late-glacial and early Post-glacial diatom succession in Loch of Park, Aberdeenshire, Scotland. Memor. Soc. F. Fl. Fenn. 44: 13–20.

Battarbee, R. W., 1979. Diatoms in lake sediments. In: Berglund, B. E. (ed.). Paleohydrological changes in the temperate zone in the last 15 000 years. IGCP 158 B. Lake and mire environments. Project guide vol. 2. Specific methods. Lund.: 177–225.

Crabtree, K., 1969. Post-glacial diatom zonation of limnic deposits in North Wales. Mitt. int. Ver. Limnol. 17: 165–171.

Florin, M. B., 1977. Late-Glacial and Pre-boreal Vegetation in Southern Central Sweden. 2. Pollen, Spore and Diatom Analyses. Striae 5: 1–60.

Hibsch, J. E., 1929. Erläuterungen zur geologischen Karte der Umgebung von Brüx. Knihovna SGU 11: 1–100.

Hurník, S., 1969. Příspěvek ke geologické problematice tzv. Komořanského jezera. Mostecko-Litvinovsko. Regionální studie. Příroda 6: 5–14.

Kabailienė, M., 1963. Die Diatomeenflora in Sedimenten einiger Seen des Riesebach-bassins. Moksliniai pranesimai 15: 253–276.

Koivo, L. K., 1976. Species diversity in postglacial diatom lake communities of Finland. Palaeogr. Palaeoclimatol. Palaeoecol. 19: 165–190.

Losert, H., 1940. Beiträge zur spät- und nacheiszeitlichen Vegetationsgeschichte Innerböhmens. 1. Der Kommerner See. Beih. Bot. Centralbl. 60 B: 346–394.

Round, F. E., 1957. The Late-glacial and Post-glacial diatom succession in the Kentmore Valley deposit. New Phytologist 56: 98–126.

Rudolph, K., 1926. Pollenanalytische Untersuchungen im termophilen Florengebiete Böhmens: Der "Kommernsee bei Brüx". Ber. d. bot. Ges. 44: 239–248.

Řeháková, Z., 1962. Subfosilni diatomové sedimenty v oblasti bývalého Komořanského jezera. Zprávy o geol. výzk. v r. 1961: 163–165.

Řeháková, Z., 1971. Changements qualitatifs des associations de Diatomées dans les sédiments tertiaires et quaternaires de Tchécoslovaquie. Etudes sur le Quaternaire dans le monde VIIIe Congrès INQUA Paris. Paléont. végétale: 275–286.

Wolf, H., 1880. Geologische Grubenrevierkarte des Kohlenbeckens von Teplitz-Dux-Brüx im nordwestlichen Böhmen. A. Hölder, Wien.

The environmental evolution and infilling process of a former lake near Vracov (Czechoslovakia)

Kamil Rybníček
Institute of Botany, Department of Ecology, Stará 18, CS 662 61 Brno, Czechoslovakia

Keywords: paleolimnology, Pannonian lakes, infilling process, environmental change, hydrosere

Abstract

Data on the stratigraphy and development of a former lake near the town of Vracov, SE Moravia, Czechoslovakia, are presented. The infilling process and vegetation succession are described. An attempt to reconstruct changes in habitat conditions during the last ca. 13 000 years has been made.

Introduction

The infilled lake near the town of Vracov in SE Moravia is ca. 0.22 km² in area. It is situated to the west of the town in a very broad and flat valley of the Vracov brook, separating a large area of drifted sands in the south from the loess with chernosem soils in the north. Under these Quaternary deposits, there are Tertiary, mostly clayey, sediments of the ancient Pannonian sea, containing high concentrations of Ca^{2+}, Mg^{2+}, Na^{+}, CO_3^{2-}, SO_4^{2-}, Cl^{-} and other ions. Due to this chemical composition spring waters, which appear in the lake, are highly saline. This environmental background is of great importance in understanding the origin, development and infilling process of the lake.

At the beginning of this century the area of the former lake was described by Čoka (1907) as wet meadows and fen with several open water pools in the central part. From the locality he reported plants such as *Carex diandra, C. appropinquata, C. davalliana, C. limosa, Blysmus compressus, Menyanthes trifoliata, Comarum palustre, Orchis incarnata, Bryum ovatum, Camptothecium nitens, Drepanocladus* species and other, less interesting plants. From the water pools *Nymphaea alba, Utricularia minor, Sparganium minimum, Potamogeton natans, Hippuris vulgaris, Lemna trisulca, L. minor* and *Myriophyllum verticillatum* were described by the same author.

During the 1920's the area was drained by several deep channels and covered over by bringing mineral soils from the sites of newly constructed houses and sands from the local neighbourhood. It was subsequently ploughed, enabling the ground to be used for growing vegetables. During the following years, the decomposition of the buried and drained fen peat led to the sinking of the land surface causing a relative rise of the water table. The Vracov lake was waterlogged again and, as a result, it was colonized by *Phragmites* stands after the Second World War.

Material and methods

We came to investigate the locality in the year 1963. By that time it had been drained before the buried peat started to be exploited for agricultural and horticultural use. We used the opportunity for describing and sampling the peat and lake sediments from the walls which were opened during the excavation of the upper 2 m of organic sediment. The excavation was carried out in strips 25 to 30 m wide. In these strips we described and geodetically

Hydrobiologia 103, 247–250 (1983).

levelled the layers every 40 m, and every 80 to 160 m we took samples for future laboratory treatment. In this way we obtained a fairly good picture about the stratigraphy, about the sequence of vegetational succession, about the filling-in process and about the general development of the lake.

The sediments and their samples were analyzed for their physical properties and for determination of their components according to Troels-Smith (1955), and for pollen and macroscopic plant remains. So far we have finished the descriptions and made lists of dominant macrofossils of about 90 sections from the open walls and many others from additional borings. For two profiles we have pollen analyses and detailed macroscopic plant analyses, and one profile has been dated by ^{14}C. The first observations were published by Rybníčková & Rybníček (1972), and by Sládková-Hynková (1974).

Results

The morphology of Vracov lake is characterized by very steep banks on almost all sides and by the division of the basin into two parts, the western and the eastern ones. Even though very many irregularities and inversions can be observed in the sediment accumulation, especially in the marginal parts of the lake, a generalized scheme of sedimentation and filling processes can be reconstructed (Fig. 1).

We suppose that the lake originated approximately at the end of the last (Weichselian) glacial period or at the very beginning of the late Weichselian when the dunes of drifted sands, which covered the region between the towns of Hodonín, Vracov and Bzenec, closed the outflow from the basin. The oldest cold periods were characterized by minerogenic sedimentation in the deepest parts of the lake, where thin layers of aeolian sand alternate with thin layers of clay. At the moment it cannot be said whether this lamination is annual or not. The characteristic plant remains are leaves of *Myriophyllum verticillatum* and *M. spicatum* always enclosed in the sand layers. The biological productivity of the lake was apparently very low owing to the low temperature and supposed oligotrophy of the water.

Organogenic sedimentation started during the Alleröd period. Highly elastic green algal gyttja is the typical sediment of this period and its deposition continued approximately to the beginning of the Boreal. In the central and deepest parts of the lake a mixture of calcareous marl and algal gyttja can be observed in the upper part of the gyttja layer. The algal gyttja contains hardly any remains of macrophytic water plants and was formed by several species of green algae like *Pediastrum, Scenedesmus, Tetraëdron, Arthrodesmus, Staurastrum, Botryococcus,* etc. The depth of this sediment (sometimes over 1 m) which accumulated over about 2000 years, and the richness of the algal flora, suggest that the biological productivity of the lake must have risen substantially in comparison to previous periods, this time owing to the supposed mesotrophy and increase of temperature of the water. At this very period the lake seems to have acquired the greatest depth in its history.

The deposition of sediments, rich in trivalent iron (*Limus ferrugineus*), is typical of the transition from Preboreal to Boreal periods. This distinctly rust-brown, sometimes jelly-like sediment, at most 2 cm thick, divides the calcareous algal gyttja from overlying lake marl in the central parts of the lake. The reasons for this characteristic type of sedimentation are at present unknown.

The Boreal period was characterized by sedimentation of lake marl (*Limus calcareus*) in the central part of the lake, while at the shore accumulation of coarse detritus gyttja can already be observed. The deposition of the coarse detritus gyttja over the whole bottom of the lake took place during the Atlantic period and during the prevailing part of the Subboreal period. This type of sediment was formed by the vegetation of macrophytic water plants, especially *Najas marina, Nuphar luteum, Nymphaea candida, Ceratophyllum demersum, Potamogeton pusillus, P. natans, P. obtusifolius,* i.e. by the vegetation close to the present phytosociological alliance *Nymphaeion*. Green algae were also an important part of the sediment, but a warmer climate caused higher humification than earlier. The lake still remained open, with initial stands of the *Magnocaricetea* communities present only near the shores. The biological productivity of this naturally meso-eutrophic water rose to the highest point we observed.

At the end of the Subboreal, probably as a result of a lowering of the water table (this period is supposed to have been warm and dry in this region),

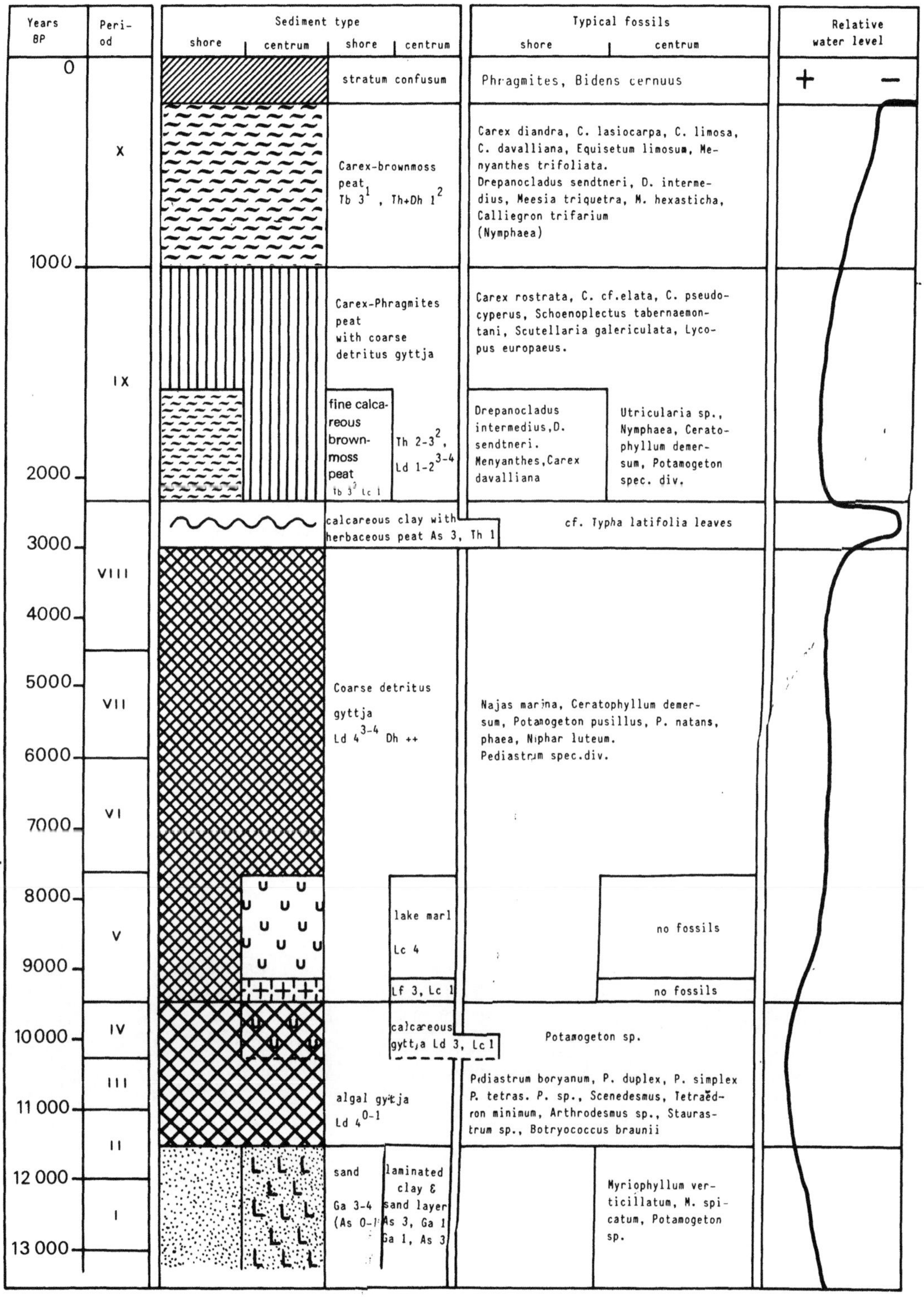

Fig. 1. Vracov (Czechoslovakia) – Scheme of the lake development.

a very distinct white calcareous clayey layer with remains of *Typha latifolia* was formed over the whole of the lake bottom. The origin of the clay is unknown; it may have resulted from agricultural cultivation of the loess soils at the northern shore of the lake, since the region has been settled since at least Neolithic times.

Higher humidity in the Subatlantic periods, both early as well as late, induced a new rise in the water level, but the lake has never become open water again since that time. Along the shores we can reconstruct the stands of tall sedges (*Carex elata C. rostrata, C. pseudocyperus*) and of *Phragmites*, while the central parts and, later on, the whole lake were covered by swim-floating carpets of brown-moss fen communities with *Drepanocladus sendtneri, D. intermedius, Calliergon trifarium, Meesia triquetra, M hexasticha, Carex diandra, C. limosa, C. lasiocarpa, Menyanthes trifoliata*, etc. The presence of many boreal and relict plants is worth stressing. Macrophytic water plants like *Nymphaea, Nuphar, Utricularia, Ceratophyllum demersum, Potamogeton* species, etc. survived only in several pools of open water. This reconstruction corresponds very well with the description by Čoka (1907) mentioned above.

The former lake of Vracov is a good example of infilling processes taking place in the small lakes of the Pannonian region during the last ca. 13 000 to 15 000 years and it demonstrates the naturally changing productivity and trophy of these former lakes in the warm regions of Central Europe.

References

Čoka, F., 1907. Příspěvky ku květeně moravské/Contributions to the Moravian Flora/. Věstn. Klubu Přírod. Prostějov 9: 80–94.

Rybníčková, E. & Rybníček, K., 1972. Erste Ergebnisse paläogeobotanischer Untersuchungen des Moores bei Vracov, Südmähren. Folia Geobot. Phytotax. 7: 285–308.

Sládková-Hynková, H., 1974. Paleogobotanická studie rašeliniště u Vracova/Paleogeobotanical Study of a Mire near Vracov/. Diploma Thesis depon. in Department of Plant Biology, Faculty of Science, University of Brno, 72 pp.

Troels-Smith, J., 1955. Karakterizing af løse jordater. Dan. Geol. Unders., IV Raekke 3/10: 1–73.

Variability of *Actinocyclus kuetzingii* (A.S.) Sim. in the Littorina sediments of the Polish Baltic Coast

Wieslawa Przybylowska-Lange[1], Irena Kaczmarska[1] & Wincenty Kilarski[2]
[1]*Botanical Institute, Polish Academy of Sciences, Lubicz 46, PL-31-512 Kraków, Poland*
[2]*Institute of Zoology, Laboratory of Scanning Microscopy, Karasia 6, PL-30-060 Kraków, Poland*

Keywords: paleolimnology, diatom flora, sediments, Vistula Lagoon, Poland

Abstract

The variability of *Actinocyclus kuetzingii* (A.S.) Sim. in the Littorina deposits of the Vistula Lagoon and Lake Druzno were studied. During the time of the Littorina Lake Druzno was a part of the Vistula Lagoon. In the different parts of the Lagoon both the abundance of *A. kuetzingii* and the diameter of its valves change in a manner associated with phases of the Littorina transgression. *Coscinodiscus* (*Actinocyclus*) *kuetzingii* fo. *subsalsa* Brockmann is probably an ecotype of *A. kuetzingii* (A.S.) Sim.

Introduction

Actinocyclus kuetzingii distinguished by A. Schmidt (1874, Pl. 57, Figs. 17 & 18) as *Coscinodiscus kuetzingii* based on samples from Cuxhaven and the Firth of Tay is a taxon widespread in the coastal waters of Northern and Western Europe, and Asia (Hustedt 1930; Proshkina-Lavrenko 1949–1950; Cleve-Euler 1951), but it is generally found as single specimens. It was found more frequently by Hendey (1964) and Paasche (1960). Paasche characterized it as accompanying the dominant diatom species in the Atlantic–Arctic floristic area, which is thought to be under Polar influences. Though most localities come from the temperate and boreal climatic zone, it has also been found by Zannon (1938) in the Gulf of Venice.

A. kuetzingii is also known as a fossil. The oldest frustules come from the Torton deposits of the Matra Mountains in Hungary (Hajoś 1968). It is also found in the Pliocene of the Far East (Proskhina-Lavrenko 1949–1950), and in the Norwegian Sea (Schrader & Fenner 1976).

In the Baltic region it is occasionally reported from Quaternary sediments (Cleve-Euler 1951; Brockmann 1954; Miller 1964; Zachowicz *et al.* 1982) as well as from contemporary waters (e.g. Cleve-Euler 1951; Molder & Tynni 1968).

An abundant occurrence of *A. kuetzingii* was observed in the Littorina sediments of the Vistula Lagoon (Brockmann 1954; Przybylowska-Lange 1974), and in the same age sediments of the adjoining Lake Druzno (Przybylowska-Lange 1976; Zachowicz *et al.* 1982). Niemelä & Tynni (1979) also found it to be common in the Eemian material in the Pohjanmaa region of Finland.

The above papers mostly concern typical *A. kuetzingii* valves, easy to identify, 30–70 μm in diameter and with characteristic ornamentation. But in the Littorina deposits in the Vistula Lagoon and Lake Druzno, smaller atypical frustules were more common. Brockmann (1954) and Przybylowska-Lange (1974 & 1976) had difficulty in identifying them. Brockmann, in spite of some doubts, described these frustules as the new taxon: *C. kuetzingii* fo. *subsalsa*.

An evaluation of the correct taxonomical position of these small frustules was necessary because of their distinctive occurrence in the sediment of both the water bodies studied.

The variability of the species was investigated in core material from both the Vistula Lagoon and the Druzno Lake (Fig. 1).

Hydrobiologia 103, 251–256 (1983).

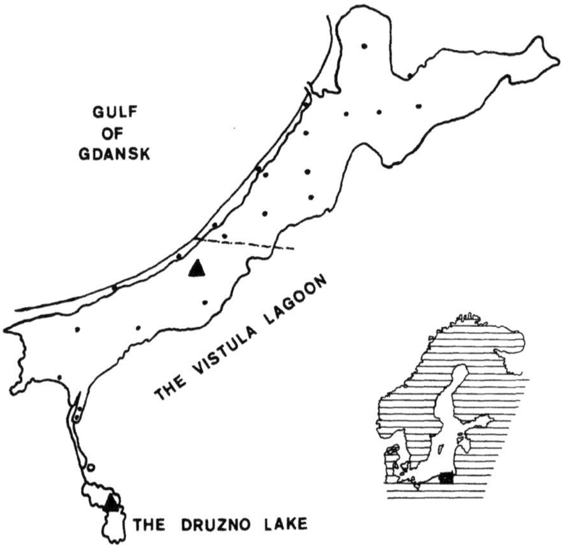

Fig. 1. Location of core sites studied by Brockmann (dots) and by Przybylowska-Lange (triangles).

Paleoecological characteristics of the material studied

The Littorina transgression of the Baltic Sea caused a rise of water level in the Vistula Lagoon, which, as a result, caused a transgression into the Lake Druzno basin. Since that time (6440 ± 50 B.P.) the lake has been a part of the Vistula Lagoon, and has had a similar diatom flora (Przybylowska-Lange 1976; Zachowicz *et al.* 1982).

Generally, the Littorina diatom flora of the Vistula Lagoon consisted of the same mesohalobous and euhalobous species as in Lake Druzno. At first, Lake Druzno was a very shallow bay of the Lagoon, as is indicated by the small number of planktonic species and their limited quantity. In the Lagoon planktonic diatoms developed more abundantly (Przybylowska-Lange 1974). Among the mesohalobous species the most characteristic was *Anomoeoneis costata*, but it was dominant only in the Older Atlantic period.

During the Younger Atlantic and at the beginning of the Subboreal period the composition of the diatom flora was very similar to that of the samples from the Vistula Lagoon and Druzno Lake: planktonic diatoms as well as mesohalobous and euhalobous species reached their maximum percentage. The percentage of meso- and euhalobous diatoms varied significantly in some samples due to the alternation of successive inflows of sea-water into the Lagoon and the more freshwater conditions in the basin. The inflows are reflected by the abundance of *Actinocyclus kuetzingii* (up to 50%). The greatest inflow occurred at the boundary between the Atlantic and Subboreal periods. The inflows of sea-water were caused by consecutive transgression phases of the Littorina Sea.

In the Subboreal period, after the culmination of meso- and euhalobous diatoms during the last phase of the Littorina transgression, their number decreased rapidly and Lake Druzno became almost a freshwater lake, although in the Vistula Lagoon meso- and euhalobous diatoms still accounted for up to ca. 10% of the population.

In spite of differences between the diatom flora of Lake Druzno and the Vistula Lagoon since the Middle Subboreal period, continuing similarities indicated that only a very slow process of isolation was taking place. This process terminated at the beginning of the Younger Subatlantic due to the rapid growth of the Nogat delta.

Description of the species under study

The specimens from the Vistula Lagoon have the same range of variability in structure as those from the Druzno Lake.

The frustules are drum-shaped, 12–66 μm in diameter. Small specimens often have a thicker frustule than the large ones. The cingulum is composed of two copulae (Fig. 2; 1 & 2). The advalvar copula (Fig. 2; 2) is narrow, open, and bears a ligula (Fig. 2; 3). The abvalvar copula is an open, wide band. In one frustule no more than one cingulum is seen (Fig. 2; 1).

The valve face is flat or, especially in bigger individuals, slightly convex in the centre, and usually has a high valve mantle seen clearly in small frustules (Fig. 2; 1).

The valve face is covered with a net of polygonal to rounded areolae (Fig. 2; 8 & 9). The areolae are arranged in radial and tangential rows. In the centre of the valve face, in a radial row, there are 8 areolae in 10 μm, or occasionally 7, especially on small valves. Toward the margin of the valve face, the areolae density gradually decreases to 10–12 in 10 μm. The margin of the valve face varies in width

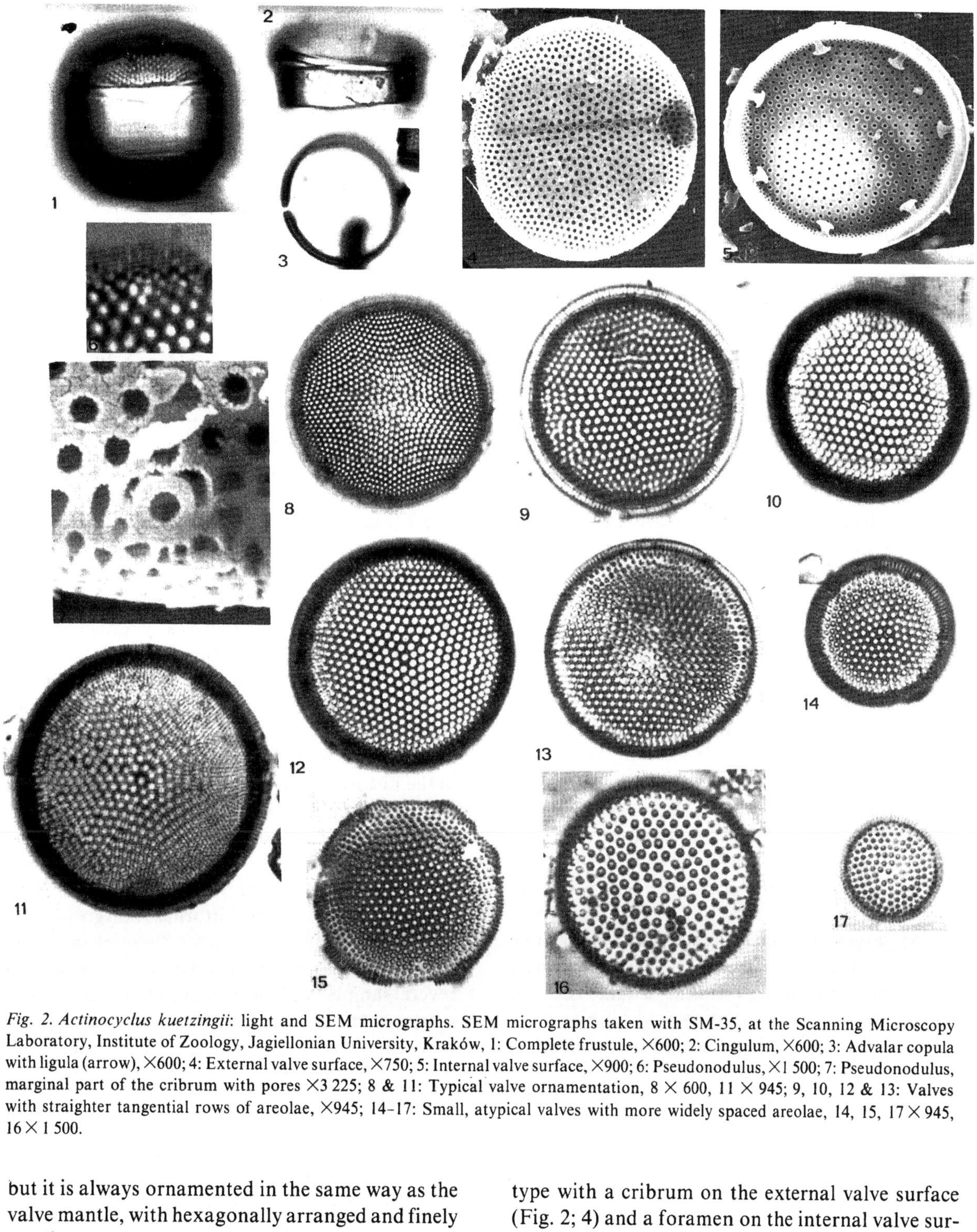

Fig. 2. Actinocyclus kuetzingii: light and SEM micrographs. SEM micrographs taken with SM-35, at the Scanning Microscopy Laboratory, Institute of Zoology, Jagiellonian University, Kraków, 1: Complete frustule, ×600; 2: Cingulum, ×600; 3: Advalar copula with ligula (arrow), ×600; 4: External valve surface, ×750; 5: Internal valve surface, ×900; 6: Pseudonodulus, ×1 500; 7: Pseudonodulus, marginal part of the cribrum with pores ×3 225; 8 & 11: Typical valve ornamentation, 8 × 600, 11 × 945; 9, 10, 12 & 13: Valves with straighter tangential rows of areolae, ×945; 14–17: Small, atypical valves with more widely spaced areolae, 14, 15, 17 × 945, 16 × 1 500.

but it is always ornamented in the same way as the valve mantle, with hexagonally arranged and finely poroid areolae, which are 16–20 in 10 μm (Fig. 2; 13 & 17).

The areolae in the valve face are of the loculate type with a cribrum on the external valve surface (Fig. 2; 4) and a foramen on the internal valve surface (Fig. 2; 5).

In the transition zone between the loculate and poroid areolae an areolate-operculate pseudonodulus is situated (Fig. 2; 4, 6 & 7). It is also found in a more central position.

The inner labiate processes are unilaterally developed and are always situated on the transitional zone between the two kinds of areolae. They are very regulary arranged, usually 6–8 on one valve, but ranging from 2 to 13 (Fig. 2; 10–17). The outer openings of the labiate processes even in SEM can hardly be distinguished from the adjacent areolae (Fig. 2; 4) since they have the same diameter as the poroid areolae. They differ by having a horseshoe-shaped area surrounding them from the admarginal side. Each labiate process is situated at the end of the radial row of the locular areolae which runs from the centre of the valve face which is also the middle row of each fascicicle. The last areola of each middle row forms a more or less regular central ring surrounding the central areola or hyaline central area. The central area is somewhat irregular and may have one or two poroid areolae. The central area, as well as the central ring of areolae, may be situated eccentrically (Fig. 2; 10, 12 & 13).

The basic ornamentation is the same for both large and small frustules. The valve face is covered with a net of areolae arranged in radial rows. The rows are gathered into fascicles, and are parallel to the middle row. The middle rows extend from the marginal labiate process to the central ring of the areolae.

The secondary ornamentation is composed of the convex tangential rows of areolae; they are clearly visible on valves bigger than 30 μm in diameter. Smaller valves, less than 30 μm in diameter, have straight or only slightly curved tangential rows of areolae.

The larger specimens, typically ornamented, were not difficult to identify. They were rarely found, however, in the Littorina sediments of the Vistula Lagoon and the Druzno Lake. Small specimens were the most abundant and are smaller than any found in other water bodies.

Such small valves were described by Brockmann (1954) as *Coscinodiscus kuetzingii* fo. *subsalsa*; he pointed out that he was not always able to distinguish the valve of this taxon from *C. eccentricus* var. *fasciculata* Hust. Hustedt (1930) similary considered *C. kuetzingii* to be related to *C. eccentricus* (now *Thalassiosira eccentrica* (E.) Cl). In the light microscope small valves of the material under study also give that impression, particularly when the central area or the central areola is placed eccentrically. When the valve margin of *A. kuetzingii* is damaged or is only faintly visible this impression is even stronger. In good conditions *T. eccentrica* is easily distinguished from this taxon in the light microscope due to its coarser ornamentation (5–7 areolae in 10 μm), and the coarser and irregular ornamentation of the valve mantle.

The ornamentation of *A. kuetzingii* valves change with a decrease in diameter. But, even in the smaller valves, the ornamentation consists of radial and more or less straight tangential rows of areolae. The straighter tangential pattern of rows causes the fasciculation to be less visible (Fig. 2; 10, 16 & 17). The areolae of the smaller valves seem to be coarser; they are probably only more widely spaced because of a thicker silica wall in small frustules (Fig. 2; 14–17). However, the number of areolae in each part of the valves is still characteristic for *A. kuetzingii* (Fig. 2; 14–17).

The *A. kuetzingii* valves have a great range of variability and thus there is no reason to classify the smaller valves separately. *A. kuetzingii* fo. *subsalsa* Brockmann must be treated as a synonym of *Actinocyclus kuetzingii* (S.A.) Sim.

The frequency of *A. kuetzingii* in the cores investigated varies greatly (Figs. 3 & 4A). In the samples from the Vistula Lagoon it is associated with other brackish diatoms which indicate successive phases of the Littorina transgression. In the samples taken from Lake Druzno sediments *A. kuetzingii* is almost the only indicator of increasing water salinity.

Both the abundance of the species under study and the diameter of its valves change in a consistent way with phases at the Littorina transgression of the Baltic Sea. To investigate these changes the diameter of 50 successive valves was measured in several samples with both high and low numbers of marine taxa. In samples where *A. kuetzingii* was very infrequent 25 individuals (Lake Druzno 65 and 99) and 40 individuals (Vistula Lagoon 34) were measured.

The measurements were grouped into 5 μm classes. Histograms in Fig. 3 & 4 demonstrate the percentage frequencies of individual classes.

In the samples from the transgression phases of the Vistula Lagoon the most numerous were valves in the 25–30 μm class and the samples contained relatively high numbers of larger, typical valves of *A. kuetzingii*. In the fresh water phases classes from

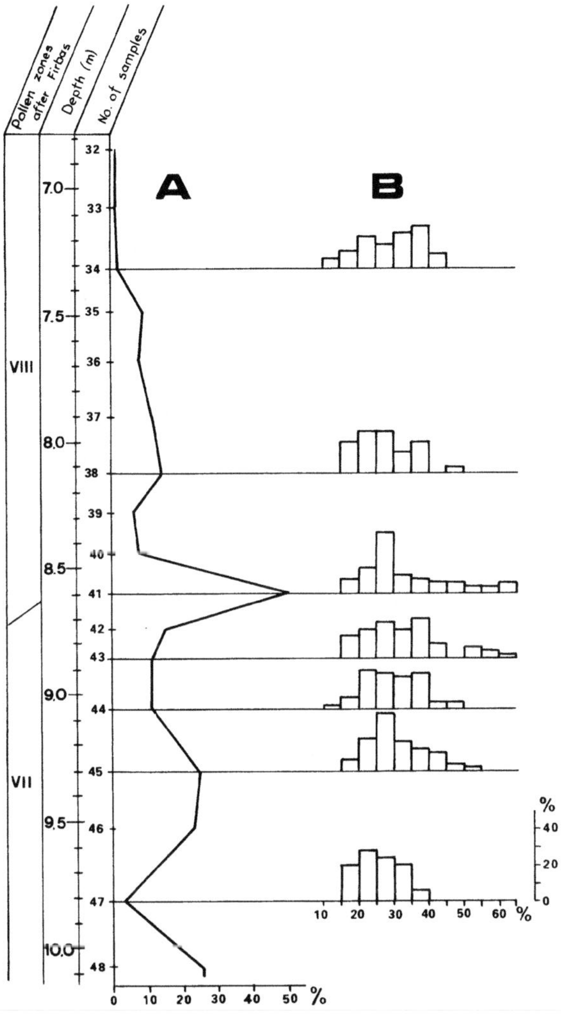

Fig. 3. A: Percentage of *Actinocyclus kuetzingii* in the Vistula Lagoon core (500 consecutive diatom valves of all species were counted); B: Percentage frequency of valve diameter size classes for *Actinocyclus kuetzingii* (n = 50).

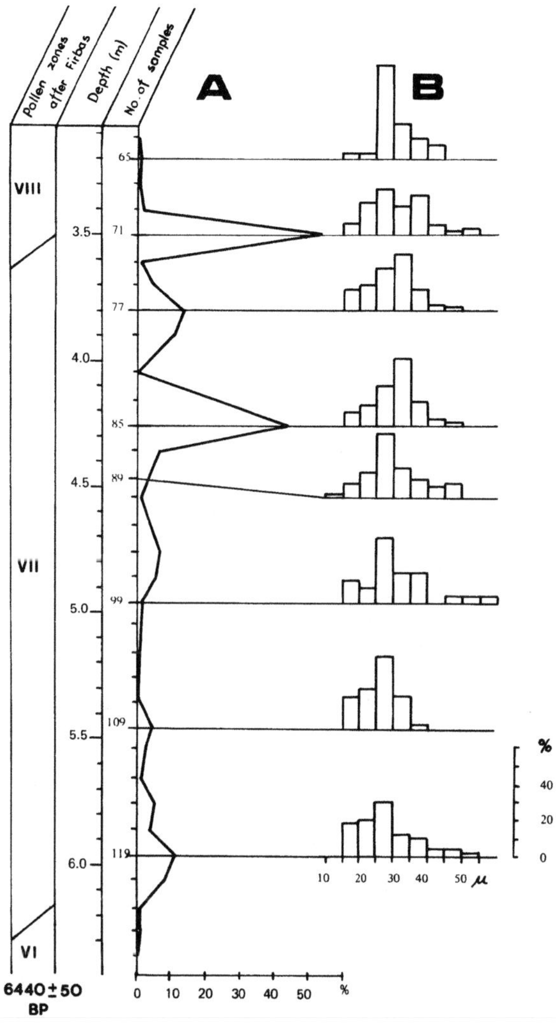

Fig. 4. A: Percentage of *Actinocyclus kuetzingii* in the Lake Druzno core (500 consecutive diatom valves of all species were counted); B: Percentage frequency of valve diameter size classes for *Actinocyclus kuetzingii* (n = 50).

15–40 become about equally numerous and larger valves are rare. In addition in the Lake Druzno samples the valve diameter measurements have a normal distribution in all transgression phases. However, the most numerous classes in the transgression phase tend to shift towards larger diameters than in the Vistula Lagoon samples.

The data obtained suggest that an ecological form of *A. kuetzingii* has developed locally and become established in Lake Druzno during the Littorina transgression phases. This form was probably better adapted to the lower salinity and greater stability of the Lake Druzno environment, but its morphological character does not differ from typical specimens of *A. kuetzingii*.

Acknowledgements

The authors would like to thank Professor Jadwiga Siemińska, head of the Phycological Department of the Botanical Institute of the Polish Academy of Sciences, for valuable support during the research, for reading the manuscript and useful criticism.

References

Brockmann, C., 1954. Die Diatomeen in der Ablagerungen der ostpreussischen Haffe. Meyniana 3: 1–95.

Cleve-Euler, A., 1951. Die Diatomeen von Schweden und Finland. Kungl. Svenska Vet. Akad. Handl., 4th, 2, 1: 163 pp.

Hajoś, M., 1968. Die Diatomeen der miozan Ablagerungen des Matra Vorlandes. Geol. Hungarica, ser. Palaeontologica, 57: 1–40.

Hendey, N. I., 1964. An introductory account of the smaller algae of British coastal waters. Part V, Bacillariophyceae (Diatoms) 318 pp. London. Her Majesty's Stationery Office.

Hustedt, F., 1930. Die Kieselalgen Deutschlands, Österreichs und der Schweiz mit Berücksichtigung der Übrigen Länder Europas sowie der angrenzenden Meeresgebiete, 1: 920 pp. In: L. Rabenhorst's Kryptogamen - Flora, 7. Akademischer Verlag.

Miller, U., 1964. Diatom floras in the Quaternary of the Göta River Valley. Sveriges Geol. Unders. Ser. C., 44: 1-67.

Mölder, K. & Tynni, R., 1968. Über Finlands rezente und subfossile Diatomeen, 2. Bull. Geol. Soc. Finland 40: 151-170.

Niemelä, J. & Tynni, R., 1979. Interglacial and interstadial sediments in the Pohjanmaa region, Finland. Geol. Sur. Finland, 302: 1–48.

Paasche, E., 1960. Phytoplankton distribution in the Norwegian Sea in June, 1954 related to hydrography and compared with primary production data. Fisk. Dir. Skv. Serie Havundersøkelser, 12: 1–77.

Proskhina-Lavrenko, A. I., 1949–1950. Diatomovyi analiz. Opredyelitel iskopayemykh i sovryemyennykh diatomovykh vodoroslyei. Gos. Izdat. Geol. Lit., Moskva. Centrales i Mediales, 2: 238 pp.

Przybylowska-Lange, W., 1974. Rozwój Zalewu Wiślanego w świetle analizy okrzemkowej. Prace Inst. Met. Gosp. Wodnej, 2: 129–164.

Przybylowska-Lange, W., 1976. Diatoms of lake deposits from the Polish Baltic Coast. I. Lake Druzno. Acta Palaeobot., 17, 2: 35–72.

Schmidt, A., et al., 1874–1959. Diatomaceen-Kunde. R. Reisland, Aschersleben Leipzig u. Akademie-Verl.

Schrader, H. J. & Fenner, J., 1976. Norwegian Sea Cenozoic diatom biostratigraphy and taxonomy. Part 1, 2. In: Talwani M., et al., Initial Reports of the Deep Sea Drilling Project, 38: 931–1099, Washington (U.S. Government Printing Office).

Zachowicz, J., Przybylowska-Lange, W. & Nagler, J., 1982. The Late-Glacial and Holocene vegetational history of the Zulawy Region (North Poland). Biostratigraphic study of Lake Druzno. Acta Palaeobot. 22, 1: 141–161.

Zannon, V., 1938. Le Alghe Della Laguna, Sezione 2, Bacillarieae, 3: 253–503.

Paleolimnological investigations in humid and arid zones of the Soviet Union

G. G. Martinson, V. I. Khomutova & V. G. Drabkova
Institute of Lake Research, Academy of Sciences of the U.S.S.R., Petrovskaya nab. 2, Leningrad 197046, U.S.S.R.

Keywords: paleolimnology, palynology, bacteriology

Abstract

There are many lake basins in the Soviet Union, both in the humid and the arid regions, including lowland and alpine, and shallow and deep-water lakes. Lithological, geochemical, radiological and paleobiological methods have been utilized in studying the history of these lakes. Lake Zaisan in Kazakhstan is described as an example of lake development in the arid zone, whilst the lake systems of the Pribaltica, where human influence on the environment is clearly recorded, exemplifies the humid zone.

Introduction

There are numerous lake systems in the vast territory of the Soviet Union, situated both in humid and arid districts. They include lowland and alpine, and shallow and deep-water lakes. The investigation of lakes in the humid climatic zone takes place in the north-west of the country, mainly in the Karelian Isthmus, Karelia, Belorussia and Pribaltica. These lake systems are of glacial origin. The water bodies in the arid zones of Kazakhstan and Middle Asia are much older. Lakes in this area include Balkhash, Issyk-Kul, Zaisan and Baikal. Particular attention has been given to the history of Lake Baikal.

Lithological, geochemical, radiocarbon, paleozoological, diatomological and palynological methods have been used to study the sediments of lake systems of different types and of terrestrial deposits formed in intermontane hollows. Special attention has been paid to the study of the human influence on aquatic systems in space and time. The history of ancient Lake Zaisan, situated in the arid zone of Kazakhstan, and the histories of young water bodies of the Pribaltica are given as examples of lake development. Ecosystem changes as a result of human influence are reflected most distinctly in them.

Lake Zaisan

The history of Lake Zaisan has been traced back 60 million years. The lake basin has its origins in the Upper Cretaceous period and has remained a lake up to the present. During this long period it has undergone repeated changes in size and has been influenced by many variations in climate. At present a marked change of lake regime has occurred as the result of human activity following the construction of the Bukhtarminsk power station on the River Irtysh, when the lake was transformed into a reservoir.

Integrated studies of the sediment accumulated throughout this long period (cf. Figs. 1, 2) have shown that numerous remains of mollusca, ostracoda, turtles and crocodiles have been preserved. Chara algae, spore-pollen assemblages, and the imprints of land vegetation have also been found in abundance. More detailed information on the sedimentology of the lake has been published in 'Paleolimnology of Zaisan' (1980).

Hydrobiologia 103, 257–260 (1983).

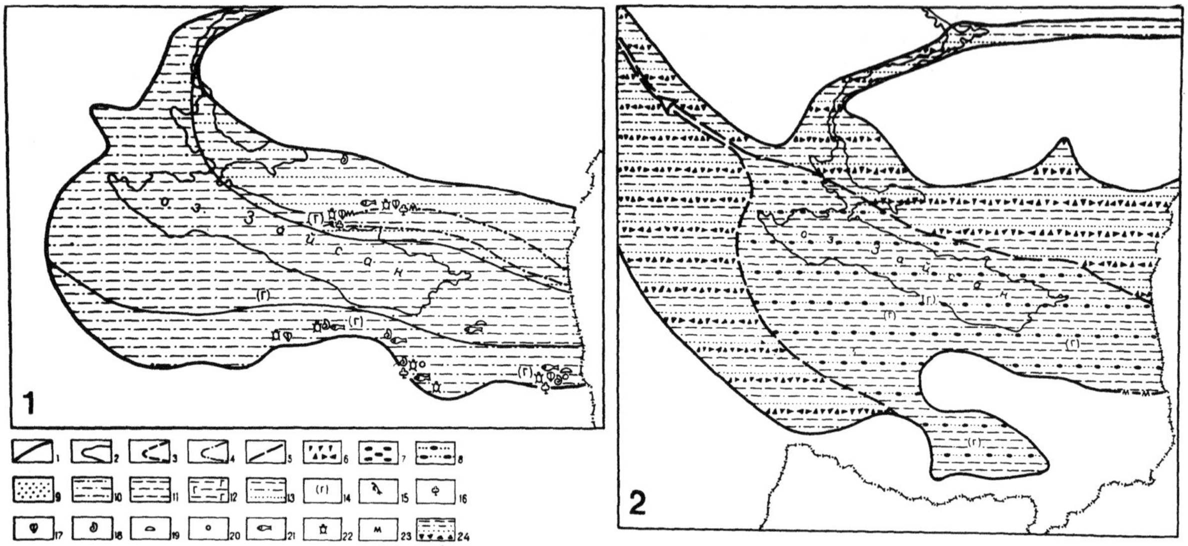

Key to 'Sediment types and location of fossil finds in paleo-Zaisan sediments'
1 – probable lake boundaries (in Fig. 2, the limit of sediment accumulation); 2 – outline of present Lake Zaisan; 3 – temporary islands; 4 – anticlinal structure, sometimes apparent both in underwater and above-water relief; 5 – probable and time variable boundary of limnic and terrestrial sediment accumulation.
Zones of preferential sediment accumulation:
6 – arboreal; 7 – gravel; 8 – sand with gravel; 9 – sand; 10 – aleurite; 11 – clay; 12 – gypsified clay; 13 – alternation of corresponding sediments; 14 – known gypsified sediments.
Fossil locations:
15 – dinosaurs; 16 – rich in vegetation; 17 – bivalve mollusca; 18 – gastropoda; 19 – ostracoda; 20 – charophytes; 21 – fish; 22 – turtles; 23 – mammals; 24 – time variable zone of terrestrial sediment accumulation.

Fig. 1. Paleolimnological map of Lake Zaisan during Late Eocene.

Fig. 2. Paleolimnological map of Lake Zaisan during Late Pliocene.

Small lakes of the Pribaltica

In small lakes of the more humid Pribaltica region human influence has become apparent to a much greater extent. From 1977 to 1980 we conducted an integrated limnological investigation of lakes of the Latgalian Elevation (Latvia). The sediments of lakes Vorkalu, Rogaizhu, Long Snidzinjas, Becheru, and Sleinovas were studied using palynological, limnological, geochemical (including trace elements), and microbiological techniques (Jakovleva *et al.* 1981).

The land surface of the Latgalian Elevation is a young one having been influenced by the last stages (probably North Lithuanian) of the Valdai glaciation; landforms produced by glaciation processes predominate. The impact of man is particularly evident in the present condition of lakes Vorkalu (eutrophic), Long Snidzinjas (mesotrophic) and Sleinovas (dystrophic).

Cores taken from the central part of the lakes show that all have similar stratigraphies. The lacustrine sediments are underlain by sands of varying coarseness with an admixture of carbonate rubble, and this is followed by a thin peat layer (1–20 cm). Above this, clayey, sometimes sandy, silts with a low but variable proportion of organic matter occur (Figs. 4, 5).

According to pollen analysis it has been determined that all lakes except Long Snidzinjas were formed at the beginning of Preboreal Time (PB), about 10 300 years ago. Lake Long Snidzinjas was formed later during Boreal Time (BO).

Pollen spectra from the lakes are shown in Figs. 4, 5 and, for Lake Vorkalu, in Khomutova (1983). The mineral component of the sediment increases in the second half of the Subatlantic period and the trend towards a gradual reduction of

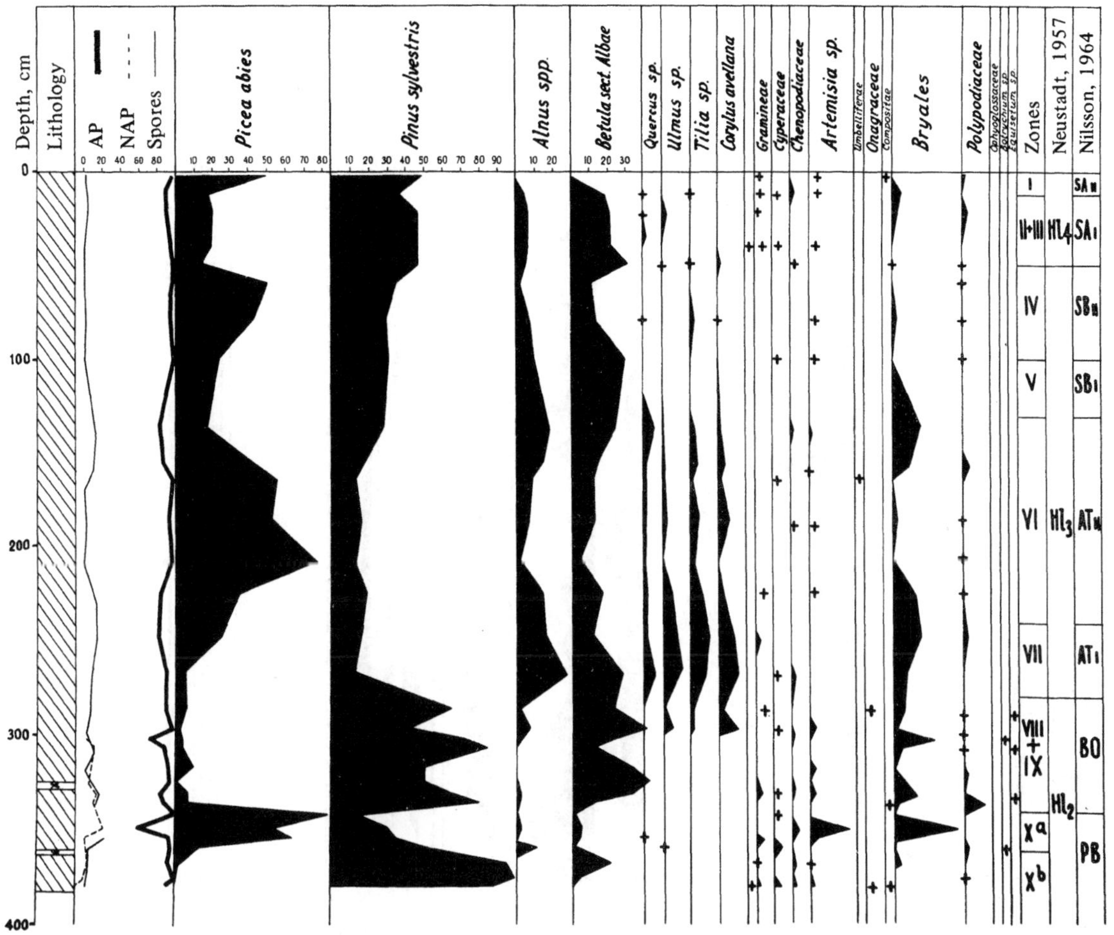

Fig. 3. Pollen Diagram for Lake Becheru (Pribaltica).

organic matter accumulation indicates the expansion of agricultural activity in the area. Judging from the present condition of lakes Vorkalu and Sleinovas, human influence has also resulted in a sharp increase in their productivity: in the first lake causing an increase in phytoplankton and in the second causing macrophyte development. Human influence on lakes Rogaizhu and Becheru is less distinct.

The redox potential of the 0–10 cm surface mud layer ranged between +180 mV and 200 mV. The lowest values occur in lakes Vorkalu and Rogaizhu. For lakes Long Snidzinjas, Sleinovas, and Becheru the minimum values of redox potentials were somewhat lower at 150 mV.

Over the growing period and average number of bottom bacteria was as follows: Long Snidzinjas 290 $\times$ 10^6 g^{-1} wet sediment, Vorkalu 580 $\times$ 10^6, Sleinovas 130 $\times$ 10^6, Rogaizhu 300 $\times$ 10^6, Becheru 365 $\times$ 10^6. The greatest number of bacteria occurred in Lake Vorkalu whilst the lowest number occurred in Lake Sleinovas. However the difference is not significant. The differences in oxygen consumption of the muds in these lakes are also small: 0.7; 1.04; 0.78; 1.05 and 0.89 g O_2 m^{-2} day^{-1} respectively.

Measurement of bottom bacteria production during June, July and August shows mean values in all lakes of 13.9–18.7 g C m^{-2}. However, this value increased to 37.8 g C m^{-2} in Lake Vorkalu and was extremely low in Lake Sleinovas, only 5.9 g C m^{-2}. These data show that in Lake Sleinovas, in spite of intensive organic matter production due to abundant macrophyte development, the decomposition of organic matter takes place extremely slowly. The lake is becoming dystrophic with considerable accumulation of organic matter taking place.

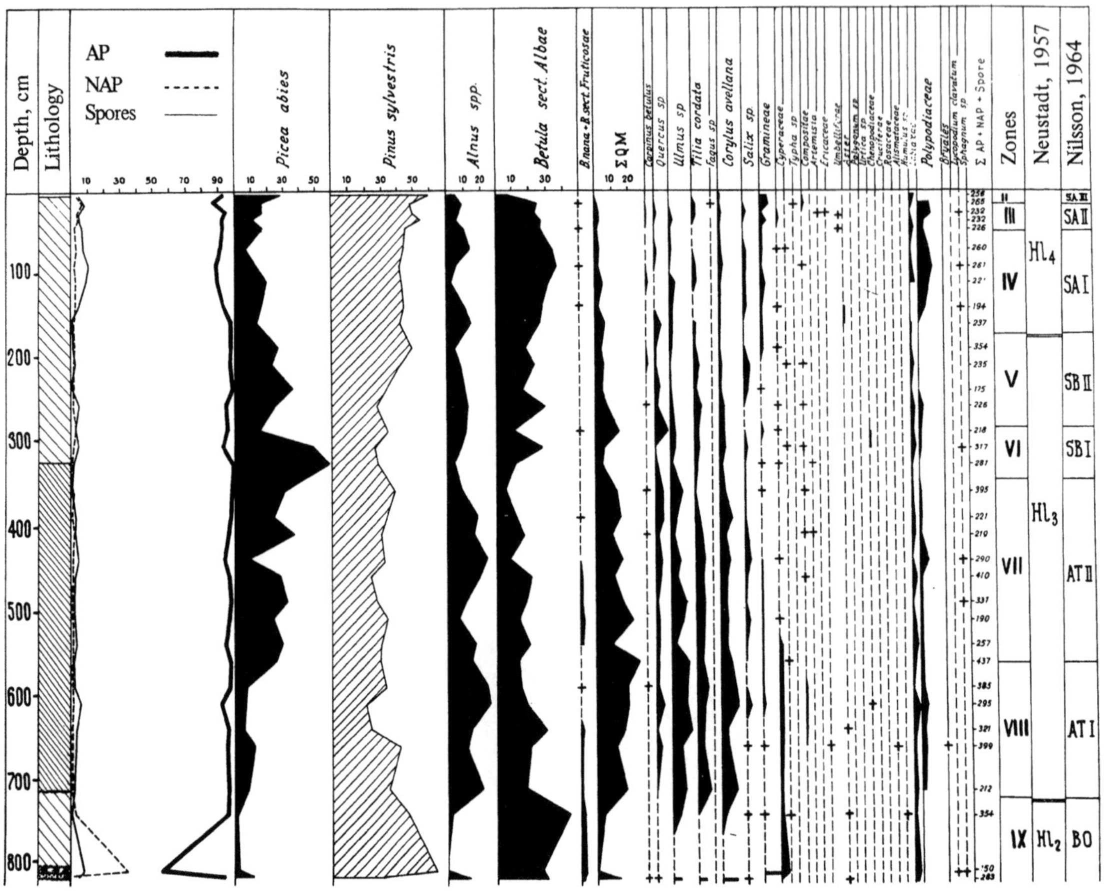

Fig. 4. Pollen Diagram for Lake V. Snidzinjas (Pribaltica).

References

Khomutova, V. I., 1983. Palynological investigations of the sediments of north-western lakes of the Russian Plain and the Baltic Sea. Hydrobiologia, this volume.

Khotinsky, N. A., 1977. Holocene of Northern Eurasia. Nauka, Moscow, 198 pp. (in Russian).

Jakovleva, L. V., Drabkova, V. G., Khomutova, V. I. & Sergeeva, L. V., 1981. Post-glacial history of lacustrine sediment accumulation in the water bodies of Latvia and its change under an influence of human factor. In: Paleolimnological Approach to Study of Human Influence on the Lakes. All-Union Geographical Society (in Russian).

Paleolimnology of Zaisan. Nauka, Leningrad, 1973 (in Russian).

Recent history of Lake Ladoga

N. N. Davydova, A. A. Kurochkina & I. S. Trifonova
Institute of Lake Research, Academy of Sciences of the U.S.S.R., Petrovskaya nab. 2, Leningrad 197046, U.S.S.R.

Keywords: paleolimnology, diatoms, eutrophication, Ladoga

Abstract

Combined paleolimnological investigations of a 1.8 m sediment core from the deepest north-western part of Lake Ladoga show stratigraphic changes in granulometric and chemical composition, organic matter content, diatom species composition and chlorophyll 'a' concentration. The sediment accumulation rate was calculated and 5 stages of lake history over the last 4 000 years were described. Changes in lake environment were mainly caused by changes in climate and lake water balance. Human impact on the lake was also traced.

Introduction

Paleolimnological methods have frequently been used to study changes in trophic state as a result of climatic and human influence. We have studied eutrophication phenomena in Lake Ladoga using a sediment core from the deepest region of the lake where the rate of sediment accumulation was highest.

Material and methods

The sediment core was taken from a depth of 230 m in the north-west part of the lake. It was 180 cm long and consisted of a fine grey silt covered by a 2 cm thick layer of brown organic sediment. The upper unconsolidated 10 cm layer of mud was collected using an Eckman sampler and subsampled at 1 cm intervals. The lower section, 10–180 cm, was taken by a JOAN sediment corer and subsampled at 5 cm intervals. For every subsample granulometric, chemical (C, N, P, Fe, S, Mn, authigenic Si) analyses, organic matter content (LOI), chlorophyll 'a' concentration, diatom concentration (valves g^{-1} of sediment dry weight) and species composition were made. The sediment accumulation rate was calculated by taking into account the area of lake bottom covered by mud, mud thickness, the period of its accumulation, and the modern accumulation rate based on sediment trap data (Semenovich 1966), and the sediment inflow into the lake.

Results and discussion

The sediment accumulation rate in the deepest part of Lake Ladoga was determined as 0.4 mm a^{-1}. The sediment in the core was divided into 5 zones according to granulometric composition (median size), chemical parameters and diatom species composition (Fig. 1). Zone I, 120–180 cm, was formed 3 000–4 000 years B.P. It has the following characteristics: median grain size, 4.6–5.2 μm; LOI, 9.09%; carbon dioxide concentration, 3.1% CO_2; phosphate, 0.45% P_2O_5; iron 9.09% Fe_2O_3; authigenic Si, 14.10–15.64% SiO_2. The diatom concentration is highest in this zone – up to 3×10^6 valves g^{-1}. The dominant planktonic diatoms are *Melosira islandica* subsp. *helvetica* (50%) and *M. distans* var. *alpigena* (18%). *M. italica* and *Stephanodiscus as-*

Hydrobiologia 103, 261–263 (1983).

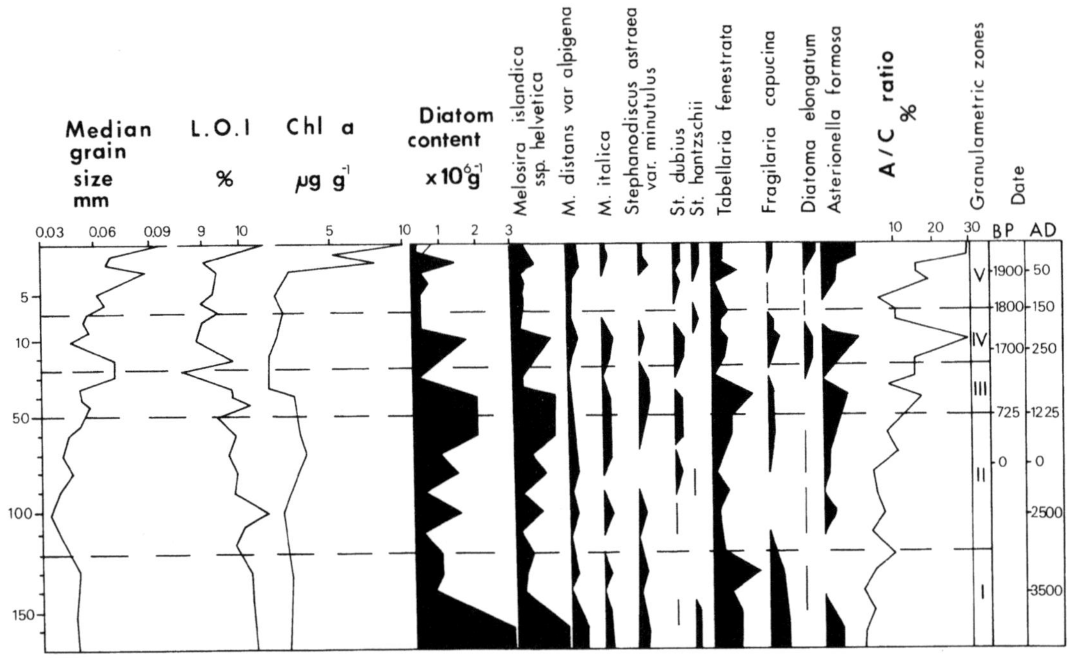

Fig. 1. Sediment Depth (cm), age, granulometric zones, median size of sediment, percent loss on ignition (LOI), percent authigenic SiO_2, chlorophyll *a* concentration ($\mu g\ g^{-1}$ dry weight), number of diatom valves $\pm 10^6\ g^{-1}$ dry weight (total and for individual species). Note change in vertical scale at 10 cm. Approximate dates are given in both B.P. and A.D.

traea var. *minutulus* are subdominants (5–10%). 89–95% of all valves comprise planktonic forms.

Zone II, 50–120 cm, was formed 3 000–1 250 years B.P. Sediments of this zone have the finest structure with a median size between 3.4 and 4.2 μm. There is a small increase in the content of biogenic elements: P_2O_5 increases to 0.52%, Fe_2O_3 to 9.81%, SO_3 to 2.49%, authigenic Si to 18.6%. The increase of chlorophyll 'a' concentration in the sediments in this zone indicates a rather high trophic status. The amount of *Stephanodiscus astraea* var. *minutulans, Melosira distans* var. *alpigena* and benthic diatoms decreases.

This zone is the period of the Ladoga transgression. It is characterized by high lake productivity and a high rate of allochthonous orgnic inflow from the catchment area. Though the main reason for the Ladoga transgression was the uplift of the northern part of the lake catchment it also coincided with a period of wet and rather warm climate.

Zone III, 15–50 cm, was formed between the years 725–1 600 A.D. The appearance of coarse sand in the sediments and the decrease of pelites indicates a marked decline in water level. The formation of the Neva River took place at the beginning of this phase and the maximum lake regression occurred in the 14–15th centuries when the climate was dry and cold. Authigenic Si values and chlorophyll *a* concentration decrease throughout the period. The maximum ratio of A/C of 15.7 indicates the low productivity of the lake. There is only one dominant diatom in the sediments, *Melosira islandica* subsp. *helvetica* (55–81%). It was an unfavourable period for other planktonic diatoms. This phase was characterized by the low productivity of the lake.

Zone IV, 7–15 cm, was formed between the years 1 600–1 800 A.D. The fine granulometric composition of the sediments of the 17th and the beginning of the 18th centuries is evidence for high water levels. The high LOI at this time is related to the influx of organic matter from newly flooded land. The high water level phase of the lake in the 17th century coincides with a wet and cool climate. The diatom content of the sediments is low, 0.5×10^6 valves g^{-1}. *Melosira islandica* subsp. *helvetica* is dominant, *Stephanodiscus astraea* var. *minutulus, Tabellaria fenestrata* and *Asterionella formosa* are subdominants. The expansion of *A. formosa* indicates the beginning of active human influence on the watershed.

Zone V, 0–7 cm, represents c. 1 800 A.D. to the present. The grain size curve suggests a decrease of lake level. During this period dry climatic phases of 50 years length alternate with wet climatic phases of 25 years. From our data a low water level occurred during the 1st quarter of the 19th century, in the second quarter (5–6 cm) there was an increase, and in the second half of the 19th century water level decreased during relatively dry and cold conditions. The sediments accumulated at this time contain the greatest sand fraction in the core.

For the 20th century we have lake level and climatic records. The mean length of wet climatic periods for this time period is 29 ± 5 years. An analysis of fluctuations of water level shows that a decrease of 1 m below the mean level causes a decrease in the littoral zone of the lake by 20%. The main effect occurs along the southern and south-eastern shores. From 1 900 to 1 925 the lake level was as high as 5.08 m.a.s.l. much higher than its mean value. The diatom concentration in the sediment is 1.4 ± 10^6 g^{-1}. *Melosira islandica* subsp. *helvetica* is the only dominant and *M. distans* var. *alpigena* and *Stephanodiscus astraea* var. *minutulus* are subdominants.

From 1 925 to the present there has been a long phase of lake level regression. This regression has been partly responsible for the appearance of coarse material in the sediments. A further reason, however, is the intensification of human activity in the catchment area of the watershed. The trophic level of the lake increases at this time. LOI increases to 10.7% and chlorophyll concentration rises to 10 $\mu g\ g^{-1}$. The diatom concentration is 1.85 ± 10^6 g^{-1}. In the uppermost 1 cm layer *Asterionella formosa* becomes a codominant with *Melosira islandica* subsp. *helvetica.* Significant amounts of the planktonic forms *Stephanodiscus dubius, St. hantzschii* and *Diatoma elongatum* indicates increasing eutrophication. Saprobic indices were calculated according to the Pantle-Buk-Sládeček technique for every sediment subsample. They are 1.19–1.33 in Zone I, 1.27–1.38 in Zone II, 1.38–1.41 in Zone III, 1.40–1.41 in Zone IV and 1.35–1.42 in Zone V. They show the oligosaprobic state of the lake but there is a tendency for the values to increase.

There are more than 10 radiocarbon dates for nearshore peat bogs and excavated wood (Liiva *et al.* 1978). The layers containing these remnants were covered by the deposits of the Ladoga transgression. Radiocarbon dates between 2 000 and 3 000 years B.P. (samples TA-362, TA-354, TA-422, TA-451, TA-603, TA-514) confirm our data concerning the maximum extent of the Ladoga transgression coinciding with the maximum water level at about 2 500 B.P. The dates $1\,710 \pm 60$, $1\,600 \pm 150$ and $1\,310 \pm 80$ (TA-450, TA-608, TA-610) which the authors considered as too young coincide with our idea of the Ladoga transgression as being long and persistent. The end of the transgression occurred about 1 250 years B.P. Differences in the position of the samples above the lake level are dependent on the varying rate of nearshore land uplift and consequently the different times of flooding of the lowlands around the lake. A very interesting date 470 ± 80 (TA-416) was obtained for charcoal from an ancient quarry near Pitkäranta. The charcoal was deposited during the lake regression after the formation of the Neva River (375–500 years B.P.). The soil horizon was later covered by sands during the rapid transgression from 1 600–1 725 A.D. The radiocarbon dates therefore confirm our stratigraphical conclusions on the recent history of Lake Ladoga.

References

Liiva, A., Ekman, J. & Rinne, T., 1978. Tartu Radiocarbon Dates VII. Radiocarbon 20 (3): 441–448.

Semenovich, N. I., 1966. Bottom sediments of Lake Ladoga-124, Leningrad (in Russian).

Diatoms in the plankton and sediments of two lakes of different trophic type

I. S. Trifonova & N. N. Davydova
Institute of Lake Research, Academy of Sciences of the U.S.S.R., Petrovskaya nab. 2, Leningrad 197046, U.S.S.R.

Keywords: paleolimnology, diatoms, plankton, eutrophication, sediment accumulation rate

Abstract

Species composition, quantity and distribution of diatoms in both the plankton and the surface sediments (0–30 cm) of mesotrophic Lake Krasnoye and eutrophic Lake Vishnevskoye (Karelian Isthmus) were studied. In the mesotrophic lake the composition of dominant diatoms (mainly *Melosira*) corresponded to those in the plankton. The Araphidineae/Centrales (A/C) ratio was 0.2–0.3%, increasing in the upper layers to 7%. Alkaliphilous and alkalibiontic forms constituted 60% of the total. In the eutrophic lake diatoms with thin valves (mainly *Synedra*) predominated in the plankton but their quantity in the sediments was insignificant in comparison with other plankton. Nevertheless, the A/C ratio was much higher, 17–35%. Alkaliphilous and alkalibiontic forms accounted for 78–90% of the total number of valves. In both lakes the highest number of diatom valves was registered in the upper layer of the sediments. From the ratio of the total number of diatoms in the upper 5 cm layer to their annual flux to the sediment from the plankton the approximate sediment accumulation rate was calculated to be 1.9 mm a^{-1} for the mesotrophic lake and 2.5 mm a^{-1} for the eutrophic one.

Introduction

The detailed study of diatom composition and distribution in the surface layers of lake sediments allows the process of lake eutrophication to be seen in a clear time perspective (Stockner & Benson 1967; Granberg 1972; Stockner 1975). Such investigations have been carried out on two lakes of the Karelian Isthmus. Lake Krasnoye is mesotrophic with an annual primary productivity of 100–120 g C m^{-2} and Lake Vishnevskoye is highly eutrophic with an annual productivity of 230 g C m^{-2} (Trifonova 1979). Judging from the thickness of their sediments (17 m in Lake Krasnoye and 12 m in Lake Vishnevskoye), both lakes have had a long evolution.

Materials and methods

Monoliths 20–30 cm long were taken in the central part of the lakes using Perfiljev's stratometer (small model). The sediment was frozen and divided into 0.5 cm sections. Every section was weighed and 1–2 g were used for quantitative diatom analysis. Besides the total number of diatom valves, the percentage relationship of the araphidinate to centric diatoms (Stockner 1975), and the percentage content of alkaliphilous and alkalibiontic forms, indicating changes in pH (Granberg, 1972), have been calculated for every level.

Hydrobiologia 103, 265–268 (1983).

Results and discussion

Diatoms in the plankton

Eighty-six species of diatoms have been determined in the phytoplankton of Lake Krasnoye. The Lake Vishnevskoye plankton is less diverse with 56 species. The main dominants and their seasonal succession are quite different in both lakes (Fig. 1). The total number of diatoms in the plankton during the annual cycle ranged from 10^5 to 10^7 cells l^{-1} in Lake Krasnoye and from 2.6×10^6 to 7×10^6 cells l^{-1} in Lake Vishnevskoye. The mean annual number was 1.2 and 2.3×10^6 l^{-1} respectively.

Diatoms in the sediments

Seventy-six diatom species have been found in the sediments of Lake Krasnoye. The dominants are euplanktonic species and *Melosira islandica* subsp. *helvetica* is the absolute dominant in all layers (Fig. 2a). The codominants are also *Melosira* spp. There is a quantitative correlation between these species in the sediments and those in the

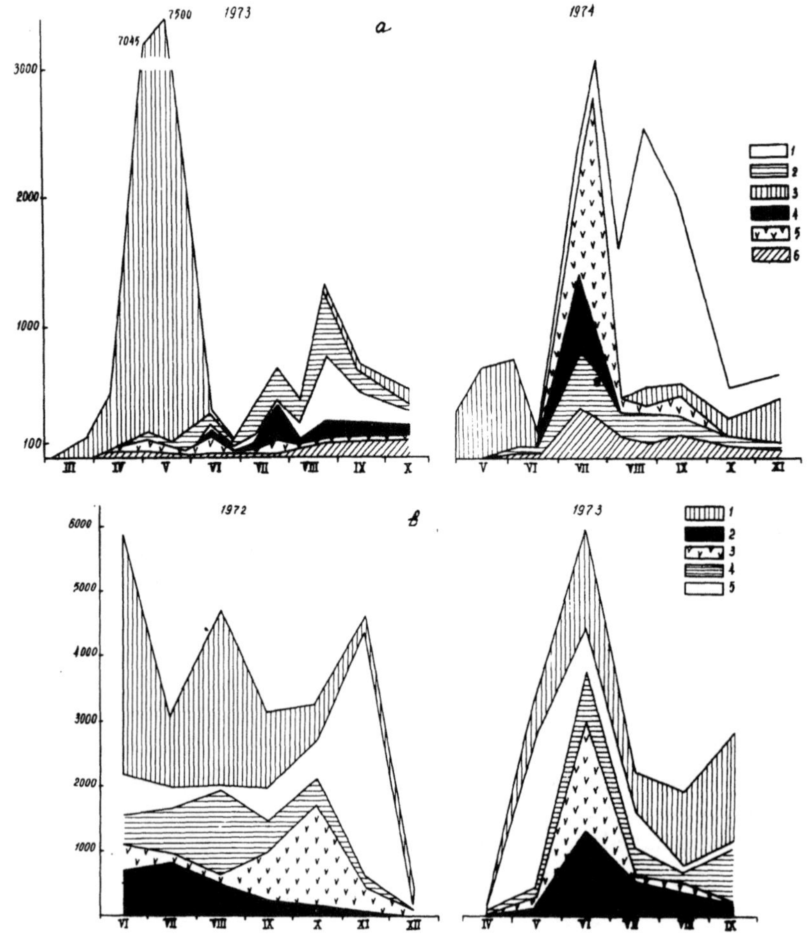

Fig. 1. Seasonal succession of dominant diatoms in the phytoplankton (cells × 10^3 l^{-1})

a. Lake Krasnoye: 1. *Melosira distans* var. *alpigena* Grün. 2. *M. granulata* var. *granulata* (Ehr.) Ralfs. 3. *M. islandica* subsp. *helvetica* O. Müll. 4. *Fragilaria crotonensis* Kitt. 5. *Asterionella formosa* Hass. 6. *Stephanodiscus astraea* (Ehr) Grün. b. Lake Vishnevskoye: 1. *Melosira ambigua* (Grun.) O. Müll. 2. *Fragilaria crotonensis* Kitt. 3. *Asterionella formosa* Hass. 4. *Synedra acus* Kütz. 5. *S. berolinensis* Lemm.

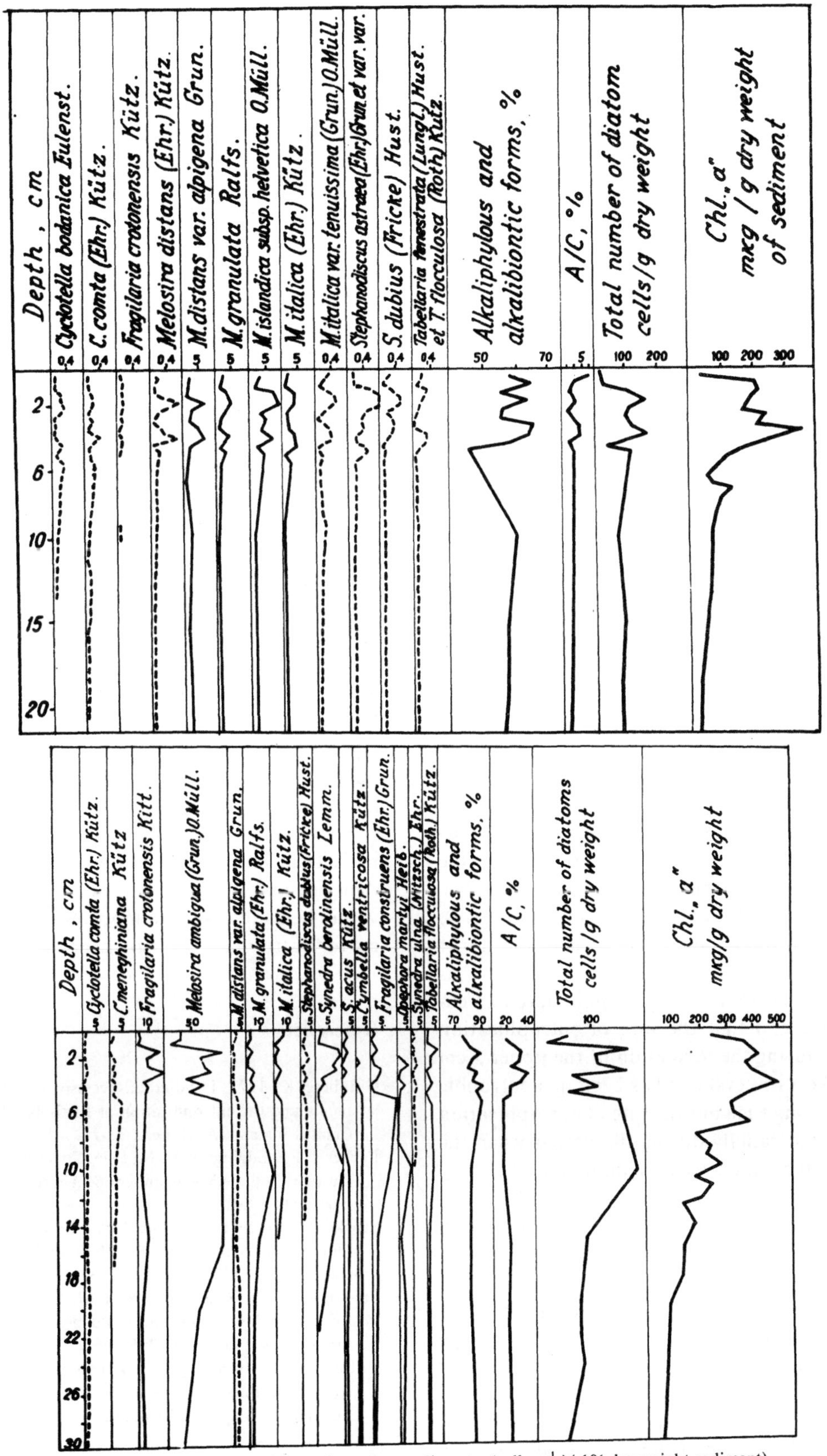

Fig. 2. Distribution of diatoms in surface sediments. (cells $g^{-1} \times 10^6$ dry weight sediment).
a. Lake Krasnoye, b. Lake Vishnevskoye.

plankton. In the upper 5 cm of core both the total number of valves and the number of valves of the planktonic dominants increase and *Fragilaria crotonensis* becomes abundant. The A/C ratio is very low (2–3%) in the lower layers of the sediments but increases up to 7% in the upper layers as the role of the pennate diatoms in the plankton becomes more important. The content of alkaliphilous and alkalibiontic forms is rather high at all levels increasing to more than 60% in the upper 5 cm layer. One hundred and thirteen diatom species have been found in the sediments of Lake Vishnevskoye, *Melosira ambigua* is the absolute dominant in all sediment layers (Fig. 2b). Subdominants are *Melosira italica, M. granulata, Synedra berolinensis* and *Fragilaria crotonensis.* A comparison of diatom composition between sediments and phytoplankton shows that only *Melosira* in the sediments corresponds closely to the plankton. The thin-walled diatoms which occur in the plankton, *Fragilaria, Synedra, Asterionella* are found in lesser quantity in the sediment. This difference is probably due to the rapid dissolution of thin-walled valves in the highly alkaline pH of the water (up to 9.4). In spite of this the A/C ratio of the surface sediment assemblage of Lake Vishnevskoye is about an order higher (17–35%) than in Lake Krasnoye and the highest values are in the upper 10 cm. The proportion of alkaliphilous and alkalibiontic forms in the sediments is also higher than in Lake Krasnoye, ranging from 78–90% of the total number of valves.

All the diatom changes observed in the upper levels of the sediment show that Lake Krasnoye has recently passed from an oligotrophic state to a eutrophic one and that the rate of eutrophication increased during the formation of the upper 5 cm layer. Lake Vishnevskoye has been in a eutrophic state for a long time but the rate of eutrophication has increased rapidly during the period when the upper 15–20 cm layer was accumulating.

The estimation of accumulation rate by diatom analysis

Having no data on the absolute age of the sediments we made an attempt to estimate the age indirectly using the method suggested by Stockner (1975).

The total number of diatom valves in the upper 5 cm layer of the Lake Krasnoye sediments has been calculated to be 37×10^6 cells cm^{-2}. The mean annual number of diatoms in the plankton is 8.4×10^6 cells cm^{-2} (in the 0–7 m water layer) and their annual production using a phytoplankton turnover of 3.3 (Trifonova 1979) is 28×10^6 cells cm^{-2}. According to the literature (Stockner 1975; Parker & Edgington 1976) the annual flux of diatom valves from the plankton to the sediments is about 5% of their annual production. Thus by dividing the total number of valves in the 5 cm layer by annual diatom flux from the plankton (1.4×10^6 cells cm^{-2}) we obtain a value of 26 years for the period of formation of this layer, and hence a sediment accumulation rate of 1.9 mm a^{-1}.

In the sediments of Lake Vishnevskoye quantitative correlations between the dominant speies are disturbed as a result of the intensive dissolution of thin-walled diatoms, making it impossible to use Stockner's method. Instead we used only species of the *Melosira* genus which are well preserved in the sediments. The total number of *Melosira* valves in the upper 5 cm layer was then 15×10^6 cells cm^{-2}. The annual production of diatom cells in the plankton has been calculated to be 14×10^6 cm^{-2} (in the 0–2 cm layer) and the annual flux is 0.7×10^6 cm^{-2}. Consequently, the 5 cm layer has taken 20 years to form at a mean rate of 2.5 mm a^{-1}.

These values are quite comparable with Stockner's data. The accumulation rate in several North American lakes ranged from 0.8 to 4.5 mm a^{-1}, the values being lowest for oligotrophic lakes and highest for eutrophic ones (Stockner 1975).

References

Granberg, K., 1972. The diatom successions in the recent sediments and the eutrophication of Ristiselkä, lake Päijänne, Central Finland. Aqua Fen.: 22–27.

Parker, I. I. & Edgington, D. N., 1976. Concentration of diatom frustules in Lake Michigan sediment cores. Limnol. Oceanogr. 21: 887–893.

Stockner, J. G. 1975. Phytoplankton heterogeneity and paleolimnology of Babine Lake, British Columbia, Canada. Verh. Internat. Verein. Limnol. 19: 2236–2250.

Stockner, J. G. & Benson, W. W., 1967. The succession of diatom assemblages in the recent sediments of Lake Washington. Limnol. Oceanogr. 12: 513–532.

Trifonova, I. S., 1979. Phytoplankton composition and productivity in Karelian Isthmus lakes of different type. Nauka, Leningrad, 168 pp. (in Russian).

Palynological investigations of the sediments of the north-western lakes of the Russian Plain and the Baltic Sea

V. I. Khomutova
Institute of Lake Research, Academy of Sciences of the U.S.S.R., Petrovskaya nab. 2, Leningrad 197046, U.S.S.R.

Keywords: paleolimnology, palynology, pollen diagram, Holocene

Abstract

In the last decade palynological investigations of the sediments of large lakes in the USSR (Ladoga, Onega, Kubenskoye, Vozhe, Lacha, Beloye) and the Baltic Sea have been carried out. Limnic and marine sediments are notable for their high pollen and spore concentrations and pollen diagrams from these deposits have a regional significance. The diagrams from Lakes Onega and Ladoga belong to the Kola-Karelian type while the sediments of Lacha, Vozhe, Kubenskoye and Beloye belong to the North-Russian type (Neustadt 1957). Three types of Holocene pollen diagrams are registered for the Baltic Sea. Modern and fossil pollen spectra reflect the composition of the zonal vegetation. Late-glacial sediments are found in several profiles of lake and sea deposits. Holocene deposits are divided into eight stratigraphic zones: Late-glacial sediments (DR_2-Al-DR_3) and Post-glacial sediments (PB, BO, AT, SB, SA). The zonations of lake and sea sediments can be clearly correlated.

The sediments of the north-western lakes of the Russian Plain and the Baltic Sea have been studied by pollen analysis for more than a decade. Investigations have been carried out both on the largest water bodies (Ladoga, Onega, Lacha, Vozhe, Kubenskoye, Beloye) and on the small lakes of the Karelian Isthmus (lakes Krasnoye, Lugovoye, Borisovskoye, Snetkovskoye and others) and of the Pribaltica (lakes Rogaizhu, Maloye Snidzinjas, Becheru, Vorkalu, Sleinovas). We have been particularly concerned with three aspects of the sediments. First, cores have been studied to determine geochronological, stratigraphical and paleogeographical problems. Second, the surface sediment samples (0–1 cm) have been analyzed in order to evaluate variations in pollen and spore sedimentation and to determine the composition of pollen spectra representing vegetation surrounding the water bodies. Finally, for lakes in the Pribaltica region the upper half-metre of sediments has been analyzed in greater detail by Müller's methods (Müller 1947) in order to assess the impact of man on vegetation in the recent past.

Lacustrine and marine sediments of the humid zone are, on the whole, notable for their high concentrations of pollen and spores. Sediments of different granulometric composition but containing extremely high amounts of microfossils are the most common type of sediments in the Baltic area and in the lakes of the North-West. The maximum pollen and spore concentration of the fine surface aleurite muds of the Baltic is 20 000 g^{-1} sediment, Lake Vozhe – 99 000, Lake Lacha – 28 000, Lake Kubenskoye – 22 000, Lake Beloye – 50 000. In small water bodies the pollen/spore concentration of bottom sediments is higher (Lake Maloye Lugovoye of the Karelian Isthmus – 400 000 g^{-1}, Lake Vorkalu in Latvia – 90 000). The high microfossil content of the sediment is due to the location of the lakes in forested areas. A detailed quantitative analysis of the microfossils has shown that their concentration varies considerably in sediments of

Hydrobiologia 103, 269–273 (1983).

the same granulometric composition. This may be due to hydrodynamic, biotic, and geochemical factors.

A comparison of the qualitative composition of microfossils and their distribution in the Baltic and in large lakes shows that the spectra primarily reflect the composition of the zonal vegetation (Fig. 1–3). For the lakes of the North-West and the Pribaltica the diagrams reflect middle and south taiga vegetation, whilst the Baltic diagrams represent three zones. In the northern part of the basin the spectra reflect the formations of the tundra zone, in the central and south-east region they reflect the taiga zone, and within the southern part of the area they reflect the nemoral vegetation of adjacent regions (Khomutova 1977).

These pollen spectra from aquatic sediments differ from spectra from terrestrial deposits in a number of ways. Pollen in the sediments of the large aquatic systems have a poor floristic composition.

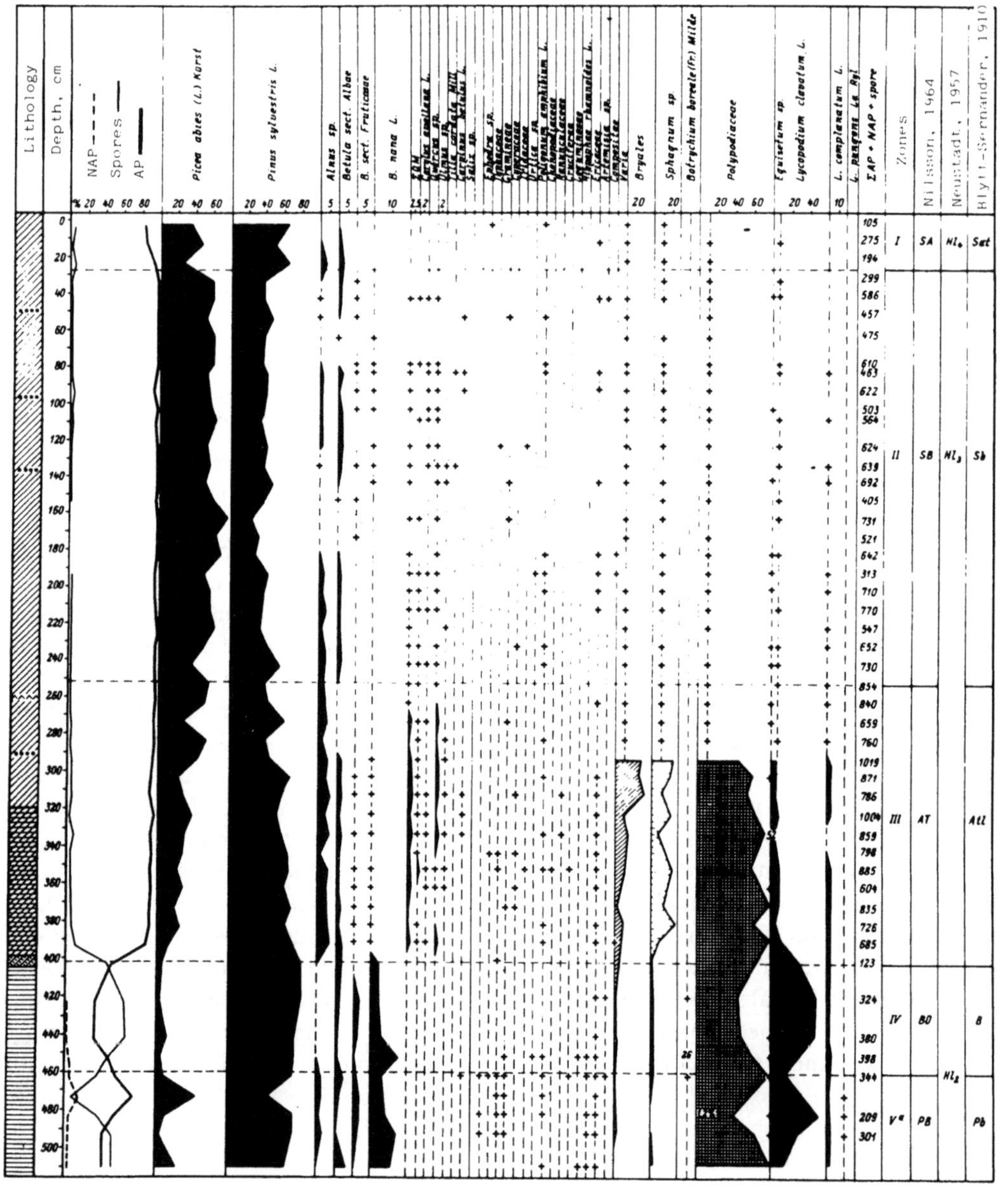

Fig. 1. Pollen diagram of a core from Lake Onega.

The content of herbaceous plant pollen is insignificant, while among spore plants only *Bryales, Sphagnum sp.* & *Polypodiaceae* are present. Local elements, which help to interpret the history of terrestrial systems (especially peats) do not occur in the sediments of large lakes although their importance increases in sediments of small water bodies. This has been found particularly in lakes of the Latgalian elevation (Fig. 4) and the Karelian Isthmus. An additional characteristic of the sediments of the Baltic and large lakes is the presence of hiatuses, probably due to different hydrodynamic processes in the water bodies in the past.

A large number of cores from both lakes and the

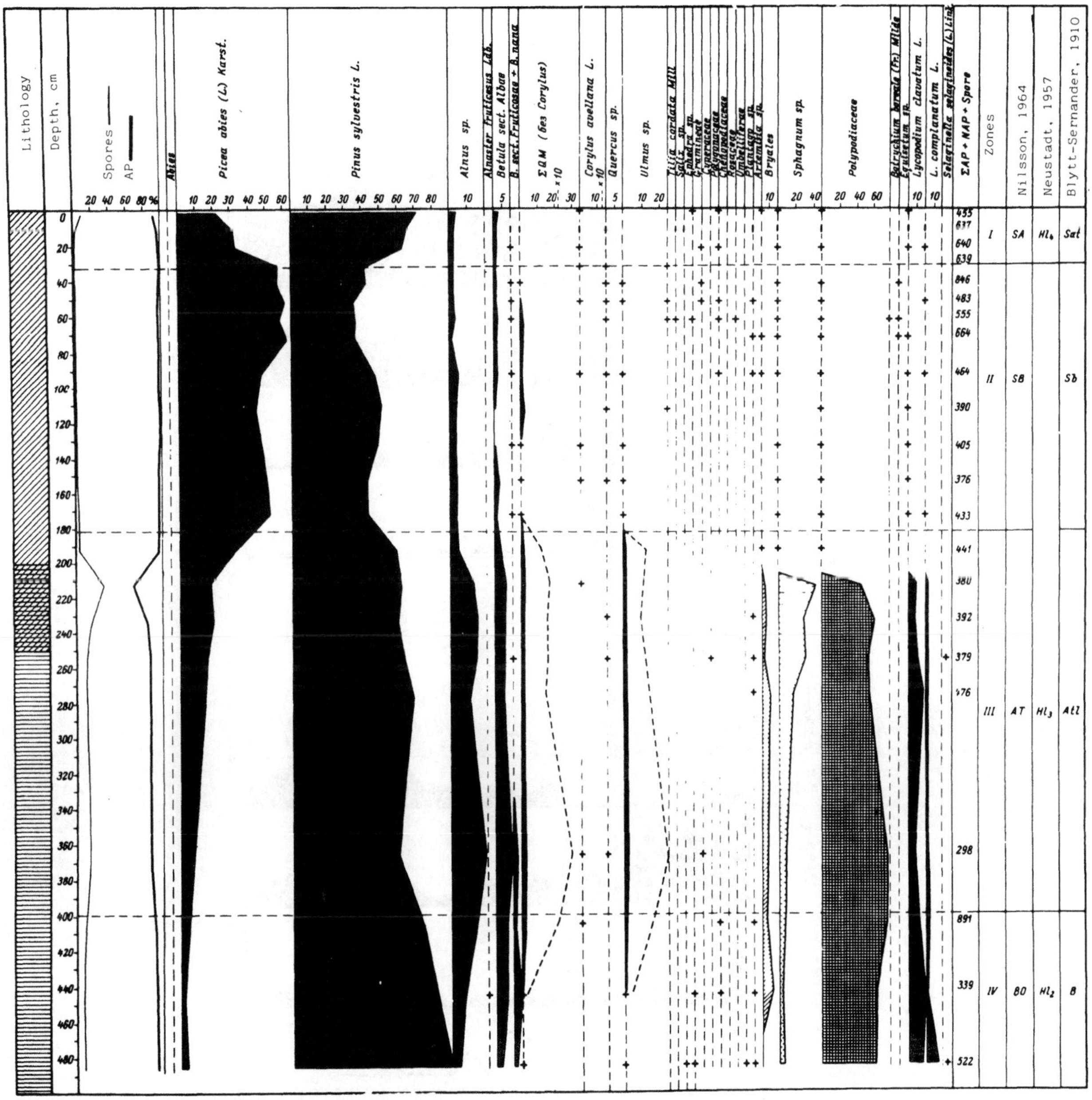

Fig. 2. Pollen diagram of a core from Lake Ladoga.

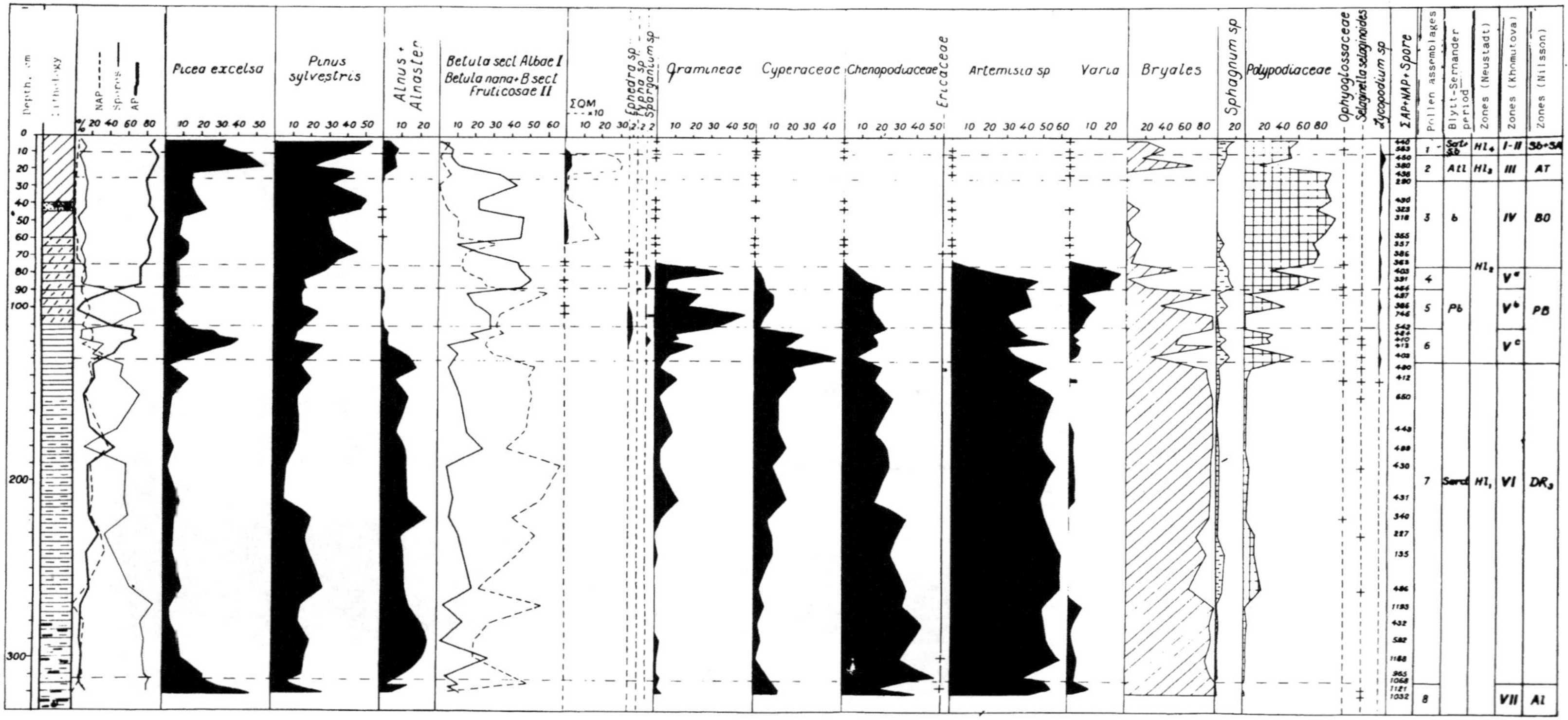

Fig. 3. Pollen diagram of a core from Lake Kubenskoye.

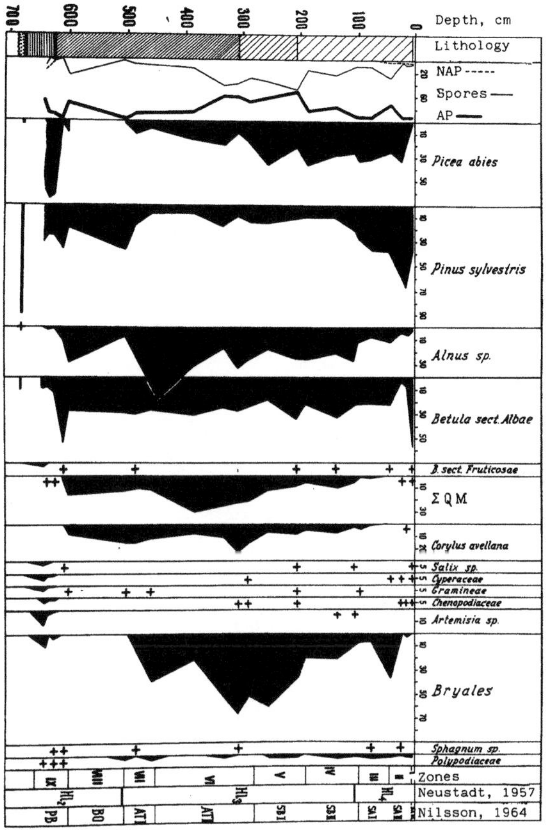

Fig. 4. Pollen diagram of a core from Lake Vorkalu (Latvia).

Baltic Sea have been analysed back as far as the Late-glacial period. The deposits have been divided into eight stratigraphic zones: Late-glacial sediments (DR_2-Al-DR_3), zones VIII–VI, and post-glacial sediments (PB, BO, AT, SB, SA) zones V–I (Khomutova 1976, 1977). There is a clear correlation between lacustrine and marine sediments of the same age in the separate zones of the Holocene.

Because of the lack of locally produced pollen in the sediments of the large water bodies, the pollen diagrams more accurately reflect regional changes in vegetation cover than pollen diagrams from terrestrial sites of the North-West. So the diagrams from the Onega and Ladoga lakes belong to the Kola-Karelian type and the diagrams of Lacha, Vozhe, Kubenskoye sediments to the North-Russian type according to the classification of Neustadt (1957). The pollen diagrams from the Baltic Sea show that the proposed regional divisions according to surface assemblages also holds for the Holocene (Kleimenova & Khomutova 1979).

Pollen diagrams of sediments from large water bodies are better for regional correlations than those from continental sites. They accord with Hafsten's proposal (Hafsten 1971) that for correlation purposes all local elements should be excluded from pollen diagrams.

References

Hafsten, H., 1971. Holocene zonal system of the world (in Russian). Theses of reports at III International Palynologic Conference, Novosibirsk.

Khomutova, V. I., 1976. Geochronology of bottom sediments according to the results of palynologic analysis. In: Palaeolimnology of Lake Onega (in Russian). Nauka, Leningrad.

Khomutova, V. L., 1977. Palynologic characteristic of Post-glacial bottom sediments of central and south-eastern regions of the Baltic Sea (in Russian). Baltica No. 6, Vilnius, pp. 17–31.

Kleimenova, G. I. & Khomutova, V. I., 1979. Some peculiarities of spore-pollen spectra of marine and lacustrine sediments (in Russian). Vestnik of Leningrad State University, No. 6, pp. 111–113.

Neustadt, M. I., 1957. Forest History and Palaeogeography of the USSR in Holocene (in Russian). Moscow, 403 pp.

Nilsson, T., 1964: Standartpollendiagramme und C14-Datierungen aus dem Ageröds Mosse in Mittleren Schonen. Lunds Univ. Årsskr. N. F. (2), 59: 7, 1–52.

The Middle-Valdaj Lakes in the Western part of the Bolshezemelskaja tundra

Emma Loseva
Komi Branch of the Academy of Sciences of the U.S.S.R., Institute of Geology, 167610 Syktyvkar, U.S.S.R.

Keywords: paleolimnology, freshwater diatoms, Middle-Valdaj

Abstract

This paper presents results of diatom analyses of dated Upper-Pleistocene lacustrine deposits from the Shapkina river in the Bolshezemelskaja tundra of the USSR. Both diatom systematics and palaeoecology have been studied.

A marine diatom assemblage was found in the lower part of the sedimentary sequence and a rich freshwater assemblage, including more than 200 taxa, was found in the upper part. Some of the characteristic diatoms are typical of standing or slowly-moving waters of more southern and western districts but are not found in the Bolshezemelskaja tundra lakes today.

According to ^{14}C dating, the age of the deposit is about 40 000 years, and represents the Middle-Valdaj (Middle-Weichselian). Palynological and diatom analysis suggest that the climate was similar to the present, that of an interglacial.

Discussion

The Bolshezemelskaja tundra is situated in the north-eastern European part of the USSR. The Pleistocene sediments of the western part are dissected by the Shapkina river. 60–170 km from the mouth of the river layers of peaty sands and aleurite are widely distributed. They have a thickness of about 15 m and are covered by a layer of till more 15 m thick. According to ^{14}C (Loseva & Arslanov 1975) and palynological analysis (Berdovskaja & Loseva 1975) the peaty sands and aleurites are about 40 000 years old and represent the Middle-Valdaj period (Middle-Wurm).

Diatom analysis of three sections indicate the characteristic assemblages of the basins in which the peaty sands and aleurites are accumulated (Loseva 1978). Characteristic species are illustrated in Figs. 1 & 2.

The first section (no. 569) is situated on the west bank of the Shapkina river, 143 km from the mouth and 4 km above the Janguta tributary. The terrace is more than 14 m high and about 32 m above sea level. The upper 11 m consist of alluvial sands and gravels, and below this layers of peat (0.25 m) and of laminated peaty clay (3 m) can be seen. 20 samples were analyzed for diatoms and the 7 samples from the lower 1.5 m contained a rich diatom assemblage (more than 180 taxa).

At the bottom of the section more than 30 brackish-water and marine forms were found together with freshwater ones. Usually the marine forms are well preserved and have an abundance of 4 and 5 ('frequent' and 'very frequent') on the 6-number scale of abundance. They constitute about 82% of the valves in the samples. The dominants are: *Paralia sulcata* (Ehr.) Kutz. + var. *biserata* Grun., *Thalassiosira excentrica* (Ehr.) Cl., *Raphoneis amphiceros* Ehr., *R. rhombica* (Grun.) Andrews, *R. surirella* (Ehr.) Grun., *Diploneis interrupta* (Kütz.) Cl., *Scoliopleura tumida* (Bréb.) Rabenh., *Pleurosigma angulata* (Queck.) W. Sm., *Nitzschia acu-*

Hydrobiologia 103, 275–279 (1983).

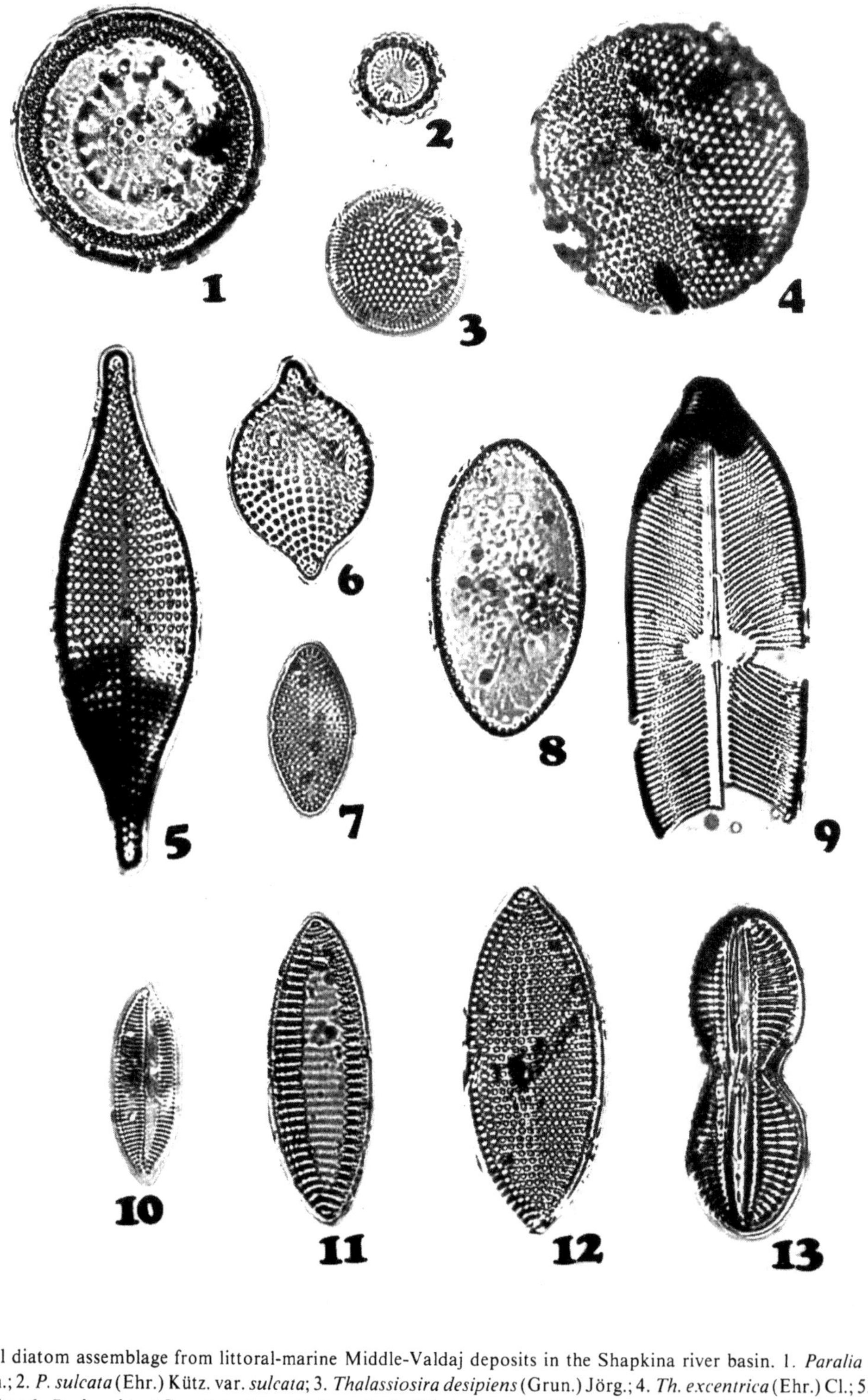

Fig. 1. Typical diatom assemblage from littoral-marine Middle-Valdaj deposits in the Shapkina river basin. 1. *Paralia sulcata* var. *biseriata* Grun.; 2. *P. sulcata* (Ehr.) Kütz. var. *sulcata*; 3. *Thalassiosira desipiens* (Grun.) Jörg.; 4. *Th. excentrica* (Ehr.) Cl.; 5. *Raphoneis amphiceros* Ehr.; 6. *R. rhombica* (Grun.) Andrews; 7. *R. surirella* (Ehr.) Grun.; 8. *Chaetoceros mitra* (Bail.) Cl.; 9. *Navicula humerosa* Bréb.; 10. *Nitzschia acuminata* (W. Sm.) Grun.; 11. *N. navicularis* (Bréb.) Grun.; 12. *N. punctata* W. Sm.; 13. *Diploneis interrupta* (Kütz.) Cl.

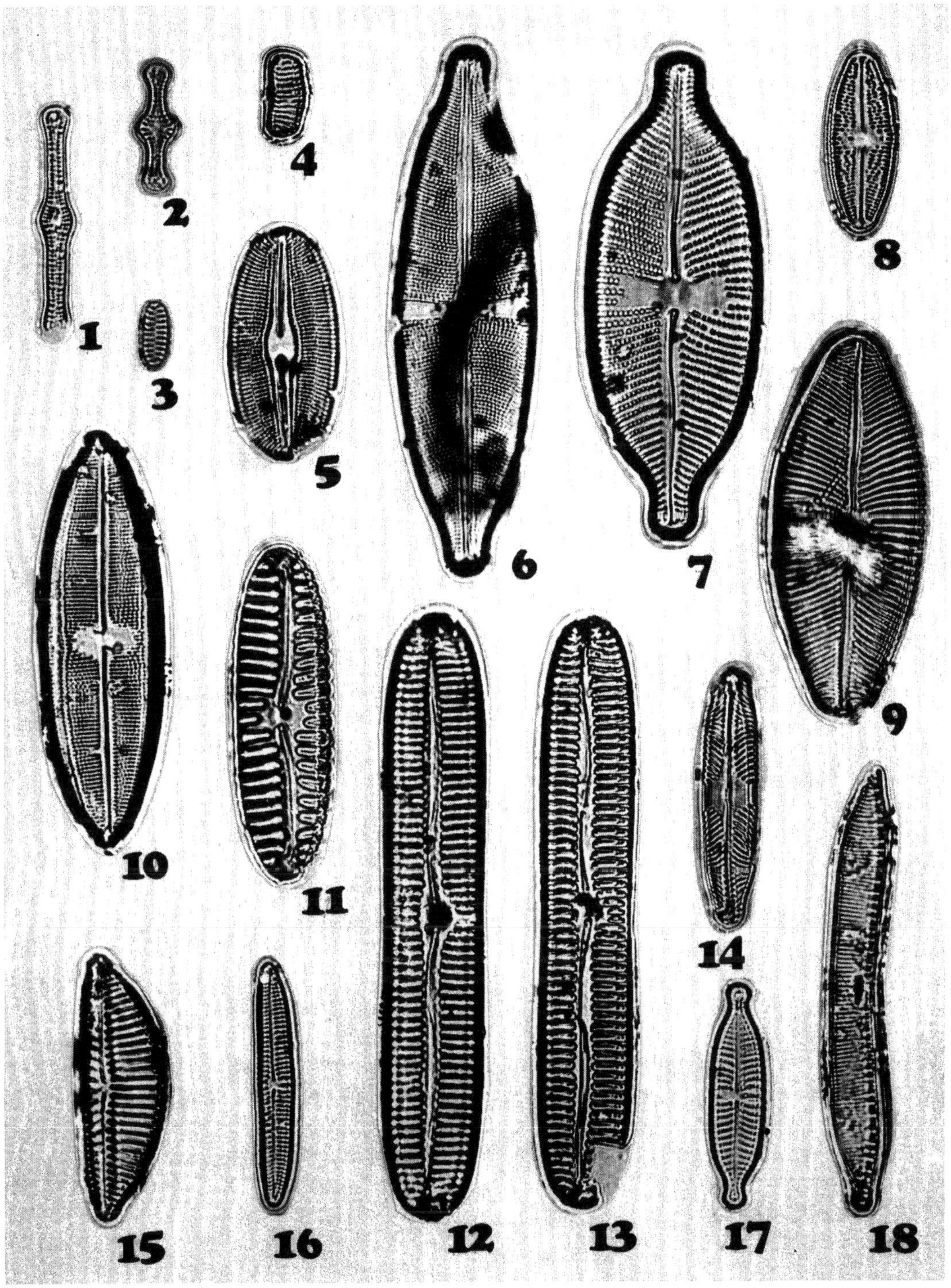

Fig. 2. Typical diatom assemblage from lacustrine Middle-Valdaj deposits in the Shapkina river basin. 1. *Tabellaria fenestrata* (Lyngb.) Kütz.; 2. *T. flocculosa* (Roth.) Kütz.; 3. *Fragilaria pinnata* Ehr.; 4. *Eunotia praerupta* Ehr.; 5. *Diploneis ovalis* (Hilse) Cl.; 6. *Stauroneis phoenicenteron* Ehr.; 7. *Navicula amphibola* Cl.; 8. *N. mutica* Kütz.; 9. *N. semen* Ehr.; 10. *Neidium iridis* (Ehr.) Cl.; 11. *Pinnularia lata* (Bréb.) W. Sm.; 12. *P. viridis* var. *intermedia* Cl.; 13. *P. viridis* var. *dispar* Schum.; 14. *P. microstauron* (Ehr.) Cl.; 15. *Cymbella turgida* (Greg.) Cl.; 16. *C. norvegica* Grun.; 17. *Cymbella* sp.; 18. *Hantzschia amphioxys* (Ehr.) Grun.

minatum (W. Sm.) Grun., *N. navicularis* (Bréb.) Grun. All are widely distributed in the littoral of modern seas and estuaries. Their good preservation, composition and abundance indicate an *in situ* assemblage.

At 0.5 m above the present river level the diatom composition is also mixed; brackish-water and marine diatoms constitute about 40% of the valves. The dominants are: *Diploneis ovalis* (Hilse) Cl., *D. interrupta* (Kütz.) Cl., *Nitzschia navicularis* (Bréb.) Grun., *Scoliopleura tumida* (Bréb.) Rabenh. and *Caloneis formosa* (Greg.) Cl.

Higher still the marine species disappear. At a height of 1–1.3 m above water-level a rich freshwater diatom assemblage is found in the laminated humic clay. The dominants are: *Fragilaria virescens* Ralfs, *Navicula cuspidata* Kütz., *Amphora ovalis* var. *libyca* Ehr. and others. In the uppermost layers of the clay and in the peat, gravel, and alluvial sands diatoms are absent.

The second section (no. 572) is situated on the west bank of the Shapkina river, 46 km downstream from the first and 97 km from the mouth. In the 25 m terrace the upper 15 m are represented by till and the lower 10 m by peaty sands and aleurite. 28 samples were analysed for diatoms. 14 samples from the sands and aleurites at 4.5–7.5 m above the water level contain a rich diatom assemblage (more than 150 taxa). The dominants are: *Tabellaria fenestrata* (Lyngb.) Kutz., *T. flocculosa* (Roth.) Kutz., *Navicula amphibola* Cl., *Pinnularia microstauron* var. *brebissonii* f. *diminuta* Grun., *Cymbella heteropleura* var. *minor* Cl., *C. norvegica* Grun., *C. turgida* (Greg.) Cl., *C. ventricosa* Kütz., *Hantzschia amphioxys* (Ehr.) Grun. + var. *major* Grun. Many other forms have an abundance of 5.

The third section (no. 573) is situated on the east bank of the Shapkina river, 23 km downstream from section 2, and 76 km from the mouth. The height of the terrace is 18 m. At a depth of 4.5 m, under boulder clay, layers of peaty sands and aleurite can be seen. 19 samples were analysed for diatoms. A rich diatom assemblage (more than 140 taxa) was found at a depth of 6–8.3 m. The dominants are: *Tabellaria flocculosa* (Roth.) Kutz., *Eunotia praerupta* Ehr., *Cymbella heteropleura* var. *minor* Cl. Many species have an abundance of 5.

Thus, the sandy-aleurite layers exposed under till in the Shapkina valley 60–170 km from the mouth of the river are characterised by rich diatom assemblages. A littoral-marine assemblage was found at the base of one section and freshwater assemblages were found in all three sections.

The freshwater assemblage includes 31 genera and more than 210 taxa. The genera *Navicula* (about 40 taxa), *Pinnularia* (30), *Eunotia* and *Cymbella* (about 20) are the most abundant. The species *Fragilaria pinnata* Ehr., *Eunotia praerupta* Ehr., *Stauroneis phoenicenteron* Ehr., *Navicula amphibola* Cl., *Pinnularia viridis* var. *dispar* Schum. + var. *intermedia* Cl., *Cymbella heteropleura* var. *minor* Cl. are well distributed and have abundances of 5 and 6. Most taxa found are widely distributed in modern basins of various types, especially in standing or slowly-moving waters, and are mainly benthic or epiphytic forms. Their predominance is characteristic of tundra lakes (Stenina 1978). Of the plankton only *Tabellaria fenestrata* has an abundance of 5. Weakly silicified valves of planktonic diatoms may not have been preserved. The peculiarity of this assemblage is the almost total absence of centric forms e.g. of the genera *Melosira, Cyclotella* or *Stephanodiscus*, and the presence of a large group of species which are unknown in modern Bolshezemelskaja tundra lakes. They are: *Eunotia monodon* Ehr., *Diploneis ovalis* (Hilse) Cl., *Navicula gibbula* Cl., *N. lagerstedtii* Cl., *N. semen* Ehr., *Pinnularia episcopalis* Cl., *P. lata* (Bréb.) W. Sm., *P. viridis* var. *dispar* Schum., *Neidium affine* var. *amphirynchus* (Ehr.) Cl., *Cymbella norvegica* Grun., *Gomphonema angustatum* var. *producta* Grun. + var. *crassula* (Naeg.) Hust. Some of them (e.g. *Navicula semen, Pinnularia episcopalis, Gomphonema bohemicus, Denticula tenuis* var. *crassula*) are typical for modern basins of more western and southern districts. Their high abundance in the terrace sequences points to warmer water conditions in the basins at the time of deposition.

In conclusion, the diatom analyses indicate that a shallow sea occupied the lower part of the Shapkina river at the beginning of Middle-Valdaj (Middle-Weichselian). The sea was then replaced by a system of freshwater lakes (not less than 40 km long). The water depth in these lakes changed repeatedly and eventually the lakes turned into a series of isolated bogs. The climate was similar to the present corresponding to interglacial conditions, a conclusion that was confirmed by palynological analyses.

References

Berdovskaja, G. N. & Loseva, E. I., 1975. Pleistocene lakes in the Shapkina river basin. The history of the lake in the Pleistocene: 114–119.

Loseva, E. I., 1978. The Middle-Valdaj sea-lake in the western part of Bolshezemelskaja tundra. Bull. Commission for Study of the Quaternary Period. 48: 103–112.

Loseva, E. I. & Arslanov, Kh. A., 1975. Middle-Valdaj horizon in the West of Bolshezemelskaja tundra. Geology and minerals of the North-Eastern european part of the USSR. Yearbook-1974, Institute of Geology, Kom Branch, Acad. Sci. USSR: 82–86.

Stenina, A. S., 1978. Diatoms. Flora and fauna of North-European lakes (Bolshezemelskaja tundra). Leningrad, Publishers 'Nauka': 21–26; 117–136.

Sediment formation in Lake Baikal

Vera A. Belova, B. F. Lut, L. P. Loginova & G. K. Khursevich
Limnology Institute, Listvenichnoye at Baikal, Irkutsk Region, P.O. 666016, U.S.S.R.

Keywords: paleolimnology, Lake Baikal, Pliocene, Pleistocene, pollen, diatoms

Abstract

The following stages in the formation of sediments in Lake Baikal have been distinguished. In the Middle Miocene lacustrine deposits accumulated in the southern depression. In Early and Middle Pliocene, as well as in Early-Middle Pleistocene the bottom of the southern and middle Baikal depressions was occupied by large lakes, but the northern basin was mainly dry with a system of small lakes, streams and rivers. This accounts for the great facies variety and the genetic heterogeneity of sediments in the northern depression. At the end of Middle and in the Late Pleistocene an abrupt warping of the depression took place, and a shore-line similar to the present one was formed. Sands in the upper part of the sediments are considered to be the product of intensive erosion which probably occurred in the pluvial epoch in the Late Pleistocene period.

Introduction

The sediments of Lake Baikal comprise a complex heterogenous and heterochronous formation. The sequence of geological events is not distinct because of the mixed facies composition of the sediments, their varying thickness, and the complex structure of the crystalline basement which influences the distribution of sediments within the lake.

An integrated study of the lake bottom, including seismographic analysis showed that the crystalline basement consisted of three grabens. Along with some smaller structures these grabens constitute the Southern-Baikal, Middle-Baikal and Northern-Baikal depressions respectively. In all these depressions unconsolidated sediments are distributed extremely irregularly. For example, from the seismic data (Lut 1964; Florensov *et al.* 1978) the thickness of sedimentary strata at the southern end of Lake Baikal is 200–300 m, whilst at the Angara River outlet it is 1 500 m and in the Selenga River delta it is 3 000 m. Tectonic processes are of great significance in the distribution of the sediments. Irregular dipping of different blocks of the depression results in stratigraphic discordances between the underlying Neogene. Pleistocene and Holocene sediments and the present-day silts. In some parts of the depressions, particularly on western slopes, outcrops of crystalline basement are superposed by a thin layer of recent mud.

Of particular importance for the formation of sediments is the character of material supply. Extremely rapid movements of debris cones, mudflows, landslide processes, and turbidity flows create a mixed facies composition and an incomplete geological record.

Stratigraphy

The oldest sediments occur on the south-eastern submarine slope of the southern edge of the lake. They are represented by dense clays of grey, blue-grey, yellow and brown colour. As a rule the clays possess insignificant intercalations of sand. In some places outcrops of carbonaceous clay schists are

Hydrobiologia 103, 281–285 (1983).

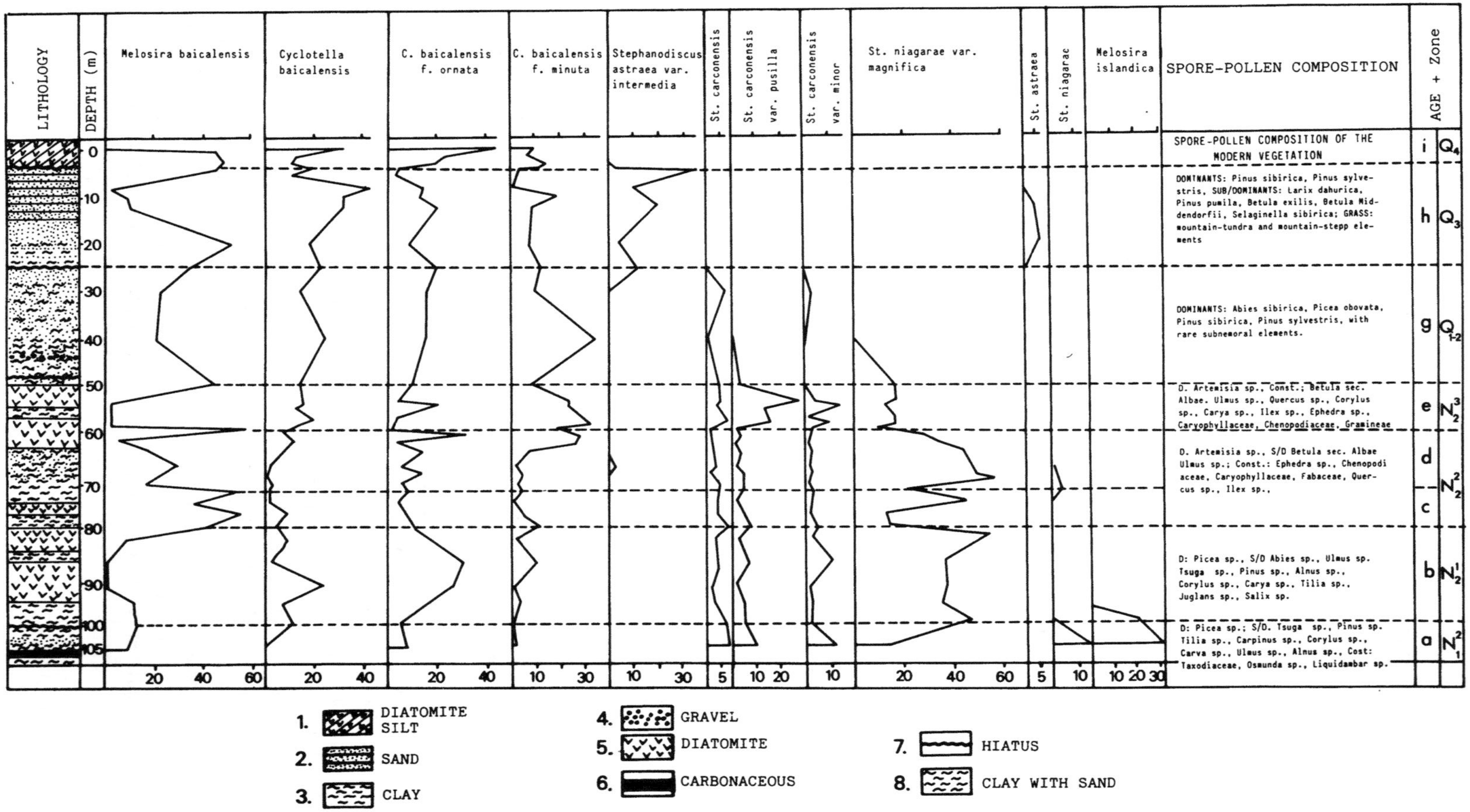

Fig. 1. The biostratigraphy of the bottom deposits of Lake Baikal.

observed. The total thickness of these sediments is about 10 m (Goldyrev *et al.* 1975). A palynological analysis of clays and carbonaceous sediments found at the base of this sequence (Fig. 1, zone a) showed similarities with lacustrine and telmatic sediments of Middle-Miocene age (K-Ar dating) from the south-eastern shore of Baikal (Belova 1975).

The next part of the stratigraphic sequence consists of dense bluish-grey clays which contain diatomite intercalations. Their thickness ranges from 0 to 5 metres. The total thickness of sediments as determined by G. S. Goldyrev is more than 12 m. They occur in the northern segment of Maloye More, along the submerged Academic Range and in several sections in the middle and southern depressions. Data presented by L. P. Loginova & G. K. Khursevich show that the main diatoms in the diatomites belong to *Stephanodiscus: Stephanodiscus niagarae* var. *magnifica, St. carconensis* and varieties especially var. *pusilla.* these species are completely absent in present day water bodies in the USSR.

The abundance of these taxa in the sediments suggests that the climate was warm at that time and consequently the sediments are probably of Pliocene age. Within the period, Early, Middle and Late Pliocene sediments can be identified.

Early Pliocene deposits are found along the south-eastern end of the South-Baikal depression. They occur with a stratigraphic discontinuity on sediments of the Middle Miocene period. The diatom and pollen content of this material is shown in Fig. 1, zone b. A similar flora was found in lake sediments of the upper horizon of the Kluevsky facies complex and the 'blue' Anosovsky clays is dated as Early Pliocene (Belova 1975).

The thickest sediments (4–5 m) of this age are found in the region of the Academic Range between the northern and middle depressions. They are represented by an alternation of clays and diatoms. The most common diatom is *Stephanodiscus niagarae* var. *magnifica* (Fig. 1c).

Similar diatom assemblages were found in blue clays in parts of the middle of Olkhon Island. These clays are covered by subaerial sediments with the remains of rodents of Middle Pliocene age (Adamenko 1975; Mats & Pokatilov 1976). Bottom sediments and coastal exposures are characterized by lacustrine deposits which contain pollen and spores with the following composition: dominants: *Picea* sp., subdominants: *Abies* sp., *Ulmus* sp., and associates: *Tsuga* sp., *Pinus* sp., *Corylus* sp., *Carya* sp. *Tilia* sp., *Juglans* sp., *Salix* sp. (Belova & Endrikhinsky 1980).

The distribution of clays of great thickness and their associated diatom assemblage make it possible to reconstruct the contours of the Early Pliocene basin of the lake. They embrace the middle and northern parts of Maloe More and extend nearly to the Barguzin Bay. An extensive lake existed in Early Pliocene time on the south-eastern coast of Baikal. The contours of this water body do not coincide with those of the present-day lake which is more to the south. The climate of the Early Pliocene was warm and humid.

A sharp change to more arid conditions accounts for the absence of diatoms in the beds of the dense grey-bluish clays, 1–1.5 m thick (Fig. 1, zone d).

Sediments contain the following pollen and spores: dominants – *Artemisia* sp., subdominants – *Betula* sec. *Albae, Ulmus* sp., associates: *Ephedra* sp., *Chenopodiaceae, Caryophyllaceae, Fabaceae, Quercus, Ilex* sp. This assemblage is similar to that from the Middle Pliocene of Priolkhonye in its floristic composition (Volkova & Belova 1981). Diatoms with this pollen flora have been found together only on the Academic Range. The Middle Pliocene desiccation was likely to cause a decrease in the area of the lake basin and a reduction in the diversity of the diatom flora. Lake deposits within the Maloe More area are disturbed by subaerial erosion and the development of gorges, slopes etc.

A marked change in the composition of the subdominant diatoms is observed (2–3 m thick) on the Academic Range (Fig. 1, zone e). The palaeofloristic assemblage represents a Late Pliocene stage of the lake at the boundary between the northern and middle Baikal depressions. The area of this basin is much smaller than that of the Early Pliocene and evolved under conditions of dry climate, steppe and wood-steppe landscape.

Analogous sediments of the same age have not been found in the southern and northern basins of the Baikal. They were probably buried beneath younger sediments of considerable thickness.

The following section of the lithostratigraphy of Lake Baikal is represented by very dense gray clays. They include an admixture of inwashed sands and gravel. In places sand and gravel make up small intercalations. In the upper segment of the section there is a distinct alternation of sands and clays with a thickness close to one metre. According to Goldyrev (Goldyrev *et al.* 1975) dense clays possess evidence of washout on the contact in which sands occur. Sediments of such a composition are widely distributed in all depressions of the lake. The total thickness of these sediments is more than 10 m.

A palynological study of sediments in this part of the section has been performed (Fig. 1, zone g) and as a result the second and the third pollen assemblage zones have been recognized (Belova 1975). They are dated as Late Pliocene and Early Pleistocene sediments. The entire stratum is characterized by the dominance of the *Melosira baicalensis,* in the lower part of the section, and *Cyclotella baicalensis* in the upper part. There is also a constant presence of *Stephanodiscus carconensis.*

Of particular importance to the general stratigraphy of the Lake Baikal sediments are sands, which are widespread in the Baikal depressions. They are of polymictic character, with components of lighter material (quartz, feldspars) predominant. In the sand strata Goldyrev found cross-bedding, essentially rounded grains and a good grading. All these factors are evidence for moving water and a current system existing during the period of formation of the sandy stratum (Goldyrev *et al.* 1975).

Diatoms are confined to the fine-aleurite sediments and have quite a different composition than in the underlying sediments (Fig, 1, zone h). There is a general predominance of *Melosira baicalensis* and *Cyclotella baicalensis* with its varieties in the upper part. A constant component of the assemblage is *Stephanodiscus astraea* and varieties, which occurs quite infrequently in the underlying horizons. This species has not yet been observed in present-day Baikal sediments. *Stephanodiscus astraea* var. *intermedia* is specific to this zone; it has been found neither in older sediments nor in recent ones. From experience so far this species, and varieties, may be considered as a representative one for sand beds.

The composition of pollen and spores is very characteristic for these sediments. Over the entire water area of Baikal the sand beds are characterized by a predominance of pollen of *Pinus sibirica* and *Picea obovata.* Among the shrubs *Betula middendorfii* and *Betula exilis* are of special interest. The grasses show that mountain-tundra and mountain-steppe elements are widespread. All these data show that a comparatively large amount of precipitation took place over Baikal. It is likely that sandy sediments in Baikal are the products of intensive water erosion, which may have taken place in a pluvial epoch of the Late Pleistocene. This epoch is associated with the culmination of mountain-valley glaciation in the region studied (Belova 1973, 1975).

Conclusion

In the developmental history of lacustrine basins within Lake Baikal the following stages are known. Middle-Pliocene time was characterized by the presence of a rather deep lake in the south-eastern part of the basin. Its deposits are only partly buried under the sediments of the present lake. They are represented by marginal lacustrine swampy formations.

In the Early Pliocene period a comparatively deep and extensive water body existed in the southern and middle parts of the basin. The diatom assemblage resembles those in the rift-valley lakes of Africa, Japan and USA, and indicates the similarity of climatic conditions that occurred in the early Baikal stage of rift genesis.

The formation of a single lake in the Lake Baikal basin probably occurred in the Late Pliocene–Early Pleistocene. The complexity of the stratigraphic structure of the sediments of this time is associated with sedimentation processes during glaciation.

With regard to the history of the Baikal depression two stages are distinguished. A slow subsidence is typical of the first (Middle Miocene–Early Pliocene) stage. It was balanced by the accumulation of lacustrine sediments in the southern depression. During the second (Neo-Baikal) stage two periods are known. In the early period (Early–Middle Pliocene) the bottom of the southern and middle parts were occupied by rather big lakes. The northern one was dry land with a system of small lakes, streams and rivers. In the second period (Early–Middle Pleistocene), due to an abrupt warping of the depression the present shore-line formed, the western side of the depression was elevated and along the eastern shore a lacustrine-alluvial plain formed that became the source of terrigenous material that was carried to the bottom of the lake through a system of submarine canyons.

References

Adamenko, R. S., 1975. Late Pliocene mammals in new discoveries of the Upper Lena. Biol. Comiss. on the study of the Quaternary period. Moscow, Nauka, N 43: 136–145 (in Russian).

Belova, V. A., 1973. Natural conditions in formation of the Baikal rift zone basins in the Cenozoic period. In: Grichuk, V. P. (ed.). Pleistocene and Pliocene Palynology. Proceedings of the III Intern. Palynological Conf. pp. 184–187. Moscow, Nauka (in Russian with English summary).

Belova, V. A., 1975. The history of development of basin vegetation of the Baikal rift zone. Moscow, Nauka (English summary).

Belova, V. A, & Endrikhinsky, A. S., 1980. Palynological evidence for the scratigraphy of the Neogene deposits of the Baikal rift zone. In: Paleopalynology of Siberia. Moscow, Nauka, 84–92.

Florensov, N. A., Belova, V. A., Goldyrev, G. S., Lut, B. F., Mats, V. D. & Popova, S. M., 1978. Some stages of Baikal Lake development in the Late Cenozoic era. Polskie archiwum Hydrobiologii. 25: 135–145.

Goldyrev, G. S., Belova, V. A., Vykhristyk, L. A., Lazo, F. J. & Fedorova, V. A., 1975. New data on the composition and age of the upper part of sedimentary rock mass in the Baikal depression. In: Florensov, N. A. & Logatchev, N. A. (eds.). Rifting Problems, pp. 40–41. Irkutsk, Publisher of Irkutsk (in English and Russian).

Lut, B. F., 1964. Geomorphology of the Baikal bottom. In: Florensov, N. A. (ed.). Geomorphology of the Baikal Bottom and Its Coasts, pp. 5–123. Moscow, Nauka.

Mats, V. D. & Pokatilov, A. G., 1976. The stratigraphy of the Quaternary sediments within the coastal zone of the Baikal. Izv. Acad. Nauk. SSSR. Ser. Geol. 3: 133–139 (English summary).

Volkova, B. C. & Belova, V. A., 1981. Basic climatic boundaries of Siberia in Late Cenozoic. Proceedings of All-Union Conf. on Quaternary period. Ufa, Bashkiria.

Recent stratigraphy and accumulation of sediment in the deep, oligotrophic Lake Pääjärvi in South Finland

H. Simola & P. Uimonen-Simola
University of Joensuu, Karelian Institute, Section of Ecology, P.O. Box 111, SF-80101, Joensuu, Finland

Keywords: paleolimnology, sedimentation, sediment laminations, diatom microstratigraphy

Abstract

Laminated sediment is deposited in the deep areas (> 30 m) of Lake Pääjärvi. In the diatom microstratigraphy a regular alteration of planktonic and littoral floras, which goes in pace with the visual structure, reflects the yearly cycle of sedimentation.

Considerable variation in sedimentation was found to have occurred during the past two decades. A sequence, dated to the early 1960's, in which the organic content was exceptionally high, could be identified in the sediment at 6–14 cm depth even in shallower areas where laminations are not formed. This level was used as a marker horizon when determining the recent distribution of sediment by means of 19 cores taken from different parts of the lake.

The average annual sediment accumulation in the profundal zone of the lake is 340 g m^{-2} a^{-1} dry weight, of which 25 g m^{-2} a^{-1} (1100 kJ m^{-2} a^{-1}) is organic carbon.

Introduction

Lake Pääjärvi is situated in South Finland, at the southern border of the Finnish Lake District (61°04′, 25°08′E). The area of the lake is 13 km^2 and its maximum depth is 86 m; it is the fourth deepest lake in Finland. Its drainage basin is 244 km^2. For hydrological and ecological background, see Ruuhijärvi (1974), Sarvala *et al.* (1981) and National Board of Waters (1976, 1977, 1980). L. Pääjärvi is mesohumic (water colour 40–60 mg Pt l^{-1}) and oligotrophic, phytoplankton primary production being about 25 g $C_{ass.}$ $m^{-2}a^{-1}$ (Ilmavirta 1981).

Laminated sediment is deposited in the deep bottom areas of the lake (below 30 m water depth). Light and dark layers form more or less regular couplets with an avarage thickness of about 2 mm (Appleby *et al.* 1979).

The aims of this paper are: 1) to present microstratigraphical evidence about the annual character of the laminations; 2) to discuss recent changes in the sedimentation; and 3) to present data about the fluxes of material into the sediment of the profundal area of the lake.

Material and methods

The stratigraphy of diatoms and Chrysophycean cysts of the five most recent laminations was studied in material from an *in situ* frozen sediment core (Huttunen & Meriläinen 1978) taken from 80 m water depth in October 1977. A tape peel preparation was made from the core (Simola 1977) and a 1.5 mm broad vertical strip was marked on it. Diatoms were analysed from this strip in the sequence 0–30 mm at 424 consecutive 70 μm intervals.

In autumn 1979 a surface sediment sample was obtained at a water depth of 55 m by *in situ* freezing. From this a sediment block was cut (height 20 cm, cross-sectional area about 1 × 5 cm) which included the undisturbed sediment surface. This

Hydrobiologia 103, 287–293 (1983).

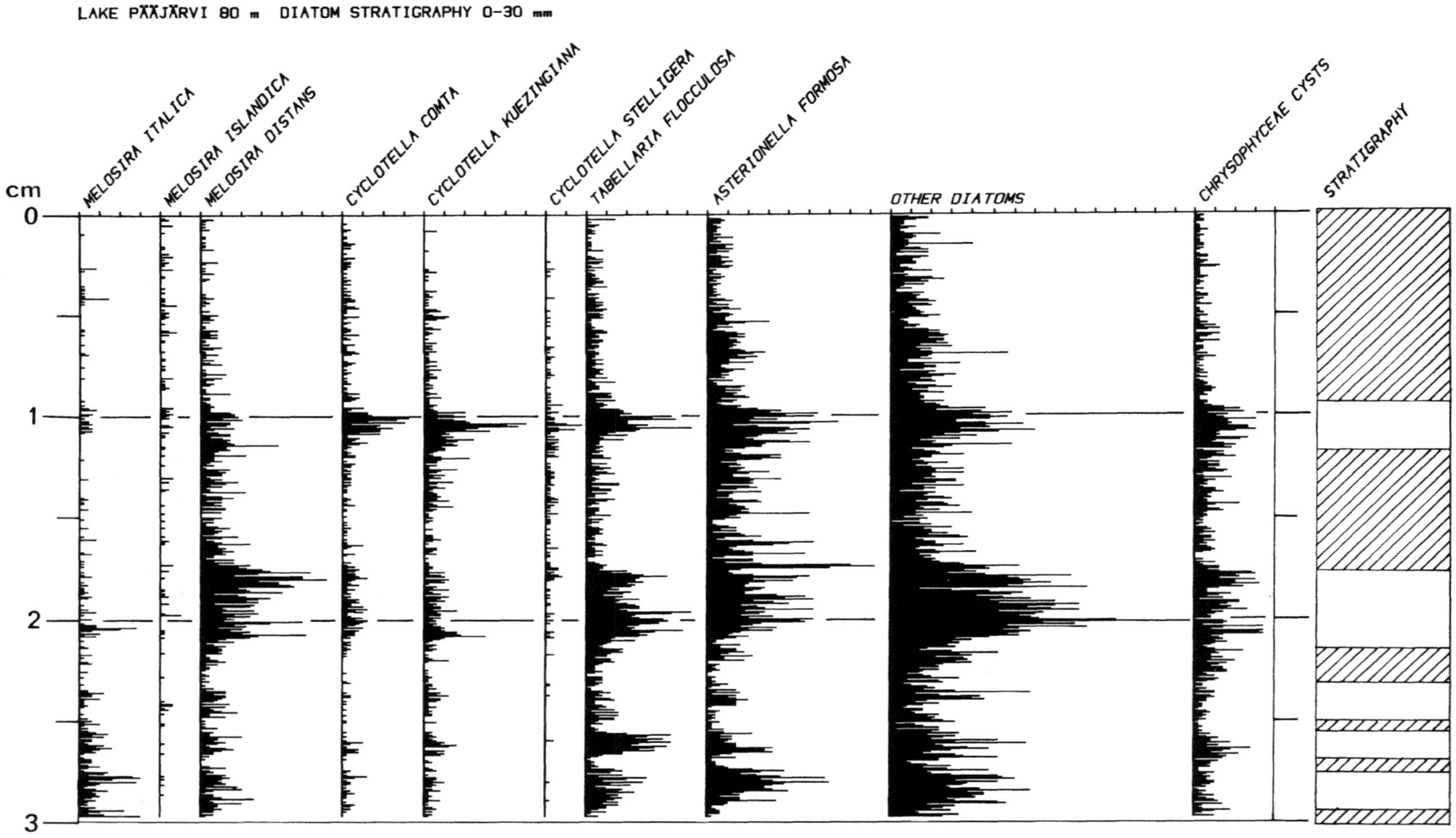

Fig. 1. Diatom microstratigraphy of the uppermost 30 mm of an *in situ* frozen core from a water depth of 80 m, taken in October 1977. The major plankton diatoms are presented as separate columns; the column 'other diatoms' mainly includes littoral forms.
Stratigraphy: hatched = dark grey (reddish brown in the oxidised surface layer); unhatched = light grey.

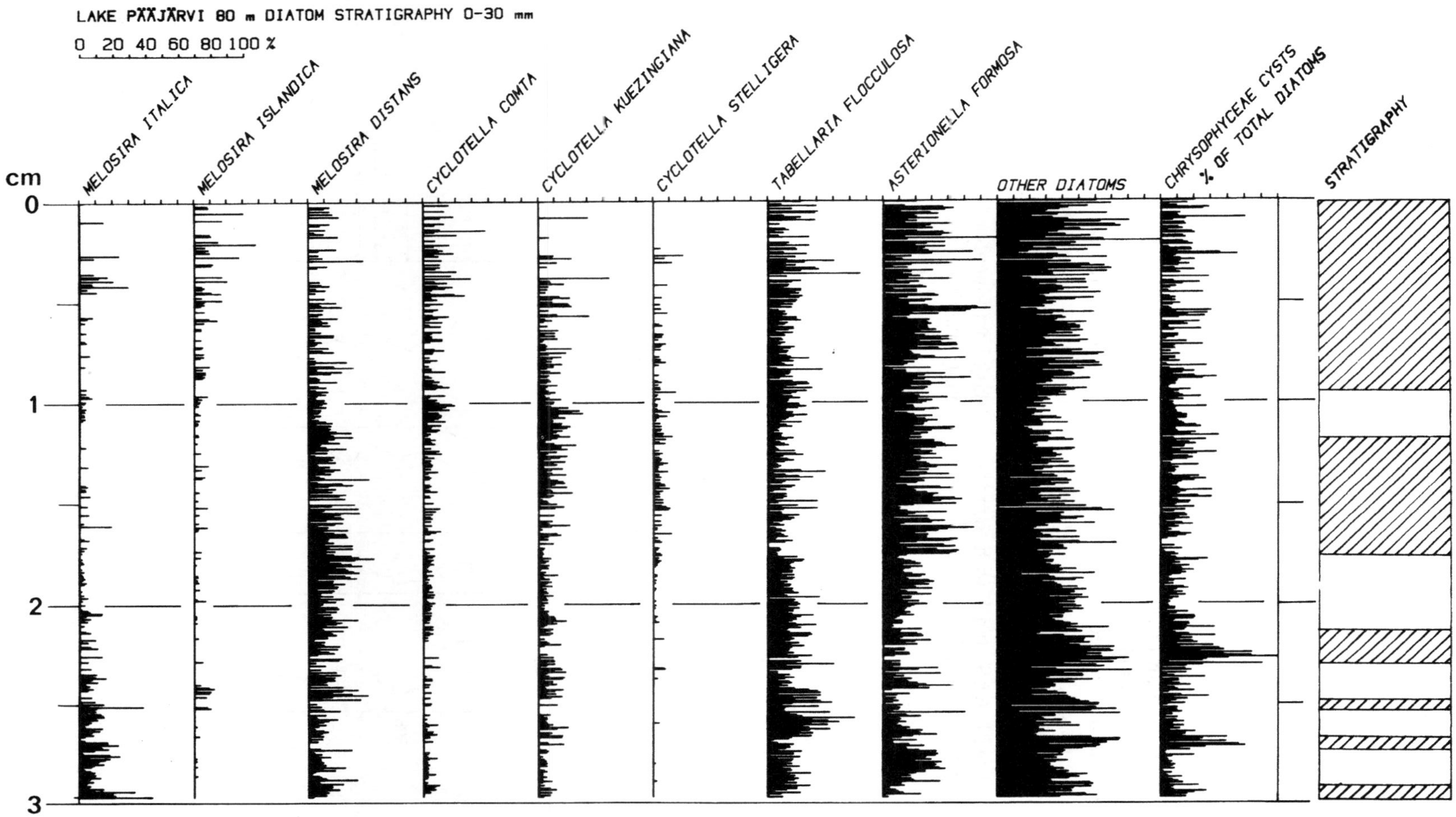

Fig. 2. Diatom stratigraphy 0–30 mm. The same data as in Fig. 1, expressed as a relative diagram. Chrysophycean cysts are indicated as a percentage of total count of diatoms.

block was freeze-dried and the 30 uppermost laminations were measured (horizontal dimensions with 1 mm accuracy), dissected and weighed individually. Organic carbon and nitrogen were determined with a CHN-analyzer from 3 subsamples of each layer.

For chemical analyses 19 sediment cores from different water depths (4–55 m) and from various parts of the lake were taken with a 15 cm^2 Kajak bottom corer in October 1979. The cores were sliced into 10–15 subsamples with an average thickness of 1 cm (2 cm in one core from 4 m depth). The size of each subsample was determined volumetrically. Fresh weight, dry weight, loss-on ignition, and (from part of the subsamples) C, N, P, Fe, Mg, Mn, K and Na were determined (C and N with CHN-analyzer, the metals with AAS, and P with the molybdenum blue method from acid extract).

Results

1. Sediment laminations

Of the five laminations in which the diatom stratigraphy was studied, the two upper ones were each about 1 cm thick. They formed the unconsolidated surface sediment. The lower ones were each about 2 mm thick.

The concentration of diatoms and Chrysophycean cysts fluctuates in the sequence studied reaching highest values in the light-coloured layers. The pattern of fluctuation is fairly similar for both planktonic and littoral forms (Fig. 1).

However, the percentages of littoral diatoms and Chrysophycean cysts are repeatedly higher in the dark components of the laminations. This is especially clear in the three lowermost, compressed laminations of the sequence (Fig. 2).

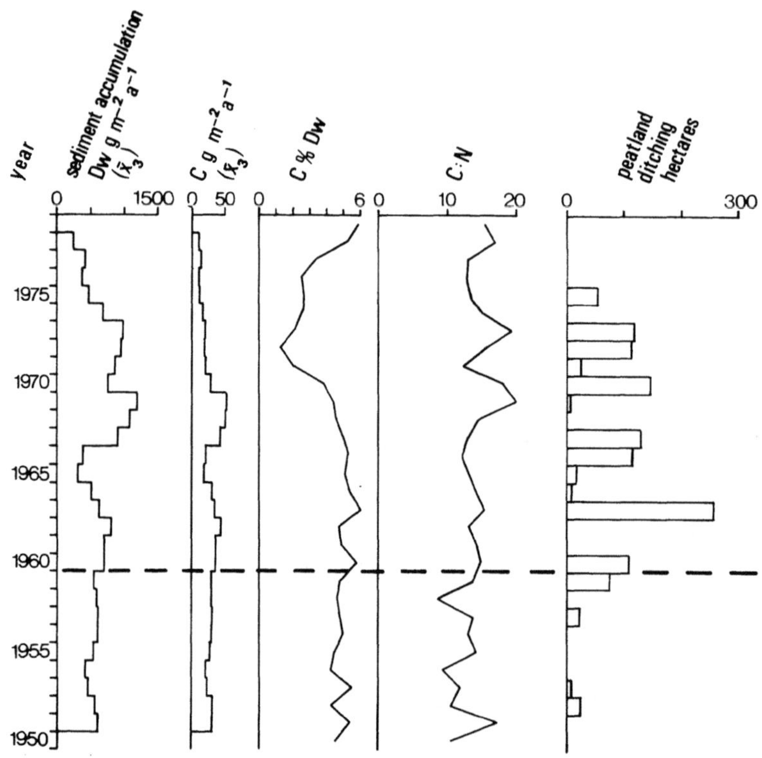

Fig. 3. Yearly accumulation 1950–1979, as determined from laminations individually dissected from a freeze-dried sample. The yearly fluxes are expressed as running averages of three successive determinations in order to smooth out any possible error caused by subjective judgement in dissecting the annual layers. The progress of peatland drainage, and the onset of lake level regulation (broken line), are also indicated.

The laminations seem to be formed by the seasonal rhythm of sedimentation so that the light-coloured diatom-rich layers represent material deposited in late spring and summer, and the dark layers, mainly resuspended and allochthonous material, sedimented during the autumn overturn (cf. Hakala 1977). The structure is thus primarily caused by physical processes, and random events like floods and windstorms may consequently create misleading structures in the laminations (cf. Lambert & Hsü 1979). Indeed, because of the irregularity of the laminations, an unambiguous identification of all the annual layers is not possible.

Within the individually analyzed 30-year sequence of laminations three periods of increased sedimentation are discernible. The last one of these contains highly minerogenic material (Fig. 3). The C/N ratio reaches its highest values (about 20) in the material deposited during the two youngest sedimentation peaks, indicating increased proportions of allochthonous organic material.

Two possible causes for the recent variation in sedimentation are a lake-level regulation and peatland drainage. The regulation of L. Pääjärvi was started in 1960 in order to prevent spring floods. The slight lowering of water level has caused a considerable reprocessing of the littoral zone especially on shallow shores. Increased spring discharge consequent on drainage operations has in some cases caused marked erosion in the lower parts of the drainage channels and the rapid formation of deltas at their inlets to L. Pääjärvi.

2. Sediment distribution in the profundal area

In each of the 19 profiles, a peak of dense (minerogenic) material is present, underlain by a stratum of looser (more organic) sediment. After comparison with Fig. 3, we judged the approximate level of the year 1962 in each profile (arrows in Fig. 4). This we used in calculating the recent accumulation rate of sediment at different water depths. The stratigraphic features thus allow sediment distribution in the whole profundal to be studied, even though annual laminations are only formed in the very deepest areas.

The average sediment accumulation rate below 10 m depth is about 350 g m^{-2} a^{-1} (Fig. 5). The rate may vary considerably between different sites due to physical factors (e.g. bottom topography, distance from inlets or the littoral zone).
In two of the shallow-water samples (4 m, 6 m) dry matter accumulation is considerably higher than in the deeper ones. These two cores seem to represent the accumulation zone at the lower edge of the littoral.

The distribution pattern of carbon and nitrogen, or organic matter, is more even than that of the total dry matter. No excess accumulation at the littoral/profundal margin is evident.

The accumulation of iron is very evenly distributed about 55 g m^{-2} a^{-1} at all depths. This uniform distribution indicates permanently good oxygen conditions even in the deep water, which prevents the escape of iron from the sediment as the ferrous ion.

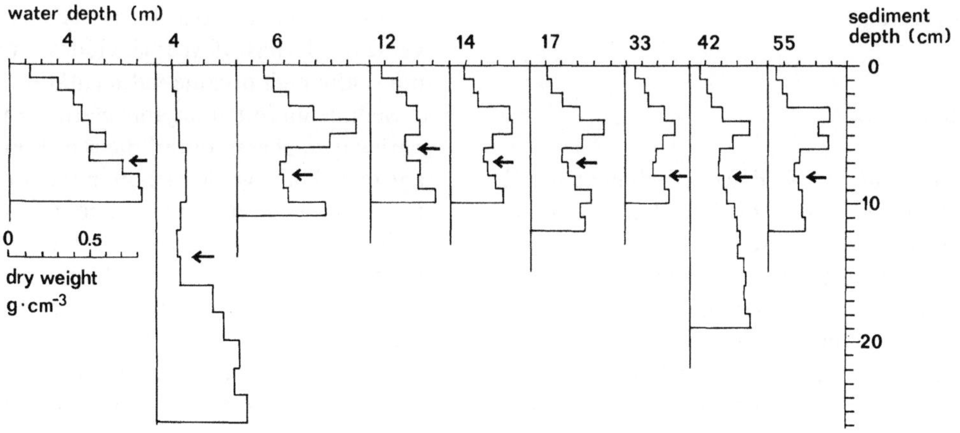

Fig. 4. Sediment dry weight profiles from different parts of the lake. Arrows point to the inferred approximate levels of the year 1962 (see text and Fig. 3). Nine out of the 19 obtained profiles are depicted.

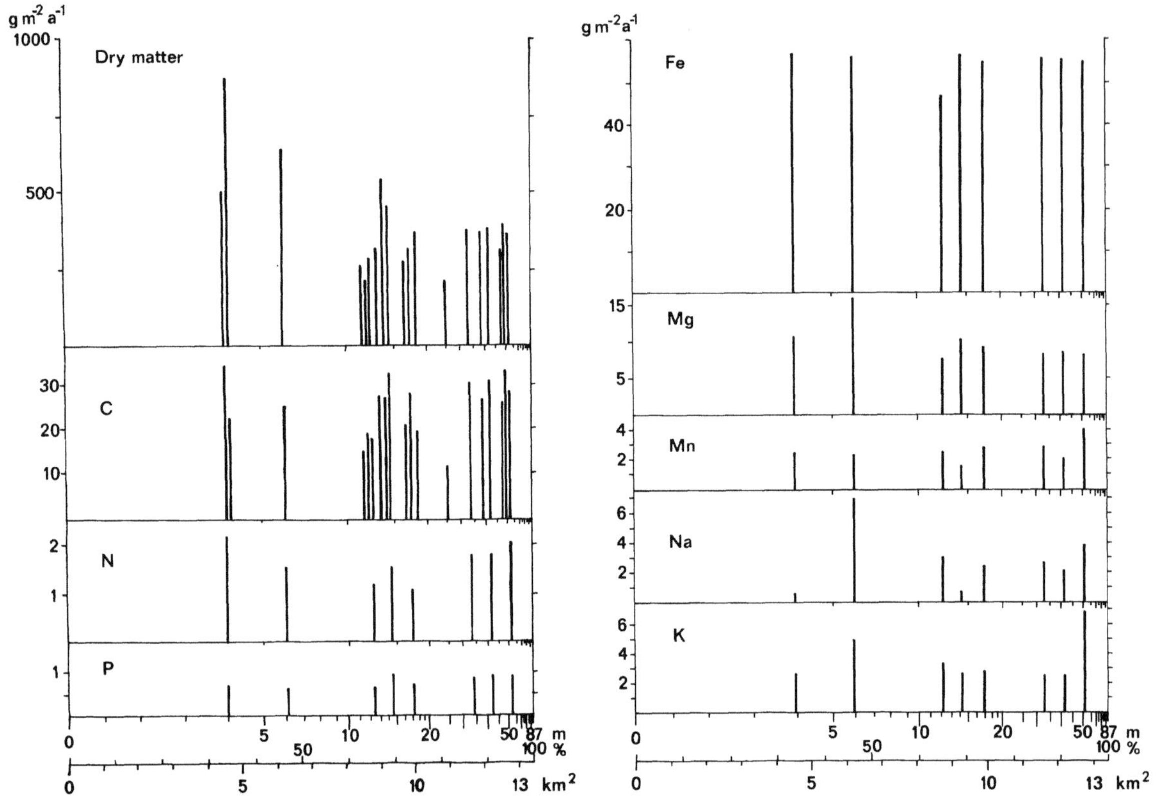

Fig. 5. Recent accumulation rates of dry matter and analysed elements in Lake Pääjärvi. In this figure, the cumulative area of the depth zones is on a linear scale on the x-axis, so the total accumulations can be determined planimetrically from the figures. The values are yearly averages, which have been obtained by dividing the increments in the sediment profiles above the 1962 level (cf. Fig. 4) by the number of years (18).

The distribution patterns of other metals are rather variable, which shows that there are great local differences in the sedimentary milieu and the quality of seston.

Discussion

Appleby *et al.* (1979) dated a sediment profile from L. Pääjärvi with ^{210}Pb. The dating was controlled by visual counting of the laminations. A perfect agreement, or a mutual support, was shown to exist between the two dating methods.

On the basis of this chronology the average yearly accumulation of carbon is about five times higher than that obtained by Hakala with sediment traps during a two-year study (Hakala 1977). It is possible that the relatively open trap model was not very suitable in the conditions of L. Pääjärvi, although the same trap design gave a yearly accumulation value quite comparable with that obtained by means of sediment varves in the small and sheltered L. Lovojärvi (Simola 1981). Also, our average value for the past 19 years includes some periods of unusually high organic sedimentation (cf. Fig. 3).

Sedimentation of organic matter is an important pathway of energy out of the ecosystem. The average burial of organic carbon in the profundal zone is, at a maximum estimate, about 25 g m^{-2} a^{-1}, or 1100 kJ m^{-2} a^{-1}, and equivalent to the primary production of the phytoplankton (Ilmavirta 1981), or about 40% of the estimated yearly input of allochthonous organic carbon (Sarvala *et al.* 1981).

The average total accumulation of inorganic matter (dry weight – 2 × carbon weight) in the profundal area of L. Pääjärvi is 2.5×10^9 g a^{-1}. A definite minimum estimate of erosion rate in the drainage area based on this figure (i.e. ignoring sedimentation in the littoral zone and in other basins, as well as outflow of inorganic matter) is 10.3 g m^{-2} a^{-1}.

Acknowledgements

This study was carried out at the Lammi Biological Station (Univ. Helsinki) and at the Karelian Institute (Univ. Joensuu). We thank the heads of these departments, Rauno Ruuhijärvi and Kyösti Pulliainen, for good working facilities. We are grateful to Jouko Sarvala and Kimmo Tolonen for their interest in different stages of the work. This study is part of a research project financed by the Maj and Tor Nessling Foundation.

References

Appleby, P. G., Oldfield, F., Thompson, R., Huttunen, P. & Tolonen, K., 1979. ^{210}Pb dating of annually laminated lake sediments from Finland. Nature 280: 53–55.

Hakala, I., 1977. Sedimentation energy flow to the profundal of the oligotrophic Lake Pääjärvi, southern Finland. Ann. bot. fenn. 14: 157–164.

Huttunen, P. & Meriläinen, J., 1978. New freezing device providing large unmixed sediment samples from lakes. Ann. bot. fenn. 15: 128–130.

Ilmavirta, V., 1981. The ecosystem of the oligotrophic Lake Pääjärvi. 1. Lake basin and primary production. Verh. int. Ver. Limnol. 21: 442–447.

Lambert, A. & Hsü, K. J., 1979. Non-annual cycles of varve-like sedimentation in Walensee, Switzerland. Sedimentology 26: 453–461.

National Board of Waters, Finland, 1976, 1977, 1980. Hydrological Yearbooks 1972–1973, 1974–1975 & 1976–1977. Publ. Wat. Res. Inst. 18: 1–192; 23: 1–195; & 35: 1–193.

Ruuhijärvi, R., 1974. A general description of the oligotrophic Lake Pääjärvi, southern Finland, and the ecological studies on it. Ann. bot. fenn. 11: 95–104.

Sarvala, J., Ilmavirta, V., Paasivirta, L. & Salonen, K., 1981. The ecosystem of the oligotrophic Lake Pääjärvi 3. Secondary production and an ecological energy budget of the lake. Verh. int. Ver. Limnol. 21: 448–453.

Simola, H., 1977. Diatom succession in the formation of annually laminated sediment in Lovojärvi, a small eutrophicated lake. Ann. bot. fenn. 14: 143–148.

Simola, H., 1981. Sedimentation in a eutrophic stratified lake in S. Finland. Ann. bot. fenn. 18: 23–36

Biostratigraphy of Lake Spitaalijärvi: an ultraoligotrophic small lake in Lauhanvuori, western Finland

Risto Salomaa & Pentti Alhonen
Department of Geology, Division of Geology and Paleontology, University of Helsinki, Snellmaninkatu 5, 00170, Helsinki 17, Finland

Keywords: paleolimnology, Flandrian biostratigraphy, *Tabellaria binalis*

Abstract

The Early and Middle Flandrian geological development and paleolimnology of Lake Spitaalijärvi was studied using pollen, diatom and cladoceran analysis and ^{14}C dating. Spitaalijärvi was isolated from the Ancylus Lake about 9000 B.P., at which time birch and pine and plants typical of open habitat communities grew on the solitary island. The rational limits for *Alnus* and *Picea* were ca. 8300 B.P. and 3700 B.P., respectively. During the first few hundred years after isolation Spitaalijärvi was probably fairly eutrophic, with a low water level. Water level began to rise before the *Alnus* rise (A°) and the lake became oligotrophic. After another transgression, which started before the *Picea* rise (P°), *Tabellaria binalis* and *Semiorbis hemicyclus* appear in the diatom stratigraphy indicating ultra-oligotrophic conditions.

The main reason for the ultra-oligotrophy of Lake Spitaalijärvi is the character of the parent material in the catchment, consisting of glacial and littoral deposits derived from the local quartz sandstone of highly siliceous nature.

Introduction

Lake Spitaalijärvi is situated in a shallow basin on the south-western slope of the till-covered sandstone hill called Lauhanvuori (Fig. 1). The top of the hill is supra-aquatic whereas the slopes emerged about 9400–8400 years ago when Lauhanvuori formed a solitary island in the open water body of the former Baltic Sea (Salomaa 1982).

Lake Spitaalijärvi is ca. 350 m long, 200 m wide and less than 1.3 m deep. The level of the lake is ca. 175 m a.s.l. Although the lake is very shallow it is poor in macrophytes and clearly shows oligotrophic features.

Spitaalijärvi receives its water supply from groundwaters and originally had no outlet channel. The shores of the lake consist of aeolian and littoral sands and gravels except in the north-western part where the lake is bordered by a small raised bog.

The sediment core taken from the deepest point of the lake does not reflect the entire postglacial history following the isolation as the youngest sediments date to ca. 3000 B.P.

For methods used see Berglund (1979).

Pollen stratigraphy

The pollen diagram (Fig. 2) starts with the *Pinus* zone, although, *Betula* is the dominant pollen and there is a comparatively high proportion of NAP. The radiocarbon date for the basal silty mud of 9020 ± 130 B.P. gives the minimum age of immigration of pine to the area. In the Early Flandrian sediments in southern Finland open vegetation cover is not usually associated with the *Pinus* zone, but with the lower part of the *Betula* zone (Donner 1971). At Lauhanvuori, however, this phenomenon is the result of the late deglaciation and emergence of an area with extensive open shore sites as a

Hydrobiologia 103, 295–301 (1983).

consequence of rapid regression. The arrival of *Alnus* starts the next pollen zone, *Betula – Alnus – Corylus – Ulmus*, and has been dated to 8250 ± 190 B.P. The immigration of spruce initiates a new stage in forest succession and has been dated at 3730 ± 150 B.P. (Hel-1172) in the bordering peat bog. The date should also represent the approximate age of Pc° in the sediments of Spitaalijärvi. The aquatic macro-vegetation was poorly developed throughout the lake history. Only *Isoetes* shows a clear maximum in the silty gyttja following the isolation.

Absolute pollen concentration is only about 5×10^4 tree pollen cm^{-3} in the lower part of the *Pinus* zone. After A° there is a strong increase in AP concentration reflecting the change from open forest vegetation to climax forest formation (Donner 1972). The high concentration figures also

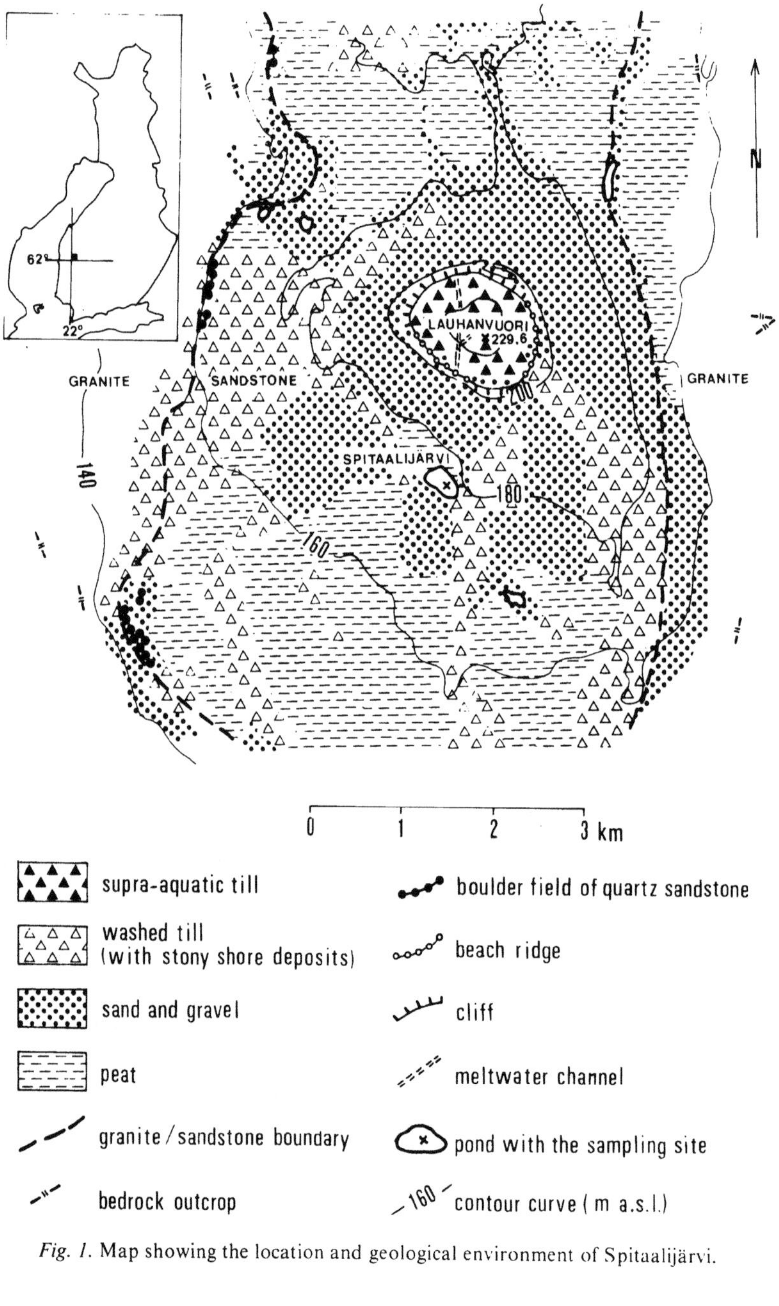

Fig. 1. Map showing the location and geological environment of Spitaalijärvi.

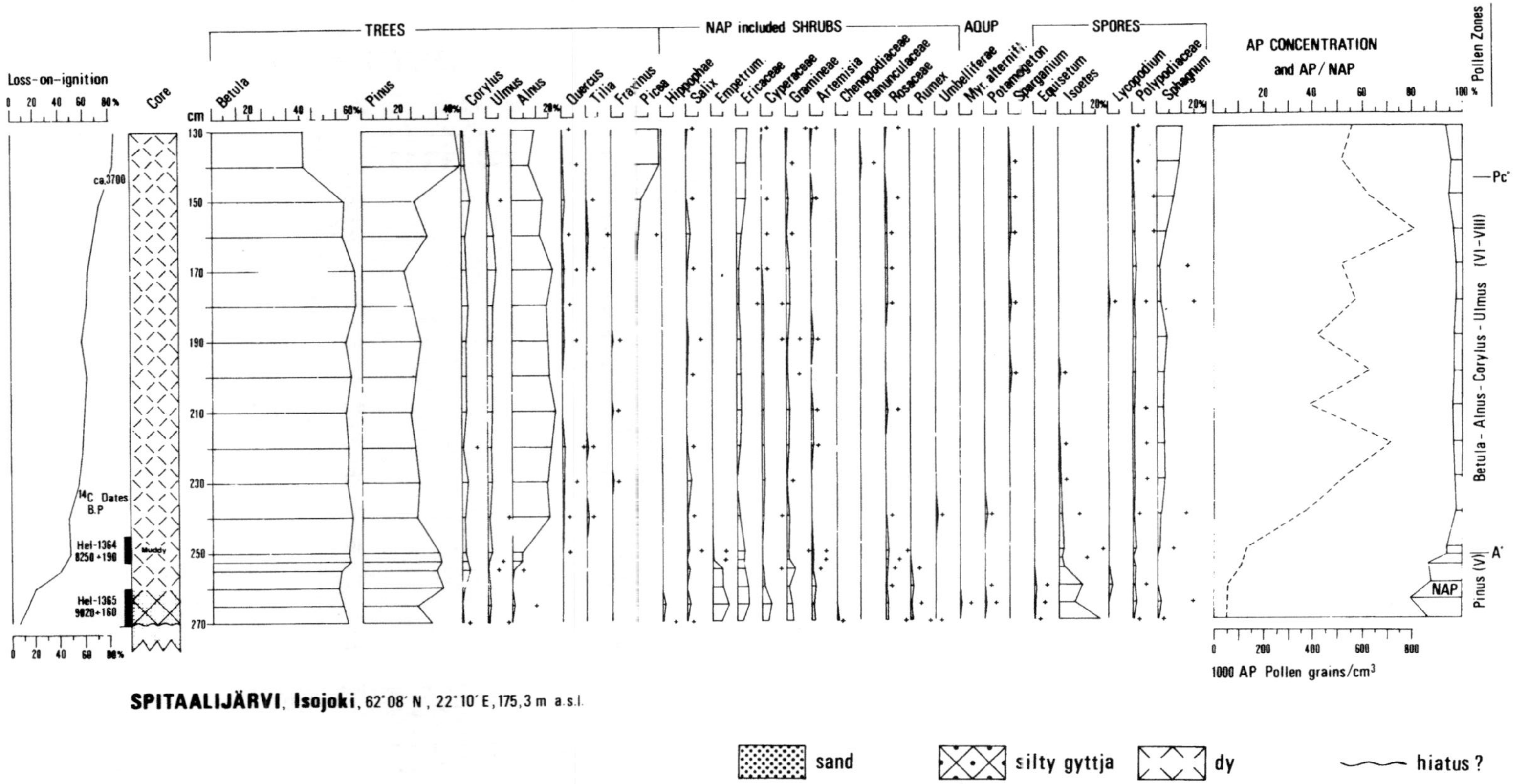

Fig. 2. Pollen diagram for Spitaalijärvi. The percentages are based on AP + NAP pollen sum. Absolute pollen concentration curve is drawn in the AP/NAP column.

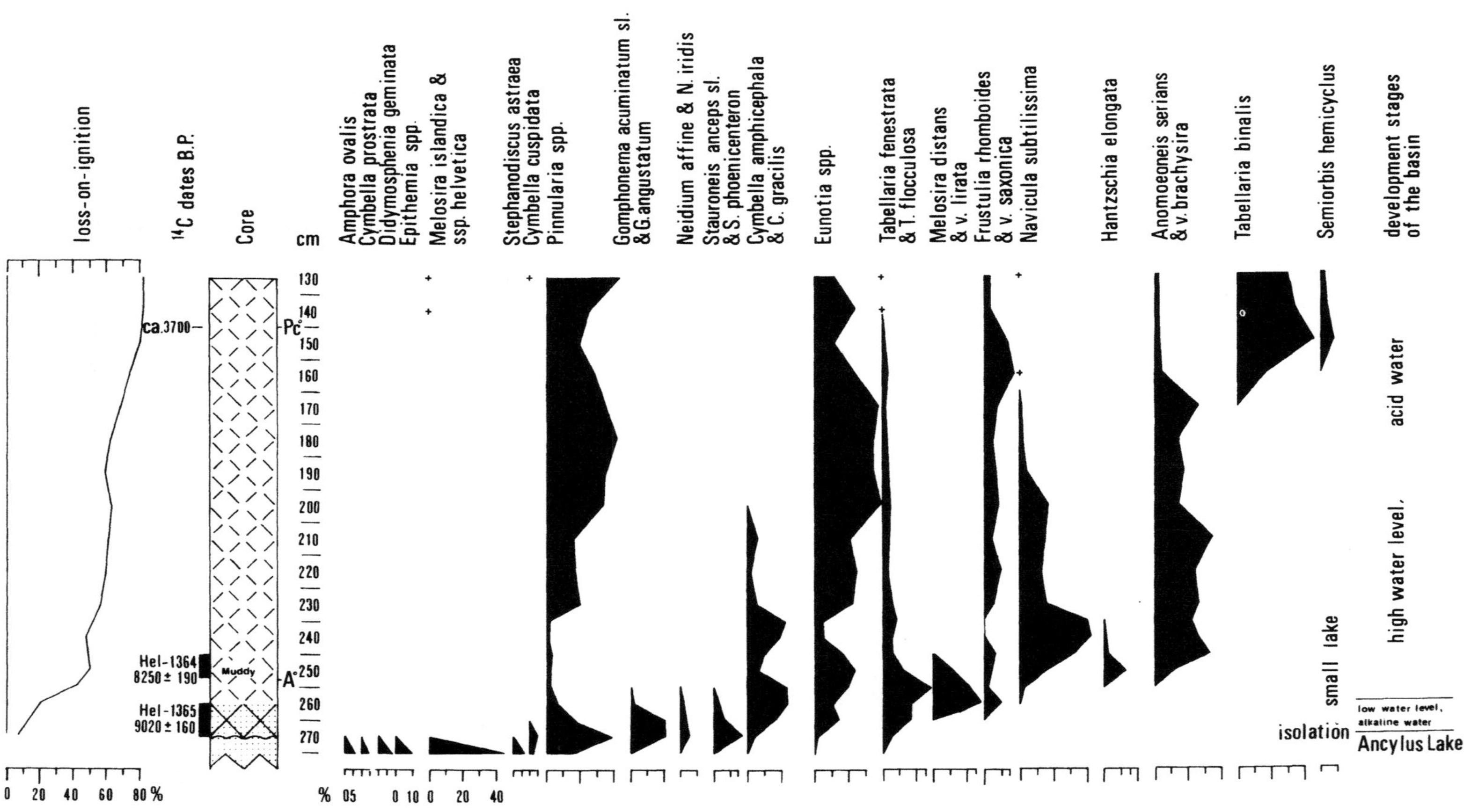

SPITAALIJÄRVI 62°08′N, 22°10′E, 175,3 m a.s.l.

Fig. 3. Diatom diagram for Spitaalijärvi. Selected species.

indicate considerable redeposition from the surroundings (Hyvärinen 1976), caused by the transgression of the lake and emphasised by the morphometry of the basin (cf. Lehman 1975).

Diatom stratigraphy

According to diatom stratigraphy (Figs. 3 and 4) Lake Spitaalijärvi was isolated from the fresh-water Ancylus Lake and the isolation took place by 9000 B.P. at the latest. The basal post-isolation sediments below the 260 cm level represent a lake phase of low water level and relatively alkaline conditions. A few hundred years after the isolation the water level had risen and the water had become more acid as indicated by the *Melosira distans* peak and the disappearance of the alkaliphilous species at the 260 cm level. Higher up in the stratigraphy, benthic species dominate again till the latter part of the Subboreal Chronozone when *Tabellaria binalis* and *Semiorbis hemicyclus* make their appearance in the stratigraphy indicating ultra-oligotrophic conditions.

Cladoceran stratigraphy

The cladoceran stratigraphy (Fig. 5) shows two planktonic *Bosmina* peaks, the first one soon after the isolation at 260 cm level and the second one in the uppermost part of the core, while littoral cladoceran dominate just after the isolation and at a depth of 250–170 cm.

The planktonic/littoral ratio (for its significance see Alhonen 1970a and 1970b) suggests a change from a littoral to a more pelagic environment in the lake development. However, the lower planktonic maximum could have been due to higher productivity of the lake at that time, and the decline in the amount of *Bosmina* could then have been caused by the dystrophication process as a result of environmental paludification (cf. Alhonen 1967; Hyvärinen & Alhonen 1970).

Conclusions

Spitaalijärvi is now very acid, pH 4.67 (20.7.1981) as indicated already by the fossil biostratigraphy.

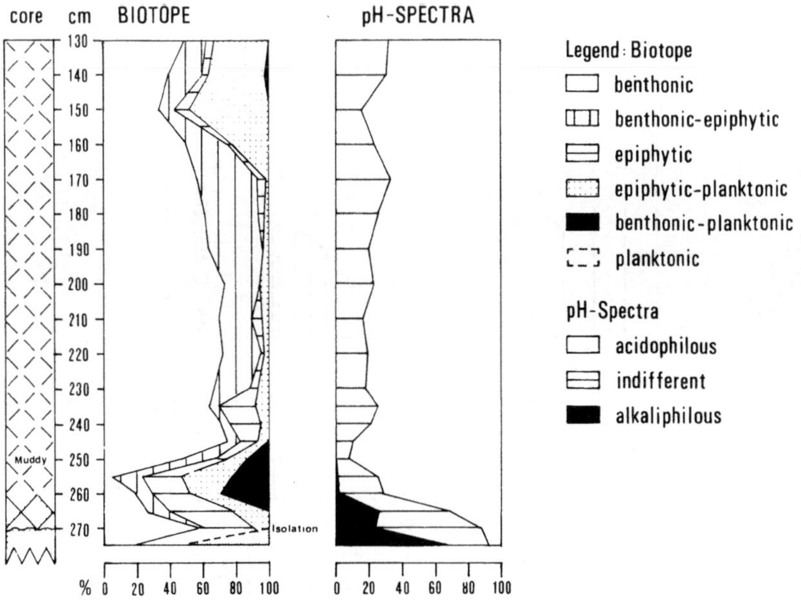

Fig. 4. Diatom diagram for Spitaalijärvi showing the ecological development of the diatom flora.

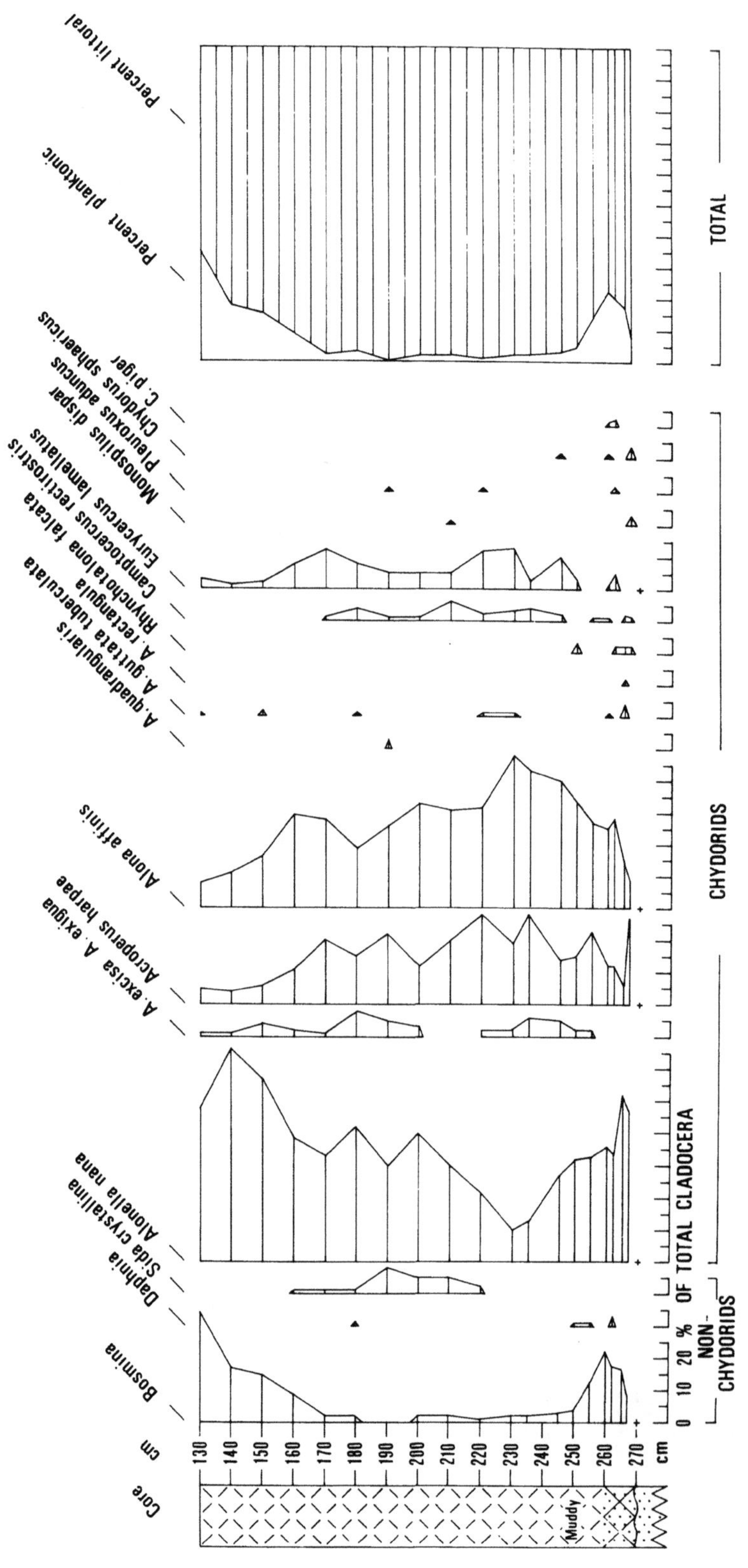

Fig. 5. Cladoceran diagram for Spitaalijärvi.

There are no other data available from Spitaalijärvi, but the average composition of an adjacent small stream, Lohiluoma, shows that silica is the main constituent (SiO_2 9.4 mg l^{-1}) and Ca, for example, is very low (1.3 mg l^{-1}).

The reason for the ultra-oligotrophy of Spitaalijärvi is the character of the local parent material of highly seliceous quartz sandstone, the SiO_2 content being about 97% (Simonen & Kouvo 1955), and the location of the basin well above the upper limits of clays and silts of the former Baltic Sea.

It is interesting to note the reason why the rare acidobiontic diatom *Tabellaria binalis* did not appear in the stratigraphy until the latter half of the Subboreal Chronozone. It is likely that the lake was becoming increasingly acidic as a consequence of the development of *Sphagnum* peat in the catchment area. This, together with an increase in water level which would extend the size of the littoral zone, may have promoted the development of *T. binalis* (cf. Quennerstedt 1949). The composition of the water could also have changed in course of time through the filtering effect of the sandstone. *T. binalis* is found in recent samples from Spitaalijärvi (Tynni, pers. commun.) and it seems to be a typical species in other small lakes on Lauhanvuori.

References

Alhonen, P., 1967. Paleolimnological investigations of three inland lakes in south-western Finland. Acta bot. fenn. 76: 1–59.

Alhonen, P., 1970a. The paleolimnology of four lakes in south-western Finland. Ann. Acad. Sci. fenn. A 3: 1–39.

Alhonen, P., 1970b. On the significance of the planktonic/littoral ratio in the cladoceran stratigraphy of lake sediments. Comment. Biol. 35: 1–9.

Berglund, B. E. (Ed.), 1979. Paleohydrological changes in the temperate zone in the last 15 000 years. IGCP 158 B. Lake and mire environments. Project guide, vol. 2. Specific methods. Lund., 340 pp.

Donner, J. J., 1971. Towards a stratigraphy division of the Finnish Quaternary. Comment. Phys-Math. 41: 281–305.

Donner, J. J., 1972. Pollen frequencies in the Flandrian sediments of Lake Vakojärvi, south Finland. Comment. Biol. 53: 1–19.

Hyvärinen, H., 1976. Flandrian pollen deposition rates and tree line history in northern Fennoscandia. Boreas 5: 163–175.

Hyvärinen, H. & Alhonen, P., 1970. The sediment history of a 100 year old pond near lake Höytiäinen, eastern Finland. Geol. Fören. Stockholm Förh. 92: 410–414.

Lehman, J. T., 1975. Reconstructing the rate of accumulation of lake sediment. The effect of sediment focusing. Quat. Res. 5: 541–550.

Quennerstedt, N., 1949. Om diatomeerna Actinella punctata Lewis och Tabellaria binalis (Ehr.) Grun. i svenska vatten. Svensk bot. Tidsk. 43: 82–97.

Salomaa, R., 1982. Erikoinen Lauhanvuori. (summary: Geological development of Lauhanvuori). Suomen Luonto 1: 30–33.

Simonen, A. & Kouvo, O., 1955. Sandstones in Finland. Bull. Comm. geol. Finlande. 168: 57–87.

Sysmäjärvi – a lake polluted by mining waste-water

Marketta Ahtiainen[1], Olavi Sandman[2] & Risto Tynni[3]
[1] *North Karelian Water District Office, Joensuu, Finland*
[2] *Mikkeli Water District Office, Mikkeli, Finland*
[3] *Geological Survey of Finland, Espoo, Finland*

Keywords: paleolimnology, acid waste-water, liming, diatoms, heavy metals

Abstract

Lake Sysmäjärvi is located inside the municipal boundary of Outokumpu in North Karelia, Finland. Since 1928 metalliferous, and at a later stage highly acid waste-water, has been discharged into the lake. Neutralization of the lake water by liming was introduced, and by the end of 1966 the pH of the water had become neutral to alkaline.

Two sediment cores (0.5 m and 5 m long, respectively) were analyzed for pollen, diatoms, loss-on-ignition, organic carbon, Fe, Cu, Ni, Pb, Zn and Hg. The maximum concentrations of organic matter and heavy metal content occur 30 cm below the sediment surface. Evidence of the liming and of accelerated eutrophication appear close to the surface of the sediment (0–10 cm), indicated by an increase in *Nitzschia plana*.

Introduction

Sysmäjärvi is a shallow lake in North Karelia, Finland, near the town of Outokumpu (Tables 1 & 2). In the 1930's, the effects of human activity on the dystrophic lake could be clearly seen since waste-water from the town made the lake more eutrophic. Copper, nickel and cobolt mining by the Outokumpu mine has also had a dramatic effect on the lake. From 1928 to 1938 the waste water from the mine was alkaline. Since 1938 sulphuric acid has been used. The mining waters made the lake water acid and in 1943 the water became clear and the fish population declined rapidly. Neutralization of the waste-water started in 1964. Today, the lake is alkaline and highly eutrophic.

The aim of this study is to investigate the evolution of Sysmäjärvi with special respect to the effects of acidification and subsequent liming recorded by the diatom flora and heavy metal content of the sediment and by historical records.

Table 1. Drainage area and water chemistry.

Sysmäjärvi data
Drainage area 114.5 km^2, water area 9.2%
Lake area 650 ha, max. length 4 km, max. width 3 km
Lake volume 6.5×10^6 m^3
Mean depth 1 m, max. depth 6.6 m
Theoretical retention time 75 d, location of the lake is N_{60} + 85.5 m
Mean discharge 1.20 m^3/s (HQ 12.0 m^3/s, MHQ 7.2 m^3/s, MNQ 0.25 m^3/s, NQ 0.1 m^3/s)
Cultivated land 12%

Division of the drainage basin.

Name of the river	Basin area (km^2)	Lake area %	HQ	MQ	Water quality
Kuusjoki	44.5	4.0	7.3	0.4	dystrophic
Kesselinpuro	21.7	1.0	3.9	0.25	dystrophic
Ruutunjoki	23.5	7.0	3.2	0.2	mining wastewater
Lahenjoki	10.3	3.0	3.0	0.10	dyseutrophic

Hydrobiologia 103, 303–308 (1983).

Table 2. Water chemistry values for Sysmäjärvi, sampling station 8.

Date	9.VIII.1979			1.IV.1980			
Depth	1 m	5 m	0–2 m	1 m	2 m	3 m	5 m
Parameter							
Temperature °C	15.3	14.2		1.6	2.7	3.0	3.2
Dissolved oxygen (mg/l)	10.1	9.3		1.9	0.3	0	0
Percentage saturation %	104	93		14	2	0	0
Conductivity (μS/cm) 20 °C	693	693		1 134			1 440
Total alkalinity (mval/l)			0.27	0.48			0.23
pH	7.66	7.37	7.77/20 °C	6.80			5.99
Chemical oxygen demand O_2 (mg/l)	9.7	10		16			4.9
Biochemical oxygen demand (mg/l)	3.2	3.4					
Total nitrogen (μg/l)	610	560		3 190			1 390
Total phosphorus (μg/l)	20	21		39			7
Fe (μg/l)	660	780		260			3 900
Mn (μg/l)	110	110		160			640
So_4 (mg/l)	340	350		380			820
Cl (mg/l)				35			44
Chlorophyll-a-conc. (μg/l)			14,8				
Primary productivity capability with C^{14} tracer technique (mgC/m^3/d)			250				

Materials and methods

The sediment samples were taken from the deepest part of Sysmäjärvi (Fig. 1). A short profile, 43 cm long, was cored with a modified Mackereth sampler in 1978. A longer core, 480 cm long, was sampled in 1980 with a Russian peat corer. The thickness of each separate subsample was 1 cm. In the long core, subsamples were taken 20 cm apart.

Pollen and diatom analyses were carried out following the methods used at the Geological Survey of Finland. The heavy metals were analyzed using standard methods (Zink-Nielsen 1975) at the laboratory of the Water Research Institute, Helsinki. The water-analysis data used were mainly based on water samples analyzed by the North Karelian Water District Office using methods of the Finnish Water Administration.

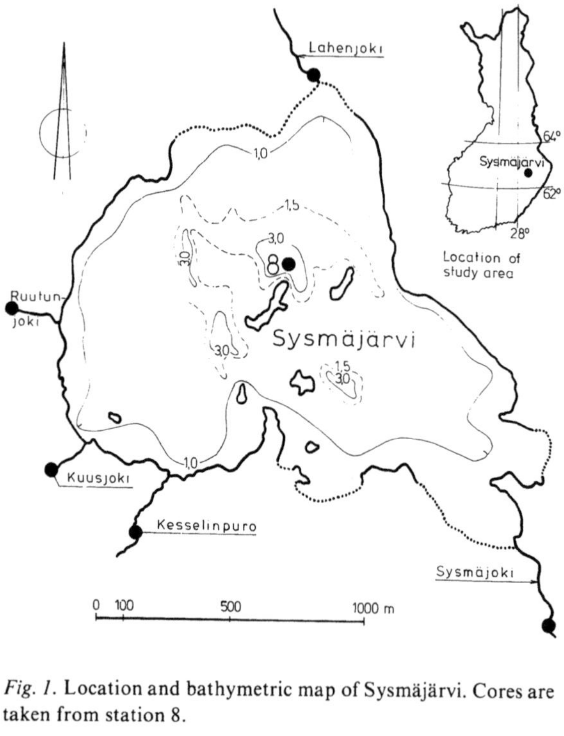

Fig. 1. Location and bathymetric map of Sysmäjärvi. Cores are taken from station 8.

Results

According to the pollen diagram and Donner's (1971) pollen zonation the base of the 480 cm sediment core from the deep part of Sysmäjärvi belongs to the Preboreal *Betula*-phase, Zone IV (Fig. 2). The whole core spans more than 9 000 years.

The lower part of the sediment was deposited in the Yoldia phase: the typical diatom at 440 cm is *Melosira islandica* ssp. *helvetica* and there is also a small quantity of *Diploneis interrupta*. The concentration of diatoms is very low.

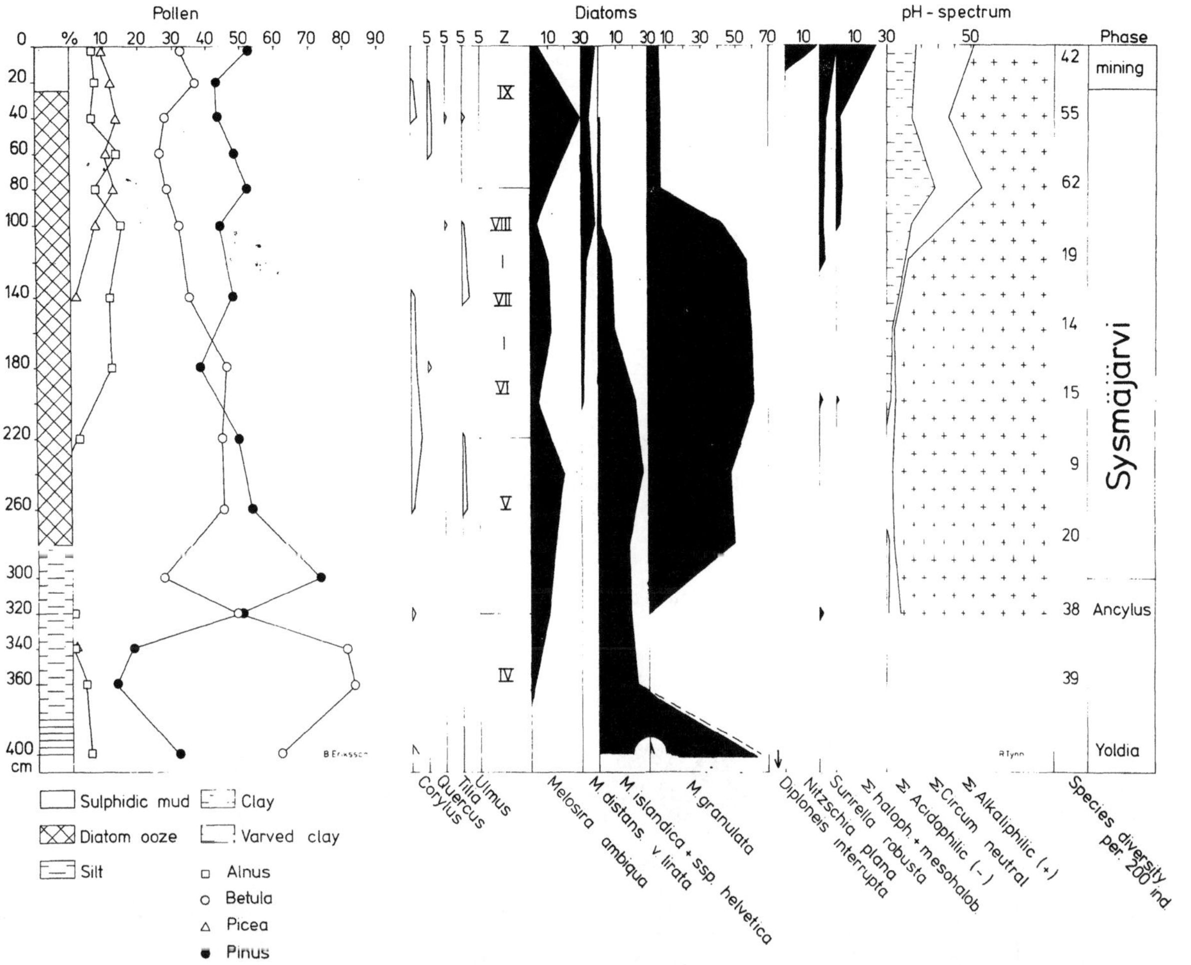

Fig. 2. Pollen and diatom diagram of the long core.

The Ancylus-lake phase extends from a sediment depth of ca. 360 cm to 300 cm. *Melosira islandica* ssp. *helvetica* is dominant at first, but then, at a higher level, taxa typical of the Ancylus flora increase. The isolation of Sysmäjärvi from the Ancylus lake took place at approximately 300 cm.

The initial and middle phases of the post-isolation lake, 300–100 cm, is indicated by a *Melosira granulata* maximum. *M. granulata* is an alkaliphilic diatom commonly found in eutrophic waters (Lowe 1974). The sediment is a diatom gyttja. This period lasted from the Boreal to the Subboreal. From 80–40 cm (the beginning of pollen zone IX) the lake became somewhat dystrophic.

In the short core, (Figs. 3 & 4), the influence of man can be detected from about 40 cm to the sediment surface, represented by increases in the proportion of organic matter, alkaliphilous and alkalibiontic diatoms, and heavy metals. This change can be attributed, directly or indirectly, to the effect of waste from the mining industry.

Sewage from the town and, from 1928–1939, circumneutral waste-waters from the mining industry made Sysmäjärvi more eutrophic. *Melosira ambigua* increases. This phase extends from a depth of 40 cm to about 25 cm (Fig. 3).

The change to an acidic pH caused by acid mine waste-waters is reflected in the diatom flora at 25–10 cm where *Melosira distans* has a maximum and corresponds to very acidic input from 1950–1966 (Fig. 5). *Nitzschia plana* appears, but it reaches its maximum later when the lake water was neutral or slightly alkaline. *Nizschia plana* is a taxon that lives in brackish waters in harbour pools

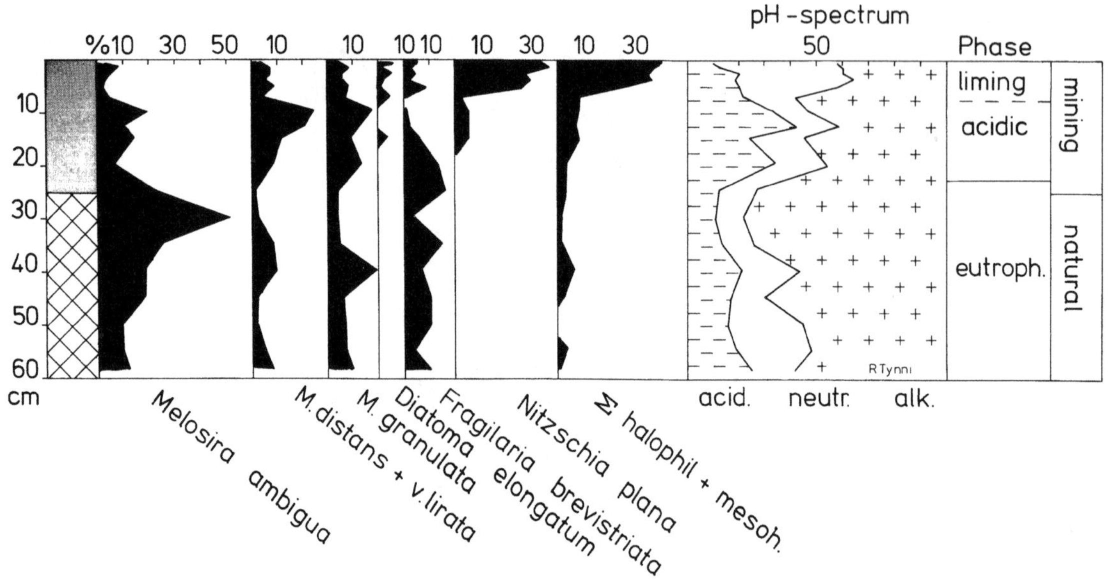

Fig. 3. Diatom diagram of the short core.

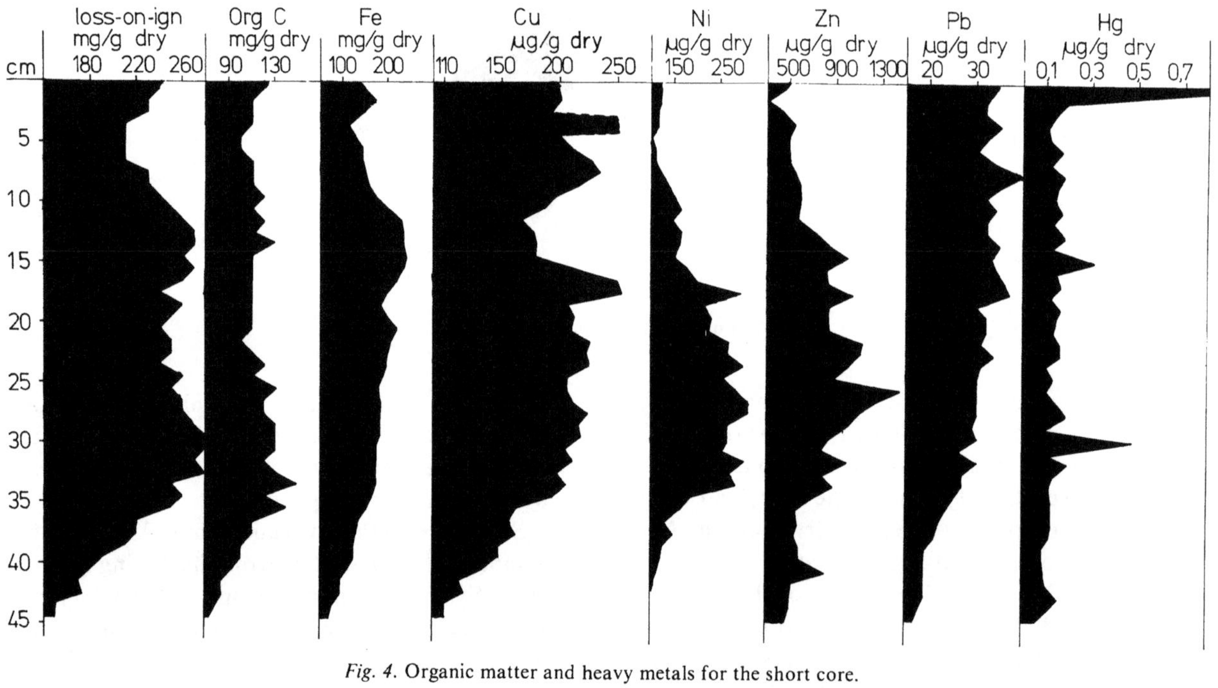

Fig. 4. Organic matter and heavy metals for the short core.

(Hendey, 1964). In Sysmäjärvi the large quantity of chemical compounds introduced seems to have favoured this species.

The radical pH changes (Fig. 5) seem to have had a smaller influence on the amounts of organic matter and metals in the sediment. The maximum loss-on-ignition and iron values at ca. 15 cm corresponds to a minimum of Cu, Ni and Zn.

Today the quality of the water in the inflowing rivers varies a great deal. In the western, deeper part of the lake, lower pH values and an abundance of iron, humus and sulphate are typical, especially in

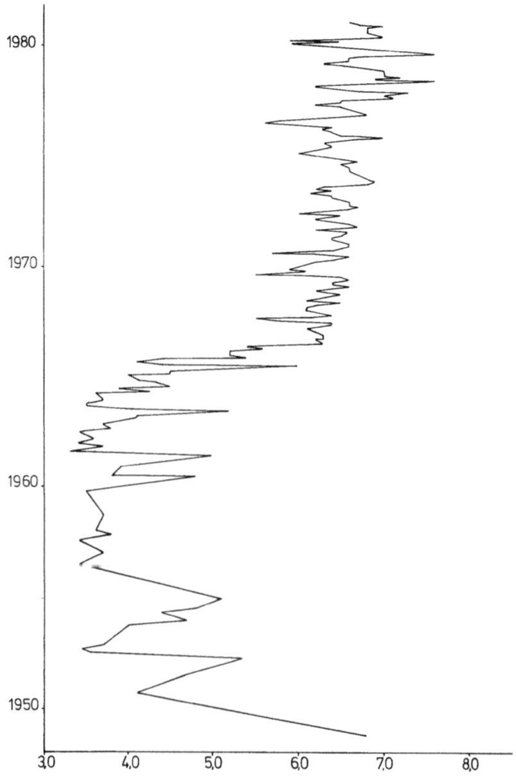

Fig. 5. pH variations at Sysmäjoki sampling station.

the winter time. On the other hand, the humus-bearing waters flowing from Kesselinpuro and Kuusjoki have some effect on the precipitation of Fe.

Some of the water polluted by the mining waste and most of the municipal sewage enters via the river Lahenjoki (Fig. 1). The effects of large amounts of nutrients can be noticed most clearly where the river enters the lake, but effects can also be observed over the whole eastern, shallow part of the lake. There are signs of increasing eutrophication. Oxygen depletion in winter has become critical in the whole water column (Table 2). Sysmäjärvi does not normally stratify in summer.

Discussion

The long core studied suggests that during the Atlantic period Sysmäjärvi was undisturbed and alkaline. The sediment was a diatom gyttja indicating water rich in oxygen and silica, and perhaps the absence of extensive macrophyte communities that are now so dominant.

A change in the pH spectrum occurred when the Subboreal period began; the climate became drier and colder and probably had an influence on the trophic level of the lake.

In the Subatlantic period the climate was wetter and colder but the lake was more eutrophic as more nutrients came into the lake from anthropogenic sources.

The more rapid change of the last 50 years is represented by the uppermost 40 cm of the sediment. The first data concerning the character of the water dates from the 1930's, when the dystrophic lake started to become more eutrophic (Järnefelt 1936) due to the effect of municipal waste-water, and the mining waste-water of the Outokumpu dressing works from 1928–1939. A clear maximum of alkaliphilous and alkalibiontic diatoms and heavy metals occurs at a depth of 35 cm.

Eutrophy and the high electrolyte content of the lake water may be the factor that diminishes the indicator value of the diatom pH-spectrum in this case. The increase in the content of halophiles, mesohalobionts and *Nitzschia plana* begins at a sediment depth of 20 cm and is clearly not an acidic phase. It therefore seems that the pH classification does not suit the diatom flora living in very strongly polluted waters. Sediment resuspension in shallow waters may also make interpretation difficult. The water level of Sysmäjärvi has been lowered several times, the first time before 1 774 (Anttila 1958), and at least twice in the early nineteenth century. The reduction of lake surface area may have accelerated the spread of macrophytes, and the decrease in water mass has increased the likelihood of oxygen depletion in deeper areas of the lake.

The amount of heavy metals in the Sysmäjärvi sediment correlates with organic matter. The peak of loss-on-ignition in the acidic phase may be associated with the sedimentation of humic colloids causing an accumulation maximum. The correspondingly high amounts of Fe in the sediment suggest that it is complexing with humus. In his extensive study, Ryhänen (1961) presents observations and analytical results that support this interpretation. Hence the general rule that metals are in solution or escape from sediments in an acidic milieu (Förstner & Müller 1979, Schindler *et al.* 1980) may not always be true. However, according to the analytical data, the water in Sysmäjärvi was very rich in iron.

When water is alkaline, copper seems to have a maximum in the sediment. As shown by Sylva (1978), the free Cu is mainly adsorbed to clay particles when pH is higher than 6,5.

The decreasing amount of organic matter and metals near the sediment surface, up to 5 cm, where the pH of the lake water increases causes difficulties in interpretation. The reduced metallic load as a consequence of a more effective dressing of the ore may have had an effect. Oxygen depletion in winter has caused the release of metals, at least Fe and Mn, from the sediment.

The rise of organic matter from a depth of 5 cm to the sediment surface may be due to the mineralization process that occurs in the surface layers of the sediment (Rich, 1975). The paleolimnology of shallow lake-basins is a difficult science. Further work is required and our next step is to date the sediments of Sysmäjärvi more accurately.

Acknowledgements

We should like to thank Dr. Pertti Huttunen, University of Joensuu, for his assistance in coring and Ms. Brita Eriksson, Geological Survey of Finland, for analyzing the pollen sequence of Sysmäjärvi.

The laboratory of the Water Research Institute, National Board of Waters, kindly made chemical analyses of the sediments.

Our special thanks are due to Dr. Boris Winterhalter for his ready help and to Mr. Paul Sjöblom for correcting the English translation.

References

Anttila, V., 1967. Järvenlaskuyhtiöt Suomessa. Kansatieteellinen tutkimus. (Referat: Die Seesenkungsgenossenschaften in Finnland. Ethnologische Untersuchung). Forssan kirjapaino. 360 pp.

Donner, J. J., 1971. Towards a stratigraphical division of the Finnish Quaternary. Soc. Sci. Fenn. Comment. Phys.-Math. Vol 41: 221–305.

Förstner, U. & G. Müller, 1974. Schwermetalle in Flüssen und Seen als Ausdruck der Umweltverschmutzung. Springer-Verlag. p. 155.

Hendey, N. I., 1964. An introductory account of the smaller algae of British coastal waters. Part V. Bacillariophyceae. Her Majesty's Stationery Office. 317 pp, 45 pl.

Järnefelt, H., 1936. Zur Limnologie einiger Gewässer Finnlands XII. Ann. Soc. Vanamo 3: 154–179.

Lowe, R. L., 1974. Environmental requirements and pollution tolerance of freshwater diatoms. EPA-670/4-74-005. 334 pp.

Rich, P. H., 1975. Benthic metabolism of a soft-water lake. Verh. Intern. Ver. Limnol. 19: 1023–1028.

Ryhänen, R., 1961. Über die Einwirkung von Grubenabfällen auf einen dystrophen See. Ann. Bot. Soc. Vanamo 22, 70 pp.

Zink-Nielsen, I., 1975. Interkalibrering of sedimentkemiske analysemetoder II. Nordforsk Miljövårdsekretariatet. Publ. 1975: 6. 19 pp, 2 app.

Sedimentary chemistry of a small polluted lake, Gallträsk, S. Finland

K. Tolonen[1] & J. Meriläinen[2]
[1] *University of Helsinki, Department of Botany, Unioninkatu, 44, SF-00170 Helsinki, Finland*
[2] *University of Joensuu, Karelian Institute, Section of Ecology, P.O. Box 111, SF-80101 Joensuu 10, Finland*

Keywords: paleolimnology, heavy metals, polluted lake, sedimentary chemistry

Abstract

Macro nutrients N, P, K, Ca, Mg, Na and S as well as heavy metals Pb, Cu, Zn, Co, Cr, Ni, Fe and Mn were analysed from a sediment core representing a 7 000-year long history of a lake including the preceding brackish (Litorina Sea) and freshwater (Ancylus Lake) stages. Sediments were also analysed for sedimentary pigments, algal growth potential (AGP) of the interstitial water, phosphorus fractions, pH, specific conductivity and redox potential. The annual accumulation values of several chemical elements were interpreted in light of other paleolimnological data from the same core. In the uppermost sediment stratum the influence of sewage pollution is clearly reflected by increases in total electrolyte content, pH, AGP and sedimentary pigments including myxoxanthophyll. Nickel, copper and manganese only slightly increase in the recent sediment, whilst other heavy metals (especially Cr and Zn) and sulphur, phosphorus, nitrogen and total organic matter are heavily concentrasted in the polluted surface sediment. The interpretation of the chemostratigraphy of the lake accords with evidence of lake development based on biological analyses.

Introduction

The studies of Alhonen (1972) and Tolonen (1980) describe the paleolimnological development of Gallträsk, a small, strongly eutrophicated South Finnish lake, on the basis of the biostratigraphy of sediments. The present paper attempts to interpret the detailed chemostratigraphy of the sediment with special reference to the recent pollution period.

The site and its sediment stratigraphy

Gallträsk is about 650 m long, 250 m wide and 11 ha in area. Its maximum depth is 1.7 m and its mean depth is 1.0 m. The topography of the lake surroundings is shown in Fig. 1.

The lake has been polluted by sewage discharge since the 1920s. At present the sediment–water contact experiences reducing conditions for long periods of the year. Except for the deepest layer, most of the water mass is alkaline. The electrical conductivity of the water varies around 15 mS m^{-1}.

The sediment stratigraphy of Gallträsk is fairly uniform throughout the basin; about a 2 m thick layer of highly organic coarse and fine detritus gyttja overlies 0.5 m of 'lagoon gyttja' of Litorina Sea age. This in turn overlies Ancylus clay and clay gyttja to a depth of at least 2.5 m. A layer of iron sulphide mud covers the whole basin; it is 0.8 cm thick close to the outlet and 20 cm thick on the opposite side near the main inlet.

Material and methods

The sediment was described in the field immediately after coring. The numbers of the strata refer to Figs. 2–4.

Hydrobiologia 103, 309–318 (1983).

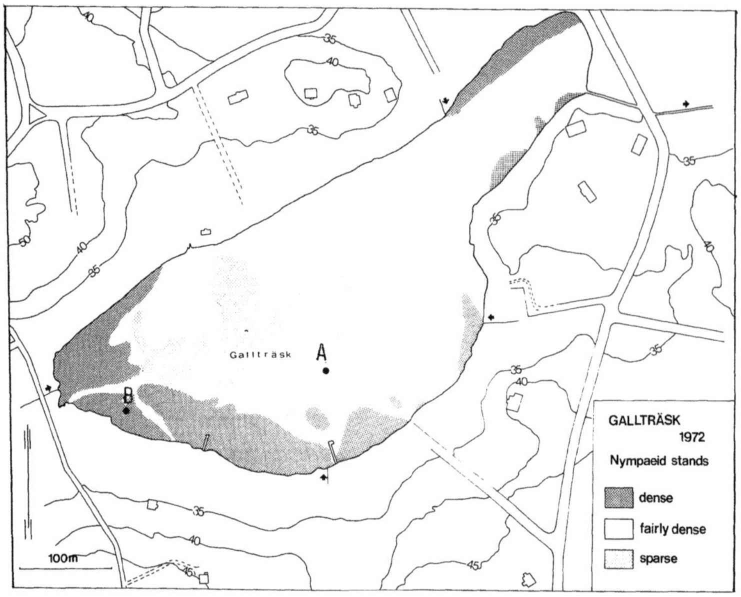

Fig. 1. Gallträsk and surroundings. Distribution and density of nymphaeid stands according to the restoration plan made by the city of Kauniainen. The study site is indicated by A. B refers to littoral sites published by Tolonen (1980).

Stratum	*Depth (cm)*	*Description*
18	150–154	Black sulphide gyttja
17	154–170	Coarse detritus gyttja, brown to greenish
16	170–185	Green detritus gyttja, very rich in moss remains
15	185–230	Coarse detritus gyttja or dy, dark brown
14	230–260	Coarse detritus gyttja, brown to greyish
13	260–350	Fine detritus gyttja, dark green
12	350–354	Fine detritus gyttja, reddish
11	354–360	Fine detritus gyttja, very rich in vascular plant remains (cf. *Naias marina*)
10	360–380	Fine detritus gyttja, pale, rich in plant remains
9	380–416	Clay gyttja and gyttja clay
8	416–450	Gyttja clay gradually becoming more bluish
7	450–460	Clay, blue
6	460–485	Gyttja clay
5	485–538	Clay
4	538–543	Sand
3	543–615	Clay
2	615–620	Sand and gravel
1	620–650	Clay

Samples for geochemical and biochemical analyses were taken from nine fresh cores from site A (Fig. 1). The specific conductivity, hydrogen ion concentration (pH), and the redox potential of the sediment were measured immediately after coring. Additional pH-measurements were also made later in the laboratory. For redox measurements, three identical platinum glass electrodes (length 40 cm), with a calomel electrode as a reference, were used (Niemistö 1973). Cut plastic syringes were used to take 3–7 ml of sediment from 10 cm intervals for density determinations.

Sedimentary pigments were analyzed and expressed as pigment units according to Vallentyne (1954) and Belcher & Fogg (1964). Myxoxanthophyll and other sedimentary pigment derivatives were analyzed by Dr. Carl Gustav Widen (the Department of Pharmacology, University of Helsinki) using thin-layer chromatography (Widen 1973). For element analysis subsamples were dried at +150° for 24 hours and the analyses were carried out using the standard methods of Viljavuuspalvelu Ltd. (Kurki 1972, Huttunen *et al.* 1978). Phosphorus fractions were made from fresh sediment samples one day after coring according to Krause (1964) and Shapiro *et al.* (1971). The test species in the AGP bioassay of the interstitial sediment water was *Ankistrodesmus falcatus* (Seppänen 1975; Tolonen & Hosiaisluoma 1978).

Sediment accumulation rates which were used for annual influx calculations are those published by Tolonen (1980, Fig. 2), and the curve of sediment accumulation rate (SR) is fitted in Fig. 5.

Results

The accumulation of sediment in the basin has varied considerably during the lake history (Fig. 5; Tolonen 1980). The rate was greatest during the Ancylus stage and during the recent polluted phase. The rate for this latter phase, however, is rather difficult to interpret because the sediment has not become fully compressed and because it is still undergoing decompositional processes.

Organic matter and sedimentary pigments

The density of the sediment (Fig. 2) decreases towards the sediment surface. Consequently, the water content increases upwards, and has been consistently higher than 90% since the isolation of the basin from the Litorina Sea 6 100 radiocarbon years ago.

Loss on ignition which approximates the organic matter content and the sedimentary pigment values increase from the basal clay. There is an abrupt rise during the late Litorina lagoon stage, and the maxima coincide with the *Naias marina* phase (Tolonen 1980). Organic matter decreases intermittently above 200 cm, probably because of the influx of eroded minerogenic matter following the first clearing and burning of the watershed (Tolonen 1978, 1980; cf. Likens *et al.* 1969).

A clear increase of the chlorophyll:carotenoid (epiphase) ratio towards a maximum of 2.9 at 190 cm is indicative of an extremely oligotrophic lake phase (Gorham *et al.* 1974; cf. Züllig 1961). Above this the ratio declines to about 1. In the uppermost black sulphide gyttja the strongly reduc-

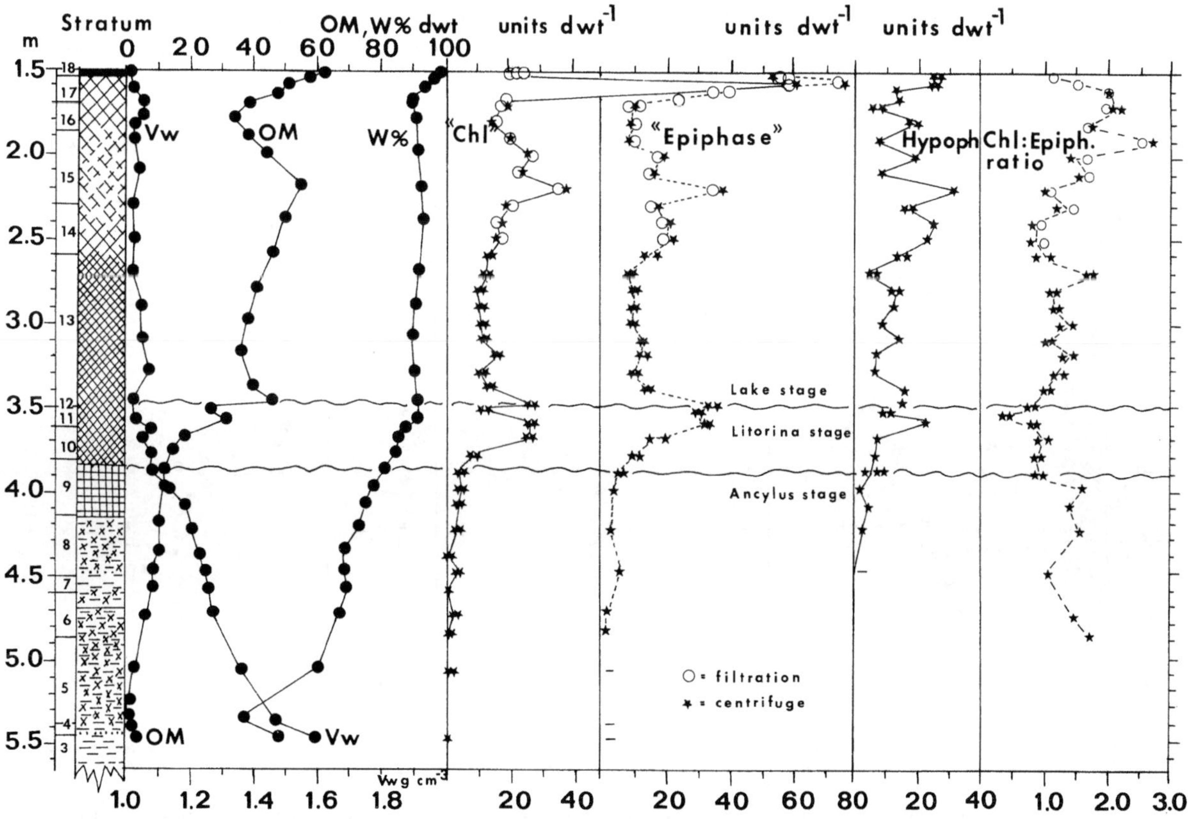

Fig. 2. Sediment sequence from site A (Fig. 1). Strata 3–18 are described in the text. Density (V_w), loss-on-ignition (OM), water percentage (W%), and sediment pigments (chlorophyll, epiphase, hypophase, chlorophyll: epiphase ratio). The brackish-water sediment (Litorina stage) is bordered by two undulating lines.

ing conditions may have enhanced the preservation of the carotenoids. Widen (1973) concluded from the abundance of mycoxanthophyll that blue-green algae were important sediment formers upwards from 166 cm in Gallträsk. This agrees well with the occurrence of several more or less eutrophic algal genera, such as *Nostoc, Gloeotrichia, Aphanocapsa* and *Microcystis* (Tolonen 1980).

Redox potential, hydrogen ion concentration and specific conductivity

The differences in the redox potential measured over the three years (Fig. 3) are probably due to methodological reasons. Nevertheless, the general trend is a gradual rise from the Ancylus stage to Litorina, an occasional fall during the isolation from the Baltic and a rather rapid decline in the uppermost one metre of sediment.

The ratios Fe:Mn and Cu:Zn have been reported to reflect paleoredox conditions in Baltic sediments (Hallberg 1974) and in freshwater sediments (Vuorinen 1978). The shape of the curves of both ratios is rather similar in Gallträsk, except in the most polluted phase at the top.

The clear rise of Fe:Mn at a sediment depth of c. 350 cm is the result of a decrease in manganese concentration. It is postulated that the fall represents the period when the lake basin was isolated from the Litorina and was due to brackish water in the hypolimnion becoming periodically anaerobic. This may have led to a migration of manganese into the water and out of the lake. The polluted sediment (150–165 cm) contains much FeS and this causes a steep rise in the Fe:Mn ratio. In the same stratum the concentration as well as the influx of zinc is many times the concentration of copper (Fig. 5). Accordingly the ratio falls to a lower level.

The hydrogen ion concentration of fresh sediment is commonly somewhat lower than in the overlying water (e.g. Digerfeldt 1972). When the pH of the topmost sediment of Gallträsk was measured in the field it agreed with the summer pH of the water. But during the storage and oxidation of the cores the sediment pH decreased considerably (Fig. 3). The diatom assemblages of the sediment

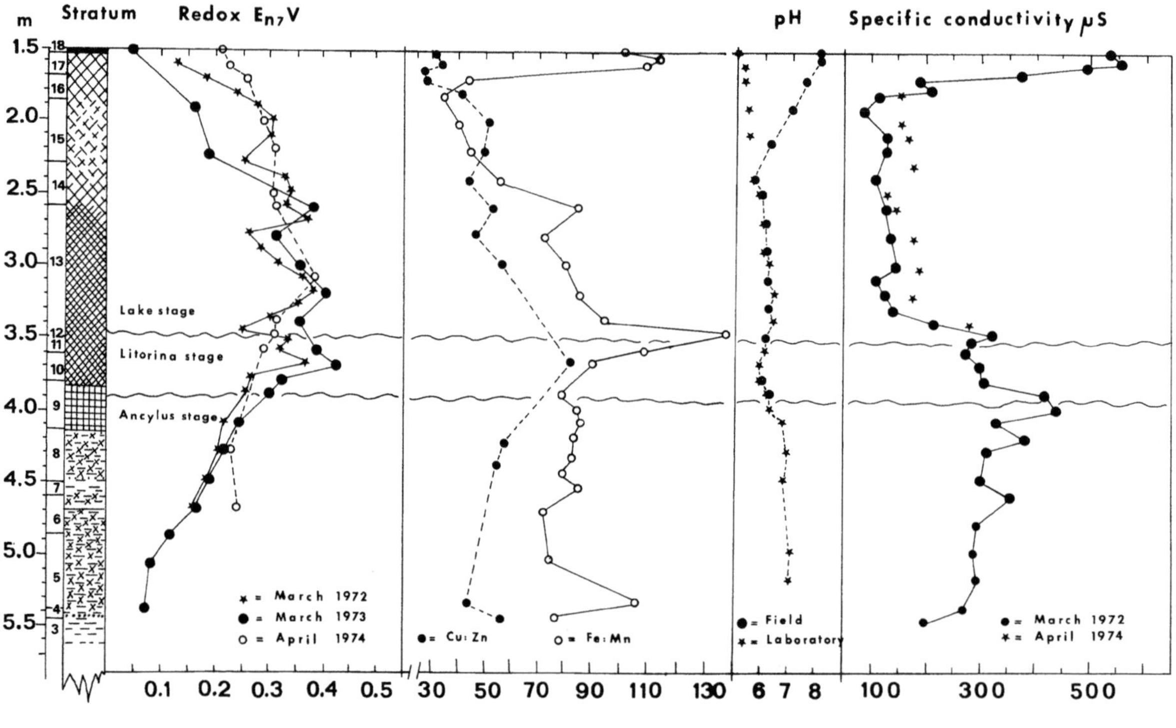

Fig. 3. Sediment redox potential for 1972–1974, paleoredox conditions expressed as ratios Cu:Zn and Fe:Mn, sediment pH, and specific conductivity of sediment (µS).

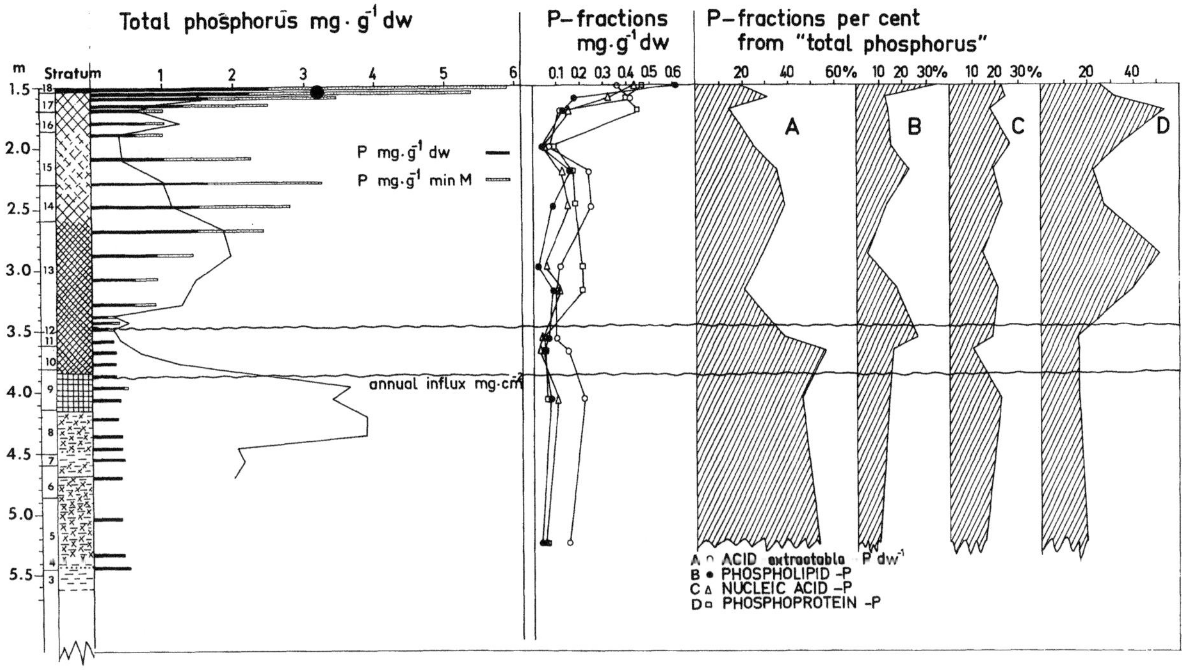

Fig. 4. Phosphorus. Total concentration expressed against dry weight (solid line) and mineral matter (dashed line) and annual influx (curve), the P-fractions: concentrations per dry weight and relative frequencies.

were used to reconstruct former pH conditions in the lake according to Meriläinen (1967) and Tolonen (1980).

It was suggested that the autumnal pH of the water had risen from 6.7 at a sediment depth of 220 cm to 7.1 (at 170 cm) and further to about 7.5 during the recent polluted period.

The specific conductivity of the fresh sediment clearly reflects the development of the lake basin: the isolation from the Baltic, a gradual and long oligotrophication and finally the cultural eutrophication and pollution during the last few decades. The electrolyte content is about five-fold in the polluted layer compared to the background (Fig. 3).

Nitrogen, phosphorus, sulphur and algal growth potential

The top layers of the sediment from the polluted period show a clear rise in nitrogen. The rise is not explained by mineral nitrogen contribution (cf. Mackereth 1966) but rather by eutrophication of the lake (Lawacz *et al.* 1978).

The total phosphorus content (Fig. 4) of the sediment closely follows that of organic matter. This might indicate that phosphorus is mainly precipitated with humus. The N:P ratio falls below 10 only in the uppermost 10 centimetres. This is a lower value than in seston (Bengtson *et al.* 1977) and probably means that phosphorus here is in excess as an algal nutrient.

Acid-extractable-P is the strongest bound phosphorus fraction (Mackereth 1967). This fraction accounts for more than 60% of the total P (Fig. 4) in the Ancylus clay, probably indicating that erosion was the main source of phosphorus during that period. It is interesting, however, that during the isolation the accumulation rate of phosphorus was small, in spite of a high sediment accumulation rate. In that period the hypolimnion may have become periodically anaerobic, as postulated earlier on the basis of sediment redox and Fe:Mn ratio, causing losses of phosphorus from the sediment.

Oxidation conditions deteriorated from after the isolation during the time that represents the sediment depth of 250 cm, when iron and phosphorus concentrations decreased and reached their minima. It is likely that these minima resulted from the losses of iron and phosphorus from the sediment and from the lake basin.

The surface maximum of phosphorus measured

from the 'pollution layer' is determined by the increase of phospholipid-P. The peak surface concentration is about 20 times greater than a sediment depth of 200 cm. The corresponding increase of the influx of total phosphorus is 10 times. The most recent influx values, however, remain uncertain because of the low redox potential and release of phosphorus from the sediment.

The phosphoprotein-P may directly measure the former productivity of the lake ecosystem (cf. Krause 1964). The assumption is supported by the biological analyses of Gallträsk cores (Tolonen 1980): the two maxima are found from the sediment depths that also indicate increased lake productivity. On the other hand, nucleic acid-P shows only a gradual increase towards the surface during several thousand years. The vertical distribution of sulphur in the freshwater sediment is roughly inverse to the redox potential. Maximum values are recorded at the end of the Litorina period. The precipitation of sulphur into the sediment is largely determined by biological activity (Mackereth 1966) which reduces sulphate to sulphide and precipitate it as ferrous sulphide on the bottom. On the basis of C:S ratios sulphate-reducing bacterial activity seems to have been greatest in Gallträsk during the isolation stage and recent pollution period. The least activity is at a sediment depth of 220–240 cm that represents the most dys-oligotrophic state in the development of the lake.

Algal growth potential (AGP) of the interstitial water was higher in the oldest minerogenic sediment than in sediments that were formed during and after the isolation. The more recent, poorly oxidized layers revealed a very distinct decrease in AGP, supporting the postulation of a loss on nutrients from the sediment and the lake basin. In the uppermost layer of the black sulphide gyttja algal growth potential may be lower because of the heavy metal concentration. In the present lake algal growth seems to be limited by the shortage of nitrogen. The addition of phosphorus alone did not influence algal growth in the culture, but additional nitrogen, especially together with phosphorus, resulted in a steep rise of algal growth.

Sodium, potassium, magnesium and calcium

Both sodium and potassium are transported to the lake mainly with mineral matter (Mackereth 1966; Koljonen & Carlson 1975). Because the concentrations of these elements are very low in Gallträsk it is assumed that leaching of sodium and potassium has been effective. On the other hand, the highest concentrations of mineral matter could indicate intensive erosion (cf. Mackereth 1966).

Before isolation the Gallträsk basin was open to intensive littoral erosion due to the fluctuation of water level and ice erosion in the Baltic Sea. This probably caused the re-sedimentation of post-glacial clay, including sodium and potassium on to the bottom (Fig. 5). After the isolation leaching of potassium was greater than sodium. The first isolated stage was rather productive (Tolonen 1980) and maybe supports the idea of a high sodium leaching efficiency in the catchment (cf. Mackereth 1966).

The distribution of magnesium (Fig. 5) is very similar to that of potassium and it could be that magnesium is also mainly brought into the basin with mineral matter.

The distribution of calcium differs from the elements discussed above. The calcium content is rather constant and very low (0.3–0.5%) in the sediment throughout the history of the lake. As in Esthwaite (Mackereth 1966) calcium is closely associated with organic matter in most Finnish lakes, including Gallträsk and it must be kept in mind that the distribution pattern of calcium against the mineral fraction (Fig. 5) only reflects changes in the sedimented organic material.

Heavy metals

Heavy metals (Pb, Cu, Zn, Cr, Co, Ni, Fe and Mn) all have a rather similar distribution from the Ancylus stage to the present. As lithophilic elements they are in general associated with mineral matter. There is, however, an exception: the uppermost sediment is rich in organic matter but also contains high contents of some heavy metals (especially Cr, Zn, and Co). For the surface sediment

---

Fig. 5. Sediment profile from site A.

Distributions of some elements: concentrations per unit dry sediment (solid line), per unit mineral matter (dashed line), and influx (curve). Recent deposition is marked by a dot. Sediment accumulation rate (SR) is added in coordinates of magnesium. Sediment depth expressed as centimetres.

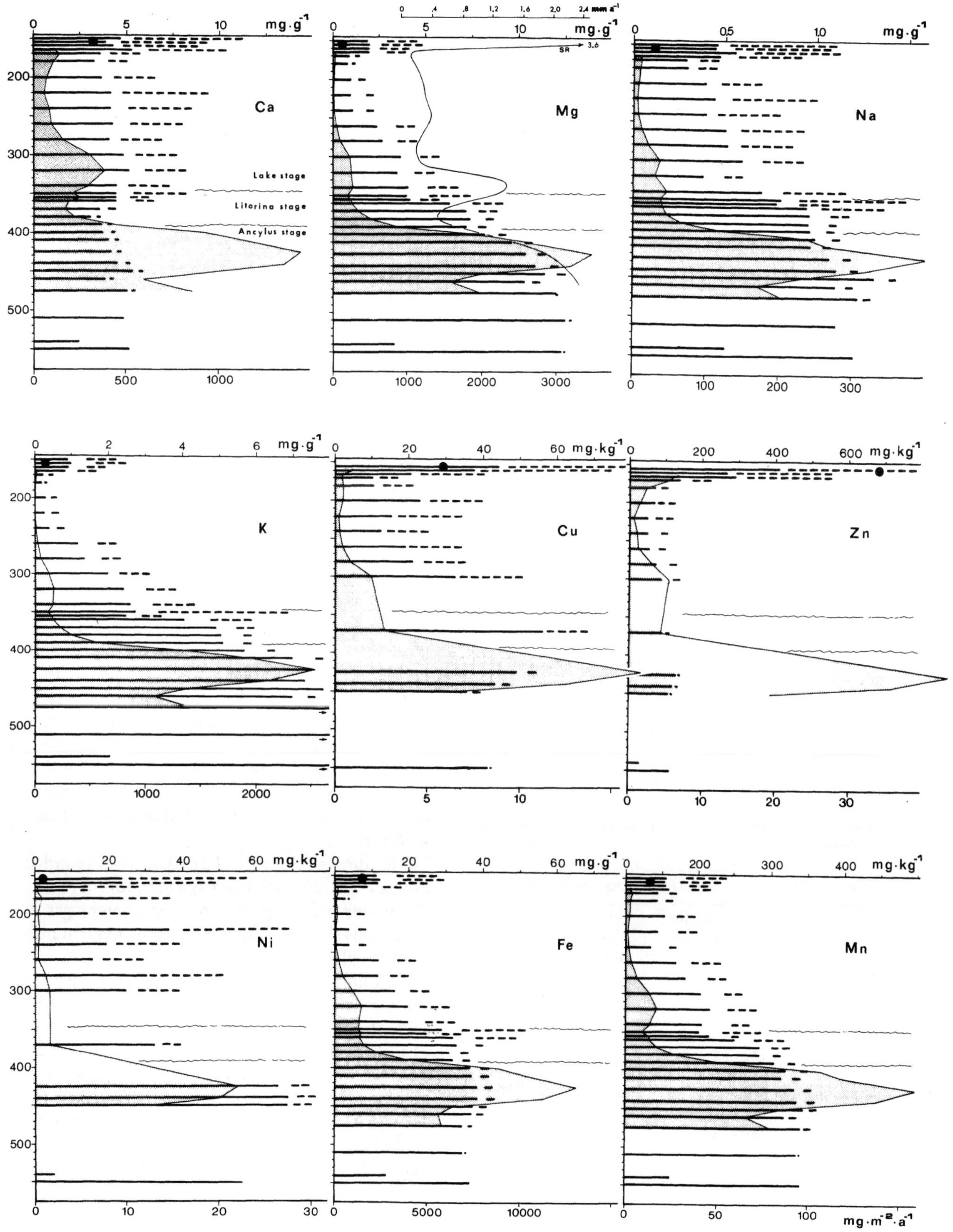

Ca
Mg
Na
K
Cu
Zn
Ni
Fe
Mn
mg·g⁻¹
mg·kg⁻¹
mg·m⁻²·a⁻¹
mm a⁻¹
SR
3,6
Lake stage
Litorina stage
Ancylus stage

Table 1. Heavy metal enrichment in the uppermost sediment of Gallträsk (concentrations in mg kg^{-1} dry sediment).

Stratum								
150–165 cm	22.0	320	10.1	11 300	145	33.5	113.7	20.9
200–240 cm	1.0	62	2.5	4 200	60	17.5	89	17.0
'Enrichment coefficient'	22	5.2	4.0	2.7	2.4	1.9	1.3	1.2

enrichment ratios are calculated from the mean contents of the 'polluted layer' (including the three topmost samples) and the mean of the 'natural background'. The rank order of their subsurface enrichment is:

Cr > Zn > Co > Fe > Pb > Cu > Mn > Ni.

Nickel can be considered perhaps as a natural calibration standard for the other heavy metals. The enrichment of heavy metals is shown in Table 1.

Discussion

The concluding remarks presented here are based not only on present results but also on Tolonen (1980) and unpublished data of the authors collected since 1962.

Six strata can be seen in the history of Gallträsk as shown in Table 2.

Stratum A (Table 2) represents the time up to the Atlantic period, and incorporates the freshwater Ancylus phase. The forest canopy closed and erosion and hence the mineral fraction of the sediment decreased. However fluctuating water-level and water movements caused erosion, re-sedimentation and leaching of post-glacial clayey sediments. The sediment accumulation rate was high, nutrient content and electrical conductivity of the sediment, and perhaps also water, was high but redox potential was rather low.

Table 2. Chronostratigraphic description of the development of Gallträsk.

Stratum	Period, BP	Depth (cm)	Characterization
F	70– 20	150–165	Pollution by sewage
E	2 300– 70	165–200	Rise of pH, slash-and-burn cultivation nearby
D	4 500–2 300	200–270	Dys-acido maximum, dys-oligotrophic phase, losses of nutrients from sediment
C	7 200–4 500	270–345	First rich in nutrients, then impoverishing
B	7 800–7 200	345–390	Brackish water (Litorina Sea)
A	–7 800	390	Fresh clear water stage (Ancylus Lake)

Stratum B (Table 2) was deposited during the Litorina stage. The Litorina regression exposed littoral sediments for erosion and re-sedimentation. The hypolimnion was occasionally anaerobic (cf. Fe:Mn ratio). The sediment was rich in nutrients, and productivity in the water column – in the warming climate – was relatively high.

The isolation caused a major change both in the sediment and in the water. There is evidence in *Stratum C* (Table 2) that even a short-lived meromixis may have taken place. First sediment accumulation increased considerably but deposition of most nutrients decreased. Then the depositional conditions stabilized. The low nutrient levels in the sediment were probably caused by leaching of soils and occasional losses to the water. Lake productivity was at its maximum. Even the demanding species *(Naias marina)* continued to live in the freshwater lake.

Stratum D (Table 2) was formed towards the end of the Subboreal and beginning of the Subatlantic period. Probably the paludification of the long western bay (Tolonen 1980; Fig. 1) was accelerated. Spruce occupied the surrounding area and the role of broadleaf trees decreased. The lake became dys-oligotrophic and periodically anaerobic. The impoverishment of the watershed was probably accelerated by the first Iron Age clearance and soil leaching.

The impoverishment of the lake was retarded about 2 300 calendar years ago (*Stratum E* (Table 2)) by intensified agriculture. The assemblages of organisms showed evidence for an increase in nutrient supply.

Stratum F (Table 2) is characterized by sewage pollution. The organic matter content reaches its highest level and includes a large proportion of autochthonous material (remains of blue-green algae). In winter the sediment is mostly anaerobic. The values of many of the sediment parameters are heavily influenced by pollution. The algal productivity especially increased to a high level. This stratum is characterized by the anomalies of Fe:Mn and

Table 3. Annual accumulation in sediment and snow (mg m^{-2}).

	Gallträsk	Snow
SO_4	1 531	954
Fe	1 746	28.9
Mn	3	8.3
Zn	34	8.0
Pb	15.4	5.5
Cu	5.9	2.5

Cu:Zn, by increases of chlorophyll content and of some heavy metals. In Fig. 5 the sedimentation curve has been completed by the dot which indicates the recent deposition. Zinc, copper and especially chromium show high influxes. In general, however, the concentrations of the heavy metals in the topmost sediment are not clearly higher than in large East Finnish lake material (Huttunen & Meriläinen, unpublished).

The present annual accumulation of heavy metals in Gallträsk (Fig. 5) can be compared with annual deposition based on snow analyses from S. Finland (in 1978–80), National Board of Waters, (Soveri, J., unpublished data). Though there are great differences in figures between the snow stations, the ratios shown in Table 3 are obtained. At the nearest snow site to Gallträsk at Vihti almost equal values are found for Pb and Cu but Zn values in Gallträsk are 2.6× higher.

The surface peats from raised *Sphagnum* bogs in the same area receive very similar fall-outs as snow for Pb, Zn and Cu. However, Fe and Mn were higher and SO_4 much lower in moss samples.

Acknowledgements

The authors wish to extend their warmest thanks to Mrs. Pirkko Dookie for technical assistance, to Licentiat of Agr.-Forest. Paavo Seppänen for putting his unpublished AGP-results at our disposal and to Dr. Carl-Gustaf Widen for his chromatographic pigment studies. Financial support for this study was received from the Natural Council of Science of the Finnish Academy and from the City of Kauniainen.

References

Alhonen, P., 1972. Gallträsket: the geological development and paleolimnology of a small polluted lake in southern Finland. Comment. Biol. Soc. Scient. Fenn. 57: 1–34.

Belcher, J. H. & Fogg, G. E., 1964: Chlorophyll derivates and carotenoids in the sediments of two English lakes. In: Miyake, Y. & Koyama, T. (eds.), Recent Researches in the Fields of Hydrosphere, Atmosphere and Nuclear Geochemistry. Tokyo: Mazaren, pp. 39–48.

Bengtsson, L., Brorson, T., Fleicher, S. & Åström, C., 1977. Beschaffenheitsänderung des Sestons in Seen nach dem Sedimentieren. Acta Hydrochim., Hydrobiol. 5: 153–165.

Digerfeldt, G., 1972. The post-glacial development of Lake Trummen. Folia Limnol. Scandinavica 16: 1–96.

Gorham, E., Lund, J. W. G., Sanger, J. E. & Dean, W. E., 1974. Some relationships between algal standing crop, water chemistry and sediment chemistry in the English lakes. Limnol. & Oceanogr. 19: 601–617.

Hallberg, R. O., 1974. Paleoredox conditions in the eastern Gotland basin during the recent centuries. Merentutk. lait. julk. 238: 3–16.

Huttunen, P., Meriläinen, J. & Tolonen, K., 1978. The history of a small dystrophied forest lake, southern Finland. Pol. Arch. Hydrobiol. 25: 189–202.

Kurki, M., 1972. Suomen peltojen viljavuudesta II. (Über die Fruchtbarkeit des finnischen Ackerbodens auf Grund der in den Jahren 1955–1970 durchgeführten Bodenfruchtbarkeitsuntersuchungen.). Viljavuuspalvelu Oy. Helsinki.

Koljonen, T. & Carlson, L., 1975. Behavior of the major elements and minerals in sediments of four humic lakes in southwestern Finland. Fennia 137: 1–47.

Krause, H. R., 1964. Zur Methodik einer differenzierten Bestimmung von Organischen Phosphorkomponenten in Süsswasserplankton. Arch. Hydrobiol. Suppl. 28(5): 282–290.

Lawacz, W., Planter, M., Stasiak, K., Tatur, A. & Wieckowski, K., 1978. The past, present and future of three Mazurian lakes. Pol. Arch. Hydrobiol. 25: 233–238.

Likens, G. E., Borman, F. H., Johnson, N. M., Fischer, D. W. & Pierche, R. S., 1969. Effects of forest cutting and herbicide treatment on nutrient budgets in the Hubbard Brook watershed ecosystem. Ecol. Monogr. 40: 23–47.

Mackereth, F. J. H., 1966. Some chemical observations on Postglacial lake sediments. Phil. Trans. R. Soc. B 250: 165–213.

Meriläinen, J., 1967. The diatom flora and the hydrogen-ion concentration of the water. Ann. Bot. Fennici 4: 51–58.

Niemistö, L., 1973. Sedimentaatio ja siihen vaikuttavat tekijät Helsingin merialueella. HKR vesiensuojelutoimiston tiedonantoja 5(5): 1–51.

Seppänen, P., 1975. On the Ankistrodesmus falcatus-AGP of some Finnish inland waters I, II. Ympäristö ja Terveys 1975: 230–238, 353–359.

Shapiro, J., Edmonson, W. T. & Allison, D. E., 1971. Changes in the chemical composition of sediments of Lake Washington, 1958–1970. Limnol. & Oceanogr. 16: 437–452.

Tolonen, K., 1978. Effects of prehistoric man on Finnish lakes. Pol. Arch. Hydrobiol. 25: 419–421.

Tolonen, K., 1980. Pollen, algal remains and macrosubfossils from Lake Gallträsk, S. Finland. Ann. Bot. Fennici 17: 394–405.

Tolonen, K. & Hosiaisluoma, V., 1978: Chemical properties of surface water in Finnish ombrotrophic mire complexes with special reference to algal growth. Ann. Bot. Fennici 15: 55–72.

Vallentyne, J. R., 1954. Sedimentary chlorophyll determination as a paleobotanical method. Canadian J. Bot. 33: 304–313.

Vuorinen, J., 1978. The influence of past land use on the sediments of a small lake. Pol. Arch. Hydrobiol. 25: 443–451.

Widen, K. G., 1973. Redogörelse för orienterande undersökning av sedimentpigment i Grankulla Gallträsk. Manuscript/-Mimeographed study, 5 pp. + 30 App. Dept. of Pharmacology University of Helsinki.

Züllig, H., 1961. Die Bestimmung von Myxoxanthophyll in Bohrprofilen zum Nachweis vergangener Blaualgenentfaltungen. Verh. int. Ver. Limnol. 14: 263–270.

GPSR Compliance

The European Union's (EU) General Product Safety Regulation (GPSR) is a set of rules that requires consumer products to be safe and our obligations to ensure this.

If you have any concerns about our products, you can contact us on ProductSafety@springernature.com

In case Publisher is established outside the EU, the EU authorized representative is:

Springer Nature Customer Service Center GmbH
Europaplatz 3
69115 Heidelberg, Germany

Batch number: 10762213

Printed by Printforce, the Netherlands